Sustainable Development in the Baltic and Beyond

Environmental Education, Communication and Sustainability

edited by Walter Leal Filho

Vol. 23

PETER LANG
Frankfurt am Main · Berlin · Bern · Bruxelles · New York · Oxford · Wien

Walter Leal Filho
Arnolds Ubelis
Dina Berzina
(eds.)

Sustainable Development in the Baltic and Beyond

PETER LANG
Europäischer Verlag der Wissenschaften

Bibliographic Information published by Die Deutsche Bibliothek
Die Deutsche Bibliothek lists this publication in the Deutsche Nationalbibliografie; detailed bibliographic data is available in the internet at <http://dnb.ddb.de>.

This book is supported by the Project BEIDS
(Baltic Environment Information Dissemination System)
and by the DG Research – EU Contract Nr.: 009244 (GOCE) – SHARING.

Front cover: Schematic view of the Baltic Sea region,
© University of Latvia, Riga.

Layout: Kumpernatz + Bromann, www.kumpernatz-bromann.de.

ISSN 1434-3819
ISBN 3-631-55282-3
US-ISBN 0-8204-9915-3

www.peterlang.de

Table of Contents

B. Science for Sustainable Development

Preface

This book documents and promotes the experiences from the "2nd Conference on Integrative Approaches Towards Sustainability" (SHARING), held in Jurmala, Latvia, from May 11 to 14, 2005. Attended by over 200 delegates from across the Baltic Sea region and beyond, the event provided an important contribution towards strengthening the integration of the region's RTD community and sharing knowledge and expertise in the Baltic, across Europe (including the Mediterranean and Black Sea regions) and elsewhere.

The first part of this book contains generalised and advanced asumptions of (some introductory remarks) sustainable development and integrative approaches in particular. It goes on to discuss the sustainability components of the economy and the challenges on new technologies for a sustainable energy system. Besides, issues related to cleaner production and moves towards zero emissions are also outlined in part 1 of the book.

Part two focuses on developments in the Baltic Sea Region, with a special emphasis on policy and implementation of technologies for sustainable development. It encompasses papers dealing with future images of sustainable urban development, environmental and thermodynamic indicators to support sustainable policy making, and the project COASTMAN, which entails integrative approaches to the Baltic Sea coastal zone management. Furhermore, it also focuses on sustainable agriculture, agroforestry and linked resources and developments, with papers outlining the use of environmental information in decision-making, a bio-based chemistry platform for sustainability and bioenergy developments in Ukraine.

In recognition of the role of education in sustainable development in the context of University-Municipality partnerships, a series of papers deal with urban sustainability strategies and teaching of sustainable development in the context of joint efforts of universities and municipalities attracting local students as the main driving force. Furthermore, a decade after the UNCED in Rio de Janeiro, the role of local Agenda 21, as a changing agent, is assessed.

The University sector is represented by integration of environmental management and sustainable development in the outlined strategies of universities and describing the role of the city as a vehicle of education. Other issues tackled in the second part are thinking in terms of systems design, the characteristics of biodiversity and possibilities for sustainable exploitation of sea areas such as the Adriatic sea and an international e-Learning didactic course of sustainable development.

In an attempt to record and disseminate the results from parallel and work sessions, part three of the book presents a wide range of papers dealing with a multitude of issues presented at those sections. Some of them are:

- research for sustainable development (EU Strategy and Policy);
- the economic and political framework
- climate change mitigation initiatives
- foresight and integrated management systems in Small and Medium size Enterprises (SMEs);
- zero emissions research;
- sustainable community development;
- integration of cultural heritage as a resource for sustainable development.

In part B of the book, readers will find a special section on science for sustainable development, where the role of Baltic Sea algae as a resource is discussed, complemented by an analysis of European rice husks studies, removal of hazardous substances in electronic processes, collaborative research on soil stabilisers, energy life cycle assessments and more technical studies with sustainability issues in the background. A special feature of this section is that papers are predominantly written by young researchers, some of whom had found the first chance of publishing results of their work in this volume.

The final section of this book deals with social corporate responsibility and the environment focusing on sustainable community development, corporate responsibility at international level and concepts of education for sustainable development.

All in all, this publication provides a good overview of sustainability issues in the Baltic and beyond, illustrating the number of excellent works being performed by municipalities, universities, research agencies and NGOs, showing the various ways by which sustainability can be delivered.

Walter Leal Filho, Arnolds Ubelis, Dina Berzina Spring 2006

Sustainable Development
–
R&D Perspectives from Sweden and the European Union

Uno Svedin, Professor PhD

Director of International Affairs,
The Swedish Research Council for Environment, Agricultural Sciences and Spatial Planning (FORMAS),
Stockholm, Sweden

Chair of IGFA (The International Group of Funding Agencies for Global Change Research)[1]

1. The Millennium Development Goals and Sustainability

In starting a reflection on the sustainability challenges for a European region such as the Baltic region it is reasonable first to reflect on the general issues of sustainability and especially in the connotation of the global context. Here we are in the lucky situation that we both have had a few decades of reflection and (partial) implementation of sustainability policies at world level. So we do not have to start from scratch. On the other side there are also divergences in the interpretations and the relative success of the concept has also at the same time eroded its precision.

In a very general sense and at global level the sustainability charge has been incorporated in the process of the Millennium Development Goals (MDG) reflection activities. Here sustainability is not the only topic involved. However there is a sort of one-to-one match between the millennium goal frame and the concerns raised by sustainability development (SD) challenges. Thus we can say that the:

- MDG – are general and targeted political goals in the long term – but that they do not always explicitly refer to SD
- MDG cover a broad range of issues – but are they really broader than SD?
- MDG – definitely have a global character – but that also holds true for SD
- MDG could be seen as a broad "container" also of SD goals
- MDG fulfillment are on their way when the SD conditions are improving distinctly
- reflections on SD challenges and problems also draw MDG issues into focus.

The to the MDG related special "Millennium Ecosystem Assessment" which delivered its results in 2005 has summed up the overriding outcome of the sev-

---

1 U. Svedin has been the EU appointed chair of the European Consultative Forum on Environment and Sustainable development and during 2003-2004 was the Chair of IGFA (The International Group of Funding Agencies for Global Change Research)

eral years of activity in a few lines – and then with a focus on the interplay between the ecosystems and the human world. Thus the "bottom line" is that:

- We are spending Earth's natural capital, putting such strain on the natural functions of Earth that the ability of the planet's ecosystems to sustain future generations can no longer be taken for granted.
- At the same time, the assessment shows that the future really is in our hands. We can reverse the degradation of many ecosystem services over the next 50 years, but the changes in policy and practice required are substantial and not currently underway (Source: Millennium Ecosystem Assessment 2005).

With special regard to the sustainability challenges we thus have to be active in securing that:

- SD integration is assured and inserted into national policy
- we turn the trend concerning vulnerable natural resources
- in order to do so we have to further develop and improve existing indicators about e.g.: land use of various kinds including the area covered by forests and agricultural practices, protected areas related to biological diversity, energy use related areas, land use for urban purposes etc.
- we also have to find indicators of quite some sophistication related to climate change, i.e. the carbon dioxide release and other connected phenomena
- we treat the SD challenges in an even wider context as we have to remember that SD is broader than just natural resources. It connects to issues about health, powerty, education etc. The connections between these items and other SD related topics reflect a broad variety of systemic causality loops
- thus it is possible to reproduce quite a bit of the mdg conceptual landscape by starting in the SD "node".

So what are our challenges in the "sustainability realm"?

- the consolidation of the SD concept at political level is historically evident – but continued problems emerge conceptually and scientifically as the concept turns more and more political
- thus we have maybe a paradoxical situation of sustainability as "political winner", but the gain may have been won through a too wide and thus thinning "retorical broadening"
- we may find ourselves to have a consensus at the surface level but carrying deep divergencies at deeper levels – showing up e.g. in terms of variations in views on strategies about action (as the climate change policy field expresses)
- we may also find that the gains have been won at a price of professionalisation and complexification of the domain which in its broader normative perspectives should have its essential roots in democracy issues as openess and public understanding and participation.

Following the analysis of the British social environmental historian Professor Susan Owens we are thus facing the following paradoxical situation that:

- the force and consensus of the SD concept came from its capacity to connect („reconciliation" of growth – with "good environment")
- but at the same time finding that "environment" and "development" diverge more and more in terms of policies and implementation
- and this opens up for an "implementation gap".

In my own perception this could be seen to be mirrored by the various steps in recent history in the last few decades in the following way:

- in the the 70's (The Stockholms Conference 1972) the environmental situation was found to be *global* in nature
- in the 80's (building up of the Brundland perspective until 1987) sustainability grew to become a *political* challenge (natural science based – but "also" social science/humanities based)
- in the 90's (the Rio-conference 1992) "environment" *and* "development" were deliberately seen to be linked (from "north" to "south") (but also through connections: ecologically-economically-socio/culturally)
- in the 2000's (The Johannesburg UN Conference in 2002) the *implementation* responsibility and attention grew, implying emphasis on more actors and their actions in partnership.

Thus the broader development from

- the 1800-1900's: emphasising *good natural resources management* ("conservation" and "maximum sustainable yield")
- turned symbolically from around 1987 (from the period of the Brundtland commission report): into "the modern times" – i.e. the promising *optimistic time* – "*it will all work*". Thus we move conceptually from environmental assessment to political challenge (pragmatic and devoted to institution building)
- Followed by a more disparate opening of the 21st century with distinct divergences – at the same time as the overriding goal structure was maintained at a rather vague global political level (and implementation schemes at regional national and sub-national levels). Thus again following Susan Owens we find from the late 90's that the operationalisation opens deeper questions (what are the pivotal critical elements? etical considerations? what is "politics"?) i.e. an "*implementation deficit*". Thus forward from now on – i.e. into the 21 century – we have to live within a "pluralistic vision" that might not converge (but that will have to operate at several levels simultaneously through "vigorous debate, argument, counter-critique" i.e. a vital and ongoing process). And we will have to stress responsibility – cooperation – structural action.

2. On the Issues of Scale: Local – National – Regional – Supraregional – Global

The SD issues are to a very large extent multi-layered in their essence. Thus we are speaking of an interlinked system not only in terms of its natural phenomena components, but also in terms of its socio-economic embedding and interwoven character between the natural and the societal aspects. Also very important is the political interpretation and institutional and legal encoding of this (temporary) understanding. This brings with strong force the issue of the knowledge production system into focus. Here we are not only speaking about research, although it is a very important part of the knowledge-knowing-understanding system. We also have all sorts of applications and also implementation measures in society. The degree of consensus formation also belongs to this arena as do the education sector aspects.

We also have to understand that the systemic nature of the planetary charge makes it very important for us in "the North" to recognise the presence of issues in "the South" – on equal par with the problems as we perceive them with a northern outlook. In this sense highlighting the perspectives of the South indeed brings new and essential elements into the overall picture, i.e. the general issue of development and what could be meant by that – also "in the North". These issues can be seen as a sort of "pointer" at pragmatic and didactic central issues for the future in relationship to SD globalised norms.

The global reach connects distinctly to "lower" scalar levels in terms of a hightened interest for regionalisation, the local impacts of globalisation and the connected issues of distribution of power and influence. We also move in terms of analysis – but also in terms of political application – towards more "place-specific" approaches of the SD-related phenomena („glocality-issues"). Related to the knowledge production system many of these topics calls for:

a) Special methodology

b) New types of networks

c) Innovation needed in research financing

d) And a break out of national confines. This especially holds true for the analytical approaches to studies:
   - Across geographical scales (multilevel)
   - Bridging Short term and Long term
   - Related to risk
   - Being "Governance" embedded.

Thus issues of concern relate to the

- Connectivity to the surrounding world
  („No One Is An Island To Himself")
  i.e. the role of regionalisation in a globalised world and the micro-macro links
- Time And Timing of integration
- Path Dependence issues –
  including those of technology innovation and investment paths
- The "Soft" Issues

- Democracy
- Inter-sectoral Coherence
- The Knowledge System Capacities
  (especially the capacity to reflect on systemic issues)
- Governance
- Identity issues.

3. The EU Approach to the SD Issues – The Göteborg Summit 15-16 June 2001 as a normative Frame of European Policy

How has all that has been said above in general terms been implemented at the European level? The key policy decision occasion for these issues was the Göteborg Summit in 2001. The major elements were stated to be in an overall SD frame:

Strategy Aim

EU Citizens shall be granted

- Economic stability
- Social supporting conditions
- A clean environment.

Means

All new major suggestions for decisions to be judged against effects:

- Economic
- Social
- Environmental.

Structurally

The environmental dimension shall be added to the Lisbon Strategy related to social and economic development.

In the follow up to Göteborg the following specifications of goals were indicated through a set of special goals e.g.

- at least 22% of electricity from renewables by 2010
- environmentally friendly transport
- the new chemicals strategy
- agricultural policy and ecologically sustainable production methods
- fisheries policies to be related to the support capacity of the ecosystems.

With special regard for natural resources the following policy norms were stated:

- change the relationship between
  - economic growth
  - consumption of natural resources
  - generation of waste
- Maintaining biodiversity
- Preserving ecosystems
- Avoiding desertification
- Add objectives to the Common Agricultural policy
- Review the Common Fisheries policy
- Implement the EU Integrated Product Policy.

In specific domains of interest the following policies were adopted:

Climate Change

- Kyoto protocol (to meet national and European commitments)
- Indicative targets for 2010 of electricity from renewable energy sources
- Invitation to the European Investment Bank to assume a more active role.

Transport

- volumes of traffic and congestion to be addressed
- decoupling of transport growth – GDP growth to be attended to
- road to rail/water/public passenger transport balances to be addressed
- infrastructure investment for SD transport systems to be developed
- better pricing policy of transport.

Public Health

- safety and quality of food (e.g. European Food Authority)
- use of chemicals (Chemicals policy)
- outbreaks of infectious diseases and resistance to antibiotics
- European surveillance and early warning networks.

4. The Baltic Sea as a System

With special regard for the Baltic Sea area we could point at the following systemic and SD related factors:

- The area is strongly historically and culturally connected („The Baltic Sea connects")
- The Geo-Systemic argument (The Baltic Sea watershed is ecologically connected)

- Geopolitical/supraregional arguments
  („a Europe built of regions as pieces in an interplay to other pieces and other global domains")
- Socio/economical/cultural matching to the ecological system.

When we look at the experiences from the Baltic region we can draw the following lessons of a more general nature:

- a lot of micro events built up macro events
- action has to be made at all levels simultaneously
- the capacity to know things has a strategic value
- regional and local phenomena are important – not only what happens globally
- it is at the lower levels that the essence of plurality is demonstrated.

Facing these experiences we have in the future to look especially towards finding appropriate understandings and solutions to the:

- vertical systems couplings
- decision system "ecology"
- democracy and the formation of identity
- knowledge formation and the creation of visions.

In terms of challenges for the knowledge production system we especially have to get to grips with the relationship between analytical models of sustainability oriented societal systems and those of the natural world systems. This could be done in terms of an expanded Earth system science with a sustainable development connotation at the normative level. We also have to improve the institutional capacity to increasingly deal with the relationship between the potential research findings and the possible design of related topics. Thus especially we have to develop our understanding of;

A. The model worlds.
   How to design the appropriate levels of multidimensional systemic "worlds" which are sufficiently convincing?

B. The policy process
   (including the feedbacks to control that perspectives and results are reasonable). That means a consultative and participatory process!

C. The decision and management system design
   Do not forget the global reach connected to perspectives and responsibility.

In terms of EU research policy – not least that which is mobilized in the 6th Framework Program and now discussed for the 7th FP you could ask if the items activated thematically match these challenges?

- What about the dynamics?
  – the R&D input to policy

  - the formulation in policy of R&D needs
- Are there other arenas in European R&D development that contribute in other ways?
- The global dimensions
  - politically
  - thematically
- The national responses.

My answer is quite positive. Our European knowledge production system is on the right way to address these issues.

5. Summary Reflections

In overriding terms there are a few points that need to be highlighted in relation to EU SD research policy:

- SD policy must analytically be supported by activities that are able to convey trustworthy images of synthesis („models")
- In order to achieve this goal processes are needed that highlight the interplay between the actors of policy and the actors of research and the stakeholder actors
- Such processes have to develop naturally over time. The activities to create the synthesis images need institutional platforms of their own and that needs investments and political courage especially with regard to setting up activities to probe long term risk potentials of finding appropriate future solutions to the development path for the combined bio-physical-chemical and socio-economic-cultural systems in Europe and its regions
- Do not forget the vulnerability aspects of the endeavour.

Sustainable development is a moving target!

References

Liljenstrom H. & Svedin U. (eds.) (2000) "Micro, Meso, Macro – Addressing Complex Systems Couplings", World Scientific, London.

Rolen M., Sjoberg S. & Svedin U. (eds.) (1977), "International Governance on Environmental issues", Kluwer, Dordrecht, The Netherlands.

Svedin U. & Hagerhall Aniansson B. (eds.) (2002), "Sustainability, Local Democracy and the Future: The Swedish Model", Kluwer, Dordrecht, The Netherlands.

Svedin U., O. Riordan T. & Jordan A. (2001), "Multilevel governance for the sustainability transition", in: O. Riordan T. (ed.) "Globalism, Localism and Identity", Earthscan, London.

EUROPEAN COMMISSION
RESEARCH DIRECTORATE-GENERAL

Directorate I - Environment
The Director

Welcome Statement

–

Integrative Approaches towards Sustainability (SHARING)

Jūrmala (Latvia) 11-14 May 2005

Pierre Valette

Acting Director,
Directorate I, DG Research, European Commission
B – 1049 Brussels, Belgium

The conference in Jūrmala this year comes at a very appropriate time.

The Commission has presented its proposal for the 7th Framework Programme which introduces new elements and aims to considerably increase the budget for research on a European level. Sustainable development is the prevailing objective and its principle cuts across the whole programme.

In addition, environmental research has been broadened and strengthened. The structure of the new Framework Programme for this area reflects the need for a more integrated approach where all three pillars of sustainable development are taken into account in our scientific research and policy considerations.

The Commission services are currently elaborating the specific programmes per area with a view to finalising them by the end of June 2005.

Therefore, the findings from this conference could produce a valuable input for the preparation of the programmes and we welcome this initiative.

Plenary Sessions

First Plenary

Sustainable Economics and "Zero Emissions" Concept

Part 1

Global Challenge – Perspectives of Implementation

The Economic Sustainability of the Economy – Assessing the Sustainability of a Complex System by Applying Orientor Theory

Joachim H. Spangenberg

Professeur invite,
Université de Versailles Saint Quentin-en-Yvelines, France

Vice President,
Sustainable Europe Research Institute, Vienna, Austria

Sustainable Europe Research Institute Bad Oeynhausen,
Hugo-Distler-Str. 1b, D – 32549 Bad Oeynhausen, Germany

Abstract

Every society can be described as comprising four dimensions, the economic, social, environmental and institutional one. Each of them is a complex evolving entity in its own right, making the coupled system one of tremendous complexity. For this system to be sustainable, each of the four subsystems must survive and evolve, and the interlinkages of the subsystems must enable a permanent co-evolution. For an adequate analysis, the appropriate level of complexity has to be respected.

It can be demonstrated that this level of complexity is beyond the scope of current economic theories, mostly dependent on assumptions of strong commensurability or at least strong comparability. Alternatively, a systems analysis perspective is presented as a new framework for discussing the co-evolution of the economy and its natural and anthropogenic environment. In this context, the economic, social, environmental and institutional sustainability of the economy must be defined and criteria for assessing it can be derived.

The analysis of the sustainability of complex systems is the raison d'être of Orientor or Orientation Theory. It application to economic systems provides the means to assess their sustainability with full respect to their complexity, development dynamics and their embeddedness in evolving socio-ecological environments.

1. The Economic Discourse: Disputing strong Comparability and Commensurability

In the economic debate, sustainable development is most often described as the need to maintain a permanent income for humankind, generated from non-declining capital stocks (Hicksian income). Thus in this perception at least constant stocks of human, man-made, natural and social capital (Serageldin 1997) are considered a necessary and often sufficient criterion of sustainable development (Pearce et al. 1990; Pearce & Atkinson 1993; Pearce & Barbier 2000). A controversy shaping much of the economic debate has arisen about the question whether each capital stock has to be maintained independently (Daly 1991), or whether the sum of all four capital stocks has to be non-declining (Pearce & Turner 1991). The dispute thus assumes strong comparability (i.e. the existence of a common characteristic like for instance 'utility') and focuses on the question of strong or weak commensurability, i.e. the existence of a common unit of measurement, like monetary value. Strong commensurability implies substitutability, but it is an open question in the economic discourse to which degree the different forms of capital can be substituted against each other, and to which degree they are complimentary.

As of today, all these debates tend to focus on increasing the stock of man made capital and the degree to which other capital stocks may be reduced for this behalf (OECD 2001). In other words, sustained growth is – often implicitly – assumed to be a part of the concept of sustainable development of the economy by most authors, and only a small fraction of ecological economists disagree. In the macro-economic debate, few other economic sustainability criteria are mentioned, like innovativeness (Rennings 2000), competitiveness (Klemmer et al. 1998) or public debt (Bundeskanzleramt 2002), while criteria like inflation or trade inbalances are politically prominent, but usually not linked to sustainable development. Again other, partly more traditional criteria like aggregate demand, consumption levels and savings rates play a minor role in the current debate (Etxezarreta et al. 2003).

In order to assess the trend in the overall wealth of nations, some studies tried to measure the development of all four categories of assets/capitals, usually in monetary terms (Serageldin & Steer 1996; Serageldin 1997), thus implying at least a certain level of substitutability. However, substitution between different capital stocks as discussed by economists mainly refers to the function of these capitals as production factors. If under changing factor constellations the product is equivalent (often in monetary terms), substitution is considered possible. Unfortunately, this analysis refers only to one criterion, utility production measured in monetary terms, and completely neglects all other aspects of the economic sphere and the unavoidable interaction of all four capital stocks (i.e. the trade offs that occur to other functions of the respective capital stocks). The restriction of economic thinking to the monetary perspective imposes a serious limitation for the analytical capacity of the discipline. As a result, for instance labour economics ignores the majority of working hours as they are unpaid (thus creating a systemic bias against sustainable development, see Spangenberg 2002), and consumption theory has no understanding of consumer satisfaction unless it is expressed on the market[1]. Due to the incommensurability of all four capital stocks, real substitution is hardly imaginable[2]. Given this, it is all the more heroic that economists still dare to make prognoses for the future development of a reality with which they are so much out of touch. In total, the weakness of strong commensurability, postulating the validity of the economic logic and numeraire outside the economic system, is obvious and has been broadly discussed (for instance in Ecological Economics 1998)[3]. Consequently from a scientific point of view, there cannot be such a thing as one comprehensive measure or index of sustainability.

These problems are specific to orthodox economic thinking; they play no significant role in ecology, sociology and political sciences (there is simply no functionally equivalent substitute for water, communication or decision making in these disciplines; see e.g. Ehrlich et al. 1999). Thus they seem to be more characteristic for the challenge that sustainability poses to economics than the economic challenges of sustainable development – much of the substitution debate results from a logical misperception, based on a rather static instead of a dynamic understanding of the systems (Coleman 1990). Consequently, Daly and Serageldin are both sceptical regarding the usefulness of monetisation and thus of the applicability of much of the apparatus of economics.

But not only the assumption of strong commensurability is the problem; it begins already when assuming strong comparability (combined with strong or weak commensurability). The assumption implies that all systems, regardless of having a common denominator or numeraire, at least have some common key

---

1 See Fisher (1906) for a different and more systematic understanding of consumer satisfaction, in particular the role for past consumption for a sustained psychic income (Lawn 2001). Although focussed on the micro level, this approach is much closer to the sustainability discourse, in particular to the analysis of sustainable consumption; the psychic income can be interpreted as a non-quantifiable measure of social sustainability.

2 However, the capital stock approach still provides a number of relevant insights. This is clearly indicated by the calculations of total wealth performed by Serageldin (1997): If development processes are increasing the man-made capital stock by depleting human, natural and/or social capital, economic development can happen to be de facto decreasing the wealth of nations, even in the sense of the weakest theories of sustainable development permitting unlimited substitution of capitals. Unfortunately, this negative development can go undetected for a long time, as long the indicators used for measuring wealth (GDP, trade balances, etc.) only take the man-made capital into account and ignore the other contributions to wealth, growth and social cohesion. Consequently, economists end up measuring wealth with "a measure of how fast resources are squandered and converted into money flows – irrespective of their effect on society" (Bossel 1996).

3 From a horizontal multi-criteria perspective it should be mentioned that indeed market value is *one* criterion amongst other which can be applied (with all the methodological difficulties known) to all four capital stocks to assess their economic sustainability, and that may permit useful answers for specific questions in this respect.

characteristics which would permit comparisons, maybe on the basis of an ordinal scale ranking. If used as a basis for comparing systems, these traits can refer to one externally defined purpose. In this case they are specific to the respective objective, but have no meaning for the sustainability of the system. Or they are intended to be sustainability criteria – this implies that the same characteristic is decisive for the sustainability of all four dimensions. However, as far as we know there is no common factor decisive for social cohesion, human satisfaction and the integrity of ecosystems. These criteria, at least as crucial to sustainable development as monetary value have to be monitored with their own yardsticks, and must be measured with their specific numeraires. An economic theory insisting on strong comparability remains helpless when trying to understand economically relevant environmental and social processes (until they are explained ex post as maximising a specifically defined kind of utility). One example shall illustrate the need for incomparable numeraires in all four dimensions when describing the sustainability-relevant aspects of a simple economic process. If a new machine (man-made capital) replaces skilled workers, this may be an effective substitution regarding production and value creation, but in terms of resource consumption (environmental capital), income generation (social capital) and skills training (human capital), the outcome is definitively different, a fact which is not captured as long as all impacts are reduced to their function in the production process.

From this extended perspective the question of substitution between two of the capital stocks simply makes no sense; it needs to be replaced by a horizontal approach referring to all four dimensions and their characteristics in order to permit an assessment of substitution impacts on all four dimensions, i.e. in a true sustainability perspective. Then the appropriate question would be whether the balance of the impacts of a certain action on all four capital stocks are considered positive, negative or neutral, according to an explicit set of multidimensional criteria (still this assessment will vary between individuals and over time). So instead of simple substitution decisions following a single criterion, a systematic or intuitive multi-criteria assessment of the trade offs would be (and in real life is) performed.

Environmental scientists and some ecological economists take a long term perspective regarding the environmental sustainability of the economy. They point out that the Earth is a closed biophysical system. In this situation, no economic system can have unlimited growth in physical terms (a different numeraire reflecting the lack of comparability) without harming its own reproduction in the long run and the viability of the social and environmental systems already in the medium term. This concern leads to the much debated demand for limits to the throughput of resources through our societies (in absolute terms, as these are what counts for nature: the argument is based on an entropic view of accumulating environmental impacts, not on resource scarcity concerns). Reducing fuel consumption by a factor of four is by now broadly accepted as a target for safeguarding the climate in the 21st century (von Weizsaecker 1997), while for material flows a factor 10 reduction is considered a first indicative goal (Schmidt-Bleek 1994). For land use, only preliminary targets have been suggested so far (WWF et al. 2000; Spangenberg 2004).

Social and institutional sustainability criteria for the development of the economy are hardly ever discussed. Economic growth is mostly perceived as the key condition for all kinds of social improvements, although this is to some degree contra factual: while empirically employment is correlated to economic growth, distributional justice is not (Alber 2002). Decreasing inequality with increasing prosperity (the 'Kuznets curve') does not emerge automatically, but results from active social, i.e. redistribution politics in affluent countries (Kuznets 1955)[4].

So whereas there are ideas to be found in the economics literature regarding the environmental, social and sometimes also the institutional sustainability of the economic system, there is hardly any information available on the economic sustainability of the economy.

This is all the more surprising, as the economic sustainability of social security systems or environmental protection legislation is a core issue of the policy debate. The understanding of capital stocks as dynamic systems and their interactions as essential for sustainable development can help to extend the perspective and to shed new light on some of these controversies. This is all the more true when the focus is on the dynamics, not in the inventory of elements of different capital stocks, as it should be.

2. An Alternative Approach to Complexity

Systems theory can help to distinguish suitable and non-adequate scientific approaches to sustainable development, by defining the four dimensions as systems and sustainability as enhancing the viability of the metasystem. The key means to assess the complexity of a system is by analysing the system rules. One way of doing so is to define rules which gradually, when applied successively, drive a system from a rather undefined to a deterministic state.

The more the rules are differentiated, the more classes of systems emerge. However, this is only useful for the analytical process, as far as the character of the resulting systems is significantly distinct between the classes defined this way. A high level of differentiation of system types with distinct behavioural characteristics can be achieved by formulating five subsequently applied rules, which

4 This suggests a new justification for the term 'Environmental Kuznets Curve' EKC as describing the effects of dedicated environmental politics in affluent countries, which is indeed the core of the EKC phenomenon (Spangenberg 2001).

together result in maximum determination (Allen 2001). Lifting them one by one changes the character of the system towards a less deterministic one, and the resulting types of systems can be compared to economic reality itself, and to economic theory.

As a result, different lines of economic argumentation can be allocated to the "mental models" of different levels of complexity. This permits to distinguish those kinds of models (and thus theories) which are capable of delivering a description of the economic reality which adequately reflects the underlying kind of system dynamics, and those which are mismatches, not suitable for relevant analyses and predictions.

The five rules could be aggregated to a smaller number and otherwise rephrased (see e.g. Bossel 1999), but this would result in a loss of information. Thus the formulation chosen permits both, the distinction of system types and the reference to key assumptions of modern economics. The five system rules as suggested by Allen 2001 are, simply expressed:

1. It is possible to distinguish between "the system" and "its environment".
2. All system components can be recognised and distinguished, permitting to understand (analytically or intuitively) their interaction.
3. The active system elements are all identical, or at least the range of their behaviour is normally distributed around the average.
4. The individual behaviour of the system elements can be described by average interaction parameters.
5. The system develops towards a stationary equilibrium, permitting to define fixed relations of system variables.

2.1 Economics

A situation where all five rules apply represents a final state system with no structural change taking place – such a system has no historical time (all interactions happen simultaneously), cannot but move towards its predetermined attractor (the equilibrium) and is unable to adapt its structure to changes in its environment. To a large degree, this is how neoclassical economics describes the economy: as a predetermined system on its way to a clear-cut and predictable future state, permitting rational decisions by allowing to compare the system state before and after a certain action is taken, e.g. by cost-benefit analysis (although strong commensurability is not a necessarily condition for such models; it might rather be a kind of sixth rule). The foundation on which this world view is built is the assumption that rules one to four are applicable.

Rule four characterises systems within which the individual behaviour of the system elements can be described by average interaction parameters. In standard economics, this is realised by restricting the analysis to market exchanges and by assuming predictable behaviour patterns based on the identical motivations of selfish actors, the individual utility maximisation of the homo economicus. Systems based on this assumption (i.e. on rules one to four, but not rule five with its predefined equilibrium) are system dynamic models, providing a mechanical description of change. Unlike equilibrium models, for which the system defines the outcome, dynamic models can develop towards different stationary states, dependant on the starting conditions. Once started in a certain "attractor basin", the model determines the development path toward the respective attractor, with no escape possible. The result is path dependent, and which path is chosen is determined by the starting conditions. The attractor itself can be constant or cyclical; Schumpeter (1928) describes such phenomena as short and medium tern innovation cycles.

Giving up rule four results in self organising systems in which according to rule three the active system elements are all identical, or at least their range of behaviour is normally distributed around the average. Standard economics makes use of this assumption, as differing individual demand and indifference curves can only be aggregated into one macro level curve (permitting to determine a clear-cut societal utility optimum) under this condition. Only if the preferences and thus the demand curves are identical (or normally distributed around the representative individual) and with proportional incomes aggregation is theoretically possible (the SMD Sonnenshein-Mantel-Debreu conditions, Keen p. 48). Giving up this assumption would imply the denial of a micro-macro-link, as macro phenomena could no longer be calculated as the aggregate of individual decisions. In such self organising systems, the possibility of the homogenous and not learning individuals to dissent from the average behaviour permits some actors to leave the attractor basin they started in, whereas the normal distribution of behaviour makes sure that this cannot happen for the majority. Nonetheless the system development is no longer determined, different attractors exhibit different probabilities, but they can be randomly chosen: self organisation is a non-equilibrium phenomenon; the development path can change over time. However, this process is undirected, not based on learning but on stochastic variation plus the positive or negative influence from the system environment. Evolution happens on the system level; amplification of processes which received a positive feedback is one of the key processes of self organisation. In this sense, the system as such – unlike the actors in it – has a learning capability, but still no reflection or anticipation capacity.

Giving up rule three lifts all restrictions for assumptions regarding actors' behaviour. It can change any time, dependent on learning, external influences or spontaneously, in any direction. Evolution happens to the system as well as to its subsystems or elements; selection processes on all levels enhance the dynamics of the system evolution. Explorative, non-average behaviour drives such processes. The outcome of the evolutionary process is unpredictable, neither the system structure nor the starting conditions determine the results. As the behaviour of individuals is not predictable, neither in a determined nor in a statistical way (Keynes pointed this out in 1937, but for obvious reasons not in a systems analysis context), no micro foundation for the macro phenomena is possible (except for ex post analysis) – the analysis of evolving systems needs a macroeconomic theory unbiased by references to micro level mechanics. The economic system evolves in co-evolution with other systems like society, the environment and other economies (Costanza et al. 2001).

2.2 Economies

The rules three, four and five can be shown not to apply or be fulfilled in real-world economic systems (in particular, if rules four and three are not valid, rule five cannot be so either). Regarding rule five, the development towards an equilibrium, Sraffa could show nearly 80 years ago that the assumed decline of marginal cost and utility leading to equilibria is rather the exemption than the rule.

As far as rule four is concerned, the economic actors are not identical homini economici acting according to standardised interaction patterns (see e.g. Ecological Economics 35/3). The diversity and dynamic development trends inherent to paid and unpaid work, production and reproduction, and the irreversibility of production and consumption processes rule out that "individual behaviour of the system elements can be described by average interaction parameters". Instead their interactions are complex, driven by changing motivations and developing highly variable patterns.

Regarding rule three, the behaviour of economic actors does not conform to the condition of being identical or normally distributed (for the consumer side see e.g. Roepke & Reisch 2004). Neither is the average constant, nor the distribution necessarily symmetrical. Innovation and structural change are driven by non-identical behaviour. Although impossible under rule three, in reality actors are learning and anticipating (despite of all the mistakes they make in trying to do so). Whereas the identity of demands may have applied to certain social groups (tribes, casts, classes, religious communities) in the past, with individualisation the range of situations where it can be assumed to be still valid is converging to nil. As a consequence, rule three does not apply to human societies, in particular not to affluent consumer societies, and models based on it are not a suitable basis for analysing the economy and a rather safe guess for wrong predictions regarding future developments.

However, rules one and two apply to economies: it is possible to define their border lines (although not undisputed, for instance in the case of unpaid labour), and their elements can be identified and their interaction understood – at least that is what much of economics is all about. Thus complex evolving systems are the only ones to match the characteristics of the real economy. Neither self organisation models nor non-linear dynamic ones stand the test, but least of all equilibrium models.

2.3 The Mismatch

As a result, standard economics is applicable only to economic systems where rules three to five apply, i.e. where consumer and producer behaviours are constant, and identical, where the interplay of actors is standardised and where the system inevitably develops towards a predetermined and unchangeable equilibrium – at best a rather specific subset of real world situations. Only for these can economic theory make useful and reliable predictions. In particular, it is not up to the challenge of understanding the evolution of economies and thus not sustainable development (even not in those derivates where the rules four and five have been modified or abandoned). Evolution is unpredictable, not even probabilities can be calculated.

Although making restrictive assumptions is a legitimate scientific procedure, conclusions reached on this basis must not be applied to any system with a different behavioural dynamic than the one used for analysis (i.e. where rules three to five do not all apply). Unfortunately, there are only rather few non-equilibrium models – many models assume rule one to five to apply, most models include at least rule one to three or four. All these are overdetermined as compared to the development dynamics of the real system, and inadequate to describe or even predict the functioning of economic systems (except under rather heroic assumptions valid in quite rare circumstances). Unfortunately, these specific assumptions are hardly ever made explicit and tested before a model is applied to reality. Before using any specific theory and applying it to a specific situation it is a matter of scientific soundness to examine the applicability of the theory to the situation. If this not done in a thorough and transparent manner, all conclusions from the model are obscure and cannot claim scientific credibility.

From these deliberations three conclusions can be drawn:

1. As for evolving systems such as the economy, the behaviour of individual actors cannot be aggregated into a macro figure either mechanically or statistically, and as the patterns of exchange between actors are variable and evolve as well, we need a revival of macroeconomics (and a microeconomic theory conforming to that, not vice versa).
2. This cannot be a mathematically formulated theory based on fixed relations of the system elements (including stochastic variations), but must permit dynamic development based on uncertainty, unpredictable and chaotic futures as part of the possible scenarios. Such models can illustrate options, but neither predict futures nor lead to the identification of unambiguously "best" politics.

3. System analysis can provide a basis for such models, as it offers the best available instruments to describe the behaviour and evolution of such complex systems. However, as the results of system analysis are necessarily on a rather abstract level, they need to be brought down to earth by interpreting them through economic theory. Evolutionary and ecological economic thinking provide a basis for this endeavour, but they should be better aligned and the possible synergies more systematically exploited.

However, if standard theory cannot deliver any solid advice regarding politics for sustainable development, on what should economists base their recommendations? The attempt to proceed in a scientifically sound way causes a serious problem: as any model making clear-cut prognoses needs to assume a fixed relation between system parameters, the alternative is between using well defined mathematical formulae to provide an unambiguous result with ambiguous relations to reality, or to restrict mathematics to system analysis, and simulation models to the role of illustrations for certain assumptions, ending with ambiguous results having a clear relation to reality. Only the second approach can be considered scientifically sound. Therefore we suggest to develop sustainability science (comprising systems analysis, post-normal science, uncertainty and transition research, evolutionary and ecological economics, plus insights from political and environmental science, sociology, psychology etc.) as a basis to derive suggestions how policies for sustainability could look like (see Funtowicz & Ravetz 1994; O'Connor 2000; Spangenberg 2003; Munda 1995; Coenen & Grunwald 2003; Hans-Böckler-Stiftung 2000; ICSU et al. 2002; Köhn 1999; Ramos-Martin 2003; Rammel 2003; Rammel & van den Bergh 2003, …). These cannot claim to represent an optimal solution, as in complex systems there is more than one criterion to be taken into account for judgements, and any multi-criteria optimisation process with incommensurable criteria must result in a range of options, not in an optimum solution. This is why decision making for sustainable development is in need of open, transparent and democratic *political processes*: they are needed to make a choice, based on informed preferences and responsibilities, for one of the sustainable options.

The following sections illustrate briefly how such a process has been conducted, resulting in criteria and indicators for an economically and otherwise sustainable economy.

3. System Analysis, Cybernetics, Orientor Theory

System analysis is a discipline analysing the common characteristics of systems in different disciplines and spheres of life in order to understand the system processes and use this knowledge to steer or guide the system development (in system theory, Luhmann with his claim that steering were rather generally impossible is a brilliant personality with an extremist view). However, there are few applications of this rather abstract theory to economic systems so far, and before using it for analysing the sustainability of economic development processes, sustainability must be defined for complex evolving systems.

Fig. 1: The Sustainability Interlinkages of the Economic System

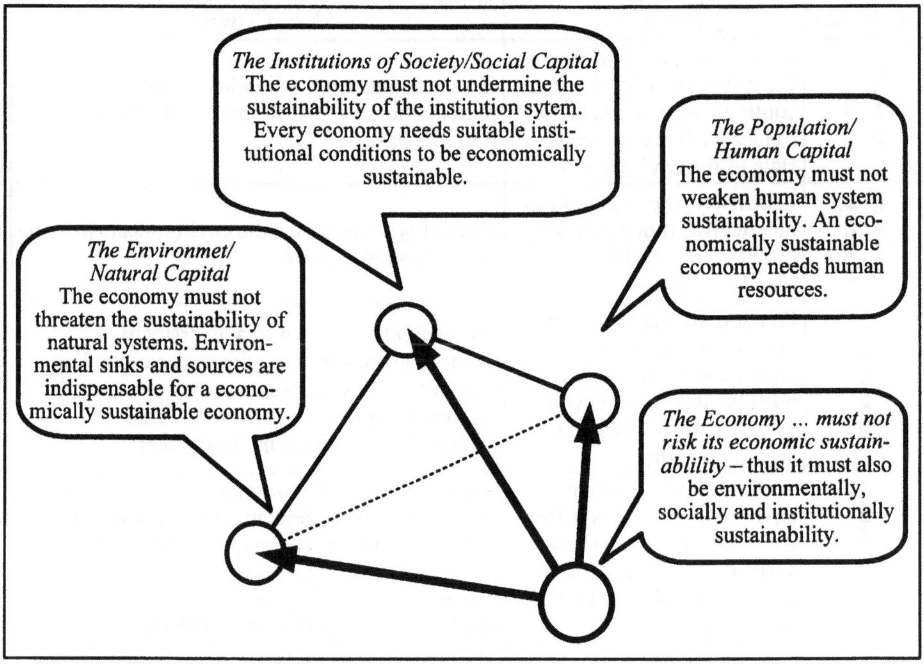

Source: Martin O'Connor, modified

The latter is the raison d'être of Orientation or Orientor Theory (Bossel 1999). According to Bossel, maintaining the viability of a system is equivalent to its sustainability, and viability is maintained if a system is able to react adequately on changes in its system environment.

For sustainable development analysis, the system environment is set to comprise population, society, the natural environment, and other economies. A sustainable economy must not undermine the sustainability of these systems it is interacting with, while defending its own viability, i.e. the economic sustainability of the economy (see Figure 1 for an illustration of this impact relationship).

The challenges posed by the complex system environment can by classified into six independent categories (none of them can be expressed by a combination of the others), which together describe the potential states. As has been shown by theoretical analysis and empirical research, the list is comprehensive. To each of the challenges the system must react by developing a specific capability, equivalent to the conditions for system sustainability (albeit still on a rather abstract level). If these characteristics of sustainable systems are interpreted in an economic way (building on the existing stock of knowledge without relying on non-applicable models), criteria for the economic and general sustainability of the economy can be derived.

Tab. 1: Challenges, Orientors, and Key Criteria

Challenges	Orientors	Key Criteria
Normal state	Existence and reproduction: the system must by able to continuously exist and re-roduce itself in the normal state of the environment	Structural integrity, reliability of conditions for reproduction, supply of goods from the given resource base.
Resource scarcity	Effectivity: the system must be effective (not necessarily efficient) in providing essential resources	Capability to influence the environment, securing access to vital resources
Diversity	Freedom of Action: As the system environment is not homogenous, the system must be able to cope with diversity of supplies, but also of challenges.	The capability to react to a diversity of environmental states requires a corresponding diversity of system structures, processes, and resources used (natural, human, intellectual, technical and institutional resources)
Variability	Security: As the system environment is not constant, the system must cope with variability, i.e. states of the environment distant from the normal state	Security needs early warning and stress detecting institutions, robustness and resistance against minor or short term pressures (enhanced by redundancies) and resilience in case of more serious pressures. Repair-, control- and steering mechanisms are part of the resilience.
Change	Adaptability: As the system environment evolves, the system must cope with changes in the normal state of its environment by adapting its own structures	The capability to evolve as a reaction of external pressures is dependent on the innovation potential within the system, and the way it is realised. A diversity of options, not just one best, but a range of second best option in any specific state is needed to be prepared when optimality conditions begin to change. Flexibility of institutions and reserves that than be managed for the change process are necessary.
Other systems	Coexistence: Every system is dependent on exchange with its environment. Undermining the viability of the system environment is thus a lethal strategy	This refers to the population, the society and the natural environment as much as to other economies. Information exchange is essential for detecting risky developments, as are adequate information processing. Coexistence in co-evolution requires a principle of reciprocity, as the right to sustainable development applies to all systems, but can only be realised if not some live on the expense of the others.

Source: fom Bossel 1996; 1998; 1999; 2000; Peet & Bossel 2000, summarised, completed and modified

These key criteria can be used to analyse by stringent or at least plausible argumentation which criteria should apply to any economy regarding its interaction with the parallel systems. For instance, resource scarcity can refer to human resources; then an adequate education system, socialisation conditions, adult education and the contribution of the economy to this, either by paying fair taxes for a public system or by providing the institutions itself would be relevant criteria.

For natural resources, the well-known management rules apply: do not use renewable resources beyond their regenerative capacity, and do not use non-renewables without developing substitutes of the same scale (for a different definition linked to a variety of measurement tools see Robert et al. 2002).

If the scarcity refers to the institutional dimension, the reliability of the legal and administrative system might be a point in case. It can be assured by combating corruption, and by enhanced distributional justice (as this strengthens the respect for societal conflict management mechanisms, including property rights and the enforceability of contracts). The challenges and the corresponding capabilities (the Orientors, as they provide an orientation for sustainable system development) are listed in Table 1.

In a similar manner, for all Orientors and all systems, criteria can be developed, and characterised by indicators which describe the sustainability of the economy. They differ from traditional economic indicators by

- Stressing the needs for diversity and redundancy of economic structures, processes and technologies;
- Emphasising the need for a balanced exchange with other economies in physical as well as in monetary terms;
- Highlighting the innovation potential in its technical, economic, social and institutional perspective;
- Demanding adequate contributions to the quality of life, the viability of the institutions, social cohesion and a sound environment as conditions of the *economic* sustainability of the economy;

Providing criteria for the social, institutional and environmental sustainability of the economy.

As a result of the system analytic process and its economic interpretation, a system of indicators has been developed permitting to assess the economic sustainability of an economy as well as its social, environmental and institutional sustainability more clearly and comprehensively than before. The difference to the usually applied indicators illustrated that a new approach was necessary, as these could have hardly be derived from standard economic thinking.

The results also emphasise the importance of a fresh effort in macroeconomic thinking, in particular in the context of sustainable development, independent of microeconomic processes and in particular of standard theories, as these are not capable to adequately deal with the problems in question.

4. Discussion and Outlook

This paper has demonstrated the need for and the possibility of a new approach to the evaluate the economic sustainability of economies. Rather a blind spot so far (most papers talking about economic sustainability do indeed address the environmental sustainability of the economy), it would be of obvious relevance for economic policies if this or a similar approach were put into practice, not based on theories which cannot cope with the complexity of reality.

The next step following the definition of criteria is the process of deriving indicators, reducing the possible plethora to a suitable number and testing this set by applying it to sustainability scenarios from different disciplines and actor groups. A second test has been conducted by analysing the theories discussed in the first section to see what lessons might be learned from them despite their weaknesses. It can be demonstrated that the criteria and indicators resulting from them are all included in the Orientor theory based analysis, which however provides a more comprehensive and complete set of criteria. This set can in turn be applied to existing sustainability scenarios to identify their weak points (missing themes, imprecise assumptions, etc.) and to help develop more comprehensive policy strategies.

As describing these steps as well would go far beyond the what can be presented in a single paper, they will be made public in future publications.

References

Alber, J. (2002), "Besser als sein Ruf – Der Sozialstaat als erfolgreiches Modell." WZB-Mitteilungen 2002 (98): 24-28.

Allen, P.M. (2001), The Dynamics of Knowledge and Ignorance: Learning the New Systems Science. Integrative Systems Approaches to Natural and Social Dynamics. H.M. Matthies & J. Kriz, Berlin, Heidelberg, New York: Springer: 3-30.

Bossel, H. (1996). 20/20 Vision. Exploratios of sustainable futures. Kassel: Center for Environmental Systems Research.

– (1998), Earth at a crossroads – Paths to a sustainable future. Cambridge: Cambridge University Press.

– (1999), Indicators for Sustainable development: Theory, Method, Applications. Winnipeg, Canada: IISD International Institute for Sustainable Development.

– (2000), "Policy assessment and simulation of actor orientation for sustainable development." Ecological Economics 35(3): 337-355.

Bundeskanzleramt (2002), Perspektiven für Deutschland. Unsere Strategie für eine nachhaltige Entwicklung. Berlin: Bundesregierung.

Coleman, J.S. (1990), Foundations of Social Theory. Cambridge, MA: Harvard Univ. Press.

Costanza, R., Cumberland, J., Daly, H.E., Goodland, R. & Norgaard, R. (2001), Einführung in die ökologische Ökonomik. Stuttgart: Lucius & Lucius, UTB.

Daly, H.E. (1991), Steady-State Economics. Second Edition with new Essays. Covelo/Washington, D.C.: Island Press.

Ecological Economics (1998), "Special issue on the Value of Ecosystem Services." Ecological Economics 25(1): 1-142.

– (2000), "Special issue 'Beyond homo economicus'." Ecological Economics.

Ehrlich, P.R., Wolff, G., Daily, G.C., Hughes, J.B., Dalton, M. & Goulder, L. (1999), "Knowledge and the environment." Ecological Economics 30(2): 267-284.

Etxezarreta, M., Grahl, J., Huffschmid, J. & Mazier, J. (ed.) (2003), EuroMemo 2002. Hamburg: VSA Verlag.

Fisher, I. (1906), The Nature of Capital and Income. New York: Kelly.

Funtowicz, S.O. & Ravetz, J.R. (1994), "The worth of a songbird: ecological economics as a post-normal science." Ecological Economics 10: 197-207.

Hans-Böckler-Stiftung (ed.) (2000), Arbeit und Ökologie, Endbericht. Düsseldorf: Hans Böckler Stiftung.

ICSU International Council for Science (ed.) (2002), Science and Technology for Sustainable Development. Science for Sustainable Development. Paris: ICSU.

Keen, S. (2001), Debunking Economics. The naked emperor of the social sciences. Annandale, NSW, Australia: Pluto Press Australia.

Keynes, J.M. (1937), "The General Theory of Employment." Quartely Journal of Economics 51: 209-223.

Klemmer, P., Becker-Soest, D. & Wink, R. (1998), Leitstrahlen, Leitbilder und Leitplanken – die drei großen "L" der Nachhaltigkeitspolitik. Zukunftsfähigkeit und Neoliberalismus. A. Renner & Hinterberger, Friedrich. Baden-Baden: Nomos.

Köhn, J. (1999), System Hierarchy, Change and Sustainability. Sustainability in Question: the search for a conceptual framework. J. Köhn, Gowdy, J., Hinterberger, F. & van der Straaten, J. Cheltenham: Edward Elgar: 85-100.

Kuznets, S. (1955), "Economic growth and income inequality." American Economic Review 49(1): 1-28.

Lawn, P.A. (2001), "Scale, prices, and biophysical assessmenets." Ecological Economics 38(3): 369-382.

Munda, G. (1995), Multicriteria Evaluation in a Fuzzy Environment. Heidelberg: Physica Verlag.

O'Connor, M. (2000), "Pathways for environmental evaluation: a walk in the (hanging) gardens of Babylon." Ecological Economics 34(2): 175-193.

OECD (ed.) (2001), Analytic Report on Sustainable Development SG/SD(2001)1-14. Paris: OECD.

Pearce, D.W. & Atkinson, G. (1993), "Capital Theory and the Measurement of Sustainable Development: An Indicator of Weak Sustainability." Ecological Economics 8(1): 103-108.

Pearce, D.W. & Barbier, E.B. (2000), Blueprint for a Sustainable Economy. London: Earthscan.

Pearce, D.W., Barbier, E. & Markandya, A. (1990), Sustainable Development. London: Earthscan.

Pearce, D.W. & Turner, R.K. (1991), Economics of Natural Resources and the Environment. Baltimore.

Peet, J. & Bossel, H. (2000), "An ethics-based system approach to indicators of sustainable development." Int. J. Sustainable Development 3(3): 221-238.

Rammel, C. (2003), "Lessons from the Red Queen." Int. J. Sustainable Development 6(4): 395-416.

Rammel, C. & van den Bergh, J. C.J.M. (2003), "Evolutionary policies for sustainable development: adaptive flexibility and risk minimising." Ecological Economics 47(2): 121-133.

Ramos-Martin, J. (2003), "Empiricism in ecological economics: a perspective from complex systems theory." Ecological Economics 46(3): 387-398.

Rennings, K. (2000), "Redefining innovation – eco-innovation research and the contribution from ecological economics." Ecological Economics 32: 319-332.

Robert, K.-H., Schmidt-Bleek, F., Aloisi de Lardarel, J., Basile, G., Jansen, J.L., Kuehr, R., Price Thomas, P., Suzuki, M., Hawken, P. & Wackernagel, M. (2002), "Strategic sustainable development – selection, design and synergies of applied tools." Journal of Cleaner Production 2002(10): 197-214.

Roepke, I. & Reisch, L.A. (2004), Sustainable Consumption and Ecological Economics. Aldershot, UK: Edward Elgar.

Schmidt-Bleek, F. (1994), Wieviel Umwelt braucht der Mensch? MIPS, das Maß für ökologisches Wirtschaften. Berlin, Basel, Boston: Birkhäuser.

Schumpeter, J. (1928), "The Instability of Capitalism." The Economic Journal 38: 361-386.

Serageldin, I. (ed.) (1997), Expanding the Measure of Wealth; Indicators of Environmentally Sustainable Development. Washington, D.C.: The World Bank.

Serageldin, I. & Steer, A. (1996), Sustainability and the Wealth of Nations; First Steps in an Ongoing Journey. Washington, D.C.: The World Bank.

Spangenberg, J.H. (2001), "The Environmental Kuznets Curve – a Methodological Artefact." Population and Environment 23(2): 175-192.

– (2002), "The changing contribution of unpaid work to the total standard of living in sustainable development scenarios." Int. J. Sustainable Development 5(4): 461-475.

– (2003), Forschung für Nachhaltigkeit – Herausforderungen, Hemmnisse, Perspektiven. Handbuch Nachhaltige Entwicklung. Wie ist nachhaltiges Wirtschaften machbar? G. Linne & Schwarz, Michael. Opladen: Leske + Buderich: 531-550.

– (2004), "Biodiversity pressure indicators – assessing the limits to sustainable use." Ecological Economics 46: accepted for publication.

von Weizsäcker, E.U., Lovins, A.B. & Lovins, L.H. (1997), Factor Four. Doubling Wealth – Halving Resource Use. London: Earthscan.

WWF World Wide Fund for Nature, UNEP-WCMC, UNEP World Conservation Monitoring Centre, Redefining Progress, Anahuac University Centre for Sustainability Studies (2000), Living Planet Report 2000. Gland, Switzerland: WWF.

Challenges on New Technologies for Sustainable Energy System – Climate Change Policy Measures in Japan

Morihiro Kurushima

Director General,
Global Environment Technology Department
Tokyo University, Tokyo, Japan

New Energy and Industrial Technology Development Organization
(NEDO), Kanagawa 212-8554, Japan

Abstract

To secure win-win situations including technology transfer for Sustainable Energy Systems, the making of alliances between the four players involved in Climate Change Policy Measures; government and its organization, academia, financial circles and industries, is essential for our common future. The "World Technology Bank for the Energy and Environment" may be the most appropriate and the shortest way to achieve our goal.

Introduction

The role of industrial technology is expected to gain further importance in the near future in providing a strong support for economic development, social development and environmental protection, which are the three main factors of sustainable development.

NEDO-Japan, an implementing agency with the Japanese government, has been carrying out technology development and promotion of new and renewable energy, energy conservation, research and development projects such as 3R (reduction of waste generation, reuse and recycling) technologies, technologies for reducing environmental pollutants, new materials, machinery, aerospace and biotechnology, thus supporting research, development and introduction of feasible technologies for industry, with subsidies from the Ministry of Economy, Trade and Industry (METI).

Particularly, with regard to global warming policies, which play an important role in environmental protection, NEDO undertakes various technology developments and promotes introduction in order to contribute to Japan's surpassing the goal obligated by the Kyoto Protocol to the United Nations Framework Convention on Climate Change to reduce greenhouse-gas emissions during 2008-2012 by 6% from the 1990 level.

1. Current Situation of GHG Emissions in Japan

Table 1 illustrates quantitative targets by sector for GHG emissions reduction in order to achieve Japan's overall target of 6% below 1990 levels, as stipulated in the Kyoto Protocol.

Tab. 1: GHG Emission Reduction Strategies in Japan

CO_2 emissions from energy use	± 0%
CO_2 emissions from non-energy use, methane emissions and nitrous oxide emissions	-0.5%
Emissions of HFCs, PFCs and SF_6	+2.0%
Reductions by innovative technologies and change of lifestyle	-2.0%
Sinks	-3.9%
Kyoto Mechanisms	-1.6%

However, domestic GHG emissions have been increasing since 1990, and were 8% above 1990 levels in 2000 (Figure 2).

Fig. 1: GHG Emissions in Japan (FY1990-2000)

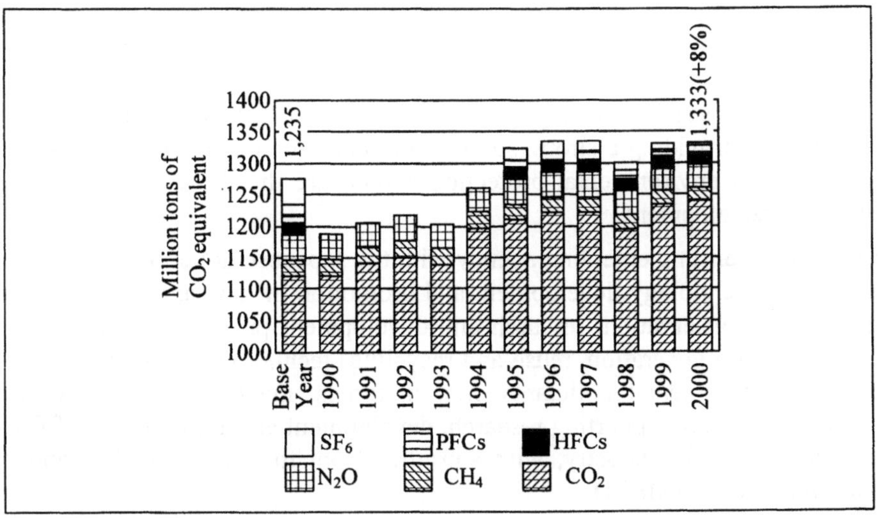

The major source of GHG emissions in Japan is energy combustion, contributing approximately 90% of total emissions. Over the decade 1990-2000, substantial increases in CO_2 emissions have been particularly evident in the transportation sector (20.6% increase over 1990) and business/home sector (21.3%). On the other hand, CO_2 emission levels from the industry sector have shown little increase since 1990 due to Japan's lengthy experience mentioned above relating to development and promotion of energy conservation technologies. Therefore, energy conservation technologies in Japan have great potential to contribute CO_2 emission reduction in developing economies and economies in transition. Because CDM/JI project activities offer the dual effects of reducing GHG emissions in such economies and transferring CO_2 reduction credits to Japan, CDM/JI utilization has become more significant.

2. Outline of Japan's Activities on CDM/JI

Japan, having already tackled development and promotion of energy conservation technologies after the second oil crisis, emits the lowest level of CO_2 of developed countries – approximately 9.4 tons per capita in the year 2000. Consequently, Japan is able to contribute to CO_2 emissions reduction in developing economies as well as in economies in transition by application of Japan's energy conservation technologies.

Tab. 2: Experiences of CDM Capacity Building

- CDM Workshop (China), Sep. 2004
- CDM Training course (Japan), June – Aug. 2004
- CDM Manuals (Vietnam and Thailand), 2003 – 04
- CDM Technology Needs Assessment (The Philippines), Dec. 2003-Mar. 2004
- CDM Training course (Japan), June – Aug. 2003
- CDM Seminar (Thailand), Mar. 2003
- Japan-ASEAN CDM Workshop (Indonesia), Jan. 2003

Because the Clean Development Mechanism (CDM) and Joint implementation (JI) of the Kyoto Mechanisms are efficient tools, the Japanese government's policy towards emission reduction makes active use of CDM/JI, thereby supporting domestic efforts in realizing Japan's reduction commitment.

METI of Japan is undertaking establishment of a system to facilitate the Kyoto Mechanisms. NEDO with METI supports CDM and JI project activities implemented by Japanese private sector enterprises.

CDM projects will contribute not only to global environment conservation but also to sustained development of host countries.

One of METI's roles is to plan, and sequentially implement, the "Asia CDM Capacity Building Program" agreed at WSSD (the World Summit on Sustainable Development) held in the Republic of South Africa in 2002.

The aim of CDM capacity building is developing incentives to accept CDM projects and assisting development of approval systems in host countries. Experiences of CDM capacity building by METI and/or NEDO are shown in Table 3.

In addition, CDM capacity building by NEDO is being planned in the Philippines, Malaysia, and Indonesia in the near future.

New Technologies against Climate Change

In the field of new/renewable energy, technology development activities such as economic and facility utilization efficiencies are now being studied. In addition, in order to expand utilization of such energy sources, financial support is provided for the introduction of highly advanced equipment and educational activities planned by domestic local government authorities.

As for energy conservation, NEDO has been involved in a series of projects aiming for basic innovation of energy use in the industrial, residential and transportation sector. In the industrial field, for example, NEDO intends to contribute to the goal by carrying out projects that concern development and introduction of high-performance industrial furnaces, development of compound semiconductors for high-efficiency conversion, etc. Furthermore, technologies to reduce standby power consumption are being developed in the residential field, whereas advanced clean energy vehicles are being developed and promoted in the transportation field.

Regarding CO_2 sequestration and utilization technology development, NEDO is carrying out projects on CO_2 separation, bio and chemical fixation, covering all levels of development ranging from fundamental studies to full-scale research. Currently, research and development projects such as predicting the environmental impact of CO_2 ocean sequestration, and geological sequestration, are under way.

NEDO is expected to play an important role in the Japanese government's future actions by making contributions to achieve the goal for greenhouse gas reduction.

CTI as the Tool Box for Energy and Environment

The Climate Technology Initiative (CTI) is a multinational initiative of 8 countries as the Implementing Agreement of the International Energy Agency (IEA). Japan serves as the Secretariat for the CTI (Figure 4).

The CTI promotes the objectives of the United Nations Framework Convention on Climate Change (UNFCCC) by fostering international cooperation for diffusion of climate-friendly technologies. The CTI provides an important mechanism to help industrialised countries fulfil their commitment to the Convention to promote the transfer of climate-friendly technology to developing countries and those with economies in transition.

As one of the CTI activities, informative seminars are held in developing and transition countries. Each seminar caters to the needs of the region, and to those of the UNFCCC process. The location of the CTI/Industry Joint Seminars on

Technology Diffusion rotates among four regions, with the objective of visiting each region once annually. For an illustration of this aspect of the joint industry seminar series, see the attached map 5.

As such, the CTI seminars provide (i) opportunity for the private sector to voice their policy recommendations to key decision-makers, helping to create a climate that attracts environmentally sound technologies; (ii) a platform to pave the way for project development; (iii) insight into financing opportunities for climate projects, including the CDM/JI, and risk management opportunities.

In addition, the CTI holds training courses in each developing region and in the countries with economies in transition. The training courses have been held in Asia, in Latin America and Caribbean, Central and Eastern Europe and Southern Africa. The training courses are designed to "train the trainers." Countries are asked to select participants who will be in a position to return to their organizations and share information and techniques that are disseminated at the events. The training course format allows for exploration of concepts in significant detail over a week long period.

Therefore, the CTI may play an important role to secure win-win situations including technology transfer in the field (Figures 7 + 8).

5. For Our Common Future

In accordance with the efforts mentioned above, the making of alliances between the four players involved in Climate Change Policy Measures, government and its organization, academia, financial circles and industries, is essential for our common future. The "World Technology Bank for the Energy and Environment" may be the most appropriate and the shortest way to achieve our goal.

From Cleaner Production to Zero Emissions

Mickael Planasch[1], Gernot Gwehenberger[1], Hans Schnitzer[1], Christoph Brunner[2]

[1] Faculty of Chemistry, Chemical- and Process Engineering, Biotechnology,
Graz University of Technology,
Inffeldgasse 21b, A – 8010 Graz, Austria

[2] Joanneum Research, Institute for Sustainable Techniques and Systems,
Elisabethstrasse 16-18, A – 8010 Graz, Austria

Introduction

This paper will discuss the idea of Zero Emissions first. Thereafter it will present different tools we have developed or were part of the developing process. We use these tools in our daily work helping companies trying to achieve Zero Emissions.

What is Zero Emissions?

To be able to discuss the question "What means Zero Emissions?" we first of all have to define the word "emission".

A production process consists of different unit operations that help in producing a product. To be able to run the process raw materials, operating materials, auxiliary material, energy and labour force are needed and represent more or less constant streams. During the process these different streams are treated mechanically, chemically or thermally. Due to losses because of efficiency factors of the single process steps (apparatuses, pipes, ...) the amount of material that will be part of the product in the end is reduced. These losses are called emissions and they can be presented in a very simple equation:

$$\text{Emissions} = \text{INPUT} - \text{OUTPUT}$$

Another name for Output is Product. Input consists of all necessary mass and energy streams for the production process and there are four different sorts of emissions: solid waste, fluidic waste, gaseous emissions and energy losses.

The reduction of mass and energy losses means a reduction of the running costs and so is not only a benefit for the environment but means also a financial benefit for companies.

Zero Emissions means a production process that produces no waste at all. All resources are used. That can be achieved on two different ways, either by the optimisation of the processes itself or by transforming the waste into a product that can be sold.

The meaning of "Zero":

Zero can be defined in different ways:

- Absolute Zero
- Analytical Zero
- Zero Impact.

Absolute Zero:

It is the strictest definition of Zero. Concerning companies it means no emissions at all: no waste, no sewage water, no polluted air and no energy loss. It may be the perfect company of the future but in our eyes it is neither necessary nor will there be the will in companies to achieve this goal. It will be nearly impossible to reach this goal with economic justifiable effort.

Analytical Zero:

Every substance has a detection sensitivity limit. If the concentration of a substance in emissions is below this limit, it will not be detected by analytical methods. To define Zero Emission by these methods requires the concentration of every substance in emissions to be below its detection sensitivity limit.

The problem is that nearly every year new detection systems are invented that make it possible to measure lower concentrations. So companies would need to optimize their plants in very short periods with increasing effort because it is getting much more expensive the lower the concentration is you want to remove from a stream.

Zero Impact:

... on the environment. As a result of our work we have discovered that zero must not be interpreted as zero in a strictly analytical sense. If the concentration and the total mass flow of a substance in emissions are smaller than natural fluctuations in geogenic flows, we assume that there is no impact on the environment. We call these emissions "Zero Emissions". Equally the input of any process has to be considered. The usage of renewable resources must not exceed the natural rate of replenishment. If non-renewable resources must be used, the yearly exploitation of the resources must not exceed the quantity that guarantees future generations the same rights of exploitation.

Why Zero Emissions?

Due to our definition of emissions, they mean lost input. Companies have either depreciations for machineries, labour force, ..., if the company is producing raw materials like stone, clay, oil, ... or costs for buying these raw materials, if they are part of the manufacturing chain. Nevertheless all companies have to deal with costs for their inputs.

To reduce these inputs by the optimisation of the production processes means to reduce their costs.

These days the most popular way companies try to reduce their production costs is firing employees. This has a short term action with fast effects but the companies have to deal with the problems, that occur due to this behaviour, in medium and long term.

Our experience shows that mostly companies believe that the highest costs during a process chain are labour costs. So their behaviour is very obvious and it is easy to understand why they choose to handle it this way. But if you take a closer look at the cost structure of producing enterprises of most branches you can identify another configuration:

In German car manufacturing companies the costs for material are 67% and for personnel just 17%. The numbers vary a bit between the branches of producing companies but the structure stays the same: Material costs take close to 2/3 of the costs in producing (no service!) enterprises and personnel costs are just close to 1/5 of the total costs.

Cost structure of the German car manufacturing industry in 2001

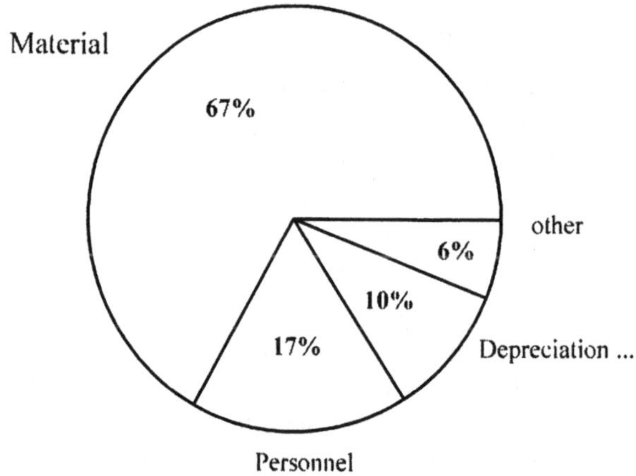

So it is very clear why it is necessary for companies to achieve Zero Emissions: They reduce their costs!

Zero Emissions is not a threat or bondage – it is the solution for a sustainable enterprise.

In the next chapter we will describe some tools we use in our daily work in companies. They have been well developed during the last few years and filled with life with a lot of case studies.

EMA – Environmental Management Accouting

In 1999 the UN Commission for Sustainable Development installed an expert group on environmental management accounting (EMA). The Austrian ministry of transport, innovation and technology (BM VIT), the Austrian ministry of agriculture, forestry, environmental and water management (BM LFUW) and the Austrian chamber of commerce (BWK) commissioned Christine Jasch to write a book on procedures and metrics for EMA. This book was published by the UN DSD and the Austrian BMVIT.

During this work the first case studies on EMA were finished with Austrian enterprises.

The methodology of EMA is used in our case studies with companies and was further developed into a specialised EMA-WATER in 2005.

What is EMA?

EMA combines financial and physical data in companies and calculates the "total" environmental costs for these companies. In this case physical data means material and energy input, material flows, products, waste and emissions, whereas financial data means expenditures, costs, earnings, savings related to company activities with potential environmental aspects or impacts.

What are Environmental Costs?

The definition of total environmental costs is not that easy anyway. Total environmental costs combine environmental protection costs and costs for wasted material and wasted capital and labour.

Environmental protection costs are defined as costs for prevention, disposal, planning, control, shifting actions and damage repair that can occur at companies, governments or people (VDI 2000). The costs for wasted material are taken into account because "waste" has a double meaning. Waste is a material which has been purchased and paid for but which has not turned into a marketable product. Waste is therefore indicative of production inefficiency. Thus, the costs of wasted materials, capital and labor have to be added to arrive at total corporate environmental costs and a sound basis for further calculations and decisions. Waste in this context is used as general term for solid waste, waste water and air emissions, and thus comprises all non-product output.

	Environmental protection costs (Emission treatment and pollution prevention)
+	Cost od wasted material
+	Cost of wasted capital and labour
=	Total corporate environmental costs

Methodology – The EMA Workshop

In each company a workshop is run on location that lasts for one or two days. The company has to take part with one member of controlling, technical and – if available – environmental management staff. The EMA team consists of two or three persons as well.

First of all the idea and methodology of EMA is described in a short presentation to help the company understand the different steps.

Thereafter the collection of data begins starting with listing all environmentally relevant plants in the company with the help of the technical stuff. The second step is to find the depreciation with the help of the controller. Thereafter the hours of work for each facility are estimated by the technical stuff and the costs for this work are calculated by the controller.

So with the knowledge of the production process by the technical stuff and the financial data the controller is responsible for all necessary data is collected.

During this step of the workshop the EMA team has the role of coordinating, asking the right questions and helping both, the controller and the engineer to communicate with each other. The calculation sheet for the workshop is based on Excel. One member of the EMA team is responsible for typing the collected data into the prepared worksheet.

All company projects are compiled and described on the basis of existing data records, balance sheets, lists of accounts, cost centre reports, list of assets etc. Each company is transformed into a neutralised description, that allows recalculation of costs from scratch.

The toolkit for environmental management accounting, structured that way, will allow environmental managers as well as accountants and controllers to actually trace the relevant costs in different records, to calculate total annual environmental costs and the costs of an investment example, and to compare their solution to the solution provided by the research team with some further remarks in the annex. Each company will be given opportunity to approve our description of their case for publication.

The experience shows that the collection of all necessary technical information takes much less time than the collecting the financial data. So after all technical data is collected, one member of the EMA staff takes a guided tour through the company to find hidden environmentally relevant facilities that where forgotten in the first place. The most important goal of this first tour is to collect technically relevant date for first optimization proposals. This information includes data in a detail that is not necessary for the EMA process itself but is very important for technical advise. As an example the total energy consumption of a single plant in a year is important for the EMA process but not e.g. the temperature or pressure of the plant. But this data is relevant for technical advice and optimization proposals.

Environmental expenditure/costs and revenue/earnings									
Environmental media **Environmental cost/expenditure categories**	**Air /Climate**	**Wastewater**	**Waste**	**Soil / Groundwater**	**Noise / Vibration**	**Biodiversity / Landscape**	**Radiation**	**Other**	**Total**
1. Waste and emission treatment									
1.1 Depreciation for related equipment									
1.2 Maintenance and operating materials and services									
1.3 Related personnel									
1.4 Fees, taxes, charges									
1.5 Fines and penalties									
1.6 Insurance for environmental liabilities									
1.7 Provisions for clean up costs, remediation									
2. Prevention and environmental management									
2.1 External services for environmental management									
2.2 Personnel for general environmental management activities									
2.3 Research and development									
2.4 Extra expenditure for cleaner technologies									
2.5 Other environmental management costs									
3. Material purchase value of non-product output									
3.1 Raw materials									
3.2 Packaging									
3.3 Auxiliary materials									
3.4 Operating materials									
3.5 Energy									
3.6 Water									
4. Processing costs of non-product output									
Σ Environmental expenditure									
5. Environmental revenues									
5.1 Subsidies, awards									
5.2 Other earnings									
Σ Environmental revenues									

It is not only the goal of the EMA workshop to identify the media or category the highest environmental costs are located in but also to show an optimisation and cost saving potential.

The following data is evaluated during the EMA workshop and analysed thereafter:

- Estimation of the total environmental costs of 200x
- Verification of material balance calculations
- Discussion about existing information systems, weak points and potential improvements.

The Following data is needed from the company

- Access to documents and IP-Structure concerning the internal information system, i.e. a list of all cost pools
- Balance list, cost centre accounting, calculation of costs for manufacturing of goods, stock list, process flow sheets, input/output material and energy balances, ...

Benefit for the companies

EMA is a very helpful tool to show companies their true environmental costs. Most of the time companies do not know that there are, e.g. labour cost or depreciations, on waste water treatment facilities that arise the costs for waste water treatment. Companies just realize the price for the discharge of the waste water. The price for discharge of waste water is around EUR 1,00.- in Austria. The true waste water treatment costs differ from company to company, depending on size, location and branches of the company, but is 5 to 6 times higher in average. During the EMA-process the perception of the companies staff can be changed significantly.

A second very important reason to run an EMA workshop is to help to identify the media a company should start with in its optimization process. It is economically not wise to reduce e.g. the energy consumption if energy is responsible for just 10% of the total costs, and to disregard e.g. solid waste, which is responsible for 36%.

The EMA workshop helps to identify what media it is best to start with and the most cost effective optimization actions.

EMA still has some more benefits which are listed below

- Critical analysis of the consistency of the existing information systems and calculation
- Exchange of experience with latest developments in cost accounting
- Weak points analysis in mass and energy balancing
- Overview of the most important environmental cost pools, divided into cost elements and environmental media
- Optimization for calculating investment appraisal
- Suggestions to optimize the environmental management system, information systems and process engineering (depending on the state-of-the-art in the company)
- More detailed information for external reporting, i.e. environmental report, sustainability report.

EMA-WATER

The EMA-Water Workshop was developed at two EMA-Workshops in food industry and set for discussion thereafter in the workshop series. With the help of a consultant specialised in slaughter houses and food manufacturing companies a model food manufacturing plant was created, that was used as a case study during the workshop.

EMA-Water Workshop means a detailed analysis of the media water in a company. It is not meaningful to make a detailed analysis of the media water without making a "Complete"-EMA Workshop beforehand. There are two reasons for this:

- Making an EMA analysis includes the media water and you recognise the importance of the media water in the company. If the environmental costs for water are very small, it is not effective for a cost reduction to set actions in the media water, i.e. reduction of the water consumption.
- The detailed EMA analysis for water is based on the data of the EMA-Workshop. It is not necessary to add additional data, so you have to do the work any way in a EMA-Workshop.

The list of the data that is evaluated during the EMA-Water workshop and analysed thereafter is increased by one:

- Estimation of the total environmental costs of 200x
- Detailed Analysis of water costs for 200x
- Verification of material balance calculations
- Discussion about existing information systems, weak points and potential improvement.

Working sheet for the EMA-Water Workshop

The environmental media water and its total environmental costs are divided into three sub-medias:

- Water supply
- Process Water
- Waste Water.

Water Supply

All costs for water supply treatment are added to this sub-media. Typical costs include:

- Depreciation for sewer, well, pipes, pumps, water supply treatment plants, e.g. DI-water plant, ...
- Costs for purchasing external water from a water supply company
- External costs for water supply activities, e.g. sewer cleaning, ...
- Personal costs for water supply activities.

If the company has an internal water supply the costs are very cheap most of the time.

Process Water

All activities concerning in-house water management are allocated in this sub-media. The block "Process Water" is most of the time very small, because most of the activities are allocated either to water supply, e.g. DI-Water plant, or waste water, e.g. Chemicals for Waste Water treatment.

Typical costs in this block are: Equipment for the treatment of water running in closed loops, e.g. a membrane plant.

Waste Water

This block is most of the time the most expensive one, because the costs for waste water treatment, including depreciation, auxiliary and operating chemicals, energy, personal costs are high. All the NPO is allocated to waste water as well because the disposal stream crosses the companies border either in the waste water or the sludge or concentrate.

Sludge and concentrate has to be added from the media "waste" in the EMA-analysis.

Example

The result of the example model food manufacturing plant is shown in the next chart. The costs for waste water are 90.1% of the total costs and 52% of the costs are based on fees and taxes. There are also a lot of costs concerning operating materials, because there was no internal recycling of operating chemicals adopted. The energy costs consist of power and compressed-air.

Possible technical advice can include an optimization of the waste water treatment plant, especially the optimisation of the energy and operating materials consumption.

The reduction of the water consumption can have a significant impact on the fees, taxes and charges.

The installation of a process water regeneration units and the closing of water loops can reduce the amount of fresh water needed as well.

The EMA-working sheet shows where to start with optimisation actions. EMA does not identify solutions. Just due to the collection of more detailed technical data engineering advice can be prepared. But at least the decisions as to what actions will be taken are still in the hands of the company.

Water expenditure/costs 2004				
	Water Supply	**Process Water**	**Waste Water**	**Total**
Environmental media		*in %*		
Environmental cost/expenditure categories				
1. Waste and emission treatment				
1.1 Depreciation for related equipment	1	2	15	17,5
1.2 Maintenance and operating materials and services			5	5,5
1.3 Related personnel			3	2,6
1.4 Fees, taxes, charges			52	52,4
2. Prevention and environmental management				
2.1 External services for environmental management				
2.2 Personnel for general environmental management activities				
2.3 Research and development				
2.4 Other environmental management costs				
3. Material purchase value of non-product output				
3.1 Raw materials				
3.2 Packaging				
3.3 Auxiliary materials				
3.4 Operating materials		6	6	12,9
3.5 Energy	1		8	9,2
3.6 Water				
4. Processing costs of non-product output				
∑ Environmental expenditure	**2**	**8**	**90**	**100,0**
5. Environmental revenues				
5.1 Subsidies, awards				
5.2 Other earnings				
∑ Environmental revenues				
Saldo Costs/Earnings	**1,7**	**8,2**	**90,1**	**100,0**

ZEPRA – Balancing Software

Mass and Energy Balancing is very important to optimise existing processes and to evaluate different optimisation actions.

So we developed a balancing tool during the last years. The focus is in the metal surface treatment industry right now, like galvanising or anodising companies, but in these days the tool is being further developed to be able to run other branches as well.

Within the framework of two projects on the reduction of water use in galvanizing plants a software tool was developed for the computer aided simulation of galvanic processes. This software, ZEPRA (Zero Emission PRogram Analysis), allows the modelling of a single galvanic process step or a sequence of several steps up to a complete galvanizing plant. It calculates the optimal water and – linked with it – chemicals use. The software is based on Visual Basic for Excel-Application (VBA), which makes it applicable on a usual PC. The tool was successfully used for the calculation of optimal water use in several case studies. A substantial reduction of water usage was possible. Several types of galvanizing processes can be simulated. Some of the processes at galvanizing plants (e.g. cleaning of parts, degreasing) are used with little or no changes in other industrial sectors. At the moment ZEPRA is adapted for the use in those other areas.

The software is linked to a technology database which allows us to find suitable technologies for reprocessing and recycling of process water. The software will also be linked to existing software for the economic and ecological optimization of the overall system.

Structure of the program

First of all the sequence of the different process baths has to be added. It is possible to fill in the physical parameters of the different baths, like geometric dimensions, temperature, weather the bath is being covered or not.

The physical parameters are listed in the screen shot below:

Beckendaten												
Postion	Typ	Länge	Breite	Tiefe	Abstand	Temperatur	Stufen	Spül-kriterium	beheizt	Galvanobad	abgedeckt	jede xte Charge
		m	m	m	m	°C			l/m³	s		
01	alkalische Entfettung	8	1,2	2	1,7	30	1	1	beheizt	nein	✓	1
02	saure Beize	8	1,2	2	1,7	20	1	1		nein	✓	1
03	Standspüle	8	1,2	2	1,7	40	1	4		nein	–	1
04	Spüle	8	1,2	2	1,7	30	2	100		nein	–	1
05	Alcotex	8	1,2	2	1,7	20	1	1	beheizt	nein	✓	1
06	Borax	8	1,2	2	1,7	60	1	1		nein	✓	1
07	Phosphatieren	8	1,2	2	1,7	20	1	1	beheizt	nein	✓	1
08	Standspüle	8	1,2	2	1,7	40	1	4		nein	–	1
09	Spüle	8	1,2	2	1,7	30	1	20		nein	–	1
10	Aktivierungsbad	8	1,2	2	1,7	30	1	1		nein	–	1
11	Seifenbad	8	1,2	2	1,7	30	1	1		nein	–	1

Thereafter it is possible to enter specific chemical and process engineering and more detailed physical parameters for every single process bath. You can find a typical input sheet below.

It is possible to enter e.g. evaporation parameters (otherwise it is calculated), rinsing criteria's in rinsing baths, recycling parameters.

Recycling parameters has to be set at "yes" if there is a rinsing cascade with regeneration of process chemicals or waste water using a process water treatment unit, e.g. a microfiltration unit. There will be the possibility to link directly to the database for experts but this functionality is not active yet. At the moment the data has to be filled in by hand.

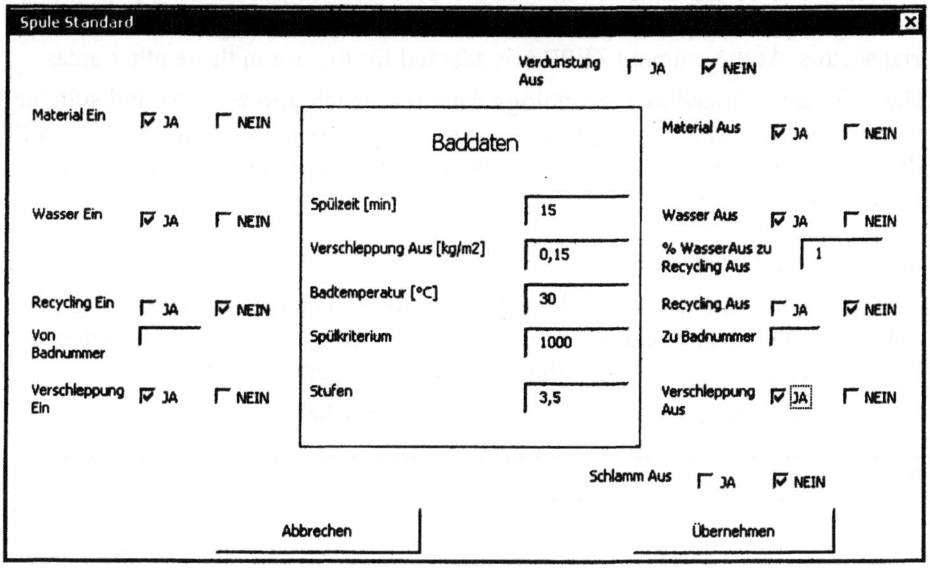

After calculating the different process baths an Input – Output sheet is created for fast and easy check of all parameters.

The Input and Output balance for the different substances is shown with a graph.

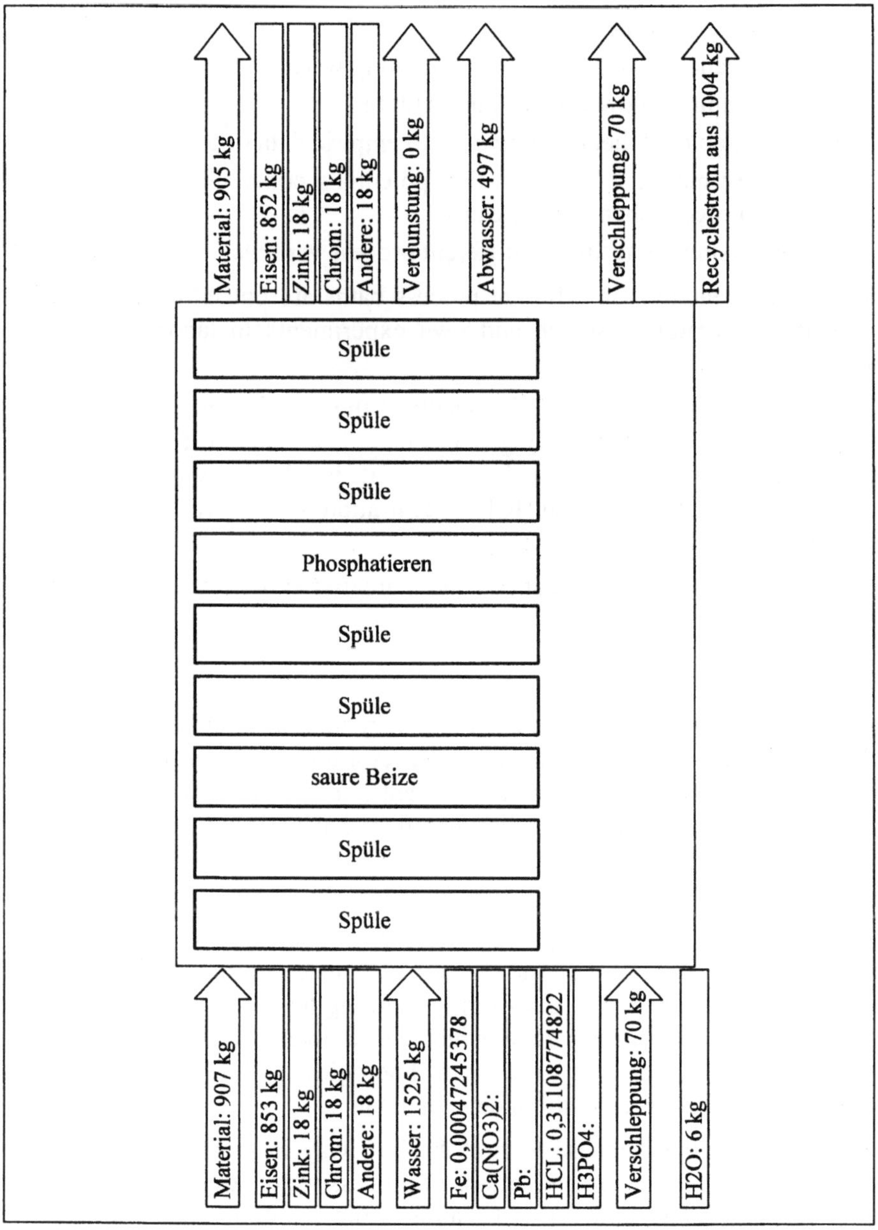

It was possible to add 13 different chemicals which can be calculated and balanced at the same time but this was not enough. So we changed the structure of the program a bit to be able to add up to 100 different substances to be balanced simultaneously.

The benefits for companies are

- Very fast balancing of the process concerning water and process chemicals
- Access to empiric knowledge of our institute
- Baths are calculated theoretically or with empiric data of the single company
- Very fast estimation of optimisation potentials concerning the process and water- and process chemicals streams
- Very fast theoretical testing of different process variants:

The chemical reactions in different process baths are put into the program via results from literature research and own experiments in laboratory and pilot scale.

It is very important to note, that the results of the calculation are just as good as the data that is put in for balancing. So, if there is no data available for a certain process it is necessary to start with experiments in laboratory scale, e.g. etching time of steel in HCl pending on HCl-concentration, Fe-concentration in the bath, temperature.

This data is the basis for calculations and thereafter experiments in pilot scale in the company.

Data base for experts

Hardly ever are companies interested in a very long developing process for occurring problems. When a company identifies a question or a problem, they either neglect solving the problem or they want it to be solved at once. So sustainable solutions are not common because they have an innovative process involved most of the time. And innovation takes its time. “Classic” problem solving including end-of-pipe technologies or processes are the norm.

We asked our selves how it is possible to help companies to solve their problems very quickly but at the same time use our experience in innovative, sustainable solutions we developed in a lot of projects during the last years. So we decided to develop a database for experts in which we compress our knowledge concerning process water treatment and its technologies. Filling the database with knowledge is a process and still in development.

It is the goal of the database to find the best way in technical and economic scales to handle certain contaminations of process water streams. It is our theory that certain technologies are used because of old traditions in certain branches. But problems are nearly the same in different branches, e.g. tensides are used in food and drink, textile, metal processing, … industries. So we try to convert the knowledge we have in single branches into “global” knowledge.

Our database has a structure that helps us to think outside of the branches and propose the “best” solution. We are not limited to solutions for certain branches but we are free to solve a certain problem, e.g. tenside contamination in a process water stream.

The data base is based on Excel VBA because nearly everybody knows how to work with Excel. The tool is based on a system of matrixes that should help the expert to find a solution concerning separability and efficiency. Using a query of process engineering and chemical parameters the expert should be able to find a solution for a certain process water stream.

Every single process or technology has a single row in the excel sheets. The different process water treatment technologies are cut into different areas because of different parameters like its selectivity, function, ... E.g. membrane technology is cut into micro-, ultra-, nanofiltration and reverse osmosis. Micro filtration again is cut into different membranes. Every membrane itself is linked to a lot of parameters like its selectivity, durability to temperature, chemicals, etc.

The next figure gives a short insight into a part of one Excel sheet. Not every process or technology needs the same information, e.g. density is very important for centrifugal separators but is not so important to ion exchange units. Cells containing information that is not necessary or not available for a technology are white.

Simplified pictured, the experts starts a query by entering some information of the process water stream he wants to handle and the tool browses through the rows of the excel sheets. The output is a list of possible processes or technologies. The more specific the input, the more accurate the result is.

The information the database is filled with consists of results of literature researches and own experiments in laboratory and pilot scale.

			ph-Wert		Konzentration [g/l]		Temperatur [°C]		Korngröße [ym]	
			min	max	min	max	min	max	min	max
1) Sedimentation	**Sedimentation**								10	
	Flotation									
	Fällung									
	Kristallisation									
2) Filtration	**Filtration**									
	Mikrosiebe	Siebkorb							10	
	Beutelfilter									
	Trommel & Scheibenfilter									
	Kerzenfilter									
	Sandfilter									
	Bandfilter									
	Plattenfilter									
	Filterpresse									
3) Membranfilter	**Membranfilter**									
	Mikrofiltration								0,05	10
	Ultrafiltration					Aufwärmung des Konzentrates!			0,001	0,05
	Nanofiltration								0,001	0,01
	Umkehrsomose							45	0,001	0,001
	Elektrodialyse				40	70			0,0005	0,1
	Pervaporation						20	60		
	Membranbioraktoren									
4) Ionentauscher	**Ionentauscher**		1	13		2 g/l			0,0001	0,1
	Kationentauscher	stark sauer	1	13						
	Kationentauscher	schwach sauer	4	9						
	Anionentauscher	stark basisch	1	10						
	Anionentauscher	schwach basisch	1	8						
	Selektivtauscher		1,5	10						
5) Adsorption	**Adsorption**									

			ph-Wert		Konzentration [g/l]		Temperatur [°C]		Korngröße [ym]	
			min	max	min	max	min	max	min	max
5) Adsorption	**Adsorption**									
	Adsorption an AK								0,001	1
	Adsorption an organischen Polymeren	Adsorberharze				5g/l				
	Adsorption an organischen Polymeren	Gelextraktion								
	Adsorption an organischen Polymeren	LPR-Retention								
6) Naßoxidation und Desinfektion	**Naßoxidation und Desinfektion**									
	Niederdrucknaßoxidation						200	250		
	Hochdrucknaßoxidation						250	310		
	Hochdrucknaßoxidation	Naßoxidation mit Sauerstoff					120	330		
	Hochdrucknaßoxidation	Naßoxidation mit Oxidantien								
	Hochdrucknaßoxidation	Schadstoffzerstörung durch UV-Bestrahlung								
	Hochdrucknaßoxidation	Schadstoffzerstörung durch Elektronenstrahlen								
7) Zentrifugen	**Zentrifugen**									
8) Hydrozyklon	**Hydrozyklon**									
9) Destillation	**Destillation**									
10) Extraktion	**Extraktion**									
11) Elektrodialyse	**Elektrodialyse**									
12) Verdampfer	**Verdampfer**									

Source: Kunz, Behandlung flüssiger Abfälle (1995); IWA, Water Recycling and Resource Recovery in Industry (2002)

The program

The starting window gives the opportunity to start the query using existing processes or to enter a new process.

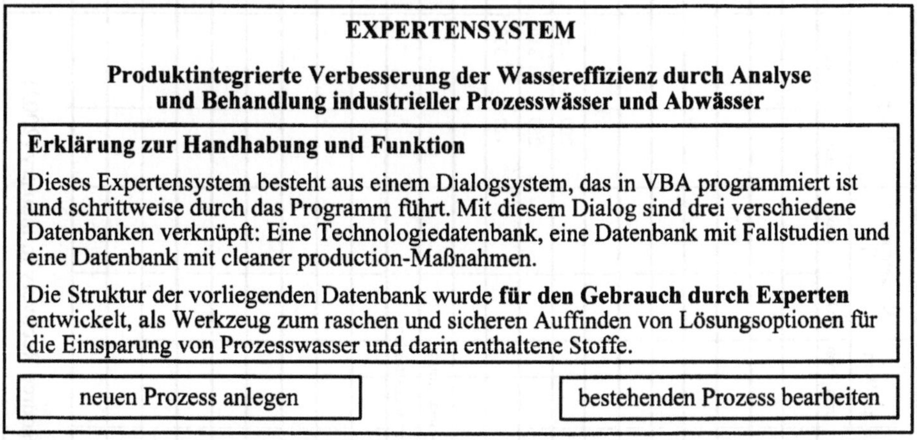

When selecting a query the user is able to enter specific information about the process and the process water stream. E.g. What kind of water is used? (freshwater, process water) How do you use the water? (cleaning, cooling) Is there a secondary use of the water planed?

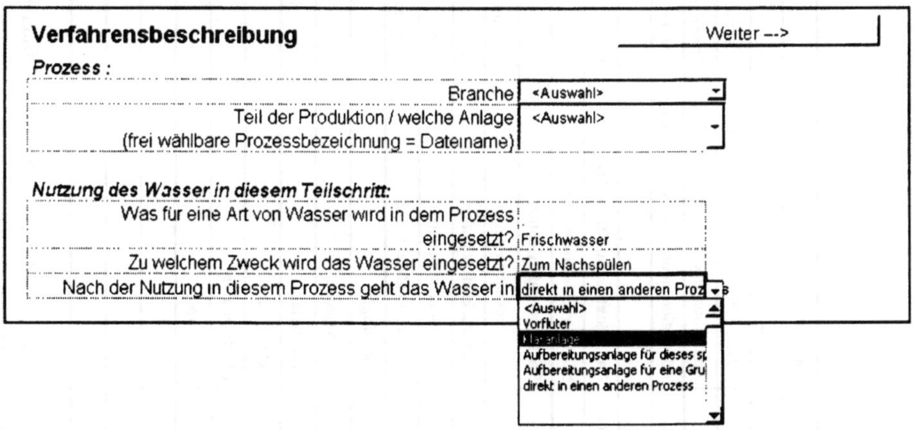

After entering this information about the process, selective information about the water stream and its components can be put in, like e.g.: organic components (CSB-scale), heavy metals (yes/no, concentration of heavy metals as sum or for each metal separeted), temperature, pH.

Thereafter the system starts browsing through exiting processes and technologies and offers a list of possible solutions to handle the process plus process water stream.

It is the always the goal of the system to offer a integrated solution first, not to offer an end-of-pipe solution.

In the next figure the system of query is displayed.

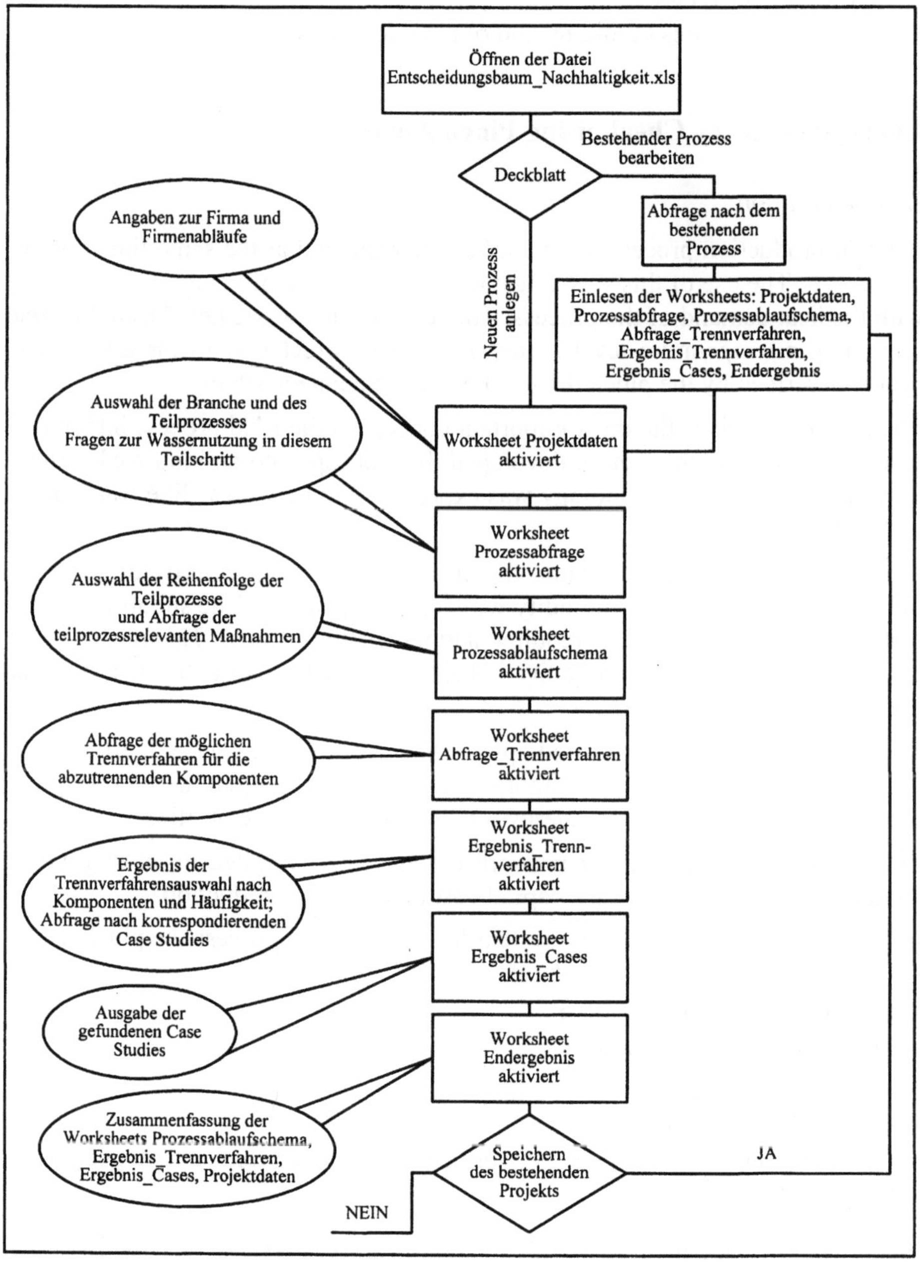

The benefits for companies are:

- solutions for waste water treatment
- decision support system
- support for consultants
- sustainable solutions (avoiding end of pipe decisions):

Energy Efficiency Check using Pinch Analysis

Heat integration

A lot of production processes have a heat demand and at the same time a lot of waste heat. The use of this waste heat has the advantage of being in concurrence with the heat demand of the process. The reuse of this waste heat has to be done at as high a temperature level as possible. The correct way to integrate waste heat is described by the pinch theory (Ferner & Schnitzer 1990).

The Pinch analysis is the most important concept in the field process integration methodologies. Linnhoff et al. developed this concept into an industrial technology and later and the technology was expanded into new areas like water pinch and hydrogen pinch.

The fundamental principal behind the pinch concept is the ability to match the individual demand for a commodity with a suitable supply. The suitability of the match depends on the quality required and the quality offered. This means in the case of energy that the commodity is heat/energy with its quality measured as temperature.

The pinch analysis is a structured approach for identifying energy inefficiencies in industrial processes. The minimum theoretical utility requirement in the processes is calculated for overall energy use as well for specific utilities.

The proposed heat exchanger network can then be compiled in an investment strategy, which details savings, capital cost and emission reduction.

In the design of a heat recovery system for an industrial process the pinch analysis can be divided into several phases:

- Building the composite curve
- Determine the energy targets
- Selection of ΔT_{min}
- Grand composite curve
- Optimization of the designed network considering economical and geographical specifications.

Pinch theory is one method of calculating a heat recovery system for an industrial process that has different material streams coming to and going from this process. These streams are the essential raw materials for production, the products, different utility materials and streams that provide heating and cooling for the process. Some of those streams have to be heated or cooled before going into the process or after leaving the process before further use. Such a stream can be described as an energy stream:

$$\dot{Q} = m \cdot c_p \cdot \Delta T \qquad \text{(kW)}$$

c_p = specific heat capacity (kJ/kg°C)
m mass
ΔT temperature difference

The pinch theory separates the production process into a cold part with surplus heat energy and therefore needs cooling and has a hot part that needs to be heated. This process is done by combining the temperature enthalpy curves of all streams that have to be heated (cold composite curve) and all streams that have to be cooled (hot composite curve) into one temperature – duty diagram (see Figure 1 for the combination of "cold" streams).

Fig. 1: Thermodynamic combination of cold streams. The composite curve is constructed by adding the enthalpy changes of the individual streams within each temperature interval

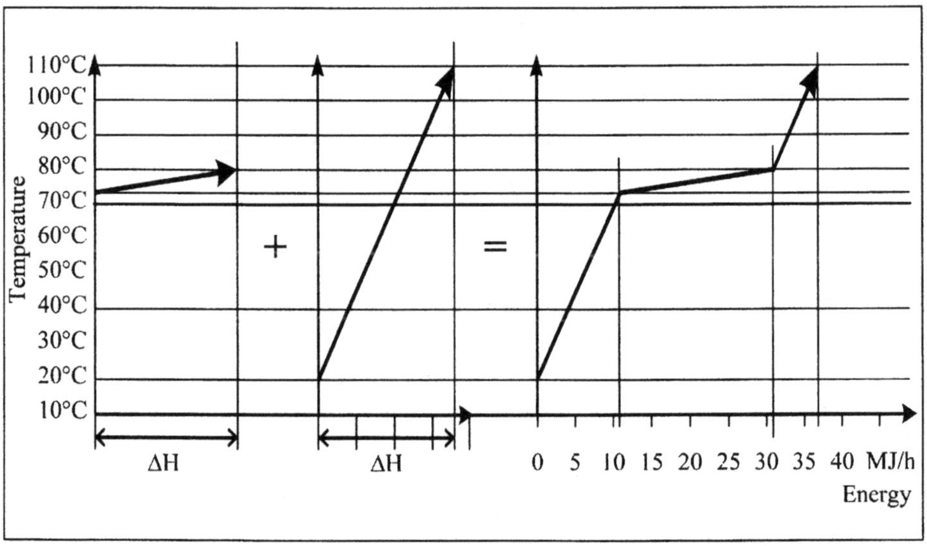

The hot streams are combined in the same way. Both curves are then drawn on the same plot in such a way that the cold streams are at a higher temperature level than the hot streams everywhere in the diagram. This can be done by moving the curves along the energy axis, as an energy value always represents a relative and not an absolute measurement. In this way the only possible energy transfer from hot to cold is provided for.

Fig. 2: Representation of the cold and hot composite curves in one diagram

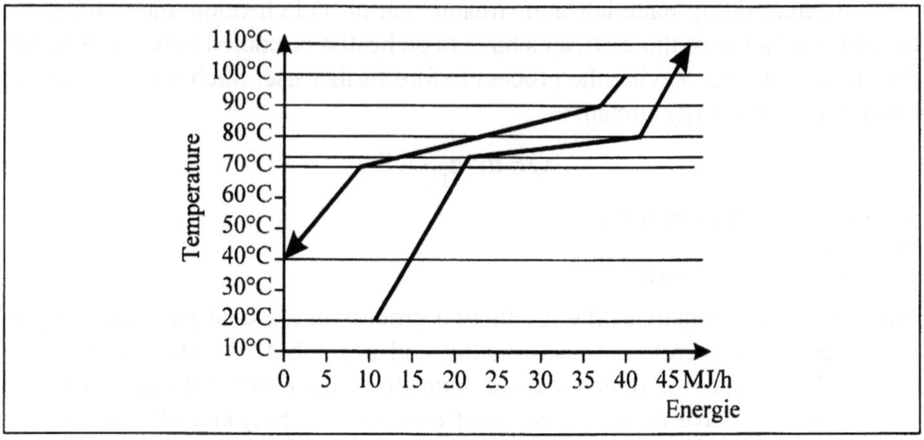

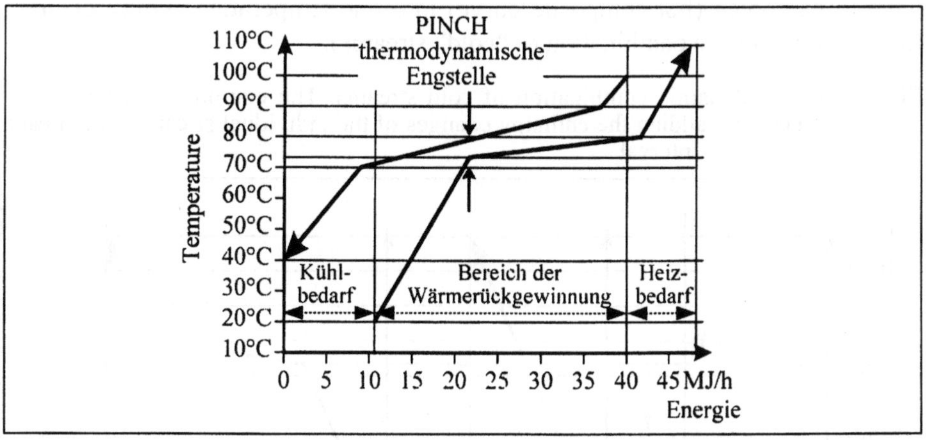

With the help of these composite curves it is possible to determine some essential facts about the process. The curves are separated by a point of lowest difference in temperature ΔT_{min}. This is the thermodynamic bottleneck (see Figure 2) of the process, the so called "pinch". In the area below the pinch temperature there is a heat surplus that has to be removed by cooling, above the pinch temperature there is an energy deficiency that has to be overcome by additional heating. Therefore external heating may only be applied over the pinch temperature. The overlap between the curves in Figure 2 shows the maximum possible process heat recovery, the minimum heating demand $Q_{H,min}$, and the minimum cooling demand $Q_{C,min}$ can also be identified from figure. The minimum temperature difference ΔT_{min} is determined by economical optimization. Typical energy differences ΔT_{min} are shown in Table 1.

Tab. 1: Typical ΔT_{min} values for various types of processes (Linnhoff March 1998)

Industrial Sector	Experience ΔT_{min} Values
Oil Refining	20-40°C
Petrochemical	10-20°C
Chemical	10-20°C
Low Temperature Processes	3-5°C

The theoretical values for $Q_{C,min}$ and $Q_{H,min}$ will be rarely achievable in praxis. The reasons for this are the difficulties of handling process streams that are polluted, corrosive or simply out of the way. But pinch analysis will give a good overview of what is possible.

Solar integration

The integration of solar thermal energy into industrial applications poses a lot of challenges. One of them is the possible temperature ranges of solar thermal applications in countries with moderate climate. In these regions the range is limited to low temperature processes from 20°C to 100°C (for short term applications), or up to 250°C (for medium term applications.

Another challenge is that all process and production specific barriers have to be identified and considered before even thinking about using solar thermal energy for industrial processes. A detailed study of the processes will most often show ways to improve the overall energy efficiency of the production. It is therefore necessary to not only compare the use of solar energy to the existing energy supply, but also to compare the solar energy supply with other alternative energy sources. This comparison will be mainly based on economic parameters, but other criteria like supply guaranty, ecological impact and sustainability should be considered as well. Some of the competing technologies are

- *Energy Efficiency Measures*: The profitability of solar thermal energy systems is mainly given by their investment costs, which are a function of size. Therefore it is necessary to identify the actual energy demand by considering all possibly ways to save energy. The insulation of hot pipes and other hot surfaces is a simple but often overlooked measure to improve energy efficiency. Another important factor may be the optimization of the control systems.
- *New technologies*: New production processes often decrease the usage of thermal energy by increasing the consumption of electrical energy. The introduction of new technologies is often not controllable by the affected companies. One example is the introduction of digital imaging technologies. This had a great impact on the consumption of water and energy as digital images don't need chemical processing, and in extension there is no need for photo development plants. A similar process can be seen with the replacement of recyclable bottles with one-way containers in the packaging of beverages. This

trend makes bottle washing machines obsolete. Other examples of new technologies with a strong impact to energy consumption are:

- o Membrane separation processes instead of evaporators
- o Microwave heating instead of steam heating
- o Sterilisation by UV light instead of heat
- o Sterilisation by membrane separation instead of pasteurisation
- o Cleaning with the help of ultrasound at lower temperatures
- o Smart biotechnological production processes instead of power-hungry chemical processes
- o Self cleaning surfaces.

- *Cogeneration*: From the point of energy efficiency cogeneration is always better then the separated production of heat and power. Large industrial plants have a steam power plant that provides heat energy (steam) and electrical power (steam turbines). These facilities are not usable for most enterprises because there is a minimum size for being economical. But there are other existing cogeneration solutions for smaller business establishments:
  - o Gas turbines in the range 1 to 5 MW
  - o Gas engines (cogeneration units) up to 2 MW
  - o Steam engines.
- *Heat pumps* are very efficient if they are able to supply heating and cooling at the same time (e.g. the usage of waste energy from refrigeration or air conditioning). The standard heat pump decreases the use of thermal energy by increasing the use of electrical energy. Absorption heat pumps are not well known but they are able to be more energy efficient without increasing the consumption if electrical energy. Absorption heat pumps, if used as absorption cooling engines, are able to be run by solar thermal energy. In accordance to the principles of pinch theory heat pumps have to be installed crossing the pinch. They are able to cool below the pinch temperature and to heat above the pinch temperature.

The integration of solar thermal energy into industrial processes can be achieved by different methods. An industrial energy system consists of several sub systems:

- the energy supply (steam, direct heat, hot and cold fluids and gases, electric power)
- the energy usage
- possible recovery systems
- possible re-cooling systems

Solar thermal energy can be integrated at the supply system or directly at the process level. Each of these integration options has different pros and cons.

- The integration into an existing energy supply allows the use of all existing control and measurement equipment already in place. The process control needs not to be changed. But the temperature levels of those energy supply systems may not be usable for solar thermal energy or the integration of solar power is done at very low efficiency. Due to the time dependent supply of solar power the conventional energy supply is still needed. In the terms of pinch theory the solar energy is to be introduced as an additional hot utility with a heat supply and energy levels depending on the collector field (see Figure 4).
- Direct heating of the processes uses solar power most efficient, but controlling and measuring may be more difficult. Due to the time dependent supply of solar power the conventional energy supply may is still needed. For pinch analysis the solar power is integrated as an additional hot stream that can be cooled (see Figure 5).
- Another possible solar option is cooling with solar power.

Fig. 3: Possibilities for the integration of solar thermal energy

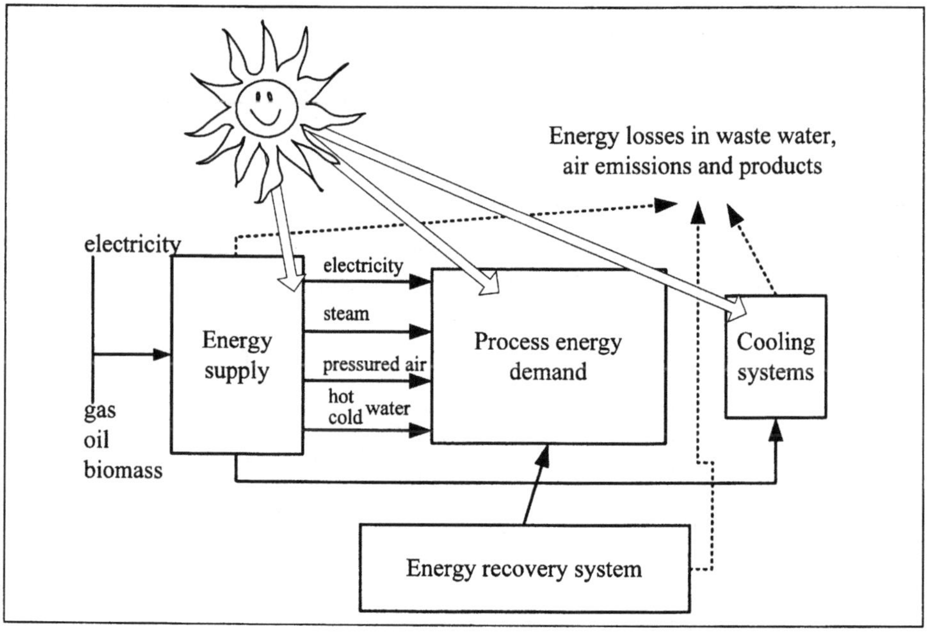

Fig. 4: Integration of solar thermal energy as a hot utility

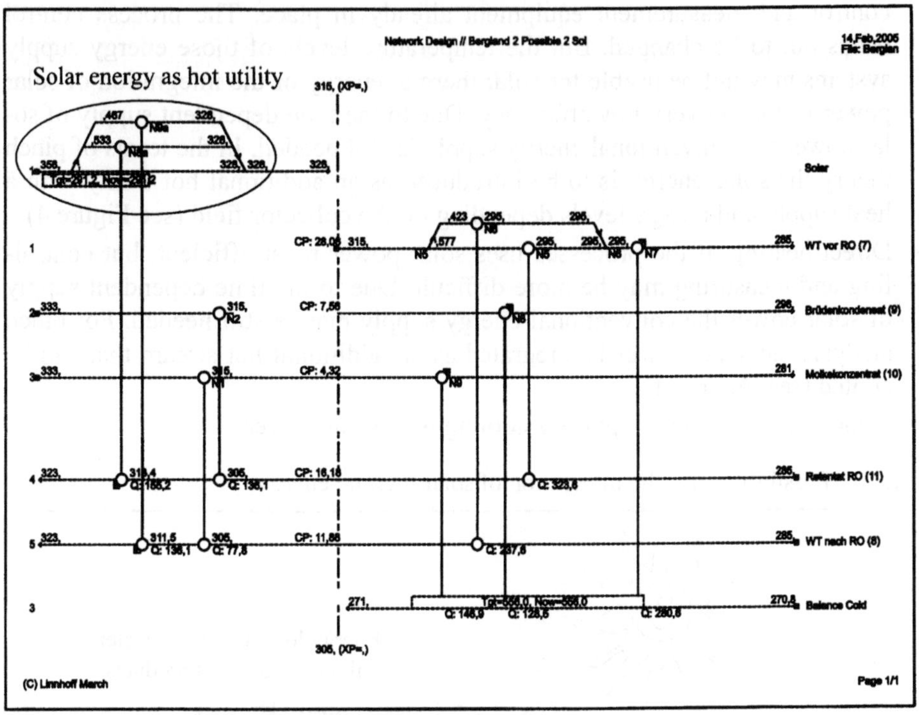

Fig. 5: Integration of solar thermal energy as an additional hot stream that decreases the minimal heating demand $Q_{H,min}$ of the system

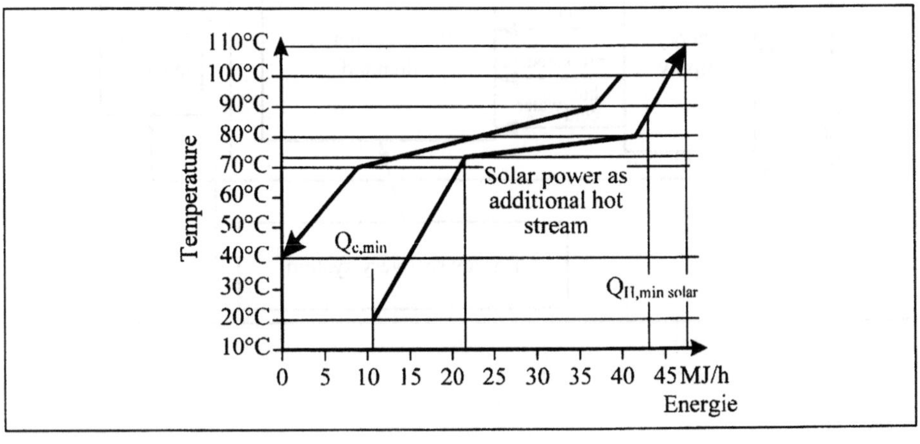

For all these cases it is essential that in accordance to the pinch theory no external heating must be done below the pinch temperature and no cooling above the pinch temperature as long as there is the possibility for internal heat integration. It is also necessary to analyse all competing possibilities to get to the best solution.

Reflections on the economics of solar thermal power

In industry the demands on the economic performance of investments usually is very high. Payback times of 3 years and less are required for production equipment, but 10 years are accepted for infrastructure as buildings and energy systems. Such numbers are difficult to obtain with solar systems if energy savings are considered as the only benefits but there exist examples that it is possible even at present energy and equipment prices. There are also new and promising considerations offering solar-contracting solutions to industry. In these contracting models, the solar energy supplier takes the investment costs of the solar equipment and the enterprise pays for the energy delivered.

Conclusion

This paper presented tools we use in companies to help them on their way achieving Zero Emissions. The main problem when working with companies is to convince them of the necessity of actions. To be able to convince them you need an economic benefit because close to no company will take actions that just "help" the environment and cause costs. So we use EMA workshops to show companies the true environmental costs of their processes and support their decision making. With the result of the EMA workshop we are able to identify the medium the most valuable actions can be taken.

Thereafter it is necessary to offer the company different technical solutions. Mass and energy balancing of the process and testing in laboratory and pilot scale conclude the technical support we can offer. But Zero Emissions does not just have an economic and technical dimension. To be able to achieve Zero Emissions the way of thinking in the company has to change.

Technical support will be able to reduce the emissions to a minimum but there is still a minimum, meaning there is still no "Zero". Economic benefits due to the reduction of input costs or the maximisation of sellable output will help convince companies to take actions but there will be a barrier of economic benefit. There will be a borderline when the actions costs are higher than the profit and the company will stop its support and change their goal "Zero Emissions".

Changing the way of thinking in the companies is as important as the technical and economic support. Changing the way of thinking means looking for new products, new processes, new resources and continuous development. Changing the way of thing means innovation.

At the moment we are developing new technical supporting systems based on system thinking and computer aided balancing software. We are improving our knowledge on waste and sewage water treatment technologies and further developing our economic supporting tool. But the main purpose of our future must be the optimisation of the cooperation of economists, engineers and innovation managers.

The main goal in the future must be to support companies in a technical innovation process because only the implementation of new products and new processes will help companies on their way achieving Zero Emissions.

The next step is to develop innovation tools that provide the cooperation of engineers, economists and companies because Zero Emissions is not achievable with continuous improvement. There will be a barrier someday.

Zero Emissions means innovation and the innovation manager of today will be the Zero Emissions engineer of the future.

References

BAT reference document (2002), European committee for surface treatment, Integrated Pollution Prevention and Control (IPPC), Reference Document on best available techniques, Surface treatment of metals and plastic materials using electrolytic or chemical process (volume of treatment vats >30 m³), May 2002.

Brezet, H. (1997), Ecodesign – a promising approach to sustainable production and consumption, United Nations Publications.

Brunner, C., Hammerl, B., Planasch, M. & Schnitzer, H. (2004), Product Service System For Water As A Guideline For Sustainable Water Management In Industry And Public Utilities. In: IWA 2004 World Water Congress: 125-137.

Linnhoff, Hindmarsh (1983), The Pinch Design Method for Heat Exchanger Networks, Chem. Eng. Sci., vol. 38, no. 5: 745-763.

Linnhoff, March (1998), Introduction to Pinch Technology, www.linnhoff-march.com.

EERE (undated), Solar Industrial Process Heat, www.eere.energy.gov/consumer-info/ refbriefs/ d102.html vom 2.6.2004 Energieverwertungsagentur, www.eva.ac.at/enz/efluss.htm vom 8.6.2004.

Fresner, J., Mayr, J., Schnitzer, H., Gwehenberger, G., Brunner, Ch. & Planasch M., Practical experiences with the reduction of the industrial use of water in the galvanising industry:

Fresner, J., Schnitzer, H., Planasch, M., Gwehenberger, G. & Brunner, C. (2005), ZERMEG II – Zero Emission Retrofitting Method for Existing Galvanizing Plants.

Jasch, Ch. (2001), Environmental Management Accounting Procedures and Principles, United Nations Division for Sustainable Development, UNITED NATIONS, New York.

Jasch, Ch. & Schnitzer, H. (2002), EMA – Environmental Management Accounting – Pilottesting, 8th European Roundtable on Cleaner Production Conference, Oct. 2002, Cork, Ireland.

Kalogirou, S. (2003), The potential of solar industrial process heat applications, Applied Energy 76, 337-361.

Pirker, U., Gwehenberger, G., Pschernig, G. & Schnitzer, H. (2002), Implementation of Zero Emissions and Waste Technologies, 8th European Roundtable on Cleaner Production Conference, Oct. 2002, Cork, Ireland.

Pirker, U., Gwehenberger, G. & Schnitzer, H. (2002), ZERIA – Zero Emissions Research in Austria, Eurosustain Conference, April 2002, Rhodos, Greece.

Planasch, M. (2003), Einsatz von wasserschonenden und wasservermeidenden Technologien in der Oberflächenbehandlung von Metallen mit Schwerpunkt auf Eloxieren, Diplomarbeit, Institut für Verfahrenstechnik der Technischen Universität Graz.

Planasch, M. & Schnitzer, H. (2005), Product Service Systems for Water – A Way To Sustainable Water Management Systems For Companies, 11th Annual International Sustainable Development Research Conference, June 6-8, 2005 Helsinki, Finland.

Schnitzer, H. (1998), Die auf eine Stoffstromanalyse basierende Implementierung von vorsorgendem integrierten Umweltschutz, Chemie Ingenieur Technik, Wiley – VCH Verlag GmbH, vol. 1+2.

Schweiger, H. et al. (2000), The potential of solar heat in industrial processes. A state of the art review in Spain and Portugal. EuroSun 2000, Copenhagen, June.

Part 2

Development in the Baltic Region – Policy, Implementation Technologies and Technologies for Sustainable Development

Factor 5 Flow City
–
Future Images of Sustainable Urban Development in the Baltic Sea Region

Per G. Berg, Professor

Department of Landscape Planning
Swedish University of Agricultural Sciences,
SE – 75007 Uppsala, Sweden

Summary

Future images with a 50-years-perspective in *Baltic Sea Region Cities*, can provide a good knowledge basis for sustainable urban development in the Baltic Sea Region. Researchers in urban planning at the Swedish University of Agricultural Sciences and Uppsala University are currently developing a new program – *Factor 5 Flow City* – tying researchers on urban and landscape planning, sociology, economy, cultural geography, soil science, engineering, ecotechnology, environmental impact assessment and communication together. International researcher-planner teams in 10 Baltic Sea cities (Uppsala, Turku, St Petersburg, Jelgava, Livani, Tartu, Kaunas, Gdansk, Hamburg and Hällefors) are invited to constitute place and situation experts on the empirical arenas for our study which runs for at least 5 years starting from 2006. To be realized – a Factor 5 Flow City requires a number of combined measures, including dematerialisation, spatial planning work, an integration of ecological and social resources and a development of cultural values in the urban settings. The main research team has been working together for more than ten years with key aspects of Sustainable Community Development but relies also on co-operation with more than 25 researchers in different fields. A high-ranking practitioners' advisory group will follow the core program and the projects linked to it. Methods used comprise systems modeling, spatial- and landscape planning tools, GIS-, EIA- and, scenario-techniques, systems ecological estimations, sociological and stakeholder communications methods.

The core research program is expected to result in a number of implement friendly place- and situation adapted strategies for achieving a Factor 5 Flow city, while keeping and developing the city's cultural and social functions and qualities. Futures images of 10 cities, well-illustrated and legibly described, is expected to be a useful tool in actual planning processes in the whole region – in short-term planning routines as well as in long-term strategic physical planning for sustainable urban-rural development.

Introduction

The untamed *urbanisation* and the continued population and consumption *growth*, are strongly linked to global change, which calls for decisive and prompt action in the world's cities (UNCHS 1996, Steffen et al. 2004). The last four decades, sprawling affluent OECD cities have appropriated ever increasing areas of life support land and waters (WWI 2004; Newman & Kenworthy 1992; WCED 1987; Vitousek et al. 1986). A cautious calculation shows that one mean square kilometre of the 29 largest *Baltic Sea Cities* requires between 500-1000 km^2 of land, inland waters and sea surface for its primary production and waste assimilation (Folke et al. 1997).

After 30 years of UN conferences the clear message during the last five years has been: we urgently need *cases* with best practices exhibiting lean and recycling city-hinterland structures outlined in different scales (WRI 2003) with inhabitants consenting to sustainable lifestyles (Spangenberg 2004; Granvik 2005).

Key factors in the resource-craving urban sprawl, seem to be the continuous quest for personal free mobility and at the same time a rational hunt for preferred positions in the city: near culturally and socially vibrant centres, near education and work opportunities and furthermore close to natural settings (Benson & Roe 2000; Orrskog & Johansson 2002; Garreau 1992).

Through its experience and competence – our proposed Research Programme Group (Table 3) can – in recurring dialogue with a national practitioners group and international research-practitioners teams (Figure 2) – efficiently investigate how to develop and reinforce resource-efficient links and slim infrastructures. We maintain that such a group is also appropriate for studying the preconditions for analysing how to create socially and culturally attractive places in dense well-positioned urban settlement nodes, as well as studying prerequisites for an appropriately designed boundary zone between townscapes and green structures.

A Comprehensive Sustainability planning Process for the City and its Local Areas

Urban structure affect citizens' lifestyles, which in turn affect cities'*sustainable development (SD)* performance. In current municipal planning one problem is therefore the complexity of the SD concept, especially when policy visions and comprehensive plans are translated into detail plans. A second problem is the growing number of stake-holders typical for SD urban conflicts between manifold interests for physical life support, economical and social aspects (Nilsson 2003). The proposed Research Program deals strongly with the *dynamics and complexity of sustainability*, with *conceivable SD stakeholders and actors* and it can deal with the *whole chain* – from theory to assessing realisation. As lifestyles with regard to typically mobility, consumption and habitation choices will have a great impact on the sustainable function of cities, one key to the Factor 5 Flow city also relates to the planning and resilient function of *local areas of cities* (Berg 2004a).

Definition and Explaination

A factor five flow city is at the same time an appropriate and radical response to global change – but also a call for finding realistic magnitudes of change within a realistic time frame. The definitions and goals presented below should furthermore not be interpreted literally and universally for all places. The objectives for any city or town must always be dependent on the local context.

> *Definition*: A *factor 5 Flow City* has reduced its physical life support areas to 20% of the present situation, without compromising with the citizens' access to culture, parks, service, education and work places.
>
> *Explanation*: A *flow city* refers to *i.a.* physical, economic and organisational and other resources continuously streaming or being exchanged in a dynamic system (*e.g.* a city). Quantifyable examples are flows of energy, money, traffic and knowledge. "Flow" could also relate to qualitative aspects, like the "light" state of mind that individuals can obtain when they "sense a good place" in a well designed city, where its inhabitants co-operate and support each other, which in turn may affect the physical flows.
>
> A *Factor 5 Flow city* is thus a semi-quantitative metaphor for a city with radically less need for physical life support and with far less cultural, social and organisational "friction" than today's situation in many OECD cities.

Summary of Quantitative Goals

The details of a factor five flow city can and must be continuously challenged and discussed. The goals below should be regarded as quantitative average goals. In some instances a factor 2 change is appropriate – in some cases it may be justified with a factor 10 goal. All is dependent on the site and situation of each city.

The general goal for a factor five flow city is thus that it is five times more efficient than today – so that it:

- Uses only 20% of the fossile fuels of today
- Has only 20% of the nutrient losses in the production-consumption-recycling chain
- Has reduced transport CO2-exhaust with 80%
- Has a five time longer "active coastline" between built and green structues
- Produces the same social, organisational and cultural resources with only a fifth of the ecosystems physical life support area.

The Need for Radical Revision of Urban Structure and Function

The world resources institute recently showed (Steffen et al. 2004) the nature of global change and its causative human activities (*the Anthropocene Era – ibid).* According to that comprehensive report, the world community have strong economic, social and ecological incentives to act radically during the coming 100 years. In his article on community building in American cities, Paterson (1997) gives a number of strong arguments why radical revision of urban culture is necessary. World demographic data, energy consumption and material flows suggest that we need new techniques and a profound change worldwide in lifestyle – particularly in the western world (WRI 2003). But why factor 5 and not – as has been suggested by the Wuppertal Institute – a factor 10? Why not less than 5? In several UN energy futures studies the target has been mentioned to make energy more efficient in the OECD countries (at least a factor 2 – Johansson & Goldemberg 2004) with more emphasis on renewable energy. Considering the possibility to substitute non-renewable energy to renewable – the total fossil fuel use can be reduced to a factor five relative to today. A factor 5 is in accordance with calculations of actual primary production areas for – *i.e.* Baltic Sea cities where the ratio 1: 200 is well established for the 29 largest cities in the Baltic Sea Region (Folke et al. 1997). This may be difficult to make much more efficient. Today cities-compared to assimilation areas have a ratio of 1: 500-1: 1000 (Folke et al. 1997). Recycling can – however – substitute a large part of such areas. The transport sector has been analysed for Swedish conditions where an 80% reduction of travel has been mentioned as a radical but necessary and realistic goal (Steen et al. 1997). It is noteworthy that the authors presuppose that fossil and other non-renewable energy need to be exchanged for renewable energy.

World Watch Institute compares different ecological footprints per capita in different parts of the world and come to the conclusion that a factor five reduction is a reasonable but also necessary goal for human culture (WWI 2004).

Why 50 Years comprehensive Plans?

In futures studies different issues have different *scenario horizons* (Wadensjö 1987) – scenarios on public school development may be appropriate within 10 years, technical innovations studies may use a horizon of 25 years, whereas political futures maps seldom exceed a few years due to the length of mandate periods. In city planning comprehensive plans are continuously reviewed every 5-10 year and is still unable to match the very long time perspective urban structures exist and develop within. In retrospect, the beginning of the 20th century is not that far away from the present. It is thus not unreasonable to have a long planning horizon for implementing a radical change of urban structures. A 100 years perspective of building sustainable cities was the scenario horizon in the international architects competition on sustainable development of nine world cities arranged by the international gas union (IGU 2003). By having a long time horizon participating local governments, companies, NGO: s and other stakeholders in nine world cities could partly or totally disregard the short-term interests and economic "realities". A drawback of a 100 year scenario horizon is loss of connection with the present – which would make a 50 years perspective more relevant. The urgency of actions in the light of global change is another strong argument – we can't wait for a whole century.

Still we wish to give a reasonable time for change – it takes more than a decade to change the infrastructure of energy flows from non-renewable to renewable energy – it will take much more than one or two decades to adapt city structures to more resilient patterns – and it will take time to dematerialise production and to reshape economy without destroying skilled companies and enterprises. It will probably also take time to change lifestyles – maybe a generation or more if we wish to implement new ways of organising everyday life, in consumption patterns and in fundamental values supporting such a life-style – without sacrificing intangible values like education, arts, traditions or just vibrant city life. Thus, things may really happen within 50 years – and we can't wait any longer.

Finding the Components of the Factor Five Flow City

Reducing radically the impact of cities on physical life support systems is a huge task. We do not believe this can be done with some simple measure like finding a new energy source or a totally new care-taking ethics. It will require hard work, co-operation and determination for many decades. And it requires a combination of measures, which may be best illustrated in a long-term comprehensive planning process. How do we find the components for radical revision of urban physical flow function? One obvious way that costs the least energy is to make an inventory of present trends and changing forces already occurring.

Forces of Change Shaping Future City-Hinterland Systems

Although it is impossible to forecast the surrounding world and future urban systems, we can make some assumptions on forces of change that we need to address when we plan for more resilient cities of tomorrow (IGU 2003; Berg 2005). Main changing forces are *knowledge*, *technologies*, *value systems* and their resulting *policies* (Wadensjö 1987). Such transformative forces of change will affect the possibility to develop *Factor five flow cities* strongly. Below the change forces (except policies) are listed in table 1 and three key issues for sustainable urban development are described briefly.

Tab. 1: Major forces of change in present human civilisation – all impacting components in factor five flow transformations with examples of plausible goals. *Modified from Vancouver CityPLUS (IGU 2003)*

Change Force	Factor Five Flow contributing knowledge on	Examples of plausible goals
Knowledge	Life support systems Economic systems Biological systems Technical systems Social Systems Demographics Globalisation	Renewable resources instead of non-renewables. Controlled global change. A high environmental quality and securing of healthy habitats Development of combined Micro-, Macro-, Common property- and Household Economics Stabilisation of global micro- and macroecosystems and their evolution Less mechanical heavy systems and more biological light systems Stable families, resilient communities, peaceful and unique cities and regions From untamed growth to controlled depopulation. Development of healthier populations Global development of networks, diversity of information, ethics, culture and markets
Technology	Transport strategy Energy principle Fuel technique Material usage Ecotechnology IT – size and volume Manufacturing	Movement of information – rather than of people, energy and materials Machine energy efficiency – from steam to combustion to fuel cell engines. From carbon to hydrogen contents of fuel Dematerialisation – from masonry to steel to polymers to carbon nanotubes to biomimics. Development of environmental diagnostics, therapy and construction Miniaturisation – from vacuum tube to solid state transistor to nano-biotransistor From large heat and pressure to nano-bio-self-assembly
Value systems	Lifestyle evolution Worldview shifts Sensory conception Cultural diversity	From social to individual to reconciled social/individual lifestyles From local to global communication and insights with/of planetary system From mainly visual to multisensory understanding of cities, towns and landscapes From mono- to multiethnicity management of the earths' cultures

Three Key Issues for the Factor Five Flow City Process

Discrete knowledges, technologies, values or policies cannot by themselves achieve the dramatic increase in cultural efficiency that is needed to curb global change. In a factor five flow city it is – as was shown above – instead the contribution or combination of different measures and events that may produce a radical result (Table 2). There are – however – some key issues that need to be emphasised – that may contribute particularly to the *factor five flow city*: future transport and communications systems, future economic systems and future lifestyles.

Key Issue 1: Well-integrated Transport and Communications Systems and their Infrastructure

Transportation can be considered as an activity in the city's public space, which is a key factor for the prospect of adjusting cities to a factor five flow city condition. Transportation is a strong localizing factor for industry and settlement. Local transport systems emerging as an effect of localisation of trade structures constitute a potential major environmental problem /load.

All movements affect the attitudes to society. The transport system is an expression for human needs and wills. Carefully designed and structured environments for transportation, mirrors/represents a society with concern for its inhabitants. If the transport system is inconvenient, sensorially unsatisfactory, expensive or environmentally tearing, it will instead signal a careless or ignorant society. Transport structures well integrated with settlements and green structures in different scales are thus imperative for a resourceful urban environment.

Communication is a basic characteristic for present urban culture. Sofar the development of communications have not substituted transportation in general – in fact their developments correlate strongly. If this is the case also in the future there may be a special role for communications to play in the quest for urban sustainability: Increasing possibilities to communicate conveniently and cheaply in *short* loops and at *short* distances in city *local* areas or regions may thus increase *local* mobility, which in turn may compete with *distant* mobility.

Planning for well-integrated transport and communications systems is therefore one main task to investigate about the *factor five flow city*.

Key Issue 2: Combined Economic Systems turning from Scale to Scope

From pre-historic cultures to last century's agrarian societies, local economic systems have played an important role in man's history. Modern economic systems comprise the public economics of the state or the municipalities *i.e.* national or macro-economics – whereas the market economy tools for companies or enterprises of different sizes is business or micro-economics. An emergent third economic principle is presently evolving (common property economic theory) dealing with common resources like *e.g.* fishing waters, recreational or housing areas. Through history up to our time, household economics has also constituted an important informal economic sphere dealing with consumer choices and preferences.

In a world community stressing *life-styles* as an important issue for sustainable development, it may be imperative to scrutinize the effectiveness in all these economic systems. To plan for the *factor five flow city* we thus need competencies on how such different economic systems – with their respective logics, actors and stake-holders – may be optimized in an appropriate future *combined economic system.*

Further changes may be a greater role for *scope* economies in relation to *scale* economies. The first is characterised by investments that can support multiple profitable activities, less expensive in combination than separately. The second refers to classical lower average cost per unit through increased production.

Planning for combined economic systems with a larger weight on scope are key economic issues for the *factor five flow city*.

Key Issue 3: Resourceful Lifestyles

For more than 30 years local and global environmental problems have been abated with industrial end-point waste treatments, introduction of critical pollution levels and – lately – increasingly strong legislation and treaties between cities or world nations. Since 1996 and the Habitat process this is considered by the international community as insufficient. It is the affluent life-style and resource-craving urban structures that need to change – especially in OECD countries. *Life-styles* affect our everyday lives' choices of transport, settlement and consumption (see *i.a.* Tynell 1997). It is even more important then, to offer human habitats which are both resourceful and at the same time attractive, affordable and reasonably convenient.

Planning for resourceful lifestyles most probably need to involve a nuanced view of in context analysed city parts, towns and rural villages. This requires recurrent citizens' involvment in the planning of future cities, first of all to assist in understanding different specific places but also to contribute to plans for the future.

Tab. 2: Summary of combined measures for reaching the goal factor five flow cities.

Resources	Measure	Example
Physical	Conserve physical resources and introduce renewable energy, materials and waters.	Save energy in homes, factories, offices and schools. Expand solar heating and local transport. Dematerialise goods and services.
Economic	Decouple economy from environmental impact	By pricing, tax policy and environmental policy implementation, make environmental consideration profitable
Biological	Fractally integrate green and built structures.	Develop the boundary zone between settlements in different scales and green wedges, courtyard green structure and entrance gardens.

Resources	Measure	Example
Organi-sational	Form an organisationally efficient city, town, community and settlement structure.	Develop comfortable and efficient public transport, gather service into nodes and develop meeting spaces and a rich small road network.
Social	Uphold strong social communities supporting a healthy and otherwise sustainable lifestyle.	Mobilise local residents, workers and school children in the inventory of place properties. Support local communities' initiatives.
Cultural	Always emphasise the local cultures over time.	Take good care of local heritage. Support local ceremonies and traditions.
Aesthetical	Design intersensory supportive human habitats.	Create or support visually, auditive, olfactorial and kinesthetically attractive settlement areas, transportation and service nodes, work places, schools and recreational areas.

Theories and Methods

The Factor five flow city research group will use planning theories dealing with the variable and complex concept of sustainable development (UNCHS 1996; Nilsson 2003; Orrskog & Johansson 2002) and a current discourse on the ever changing subject of sustainable community development (WRI 2003; UNCHS 1996; Berg 2002). The research will also be based on landscape analysis theory (Benson & Roe 2000), on findings about the importance of green structure (Florgård 1981; Lundgren Alm 2001) on outdoor planning theory and practice (Berglund 1996) and on structure and function of infrastructure (Steen et al. 1997). The structure and function of local areas (Franzen & Sandstedt 1993; Berg et al. 2004a; Berg et al. 2005), sociological theory (Giddens 2001) and architectural pattern language theory (Alexander et al. 1977) will be important for understanding the nature of sustainable human habitats. The management of human resources will rest on combined economic theories for macro-, micro- (*i.a.* Lindblom 1977) and common property (Ostrom 1990) economic systems and also consider household economic behaviour. Communication activities in the research process will rest on theories of human interaction and communication (Ljung 2003).

As was discussed earlier – a number of measures contribute to a factor 5 flow city. Examples are *dematerialisation*, use of *renewable* energy, planning for *short distances* between human settlements and primary production and assimilation areas, *integration* of *green* structure at all scales with *settlements*, changes in *behaviour* with regard to *consumption*, *travel* and *housing*. It is reasonable to believe that some cities can achieve a better result than factor 5 and some need to set a more modest goal.

Our Research Group will use current planning methods to understand cities as *systems*. For analysing sustainability according to the Habitat agenda – *resource analysis* will be used (Berg 2004a) (Figure 1), as well as SWOT practice methods[1]. An overall sketch of the core program method is shown in Figure 2. Working with future images of urban systems will require systems modelling (Nybrant, 1999) and scenario techniques (Cole 1988), Environmental Impact Assessment (Petts 1999), GIS (Haining 2003) and other IT based methods (Eckerberg 2004).

Fig. 1: Model for analysing site and situation contextual sustainability properties of communities, cities and regions. This simple model is developed from the final texts in the UN Istanbul conference (Habitat II – UNCHS 1996) and is now used for comparing the sustainability status of Swedish and Baltic City local areas (Berg 2004a; Berg 2004b). P= Physical; E= Economic; B= Biological; O= Organisational; S= Social; C= Cultural and A= Aesthetic resources. Irrespective what initial focus any program researcher will have on Sustainable Development, it is imperative to be able to relate to all resources.

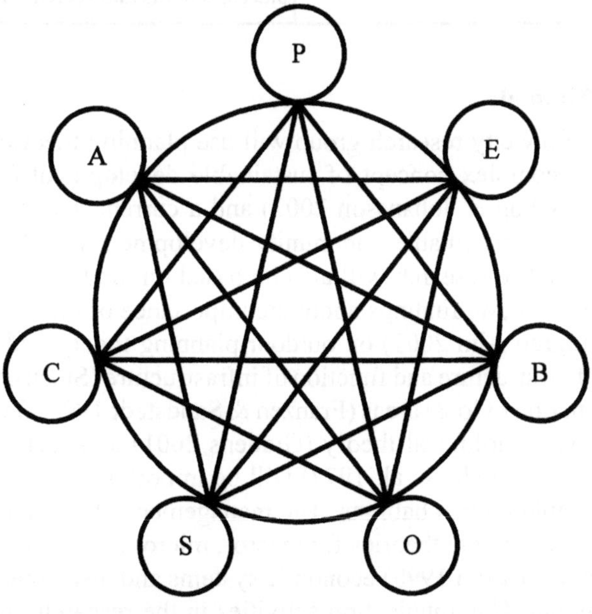

Formative evaluation of the Research Program will be carried out to monitor a changing perspective with methods developed in our communications research team (Ljung 2003). The ever-changing perspectives and concept development within the network will be observed. Comparative research in the Baltic Sea Region – will rely on *teams* of researchers and practitioners (Table 2) and on seminar workshops. *Scientific mediation* (FRN 1980) will be used for settling different opinions on *scenarios*.

The Research Arena – The Baltic Sea Region

The Baltic Sea Region has a growing power and will (Maciejewski 2002; Ryden & Migula 2003) to change towards a greater degree of Physical, Economic and Social sustainability. But the new EU-acquisition states and their cross-the-Baltic-neighbours have different foci on SD. The main applicants of this research program have a decade-long and strong relation with the international teams of planners and researchers we propose in this program as well as large number of universities in Sweden and the Baltic Sea Region (see Table 4, Figures 2 and 4). The main research program process is described in Figure 3 below.

Fig. 2: Ten cities in the Baltic University Program network active in the Sustainable Urban Planning field (Copenhagen, Lodz and Rostok have also been mentioned). They are now preliminarily proposed as partners in a new research effort during the years 2006-2010. Factor five flow city future images will be constructed based on 10 current comprehensive plans developed by researchers and planners into 50 years' scenarios. The process is led by four main researchers/planners and involves about 50 senior and junior researchers. National and international teams of practitioners and a communication process team successively follow the project.Ill: Per G Berg

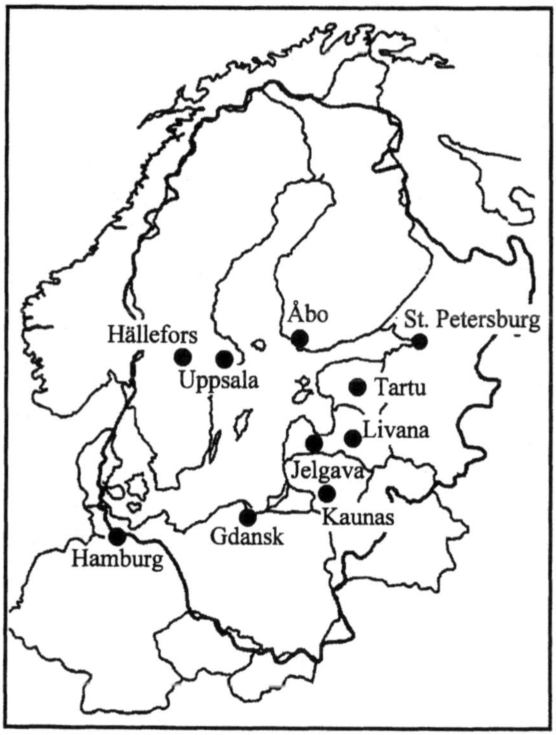

Fig. 3: Factor 5 Flow City Core Program Process

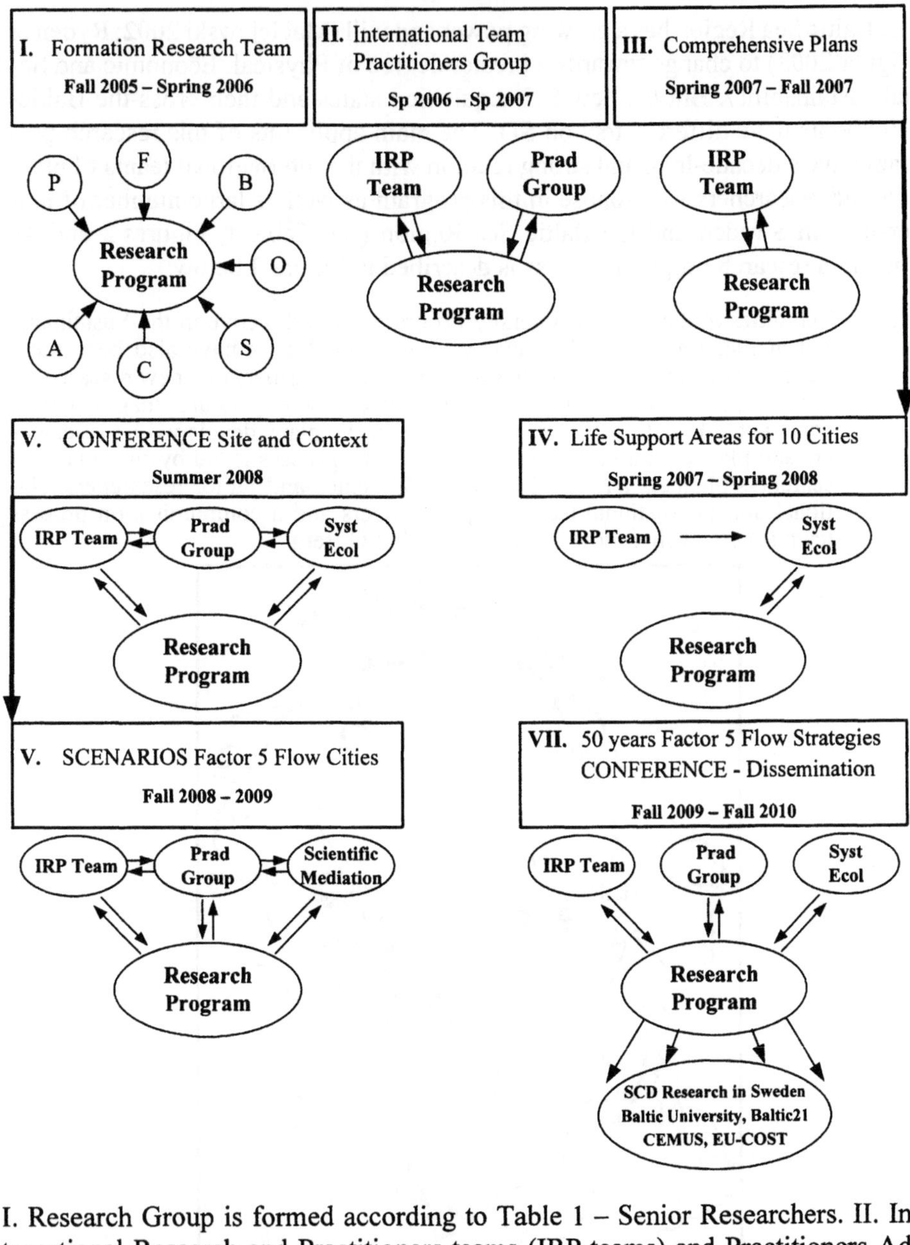

I. Research Group is formed according to Table 1 – Senior Researchers. II. International Research and Practitioners teams (IRP-teams) and Practitioners Advisory Group (Prad-group) are organised and communication starts. III. Comprehensive plans from 10 Baltic Sea Cities are translated. IV. Life Support Areas for 10 cities are estimated with e.g. Systems Ecological Methods and criteria for decreasing them are identified. V. Research Group Conference with IRT-teams,

Prad Group and Invited Systems Ecologist researchers, where 10 unique cities are described and the specialised options for a Factor 5 Flow City process is analysed. VI. Scenarios for 10 Factor 5 Cities are constructed and discussed under Scientific Mediation. VII A 50-year Strategy is formulated for achieving a Factor 5 Flow stage. A final conference starts the Dissemination to Political, Practitioners, University and Research Networks. *) Scientific Mediation enhances several partly conflicting solutions given by different researchers and/or practitioners. P,E,B,O,S,C,A refer to the seven resources or aspects considered in every city (Figure 1) and also refer to the combined competence of the main research group (Table 3).

Tab. 3: List of proposed participating Swedish co-ordinating senior researchers (and key experienced PhD-students) in the Factor 5 Flow City research program

Program focus[1]	**Senior Researchers, Position, Academic grade, (Exam Year)**	**Professional foundation**	**Strategic Focus in Research Program**
Physical resources	Per Anders Hansson, Prof, PhD (93)	Engineer	Energy technology
	NN	Engineer	Waste water technology
	Harry Linnér, Prof, PhD (84)	Agronomist	Soil and hydrology
	NN Crop Production	Agronomist	Plant-soil-water systems
	Hans-Georg Wallentinus Assoc Prof, PhD (73)	Biologist	Environmental impact assessment
	Ulla Myhr, PhD-stud	Landsc Arch	Eco-effect: outdoor sustainability Indicators in housing areas
Economic resources	NN	Economist	Micro and macro economy
	Markus Larsson, PhD-stud	Macro Economist	Common property economy
	Leif Bratt PhLic	Business Economist	Economic systems
	Kåre Ohlsson, PhD	Social Economist	Social economy
Biological resources	Clas Florgård, Prof, PhD (81)	Landsc Arch	Indigenous vegetation
	Urban Emanuelsson, Associate Prof (84)	Biologist	Biodiversity
	Oskar Forsberg, PhD-stud	Forester	Indigenous forest biotopes in Urban environments
Organisational resources	NN	Engineer/Archit	Communication/transport systems
	Magnus Ljung, PhD (01)	Agronomist	Environmental communication
	Klas Eckerberg, PhD (04)	Landsc Arch	IT in planning
	Kerstin Nordin, Assist Prof	Landsc Arch	GIS in planning

Program focus[1]	Senior Researchers, Position, Academic grade, (Exam Year)	Professional foundation	Strategic Focus in Research Program
Social resources	Ulla Berglund, PhD (96)	Landsc Arch	Outdoor urban environments
	Tuula Eriksson, Senior Res	Sociologist	Acceptance and preparedness for sustainable lifestyles
	Madeleine Granvik, PhD-stud	Geographer	Individuals' perspectives in S.D.[2]
	Eva Sandstedt, Assoc Prof, PhD (81)	Sociologist	The function of local urban areas
	David Kronlid, PhD (03)	Theologist	Context dependent value systems
Cultural and Aesthetic resources	Katri Lisitzin, Lecturer	Architect	City culture
	Nils Ahlberg, PhD (05)	Urban Historian	The urban heritage
	Per Hedfors, PhD (03)	Landsc Arch	Intersensory experience of cities and their hinterlands
	Per G Berg, Assoc Prof, PhD (86)	Syst Biologist	Human aesthetic experience
Integrative aspects and co-ordination	Per G Berg, Assoc Prof, PhD (86)	Landsc Planner	Integration of Habitat aspects for urban and rural development
	Thomas Nybrant Prof, PhD (86)	Engineer	Modelling and scenarios of urban and rural sustainable systems
	Lars Rydén Prof, PhD (72)	Envir Scientist	BSR[2] cases on social, ecological and economic connections
	Kristina Nilsson, Assist Prof, PhD (03)	Urban Planner	Integrated planning for sustainable urban and rural development

1) Factor five flow – researchers typically relate to – or work with – several or integrated aspects of the Factor 5 Flow City. The program focus should be considered as an entry point for the respective researcher and constitute a special area of surveillance – as a service to the whole Research Group.
2) S.D.= Sustainable Development
3) BSR= Baltic Sea Region

Tab. 4: External Key National and International Partners for the Factor 5 Flow City Programme

Partner	Research Contact/University	Importance for the Research Program
International Re-search and Practitioners Teams (IRP-teams)	Per G Berg/Uppsala Paula Lindroos/Turku Ksenia Shelest/St Petersb Karin Bachman/Tartu Iveta Lacauniece/Jelgava Arnolds Ubelis/Livani Linas Kluichininkas/Kaunas Tomasz Rozwadowski/Gdansk Walter Leal Filho/Hamburg Anders Bro/Hällefors	IRP-teams consist of 1 high-rank planner, 1senior researcher, 1 PhD- and several younger students. IRP-teams translate and interpret comprehensive plans and related documents. They further co-operate in developing scenarios and strategies for realizing an appropriate Factor 5 Flow process for each city.
Swedish Practitioners Advisory Group (Prad Group)	Carl-Johan Engström, Uppsala Municipality Torbjörn Tiren, Uppsala County Council Torbjörn Sunesson, Swed National Road Administration Åsa Hedenberg, Uppsalahem Housing Company Anders Sandberg, Strategic Planner, Stockholm	The Prad-group (here only a selection of names) contain high-ranking planners of Uppsala and Stockholm cities, Swedish roads, county council and residential company, will recurringly have a dialogue with our Research programme.
Baltic University Urban Forum (BUUF)	Dorota Wlodarzcyk, KTH (Technology), Stockholm	20 Baltic Sea Cities highlights 10 key topics for SCD. The BUUF cases, its local planners & re-searchers are the basis of IRP-teams (see above).
Baltic Recycling Agriculture and Society (BERAS)	Artur Granstedt, Biodynamic Research Institute in Järna	BERAS investigates primary produ-cers-distributors-retailers-consumer networks in local areas outside larger cities in the Baltic Sea Region. Our Program will able to match the BERAS and the BUUF perspectives, with regard to town-country relations.
Systems Ecology Department	NN, Stockholm University (Not yet asked)	This department can – together with our Research group – co-operate on assessment and interpretation of cities' Life-Support Areas.
Assoc of Euro-pean Schools of Planning (AESOP)	(Kristina Nilsson, The Research group)	This international network of planning researchers and educators and this Research Group will have a recurring fruitful information exchange.
Nordic Network of Local Authorities	(Kristina Nilsson, The research group)	This Nordic network of planning researchers will offer further cases and information exchange.

Fig. 4: Selected research links where the core Research Team have strong professional bonds (to other researchers or teams) to departments at SLU and UU (inner ring), to Swedish Universities (middle ring) and to International Universities of great relevance to our Research Program

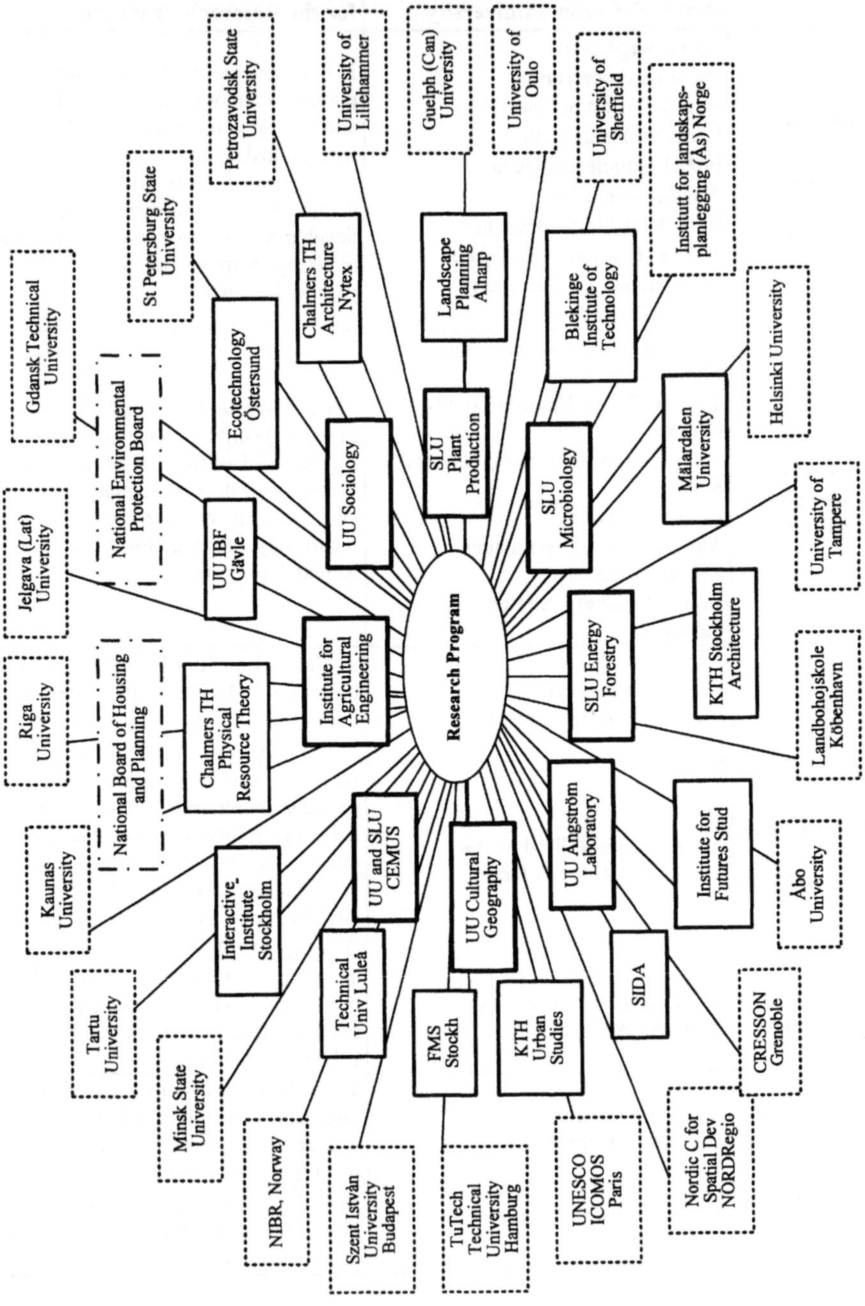

Strategic Relevance of the Factor 5 Flow City Program

This Research program embrace a broad array of aspects on sustainable urban development, mostly of direct relevance to Baltic University Program projects – *e.g. Baltic University Urban Forum* (BUUF) and in basic and master university training on Sustainable Community Development. The results will also be relevant for planners, policymakers, constructors, city and countryside inhabitants and researchers. *Life support areas* for and *technical performance* of city functions, *management* of *flows* in *city-hinterland systems*, are dealt with by our researchers. Other parts of the program focus on *public space*, local *communities* and inhabitant's *lifestyles*. Urban *green* structure and *biodiversity* are basic topics as well as *management* of *landscape values* and particularly *cultural landscapes assets* in cities and towns. Methods for dealing with complex *spatial planning* and its *environmental impact*, *design* and *visualisation* as well as *communication* between *actors* and *policy makers* are also important.

Dissemination of Popular Scientific Results

We have earlier and will publish results on different media and levels depending on the target groups within the Baltic University program and Baltic21 school networks and also as a basis for cross-scientific communication within EU COST researcher networks, UNESCO and AESOP european planners networks.

With such a Factor-five-flow-city research group we also expect to be able to develop a system for research-team-generated re-education of practitioners teams at site in local communities and local authorities within Sweden and in the extention within the whole Baltic Sea region.

References

Alexander, C., Ishikawa, S. & Silverstein, M. (1977), A Pattern Language – Towns, buildings, construction. Oxford University Press, N.Y.

Axelsson, A. (1997), Miljöekonomi för hus och hem (Environmental Economics for house and home – in Swedish) Larsson (Täby).

Benson, J.F. & Roe, M.H. (2000), Landscape and Sustainability. Spon Press, London.

Berg, P.G. (2002), The City as a Sustainable Living System (Chapter 4); Demonstrating sustainability in human habitats (Chapter 5); Developing sustainability in Hågaby village (Chapter 6). In: Ryden L. (ed.) Basic Patterns of Sustainability. Case Studies I. Reports from the SUPERBS (Sustainable Urban Patterns around the Baltic Sea) project. Baltic University Press. Uppsala University p. 24-56.

– (2004a), Sustainability Resources in Swedish Townscape Neighbourhoods. Landscape and Urban Planning 68: 29-52.

– (2004b), The Sustainable Neighbourhood – from Model Area Hågaby to Common Neighbourhoods in a Baltic Sea Perspective – In: Leal Filho, W. & Ubelis, A., Integrative Approaches towards Sustainability in the Baltic Sea Region. Peter Lang Verlag (Frankfurt am Main).

Berg, P.G., Eriksson, T. & Granvik, M. (2005), Building the Sustainable City from within – Implementing the Habitat Agenda in three Swedish Local Townscape Type Areas (in manuscript).

Berglund, U. (1996), Perspektiv på stadens natur (Perspective on city nature) Dissertation KTH (Stockholm).

Cole, S. (1988), Global Models and Futures studies. Institute for Futures Studies (Stockholm).

Eckerberg, K. (2004), Etta eller nolla. Dissertation Landscape planning Ultuna. Acta universitates Agriculturae Sueciae. Agraria vol. 463.

Florgård, C. (1981), Natural Vegetation and Settlements (Dissertation in Swedish) Konsulentavd Trädgård SLU (Alnarp).

Folke, C., Jansson, Å., Larsson, J. & Costanza, R. (1997), Ecosystem Appropriation by Cities. Ambio 26: 167-172.

Franzen, M. & Sandstedt, E. (1993), Välfärdsstat och byggande – Om efterkrigstidens nya stadsmönster i Sverige (Welfare State and Building – on post war city patterns in Sweden (only in Swedish) joint Dissertation. Department of Urban Design and Planning School of Architecture, Chalmers (Göteborg).

FRN (1980), Scientific Mediation – a practical method used for describing scientifically controversial problems about e.g. energy, pollution or genetic engineering. Swedish Research Council's periodical Källa 1980-1993.

Garreau, J. (1992), Edge City – life on the new frontier. Anchor books (New York).

Giddens, A. (2001), Sociology 4th Edition. Blackwell (Oxford).

Granvik, M. (2005), Late modern lifestyle and implementation of the Habitat-agenda in local communities – dwellers' real action, time and interest in public-participation processes in a Swedish and Russian context. Dissertation Department of Landscape planning Ultuna SLU Uppsala.

Haining, R.T. (2003), Spatial data Analysis – theory and practice. Cambridge University Press.

IGU (2003), International Competition for Urban Sustainability. 100 year future plans for 9 world cities. World Gas Conference in Tokyo – June 2003. Final Report from the 9 teams. (International Gas Union).

Johansson, T.B. & Goldember, J. (eds.) (2004), World Energy Assessment – Overview – 2004 Update. U N Develoment Program (UNDP – N.Y.).

Lindblom, C.E. (1977), Politics and Markets – The world's political-economic system.

Ljung, M. (2001), Collaborative Learning for Sustainable Development of Agri-Food Systems. Dissertation at the Department of Landscape Planning Ultuna, Swedish University of Agricultural Sciences.

Lundgren, Alm E. (2001), Stadslandskapets obrukade resurs – Om grönstrukturens potential och synliggörande i en hållbar stadsutveckling. (On green structure – in Swedish) Dissertation CTH (Göteborg).

Lundgren, L.J. (ed.) (1992), Livsstil och miljö – På väg mot ett miljövänligt beteende? (Lifestyle and environment – in Swedish). Report from a conference in Hässelby 18-19 March. Forskningsrådsnämnden (Stockholm).

Maciejewski, W. (ed.) (2002), The Baltic Sea Region – Cultures, Politics, Societies. Baltic University Press (Uppsala) 686 pp.

Newman, P. & Kenworthy, J. (1992), Cities and Automobile Dependence – Avebury (Hampshire, England).

Nilsson, K. (2003), Planning in a Sustainable Direction – The art of Concious Choices. Trita Infra 03-58. Dissertation. KTH (Stockholm).

Nybrant (ed.) (1999), Proceedings of the 3rd International Symposium on Mathematical Modelling and Simulation in Agriculture and Bioindustries, Uppsala/ Sweden Published for IMACS and IMAC by SLU, the Swedish University of Agricultural Sciences.

Orstrom, E. (1990), Governing the commons – The Evolution of Institutions for Collective Action. Cambridge University Press.

Orrskog, L. & Johansson, B. (2002), Att bygga ett hållbart samhälle. Natur & Kultur (Stockholm).

Paterson, D.D. (1997), Comunity building and the necessity for radical revision. Landscape and Urban Planning 39: 83-98.

Petts, J. (1999), Handbook of Impact Assessment Blackburn (Oxford).

Ryden, L. & Migula, P. (2003), Environmental Science – A comprehensive Textbook. Baltic University Press (Uppsala) 824pp.

Spangenberg, J.H. (2004), Environmentally Sustainable Domestic Consumption-Material-Flow-Based Indicators for Priority Fields of Action. In: Leal Filho, W. & Ubelis, A. Integrative Approaches towards Sustainability in the Baltic Sea Region. Peter Lang (Frankfurt a.M).

Steen, P. et al. (1997), Färder i framtiden – transporter i ett bärkraftigt samhälle (– Travel in the future only in Swedish) Forskningsgruppen miljöstrategiska Studier – FMS, KTH (Stockholm).

Steffen et al. (2004), Global Change and the Earth System: A Planet Under Pressure. Int Geos and Bios Program.Springer-Verlag (Berlin).

Tynell, M. (1997), Life-style and Mobility. In: Tengström and Tynell (eds.), Towards Sustainable Mobility – Transporting people and goods in the Baltic region. A sustainable Baltic Region, vol. 6. Baltic University Program. Uppsala University.

United Nations Conference on Human Settlements (1996), The Habitat Agenda and the Istanbul Declaration. UNCHS (New York).

Villee, C.A., Solomon, E.P. & Davis, P.W. (1985), Biology. Saunders publishing (New York).

Vitousek, P.M., Ehrlich, P.R., Ehrlich, A.H. & Matson, P.A. (1986), Human Appropriation of the Products of Photosynthesis. BioScience 36: 368-373.

Wadensjö, C. (1987), Formbar framtid (Shaping the Future – in Swedish) Secretariat for Futures studies (Stockholm).

WCED (1987), Our Common Future – World Commission on Environment and Development (The Brundtland report). Oxford University Press.

World Resources Institute (2003), World Resources 2002-2004. WRI (New York).

World Watch Institute (2004), State of The World Yearbook WWI (New York).

Environmental and Thermodynamic Indicators in Support of Fair and Sustainable Policy Making

–

Investigating Equitable Trade among Latvia, Denmark and Italy

Sergio Ulgiati[1], Catia Cialani[2]

[1] Department of Chemistry, University of Siena,
Via Aldo Moro 2, I – 53100 Siena, Italy

[2] Department of Economics, University of Foggia,
Largo Papa Giovanni Paolo II, I – 71100 Foggia, Italy

Abstract

A theoretical framework and a case study for equitable trade policies based on Emergy Synthesis (Odum 1996, Brown & Ulgiati 2004)) is presented in this paper, aimed at integrating monetary evaluation and measures in support of more comprehensive trade balance between countries, for equity and international stability. The paper stems from the recognition that the economic growth of developed countries is very often based on the depletion of primary resource storages of developing countries, not adequately compensated for in trade dynamics in terms of real wealth exchanged. One of the basic principles of the emergy systems perspective is that wealth relies on availability and good use of resources, not on money in itself. Evaluating international trade and net benefit from investments using only the inflows and outflows of currency often shows a monetary balance of payments but disregards the inflows and outflows of resources.

Since resources have different qualities and therefore different abilities to drive an economic process and provide wealth, accounting for quality and ensuring a balanced resource exchange is very important for trade to be beneficial to both partner countries. In this paper, the Authors investigate in emergy terms the economic performance of Latvia, Denmark and Italy, and then calculate the emergy exchange ratios associated to global trade as well as to selected products traded among these countries. Results show that Denmark has the highest advantage from bilateral trade with Latvia and Italy, and Latvia the lowest. The reasons can be identified in the high emergy/GDP ratio of Latvia as well as in the fact that Latvian exports mainly rely on primary resources, while Italian and Danish exports basically consist of manufactured goods and commodities. The Authors underline the importance of good trade partnership and call for policy measures for implementation of equitable trade pattern.

1. Introduction. Impact of Economic Investment

A systems overview of a region (Figure 1) suggests that its ecological, economic, and cultural systems are closely intertwined[1]. As a region's economic system develops, there are resulting changes in its ecological and cultural systems because the increased economic activity affects wider and wider spatial area and may cause changes in values and ethics. The extent of change in each of these systems is more or less dependent on the extent of change in the other. Figure 1 illustrates the interconnections between environmental, cultural, and economic systems of regions. A balanced and well-adapted subsistence economy might have the organization depicted in Figure 1 a.

Fig. 1: Systems diagram of a regional economy (a) having no trade with external markets and (b) an economy that has developed trade. Money is shown as dashed lines, and energy and information flows as solid lines. While invested money can circulate within the economic system, eventually, like income from exports, it is used to purchase goods and services from external economies. Energy systems symbols from Odum (1996).

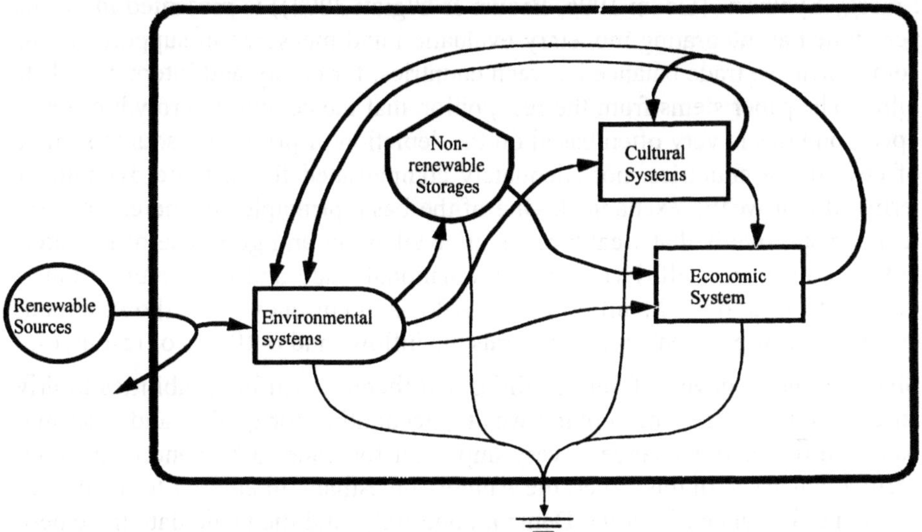

Ecological resources are extracted by the economic system, converted to goods, and consumed by cultural components which, in turn, provide the necessary organizational structure and "manpower" for the economic system. Byproducts of the economic system are recycled back to the environment and information and "good stewardship" are fed back from culture. The driving forces are renewable resources shown coming from the left side of the diagram and the nonrenewable resource storages from within. The overall system that develops (i.e., the levels of ecological productivity, economic activity, and cultural organization) is, to a

1 Systems symbols and definitions can be found in Odum (1996).

large degree, dependent on the magnitude of renewable emergy flow and the nonrenewable storages that are available.

Economic investment from outside can be depicted like that in Figure 1 b. Investment dollars are used to purchase fuels, goods, and services from outside the local economy. A second outside source of resources now influences the system. As a result of the connections between components of the regional system, any increase in one compartment affects the other two compartments (whether they increase or decrease depends on the nature of the interconnections and is not necessarily important at this point). The bigger the influence from outside investment, that is the bigger the magnitude of the flows coming from the top right (Figure 1 b) compared to the flows coming from the left, the greater the impact.

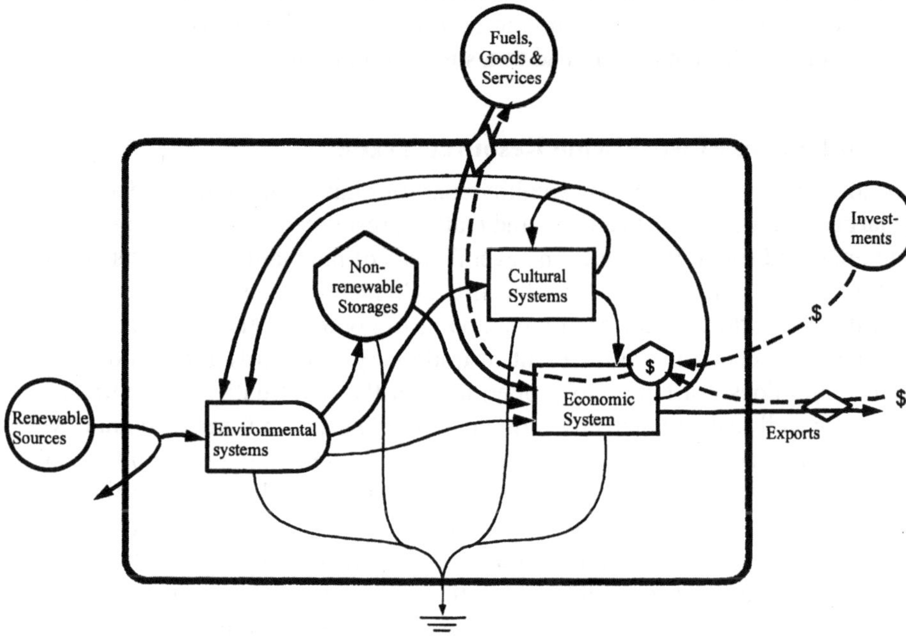

Secondary impact of economic investments is also illustrated in Figure 1 b. Economic investments from outside are made as a means of financing enterprises that either directly extract natural resources (wood, minerals, fuels, or fish, for instance) and sell them to outside markets, or to develop enterprises for the conversion of resources within the local economy (hydroelectric projects or tourist developments, for instance). In either case, the "attracted" investments carry with them a significant debt that must be repaid and which is financed through the export and sale of resources. The net benefit of investments from outside to the local economy, then, becomes a matter of determining the balance between what is purchased with the investment, and the resources that are exported over the long term.

Additional insight related to the net benefit from investment is gained using emergy analysis (Odum 1996), as described in this paper. One of the basic principles of the emergy systems perspective is that true wealth comes from resources, not from money. Money can be used to purchase resources, but the money in itself is not representative of wealth. Evaluating international trade and net benefit from investments using only the inflows and outflows of currency often shows a monetary balance of payments but does not take into account the inflows and outflows of wealth. Often the investing economy receives double benefit, namely the resources extracted directly, and the resources that must be extracted and sold by the developing economy in order to pay interest on outside loans. Most developing economies seek money from outside sources instead of seeking resources (the true basis of wealth). Thus they often sell their potential wealth cheaply to purchase economic goods that have less effect in stimulating their economy and that do not lead to a sustainable future.

2. Unbalance of worldwide Resource Trade

It is nothing new that very few wealthy countries have access to and actually use the largest part of the world energy and material resources. The generation of environmental and social instability in several areas of the planet can be discussed in relation to the existence of this disparity. Conventional economic approaches quantify traded flows in terms of the amounts of goods traded and the money paid for them. The economic assessment of trade very often only focuses on money balance and does not take into proper account the real quality of the traded resources as well as the related environmental problems, both from the point of view of the depletion of resources and of the pollution generated in the exporting country. Resources are very often mined and partially processed in the exporting country, then refined and used in the importing developed countries. The price of exported resources is very often inadequate to compensate for the depletion of local storages and the environmental burden that is generated by resource extraction and primary processing. Instead, resources drive significant economic and environmental benefits in technologically and economically developed countries.

The unbalance of resource trade among countries is of paramount importance to environmental and political stability, as clearly pointed out by H.T. Odum (1994):

> "Trade and projects that unbalance local economies [...] increase emergy inequity between countries, do not maximize the world economy, because they leave major sectors of the world's population in poverty, essentially outside the world economy. This pattern wastes resources into luxury and excess of the developed countries, diverting resources that used to go directly to population support (without payments). This pattern is not sustainable, does not maximize world wealth and emergy, does not reinforce world production, and will not last. These patterns will become discredited as world opinion changes, as revolu-

tions occur, and worldwide resource depletion soon cuts off the largesse of the overdeveloped countries."

Brown (2003) clearly identified the "advantage to the buyer" in the relations between industrialized countries and countries the economies of which are mainly based on exports of primary resources, for a large number of countries worldwide. This happens in spite of an apparent balance of money flows between trading countries. The situation appears even worse when the environmental burden associated with the traded resources is also taken into account and its geographical distribution is carefully investigated.

Sampat (2003) provided a vivid picture of the inequality of the present primary resource exchange worldwide, by showing an unequivocal set of data about polluting emissions and tradeoffs in mineral extraction and trading. According to Sampat's report, the mining industry was responsible in the late 90's for about 5.3 million square kilometers of threatened forests (39% of the total damaged forest area worldwide), water and air pollution, loss of biodiversity, harm to the people living near the extraction sites, as well as for about 7-10% of the total world energy use. Based on World Bank data, Sampat also calculated a significant decline (about 40%) in the market price of metals and minerals in the years 1960-2001, which decreased the economic advantage of the exporting countries in spite of the huge environmental problems suffered by them. Finally, analyzing data from UNCTAD (2001), World Bank (2001) and UNDP-United Nations Development Program (2001), Sampat ascertained that several countries with high export of minerals still have large fractions of population below the national poverty line, thus confirming the lack of any real economic advantage compensating for the depletion of local resource storage and the related pollution.

3. The Emergy Approach. Assessing the Donor-Side Resource Quality

The same product may be generated via different production pathways and with different resource demand, depending on the technology used and other factors, such as boundary conditions that may vary from case to case and process irreversibility. In its turn, a given resource may require a larger environmental work than others for its production from nature. As a development of these ideas, Odum (1988, 1996) introduced the concept of *emergy*, i.e. "the total amount of available energy (exergy) *of one kind* (usually *solar)* that is directly or indirectly required to make a given product or to support a given flow". We may therefore have an oil emergy, a coal emergy, a solar emergy, etc. according to the specific goal and scale of the process. In some way, this concept of embodiment supports the idea that something has a value according to what was invested into making it. This way of accounting for required inputs over a hierarchy of levels might be called a "donor system of value", while exergy analysis (Szargut et al. 1988) and economic evaluation are "receiver systems of value", i.e. something has a value according to its usefulness to the end user. *Solar emergy* was therefore suggested as a measure of the total environmental support to all kinds of processes in the biosphere, including economies. Flows that are not from solar source (like deep heat and gravitational potential) are expressed as solar equivalent energy by means of suitable transformation coefficients (Odum 1996).

The amount of input emergy dissipated per unit output exergy is called solar transformity. The latter can be considered a "quality" factor which functions as a measure of the intensity of biosphere support to the product under study. The total solar emergy of a product may be calculated as: (solar emergy) = (exergy of the product) * (solar transformity). Solar emergy is usually measured in solar emergy joules (seJ), while the unit for solar transformity is solar emergy joules per joule of product (seJ/J). Sometimes emergy per unit mass of product or emergy per unit of currency are also used (seJ/g, seJ/$, etc). In so doing, all kinds of flows to a system are expressed in the same unit (seJ of solar emergy) and have a built-in quality factor to account for the conversion of input flows through the biosphere hierarchy.

Values of transformities are available in the scientific literature on emergy. When a large set of transformities is available, other natural and economic processes can be evaluated by calculating input flows, throughput flows, storages within the system, and final products in emergy units. As a result of this procedure, a set of indices and ratios suitable for policymaking (Ulgiati et al. 1995; Ulgiati & Brown 1998; Brown & Ulgiati 1999) can be calculated. Such indicators expand the evaluation process to the larger space and time scales of the biosphere. While the emergy approach is unlikely to be of practical use in making decisions about the price of food at the grocery store or the way a process should be improved to maximize exergy efficiency at the local scale, its ability to link local processes to the global dynamics of the biosphere provides a valuable tool for adapting human driven processes to the oscillations and rates of natural processes, towards sustainable patterns of human economies.

3.1 Emergy Method in Practice

A short summary of the emergy method is provided in the following. Then two emergy intensity indicators (emergy per unit of product and emergy per unit money) are discussed due to their relevance for the evaluation presented in this paper. Further details and concepts can be found in Odum (1996) and Brown & Ulgiati (2004).

3.1.1 Drawing a Diagram

Systems diagrams are preliminarily drawn to show the inputs that support and drive a process or a system. The purpose of the system diagram is to conduct a critical inventory of processes, storages and flows that are important to the system under consideration and are therefore necessary to evaluate. Components and flows within diagrams are arranged from left to right reflecting more available energy flow on the left, decreasing to the right with each successive energy transformation. For example, abundant solar energy is utilized in successive transformations in ecological, agricultural, and techno-economic subsystems to support a small amount of high quality energy of humans, their government and information. A simple diagram of a national environmental, economic and social system is shown in Figure 2. The left to right organization also corresponds to increasing scale of territory and turnover time. As illustrated in Figure 2, every energy transformation box has more than one input, including larger energy flows from the left, lesser amounts from units in parallel, and small but important controlling energies feeding back from the right.

Fig. 2: Generic systems diagram showing driving energies, production, cycling, and the hierarchy of ecological and human-dominated components. Energy systems symbols from Odum (1996).

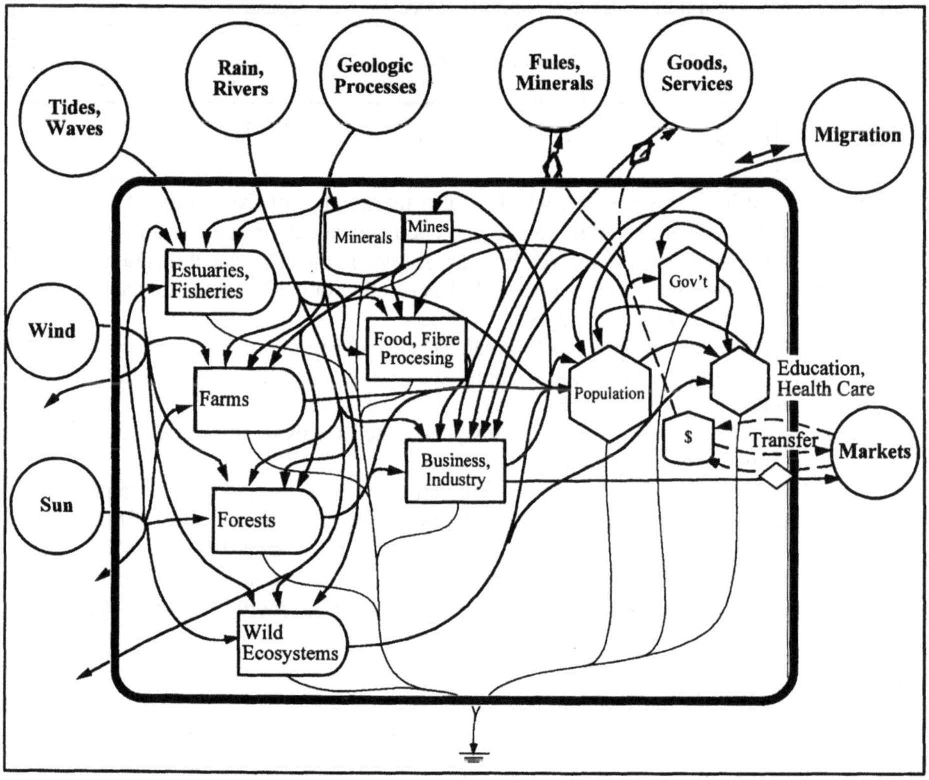

3.1.2 Preparation of an Emergy Evaluation Table

Tables of the actual flows of materials, labour and energy are constructed from the diagram. Raw data on flows and storage reserves are converted into emergy units, and then summed for a total emergy flow to the system (Table 1). Inputs that come from the same source are not added, to avoid double counting. Only the larger input is accounted for. If the table is for the evaluation of a process, it represents flows per unit time (usually per year). If the table is for the evaluation of reserve storages, it includes those storages with a turnover time longer than a year.

Tab. 1: A generic emergy evaluation table for processes, systems, and economies.

1	2	3	4	5	6	7
Note Value*	**Item**	**Data**	**Units**	**Emergy/unit (sej/unit)**	**Solar Emergy (E+xx seJ/yr)**	**em$ (em$/yr)**
1.	First item	xx.x	unit/yr	xxx.x	xxx.x	xxx.x
2.	Second item	xx.x	unit/yr	xxx.x	xxx.x	xxx.x
...						
n.	n^{th} item	xx.x	unit/yr	xxx.x	xxx.x	xxx.x
	Output	xx.xx	unit/yr	xxx.x	$\sum_{n}^{1} Em$	xxx.x

Column #1 is the line item number, which is also the number of the footnote found below the table where raw data sources are cited and calculations are shown.

Column # 2 is the name of the item, which is also shown on the aggregated diagram.

Column # 3 is the raw data in joules, grams, dollars or other units. The units for each raw data item are shown in column #4.

Column # 5 is the emergy per unit used for calculations, expressed in solar emergy joules per unit. Sometimes, inputs are expressed in grams, hours, or dollars, therefore an appropriate conversion ratio is used (sej/hr; sej/g; sej/$).

Column # 6 is the solar emergy of a given flow, calculated as raw input times the transformity (Column 3 times Column 5).

Column # 7 is the "emdollar value" of a given item for a given year. This is obtained by dividing the emergy in Column # 6 by the "emergy-to-money ratio" (EMR) for the country and selected year of the evaluation (units are sej/$). The EMR is calculated independently. The resulting values in this column express the amount of economic activity that can be supported by a given emergy flow or storage.

3.2 Emergy Intensity

The total emergy supporting a process is assigned to the process product(s) according to emergy algebra rules which are not described in this paper (Brown & Ulgiati 2004). The ratio of total emergy to the available energy or mass of a product is an "emergy intensity", respectively called transformity if the product is measured in terms of its available energy (seJ/J) and specific emergy is the product is measured in terms of its mass. For example, if 4000 solar emjoules are required to generate a joule of wood, then the solar transformity of that wood is 4000 solar emjoules per joule (abbreviated sej/J). Solar energy is the largest but most dispersed energy input to the earth. The solar transformity of the sunlight absorbed by the earth is 1.0 by definition. Solids may be evaluated best with data on specific emergy for its concentration. Because emergy is required to concentrate materials, the unit emergy value of any substance increases with concentration. Elements and compounds not abundant in nature therefore have higher emergy/mass ratios when found in concentrated form since more work was required to concentrate them, both spatially and chemically.

Another, very important, indicator of emergy intensity is the ratio of total emergy driving a process to the money value of the product. If the process is a whole national economy, its product can be grossly identified as its Gross Domestic Product (GDP) or any other equivalent measure of the economic activity. Therefore, the total emergy supporting a Nation also supports its economic system as a whole and its GDP, as well. The emergy/GDP can be regarded as a measure of the efficiency with which resources are converted into money or, in other words, of how much environmental support is needed to generate one unit of GDP (seJ/$; seJ/€). It can be used to convert money payments into emergy units. Since money is paid to people for their services and not to the environment, the contribution to a process represented by monetary payments is the emergy that people purchase with the money. The amount of resources that money buys depends on the amount of emergy supporting the economy and the amount of money circulating. The emergy/GDP ratio varies by country and has been shown to decrease each year at the present trends of country development worldwide, increase of resource consumption and, sometimes, increase of inflation.

In this paper, as better explained in the next Section, emergy intensities from the present and previous works are used to calculate the emergy of selected products traded among Latvia, Italy and Denmark, in order to assess the emergy balance among these countries, and then suggest policy measures for fair international trade.

3.3 Calculating Emergy Intensities of individual Products

Fig. 3 shows the main input flows of energy, matter and labour supporting a generic agricultural process. Some flows are locally renewable, others are nonrenewable or imported from outside the system. Each of them is characterized by a different quality, both from the point of view of the environment (i.e., how much environmental work supports the flow) and from the point of view of the user (i.e., what the user can extract out of each flow or product). These characteristics cannot be properly accounted for by their money cost, which only measures the cost of the applied human labor and of the willingness to pay for a given product. Before converting all these input flows to emergy units, it is very useful to consider them as such, in terms of their physical nature.

Fig. 3: Energy systems diagram of a generic agricultural production process. Energy systems symbols from Odum (1996).

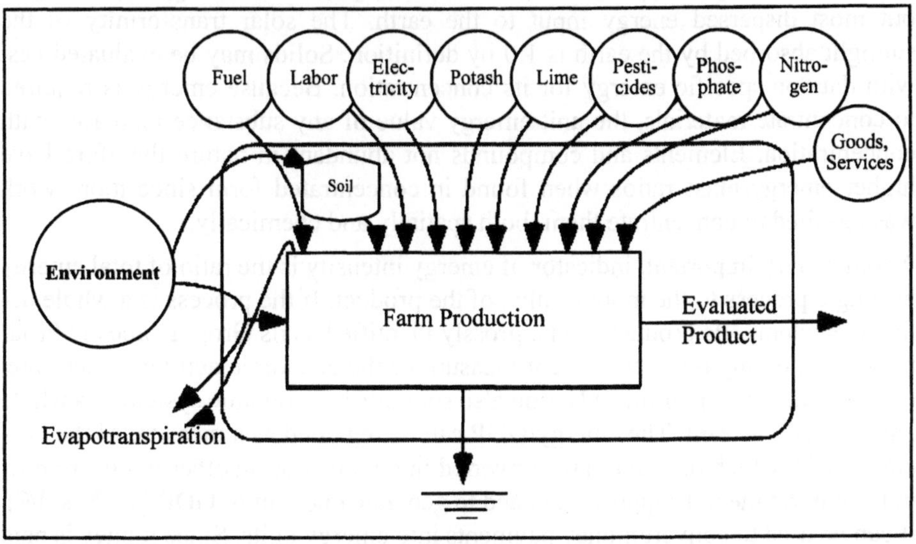

An input of topsoil degraded due to intensive agriculture means a decrease of the physical support to further plant growth; an input of water for irrigation means a depletion of underground water storages, with lowering of the water levels and possible consequences to the surrounding ecosystem integrity; an input of pesticides means an increase of the risk of chemical pollution; and so on. Resource depletion and pollution problems individually associated to each input or output flow of the process should be charged to the product, since they are "embodied" in it. When the product is sold, its price should not only reflect the labor and services invested for its production, but also the associated problems generated in the ecosystem dynamics, which are generally outside of the market system of values. When a product is exported, it contributes to the economy of the importing country, without delivering to it the problems generated in the exporting country. In other words, we could identify a "hidden, virtual export" of land, water, chemicals, labor, fuels, electricity, and soil, diverted or degraded in order to make the exported products. We use the word "virtual export", because the input supporting the production process is no longer available to the system, as if it were exported together with the product.

Converting each input flow into emergy units and then calculating the total emergy input to the process, i.e. aggregating all data into a total, hides the individual problems but offers a global view of how these problems are embodied in the product. The emergy intensity (calculated by dividing the total emergy by the exergy or mass of the product) expresses in a quantitative and scientifically based way the convergence of inputs and problems generated during the process. The emergy intensity is a unit measure of cost, and provides much more information (although aggregated) than the actual economic cost of an item. When people go to the grocery store, the price of commodities provides them only little information about the product, yet enough to understand how much is required to get the product and bring it home. Policy makers and managers in charge of national welfare should know much better the individual problems involved in a given production process before deciding to sell at low cost the potential wealth of their country. Once the process is well known, an aggregated indicator of true environmental cost such as the transformity or other emergy-based indicators may be coupled (as indicated in the following) to traditional economic indicators for the implementation of more equitable trade.

4. Emergy Evaluation of Trade

Economists define the so-called "terms of trade", i.e. the relationship between the price received for exports[2] and the amount of imports[3] a country is able to purchase with that money:

$$\text{Term of Trade} = \frac{\text{Average Price of Exports}}{\text{Average Price of Imports}}$$

The terms of trade is generally equal to 1 for a country, for the obvious reason that it is impossible for any country to purchase commodities from abroad without earning the same amount of money by exporting its own local resources. The terms of trade fluctuate according to changes in export and import prices. Clearly the exchange rate and the rate of inflation can both influence the direction of change in the terms of trade. Table 2 shows the terms of trade for Italy in selected years. When a complete factor price equalization is not observed because of wide differences in resources, barriers to trade, technology, and purchasing power of a country's currency, the result is almost always an increase of the debt for the developing countries or weakest economies.

Tab. 2: Terms of Trade of Italy in selected years (Cialani et al. 2005)

	Selected years				
Term of trade	**1989**	**1991**	**1995**	**2000**	**2002**
Exports/Imports	0.97	0.97	1.12	1.01	1.03

Since money only pays for the human labour and services, it is highly unlikely that market prices take into account the "hidden imports" embodied in each product.

---

2 Exports: The sale of goods, services and energy to buyers from other countries leading to an inflow of currency to a country.

3 Imports: The purchase of goods, services and energy from abroad that leads to an outflow of currency from a country.

Hidden imports are the resources of water, fertile land, constrained landscape, fertilizers, indigenous minerals, environmental services which support the production of traded commodities as well as labour and services. Since the emergy synthesis takes all these driving forces into account, a new form of more comprehensive "terms of trade", also called *emergy exchange ratio*, can be accomplished by using the emergy synthesis procedure. Figure 4 describes trade of commodities to Italy in emergy terms.

Fig. 4: Definition of trade in emergy terms. The emergy associated to the traded resource is compared to the emergy associated to the money received.

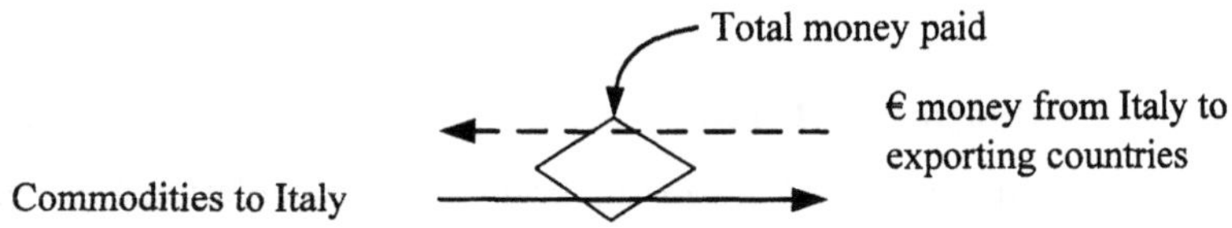

$$\text{Emergy benefit to buyer} = \frac{\text{Emergy of traded products}}{\text{Emergy of money paid}}$$

Each traded product is multiplied by a suitable emergy intensity (transformity or specific emergy), so that the emergy supporting its manufacture and transport is calculated. The total emergy (embodied resources) associated to the traded product flow is then compared to the total emergy associated to the commodities which can be purchased on the international market thanks to the money received (i.e. the emergy embodied in the money paid for). Emergy terms of trade depend on two main factors: the actual money cost of a commodity or resource, and the emergy/GDP ratio of the importing country (i.e. the buying power of its currency on the international market). In general, a purchase of primary, non manufactured resources (minerals, oil, wood, etc) gives an advantage to the buyer, because these resources are generally purchased by industrialized countries (low emergy/GDP ratio) and sold at low cost on the international market. The combined effect is that the exporting country buys less emergy than supplied, when it invests on the international market the money received in payment for the resource exported. When a country's economy develops, its emergy/GDP ratio generally decreases, resources are more likely to be processed at home and sold as manufactured goods at higher price: the combined effect of this new situation is better emergy terms of trade. A general policy for sustainable development of countries rich with primary resources is to process these resources at home, thus creating jobs and improving the local economy, and sell them as manufactured goods at higher price.

The emergy terms of trade of a given product can vary from a few units up to a very high value of the ratio, depending on the resource and market price. Instead, if the analysis is performed at the national level and extended to the whole import-export relationship between two countries, the advantage to the buyer may be higher for some items and lower for others, and therefore generally translates into a global emergy terms of trade of a few units. On the average, the Italian *import/export ratio* was ≈1 between 1989 and 2002 on a money basis (Table 2), while it was respectively 2.5 in 1989, 2.2 in 1995, and 1.7 in the year 2002 on an emergy basis. The area receiving the larger emergy receives the larger potential value and has its economy stimulated more. Raw products such as minerals, rural products from agriculture, fisheries and forestry, all tend to have high emergy exchange ratios when sold at market price. This is a result of money being paid for human services and not for the extensive work of nature that went into these products.

In the following we investigate and compare the economic performance of Latvia, Denmark and Italy in the year 2002 and provide an evaluation of trade equity-patterns among these countries.

5. National Analyses of Denmark, Latvia, and Italy

Tables 3 and 4 show respectively the emergy synthesis evaluation of Denmark and Latvia for the year 2002. The same evaluation for Italy was previously performed by the Authors of this paper and can be found in Cialani et al. (2005). Source of National data for Italy, Latvia and Denmark were the 2002, 2003, 2004 online Statistical Yearbooks of each country (www.coeweb.istat.it, http:// data.csb.lv/ EN/, www.dst.dk/) as well as other national and international databases (INEA 2003; United Nations 2004, 2005; UNCTAD 2004). Transformities used in Table 3 and 4 are from Brown & Ulgiati (2004) and Cialani et al. (2005).

The main energy and matter flows to each system (with reference to Figure 2) are listed and their emergy value quantified by multiplying each input flow by a suitable transformity from published literature in the field. A rough estimate of the amount of national GDP "generated" by each flow is provided in the last column of Tables 3 and 4, in order to show the relative importance of direct and indirect environmental flows and services supporting an economy. It clearly appears that when a nation develops an industrialized structure for its economy the most concentrated sources support the larger fractions of a nation's GDP. A clear picture of the main driving forces is provided by Figures 5 a,b,c respectively for Denmark, Latvia and Italy. Latvia, less industrially developed, relies to a larger extent on local renewable and nonrenewable sources (Figure 5 b), while the amount of imports still accounts for only 45% of the total emergy driving the country's economy. Instead, when a country develops, its main driving forces are fossil fuels and imported goods and commodities, as Figure 5 a (Denmark) and 5 c (Italy) clearly show. The fraction of imports to total emergy used is respectively 68%

for Denmark and 78% for Italy. Emergy flows imported at low cost, further contribute to the development of already industrialized countries. The emergy/GDP ratio for Italy was calculated as 2.42E+12 seJ/$ after Cialani et al. (2005, updated and converted from seJ/€ to seJ/$). The ratio of the total emergy supporting Denmark and Latvia to their respective GDPs yields emergy/GDP values equal to 1.89E+12 seJ/$ for Denmark and 4.47E+12 seJ/$ for Latvia.

Tab. 3: Emergy Synthesis of Resource Basis for Denmark (2002)

Note	Item	Raw amount	Unit	Transformity (sej/unit)	Solar Emergy (E20 sej)	Economic product of emergy flow (E7 Em$)
Renewable resources						
1	Sunlight	1.14E+20	J	1.00E+00	1.14	6.20
2	Rain, chemical	1.22E+17	J	3.06E+04	37.18	202.19
3	Rain, geopotential	3.72E+15	J	4.68E+04	1.74	9.48
4	Wind, kinetic energy	1.21E+18	J	2.51E+03	30.48	165.73
5	Waves	8.73E+16	J	5.13E+04	44.79	243.59
6	Tide	3.76E+14	J	2.83E+04	0.11	0.58
7	Earth Cycle	8.83E+16	J	3.44E+04	30.37	165.16
Nonrenewable sources from within system						
8	Natural Gas	3.16E+17	J	8.06E+04	254.53	1384.17
9	Oil	7.61E+17	J	9.07E+04	690.07	3752.69
10	Stone, gravel, sand	4.70E+07	J	6.72E+04	0.00	0.00
11	Calcium Carbonate, Lime	4.86E+06	g	1.68E+09	0.00	0.00
12	Clay	9.95E+04	g	1.68E+09	0.00	0.00
13	Top Soil	1.35E+16	J	1.24E+05	16.80	91.37
Imports and outside sources						
14a	Coal and coal products	1.16E+17	J	6.72E+04	78.08	413.67
14b	Oil derived products	3.29E+17	J	1.11E+05	364.92	1984.46
15	Metals	7.54E+11	g	3.02E+09	22.80	124.02
16	Minerals	1.51E+12	g	1.68E+09	25.41	138.19
17	Food & agricultural products	5.14E+16	J	3.36E+05	172.62	938.76
18	Livestock, meat, fish	4.45E+15	J	3.36E+06	149.62	813.68
19	Plastics & rubber	7.24E+15	J	1.11E+05	8.03	43.64
20	Chemicals	3.32E+11	g	3.80E+08	1.26	6.85
21	Wood, paper, textiles	1.15E+17	J	5.86E+04	67.42	366.62
22	Mechanical & transport equipment	1.15E+12	g	1.13E+10	129.14	702.30
23	Service in imports	2.51E+10	$	3.11E+12	780.61	4245.03
24	Tourism	5.79E+09	$	3.11E+12	179.80	977.76

Note	Item	Raw amount	Unit	Transformity (sej/unit)	Solar Emergy (E20 sej)	Economic product of emergy flow (E7 Em$)
Exports						
25	Food & agricultural products	1.43E+16	J	3.36E+05	48.11	261.62
26	Livestock, meat, fish	1.76E+16	J	3.36E+06	592.09	3219.84
27	Wood, paper, textiles	7.02E+15	J	5.86E+04	4.12	22.39
28	Oil derived products	7.30E+17	J	8.90E+04	650.17	3535.73
29	Metals	4.06E+11	g	1.68E+09	6.82	37.11
30	Minerals	2.50E+11	g	1.68E+09	4.19	22.80
31	Chemicals	2.86E+11	g	6.38E+08	1.83	9.93
32	Mechanical & transport equipment	1.84E+12	g	1.13E+10	206.80	1124.62
33	Plastics & rubber	7.84E+15	J	1.11E+05	8.69	47.27
34	Service in exports	2.72E+10	$	1.84E+12	499.84	2718.20
Production sectors based on indigenous resources						
35	Hydroelectricity	1.14E+14	J	2.77E+05	0.32	1.72
36	Agricultural Production	2.48E+17	J	3.36E+05	832.78	4528.77
37	Livestock Production	9.56E+15	J	3.36E+06	321.10	1746.20
38	Fisheries Production	5.56E+15	J	3.36E+06	186.89	1016.32
39	Fuelwood Production	4.88E+15	J	3.14E+04	1.53	8.34
40	Forest Extraction	2.35E+16	J	3.14E+04	7.38	40.13

Tab. 4: Emergy Synthesis of Resource Basis for Latvia (2002)

Note	Item	Raw amount	Unit	Transformity (sej/unit)	Solar Emergy (E20 sej)	Economic product of emergy flow (E5 Em$)
Renewable resources						
1	Sunlight	3.04E+20	J	1	3.04	680.70
2	Rain, chemical potential	2.62E+17	J	30574	80.22	17944.22
3	Rain, geopotential	1.37E+17	J	46828	64.24	14369.84
4	Wind, kinetic energy	3.67E+18	J	2513	92.27	20639.96
5	Waves	9.86E+16	J	51324	50.62	11322.76
6	Tide	1.03E+17	J	28295	29.06	6499.94
7	Earth Cycle	9.37E+16	J	5.77E+04	54.09	12099.56

Note	Item	Raw amount	Unit	Transformity (sej/unit)	Solar Emergy (E20 sej)	Economic product of emergy flow (E5 Em$)
Nonrenewable sources from within						
8	Coal	1.34E+15	J	6.72E+04	0.90	201.37
9	Dolomite	2.19E+06	g	1.68E+09	0.00	0.01
10	Clay	8.25E+07	g	1.68E+09	0.00	0.31
11	Sand	3.26E+06	g	1.68E+09	0.00	0.01
12	Gypsum	1.23E+06	g	1.68E+09	0.00	0.00
13	Limestone	3.03E+08	g	1.68E+09	0.01	1.14
14	Peat	1.11E+15	J	1.68E+04	0.19	41.55
15	Top Soil	8.66E+14	J	1.24E+05	1.08	240.92
Imports and outside sources						
16	Oil derived products	5.71E+16	J	1.11E+05	63.33	14166.41
17	Natural Gas	5.34E+16	J	8.77E+04	46.83	10476.97
18	Coal	1.85E+15	J	6.69E+04	1.24	277.31
19	Metals	6.49E+11	g	3.02E+09	19.61	4387.61
20	Food & agricultural products	1.83E+15	J	3.36E+05	6.15	1376.87
21	Livestock, meat, fish	3.20E+14	J	3.36E+06	10.76	2407.55
22	Plastics	9.98E+10	J	1.11E+05	0.00	0.02
23	Chemicals	1.16E+11	g	6.38E+08	0.74	165.95
24	Paper	1.39E+11	g	6.38E+06	0.01	1.99
25	Textiles	7.33E+09	g	6.38E+06	0.00	0.10
26	Mechanical & transport equipment	8.01E+10	g	1.13E+10	9.01	2016.11
27	Services in imports	7.08E+08	$	3.11E+12	22.00	4922.43
28	Tourism	1.61E+08	$	3.11E+12	5.00	1119.37
Exports						
29	Food & agricultural products	2.43E+15	J	3.36E+05	8.18	1829.90
30	Livestock, meat, fish	1.14E+14	J	3.36E+06	3.84	859.77
31	Wood	4.83E+16	J	5.86E+04	28.29	6328.63
32	Paper and paperboard	3.51E+10	g	6.38E+08	0.22	50.10
33	Textiles	1.42E+10	g	6.38E+06	0.00	0.20
34	Peat	7.55E+14	J	1.68E+04	0.13	28.37
35	Metals	6.78E+11	g	1.68E+09	11.40	2549.50
36	Chemicals	1.66E+10	g	6.38E+08	0.11	23.76
37	Mechanical & transport equipment	2.76E+10	g	1.13E+10	3.10	694.51
38	Plastics & rubber	4.94E+14	J	1.11E+05	0.55	122.61
39	Services in exports	1.25E+09	$	4.47E+12	55.97	12520.00

Note	Item	Raw amount	Unit	Transformity (sej/unit)	Solar Emergy (E20 sej)	Economic product of emergy flow (E5 Em$)
Production sectors based on indigenous resources						
40	Hydroelectricity	7.09E+15	J	2.77E+05	19.66	4398.61
41	Agricultural Production	2.99E+16	J	3.36E+05	100.59	22502.55
42	Livestock Production	2.65E+14	J	3.36E+06	8.90	1991.63
43	Fisheries Production	5.72E+14	J	3.36E+06	19.23	4301.91
44	Fuelwood Production	7.74E+16	J	3.14E+04	24.32	5439.73

Fig. 5 a: Emergy signature of Denmark in the year 2002, showing a large dependence on fossil fuels (local and imported) and imports of goods and commodities other than fuels.

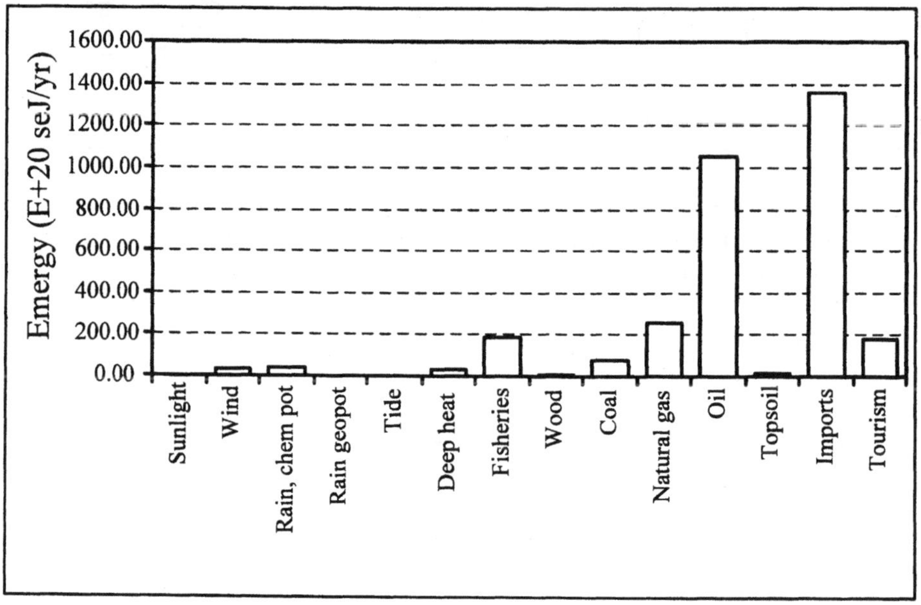

Fig. 5 b: Emergy signature of Latvia in the year 2002, showing a large dependence on local renewable inputs and comparable amounts of imported fossil fuels and commodities other than fuels.

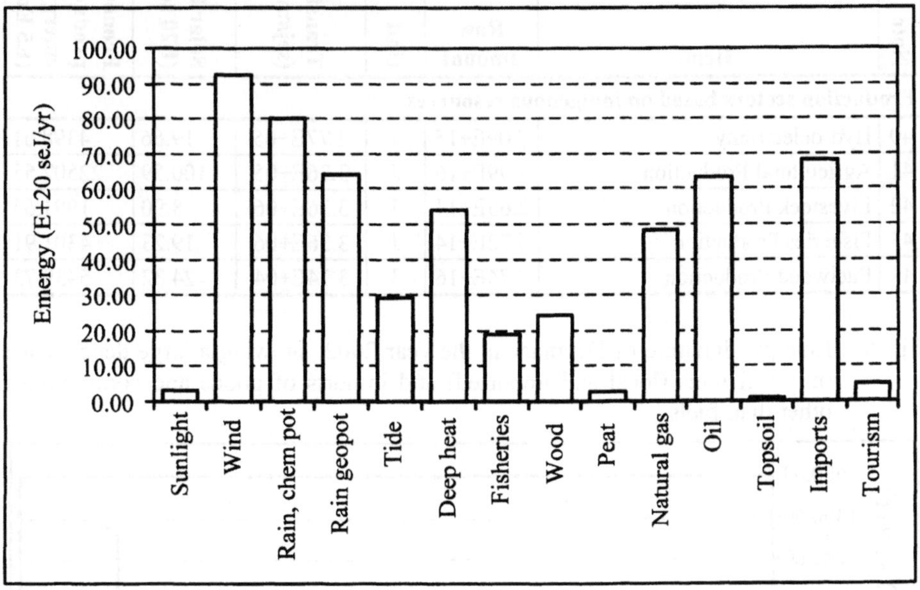

Fig. 5 c: Emergy signature of Italy in the year 2002, showing large dependence on fossil fuels (local and imported), imports of goods and commodities other than fuels, indigenous minerals and tourism.

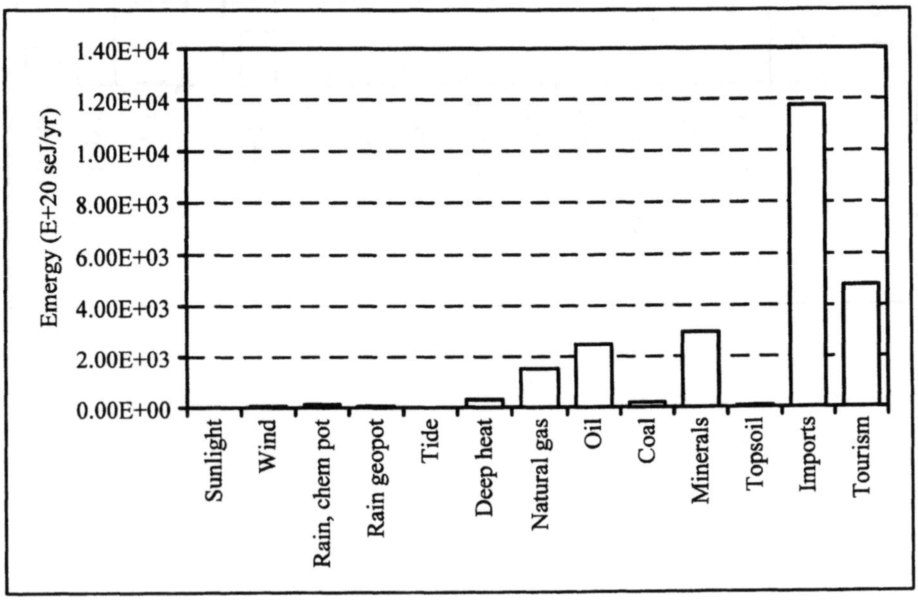

The discussion of the individual performance of each country is not, however, the main goal of this paper, which is instead concerned with resource trade and trade equity. We already pointed out that several aspects of a country's imports and exports are not accounted for by monetary exchange indicators. Table 5 lists the emergy value of selected products traded among the three investigated countries. Raw amounts are multiplied by their transformities in order to calculate the emergy which is associated to each flow of resources. Their monetary value (i.e. the money paid for by the importing country) is also indicated and multiplied by the country's emergy/GDP ratio, in order to estimate the emergy associated to the money flow. Finally, an emergy exchange ratio for each traded commodity is calculated. When the ratio is higher than one, it indicates that the importing country gets more emergy than it exports in the money paid for (or, vice versa, that the money paid is too small). This unbalance favours the economy of the importing country. Of course, since a large number of resources, goods and commodities are traded between two countries, the global advantage must be calculated by means of a weighted average of the ratios calculated for all traded items.

Table 5 also shows cases where the emergy exchange ratio is lower than one, clearly indicating that the importing country loses emergy in the trade. In general, while the emergy exchange ratio may also have high values for individual items traded, it averages and becomes lower when calculated over the whole set of traded commodities and reflects the different emergy-based buying power of each country's currency, as described in Figure 6. In this Figure, the buying power of one international dollar in each country is converted to emergy amounts: one dollar buys more emergy in Latvia and less in Denmark and Italy. This is because the Latvian economy is driven by large flows of primary emergy sources (peat, wood, fishery, local agriculture), which are sold at a low price but provide a huge amount of emergy to the importing country. Since money only pays for human labour and services, the unmonied emergy fraction is not paid for and contributes "for free" to the economy of the importing country. The emergy exchange ratio of Italy relative to Latvia is 1.85: 1 (Latvia sends 1.85 seJ to Italy for each seJ received); the same ratio calculated for Denmark relative to Latvia is 2.37: 1, while it is only 1.28: 1 for Denmark relative to Italy. Latvia loses significantly in emergy terms when trading with Italy and Denmark, although the global Latvian terms of trade in economic units is almost equal to one so that the trade seems to be quite fair and balanced.

Tab. 5: Emergy evaluation of trade for selected products exchanged by Italy, Denmark and Latvia in the year 2002.

	Total imports (g)	Energy Flow (J)	Transformity (seJ/unit)	Emergy of product (seJ)	Monetary value (US $)	Emergy of money paid (seJ)	Emergy Exchange Ratio
ITALY's imports from Latvia and Denmark (2002)							
Imports from Latvia							
Wood	9.39E+09	1.42E+14	8.85E+04	1.25E+19	3.90E+06	9.44E+18	**1.33**
Cereals (mainly wheat)	1.49E+10	3.12E+14	1.14E+05	3.56E+19	2.49E+06	6.01E+18	**5.92**
Peat	4.92E+10	5.15E+14	1.87E+05	9.63E+19	5.37E+06	1.30E+19	**7.41**
Imports from Denmark							
Wood	7.50E+08	1.13E+13	2.30E+05	2.60E+18	1.51E+06	3.66E+18	**0.71**
Cereals (mainly rice)	2.35E+06	4.91E+10	9.00E+04	4.42E+15	2.51E+03	6.06E+15	**0.73**
Peat	7.61E+09	7.96E+13	1.87E+05	1.49E+19	1.64E+06	3.98E+18	**3.74**
DENMARK's imports from Italy and Latvia (2002)							
Imports from Italy							
Pasta	2.20E+10	4.61E+14	4.00E+05	1.84E+20	1.83E+07	3.36E+19	**5.49**
Wine and liquors	3.83E+10	1.34E+14	4.20E+05	5.63E+19	7.91E+07	1.45E+20	**0.39**
Imports from Latvia							
Textiles	8.70E+08	1.38E+13	3.25E+06	4.48E+19	1.71E+07	3.14E+19	**1.43**
Plastics and rubber	8.25E+06	2.42E+11	1.33E+06	3.22E+17	2.84E+05	5.23E+17	**0.61**
LATVIA's imports from Italy and Denmark (2002)							
Imports from Italy							
Pasta	3.79E+09	7.93E+13	4.00E+05	3.17E+19	2.17E+06	9.69E+18	**3.27**
Wine and liquors	1.45E+09	5.09E+12	4.20E+05	2.14E+18	3.26E+06	1.46E+19	**0.15**
Imports from Denmark							
Textiles	1.15E+09	1.82E+13	3.25E+06	5.90E+19	1.24E+07	5.56E+19	**1.06**
Iron and steel	5.42E+05		5.46E+09	2.96E+15	2.56E+03	1.14E+16	**0.26**

Note: Mass of items is converted to energy (when applicable) by means of their available energy factors and then converted to *em*ergy, by means of average transformity values (seJ/J). Mass of steel is directly multiplied by a specific emergy factor (seJ/g).

Fig. 6: Emergy (x 10^{12} seJ) traded on the basis of 1 \$ paid. One \$ paid to Latvia buys on the average 4.47 x 10^{12} seJ, while one \$ to Italy buys only 2.42 x 10^{12} seJ. Therefore, even if the economic terms of trade are balanced, the actual trade of resources is uneven (advantage 1.85: 1 for Italy). Denmark gains from both Italy and Latvia (respectively 1.28: 1 versus Italy and 2.37: 1 versus Latvia), while Latvia loses in both trade relationships.

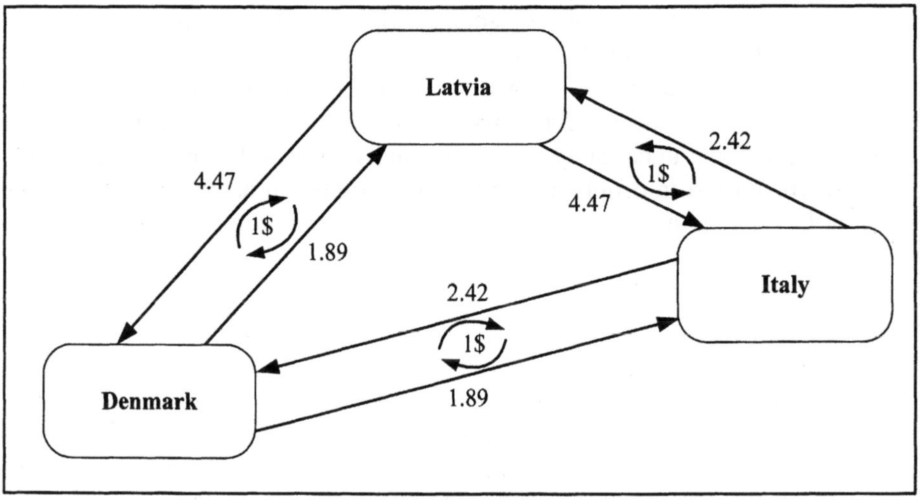

(x 10^{12} seJ)

One of the reasons of the trade advantage for the two importing countries relies on their low emergy/GDP ratios, typical of developed economies characterized by large industrial, information, and service sectors. These countries generally sell processed items with high labor content and consequent high price, while Latvia sells primary resources with low labor content and consequent low price (e.g., wood). The combined effect of paying a low price to a high emergy/GDP country is that the importing country gets the most emergy and its economy is supported more. A good policy for Latvia would be to process its primary resources at home by developing a manufacturing sector within the country, and export manufactured goods with high labor and information content at higher prices. By making its economy more efficient (i.e. able to generate a larger GDP due to better use of resources), Latvia would finally reach a lower emergy/GDP ratio, which in turn would have feedback effect on balancing its exchange with trade partner Nations.

Equitable trade policies suggest emergy exchange ratios (= $Emergy_{out}/Emergy_{in}$) to be equal to 1. By setting equal to one the emergy terms of trade of Figure 4, it is possible to calculate a reference price of the exported commodity, for equitable trade. For example, Italy pays 0.42 \$/kg for semi-processed wood from Latvia, while it pays about 2 \$/kg for processed wood for forniture from Denmark. Taking into account two different transformities for wood imported, the requirement for the Emergy Exchange Ratio to be equal to 1 would translate into an average price of about 0.55 $/kg of wood. This "virtual" price is higher than for the wood imported from Latvia and much lower than for the wood imported from Denmark. As a consequence, Italy gains emergy in the import from Latvia and loses emergy in the trade from Denmark (Table 5, next Section), due to the fact that Denmark partially processes its wood within the country. Since emergy accounting methods are unlikely to be accepted in the present system of market economy, an emergy-based price can be considered as a reference point, in order to compare the fairness of trade and act for improvement. Our figures are very likely affected by significant uncertainty, due to the uncertainty and sometimes huge differences which exist in official national and international databases. Much work is still to be done in this regard: however the calculated orders of magnitudes are reasonable and provide interesting hints for trade evaluation.

5. Conclusions

We have shown in this paper that fairness of trade and resource exchange among financially and industrially developed countries and less developed countries is hardly expressed by monetary indicators as well as by the conventional terms of trade. This is because money does not measure a large set of free environmental services, which are embodied in the exchanged resources and goods, and therefore it is unable to assess the existence of unfair trade and unbalanced resource flows.

This paper investigates the hidden flows of matter, energy and environmental services which support international monetary transactions. We maintain that these indicators should be first identified and individually calculated, and then aggregated into larger-scale indices (such as the emergy exchange indicator investigated in the paper), in order to provide a more comprehensive picture to complement monetary evaluations.

We have pointed out elsewhere that the multifaceted complex systems such as economies and human societies cannot be dealt with by means of only one evaluation approach (Ulgiati et al. 2005a), because their dynamics (the so-called societal metabolism) develops on multiple scales according to multiple goals, driving forces and constraints. Multi-criteria and multi-scale methods are needed for such an evaluation. However, societal metabolism (the way societies grow on available resources) is embedded in the larger-scale dynamics of ecological metabolism (the way ecosystems grow on available resources) and shows like patterns. Exploring societal and ecological metabolisms requires the evaluation of unmonied flows of resources and environmental services which are very often neglected and undervalued and for which the emergy method is on our opinion the best-suited tool. In particular, we explored in this paper a specific aspect of the economic process, i.e. the support it receives by the larger scale of biosphere and how this support can be accounted for in trade relationships. This donor-side perspective is capable of complementing pieces of information provided by different approaches (Bargigli et al. 2005; Ulgiati et al. 2005b), towards a more comprehensive understanding of the whole system's and trade dynamics.

It is therefore of paramount importance that governments, trade operators, NGOs, and consumers realize that there are other parameters and performance indicators which must be taken into account in order to better gauge the real value of resource exchange and the dynamics of the transfer of wealth among countries. The environmental and social stability of the world depends on balanced exchange of resources and equitable trade. This issue is equally very important in the trade relations among European countries, whether they are Members of the European Union and of the Euro monetary area or not.

References

Bargigli, S., Raugei, M. & Ulgiati, S. (2005), Mass flow analysis and mass-based indicators. In: Handbook of Ecological Indicators for Assessment of Ecosystem Health. Sven E. Jorgensen, Robert Costanza & Fu-Liu Xu Editors, pp. 353-378. CRC Press, Boca Raton, Florida.

Brown, M.T. (2003), Resource Imperialism: Emergy Perspectives on Sustainability, International Trade and Balancing the Welfare of Nations. In: Book of Proceedings of the International Workshop "Advances in Energy Studies. Reconsidering the Importance of Energy". Porto Venere, Italy, 24-28 September 2002. S. Ulgiati, M.T. Brown, M. Giampietro, R.A. Herendeen & K. Mayumi, Editors. SGE Publisher Padova, Italy, pp. 135-149.

Brown, M.T. & Ulgiati, S. (1999), Emergy Evaluation of the Biosphere and Natural Capital. Ambio, 28(6): 486-493, September 1999.

– (2004), Emergy Analysis and Environmental Accounting. In: Encyclopedia of Energy, C. Cleveland Editor, Academic Press, Elsevier, Oxford, UK, pp. 329-354.

Cialani, C., Russi, D. & Ulgiati, S. (2005), Investigating a 20-year national economic dynamics by means of emergy-based indicators. In: Brown, M.T., Campbell, D., Comar, V., Huang, S.L., Rydberg, T., Tilley, D.R. & Ulgiati, S. (eds.) (2004), Emergy Synthesis. Theory and Applications of the Emergy Methodology – 3. The Center for Environmental Policy, University of Florida, Gainesville, FL, 2005.

INEA, Istituto Nazionale di Economia Agraria (2003), Annuario dell'Agricoltura Italiana 2002. Il commercio con l'estero dei prodotti agroalimentari, vol. LVI. Edizioni Scientifiche Italiane, Napoli. ISBN: 884950750X, pp. 622.

Odum, H.T. (1988), Self organization, transformity and information. Science, 242: 1132-1139.

– (1994), Emergy and Policy, vol. 1, pp. 25-29. Environmental Engineering Sciences, University of Florida, Gainesville, FL.

– (1996), Environmental Accounting: Emergy and Environmental Decision Making. John Wiley & Sons, N.Y.

Sampat, P. (2003), Scrapping Mining Dependence. In: The State of the World, 2003, The World Watch Institute, pp. 110-135.

Szargut, J., Morris, D.R. & Steward, F.R. (1988), Exergy analysis of thermal, chemical and metallurgical processes, Hemisphere Publishing Corporation, London.

Ulgiati, S., Bargigli, S. & Raugei, M. (2005a). Dotting the I's and Crossing the T's of Emergy Synthesis: Material Flows, Information and Memory Aspects, and Performance Indicators. In: Brown, M.T., Campbell, D., Comar, V., Huang, S.L., Rydberg, T., Tilley, D.R. & Ulgiati, S. (eds.). Emergy Synthesis. Theory and Applications of the Emergy Methodology – 3. The Center for Environmental Policy, University of Florida, Gainesville, FL, 2004.

Ulgiati, S. & Brown, M.T. (1998), Monitoring patterns of sustainability in natural and man-made ecosystems. Ecological Modelling, 108: 23-36.

Ulgiati, S., Brown, M.T., Bastianoni, S. & Marchettini, N. (1995), Emergy based indices and ratios to evaluate the sustainable use of resources. Ecological Engineering, 5(4), 519-531, 1995.

Ulgiati, S., Raugei, M. & Bargigli, S. (2005b). Overcoming the inadequacy of single-criterion approaches to Life Cycle Assessment. Ecological Modeling, forthcoming, 2005.

UNCTAD (2001), Handbook of World Mineral Trade Statistics 1994-1999. New York.

– (2004), Handbook of Statistics, Part 5, International Trade in Services.

UNDP (2001), United Nations Development Program, Human Development Report 2001. New York.

United Nations (2004), International Trade Statistics yearbook, vol. 1, Trade by country. United Nations, New York.

– (2005), http://unstats.un.org/unsd/comtrade (data 2002).

World Bank (2001), World Development Indicators 2001. Washington, DC.

Coastman
–
Integrative Approaches to Coastal Zone Management in the Baltic Sea

Walter Leal Filho, Professor

TuTech Innovation,
Harburger Schlossstrassse 6-10, D – 21079 Hamburg, Germany

Abstract

One of the priorities in respect of sustainable development in the Baltic Sea Region is coastal zone management. Due to its economic and ecological relevance, the proper management of the Baltic coast poses a major challenge to all countries in the region.

This paper will present some examples of initiatives on coastal zone management in the Baltic Sea Region and will pay a particular attention to Coastman ("Coastal Zone Management in the Baltic Sea" – "Coastman"), an Interreg IIIB Project which aims at analysing the interactions between environment conservation, conflict management and sustainable development in seven coastal zones of the Baltic Sea. The project partners Sweden, Germany, Finland, Estonia, Lithuania, Latvia and Russia will analyse environmental problems and assess the risks of the Baltic Sea coast with reference to transportation, fishing and tourism, suggesting measures to improve the current situation. Elements to be considered in order to establish a better coastal zone management in the Baltic Sea, North Sea and Mediterranean, Sea, via exchanges through seminars, inspections and dialogues with regional decision-makers will be discussed.

Introducing the Project "Coastman"

Today around 60% of the earth's population lives within 60 km of a seacoast. To this must be added coastal tourism as practised in many countries. The burdens on coastal ecosystems are accordingly very heavy.

At present, there are many pressures and threats to coastal regions in the Baltic Sea Region: cargo shipping is growing steadily; commercial fishing is now exceeding fishing limits; and now more and more exploration and pumping platforms are being built at sea, whilst wind parks are already established along the coasts of Denmark and Sweden and the construction of such parks has begun in German waters. Additionally, there are plans for many other exploitative uses of the maritime areas of the North Sea and the Baltic. Finally, many coastal areas suffer from problems (especially pollution) deriving from agriculture (Leal Filho 2004).

All these developments increase pressure on coastal ecosystems.

Furthermore, every use of the sea has its counterpart on land: harbours, wharfs, company offices, processing plants, service and secondary companies.

In short, one can say that coastal areas are being subjected to immense and growing burdens.

This developmental process is not happening within a comparatively resilient and stable system but rather in areas characterised by their frailty and sensitivity. Two different ecosystems are not only juxtaposed in coastal areas; they actually constitute a continuum: the habitats and favourable sites of many biotic elements in the two ecosystems – 'land' and 'sea' – overlap and impinge on each other and thus respond particularly sensitively to disruption.

By reason of its dynamic nature and suitability as a transport medium, the sea itself is another important sensitivity factor.

Since the 1970s, the development of coastal zones has demonstrated that construction projects, sealing of the natural surface to rainfall, unsecured waste, waste water and litter disposal, release of pollutants (either in major accidents such as oil spills or from small but persistent leaks) have all seriously damaged the capacity of coastal ecosystems to sustain life. A conclusion that can be drawn from this is that the principle of 'Sustainability' must be the more widely used and serve as a continual basis for action – although at first 'Sustainability' was taken to refer only to the retention of natural systems. 'Sustainability' may be attainable through holistic, literally 'ecological' thinking and acting.

The beginnings of 'Integrated Coastal Zone Management' (ICZM) can be traced to the passing of the 'Coastal Zone Management Act' in the USA in 1972 to co-ordinate development planning along the American coastline. The four aims that it established are still valid today:

1. to preserve, protect, develop, and where necessary and possible, to restore or enhance, the nation's coastal resources in the widest sense, although this was mainly taken to refer to the ecological potential.
2. to implement national environmental protection policy by supporting the coastal states in the creation of coastal zone management plans
3. to improve the quality of life by protecting significant ecosystem types and by the creation of planning security and long-term reliability
4. to involve governmental and local administrations and to work with regional and national authorities.

Based on the need to analyse current problems seen in coastal areas in the Baltic and identify ways to address them, the project Coastman ("Coastal Zone Management in the Baltic Sea" – "Coastman") has been created. It is an Interreg IIIB Project involving Estonia, Germany, Latvia, Lithuania, Russia and Sweden, which aims at analysing the the interactions between environment conservation, conflict management and sustainable development in six coastal zones of the Baltic Sea.

The Aims of Coastman and Problems to be Addressed

From a perspective of sustainable development, integrated coastal zone management needs interdisciplinary and spatially integrative approaches, where all important stakeholders –especially those who are responsible for planning and realization of exploitation of coastal zones – are involved.

Therefore, the central objectives in the project *Coastman* are:

1. to demonstrate the conditions under which spatial conflict resolution in coastal zone management can be handled in a bottom-up perspective, starting from concrete case studies covering the problem dimensions and different cultures around the Baltic Sea;
2. to identify legal, organizational, economic, methodological and cultural frameworks for CZM in participating countries;
3. to find out how differences in these frameworks influence strategies for conflict resolution on local and regional levels in CZM and what adaptations of methods are necessary for each participating country.

Furthermore, *Coastman* intends to promote the sustainable development of coastal areas through an information exchange platform, information events and specialist workshops, combined with and exchange of experience and best practices in conflict resolution in CZM with related Interreg projects.

A further objective of the project is the development of joint educational programmes for coastal zone management in the Baltic, with a focus on conflict resolution in spatial planning. This will be combined with initiatives to create awareness among the various stakeholders in all Baltic countries, that in order that problems are fully solved, a mixture of facts and values are needed and that solutions very much have to be found in a participatory processes where all stakeholders are involved.

As far as the problems to be addressed are concerned, coastal zones in the Baltic region have a great potential for economic growth. There are however several serious threats to a sustainable growth of coastal zones in and around the Baltic Sea. Two of the main threats are conflicting interests and goals, and environmental pollution. In addition, potential conflicts occur which are multidimensional, including cultural and spatial dimensions. Environmental threats to coasts are also very complex and versatile. Several of the existing problems have historical roots but the rapid growth of the economies in the region (the rate of economic growth in Estonia, Latvia and Lithuania for the years 2005 and 2006 is well above EU average) will no doubt create new problems which might turn out to be even worse than we can imagine today. One important example is oil transports from different terminals around the Baltic Sea. This has already created conflicts at both the local and regional levels. There is often a lack of system-view in the handling of this problem. This is the case for oil transports, as well as for the investments in environmental technologies. The solutions have to be found in a process where all spatial aspects and stakeholders take part in order to prevent that intractable conflicts appear.

Many on-going and planned projects in the region are aimed at handling either environmental problems or problems in respect of economic growth (usually in different sectors such as fisheries or tourism) separately. The problem with this approach is that conflicting interests and goals cannot be properly handled, with a risk that the results are jeopardized. In specific cases it is often obvious that there is a mix of basic facts and values which confuses the struggle for a practical solutions. The *Coastman* project takes on the task of analysing concrete approaches in specific cases studies. In this way there is the potential for the exchange of "hands on" experience between the different countries thus contributing to a necessary harmonisation of conflict resolution strategies and methods in the Baltic.

The Approaches Used

The main approach of the *Coastman* project, which involves universities, municipalities and companies in the Baltic, is to apply existing methods of system analysis and conflict resolution on coastal zones in order to handle the multidimensional aspects of sustainable development.

Many of these methods are in use today in educational programmes in order to get students to realize that there are often many dimensions and conflicting interests in problems related to spatial planning. Conflicts can often be resolved if stakeholders can shift their perspectives of the problem. The methods have to be simplified, validated, and integrated in order to be of value in a process where stakeholders work together in a process for finding solutions. In many of the countries participatory processes are not common and it will be an important part of the project to harmonize the way in which all stakeholders can be involved in conflict resolution. The process of developing indicators maybe useful in this process (Bach 2004), whilst environmental information plays an important role (Langaas, Aliakseyeva, Gooch, Lopman, Nilsson & Timmerman 2002):

To demonstrate and evaluate the available methods, the project uses selected case studies where coastal zones are exploited for oil terminals and waste water treatment plants. In some of these cases, there are intractable on going conflicts which have developed during several years, conflicts which tends to threaten economic growth in the area. Joint workshops focus on benchmarking and evaluation of common methods and to see which role cultural, social and economic differences play.

The project partners involved come from several important coastal zone areas around the BSR. That is south-east coastal area of Finland, the St Petersburg area, coastal zone areas in Estonia, Latvia and Lithuania, Hamburg, and the Stockholm coastal zone.

The central part of the project is the case studies carried out in several partner countries.

The case studies take place in:

- Loudden oil port located in the central part of the City of Stockholm
- Ventspils oil port located on the coast line 200 km West of Riga
- Klaipeda oil port located in the central part of Klaipeda
- Haapsalu waste water treatment plant, located on the coast line 100 km SW from Tallin
- Primorsk oil port west of St. Petersburg (case study work within Tacis project).

As far as dissemination is concerned, the project uses a number of means such as:

1. a project website, which is upkept and marketed. It contains a description of the projects, activities and case studies, as well as guidelines and other developed materials as well as information concerning dissemination workshops and other project events;
2. a project newsletter, published on a regular basis;
3. dissemination workshops and training courses organised for different stakeholders.

The developed guidelines and recommendations will be disseminated through networks established during the project. In addition, the mass media will be involved into the dissemination activities. Moreover, the establishment of an educational programme in the Baltic will start a long-term dissemination process.

Fig. 1: Coastman Home Page (http://www.coastalmanagement.net/en/)

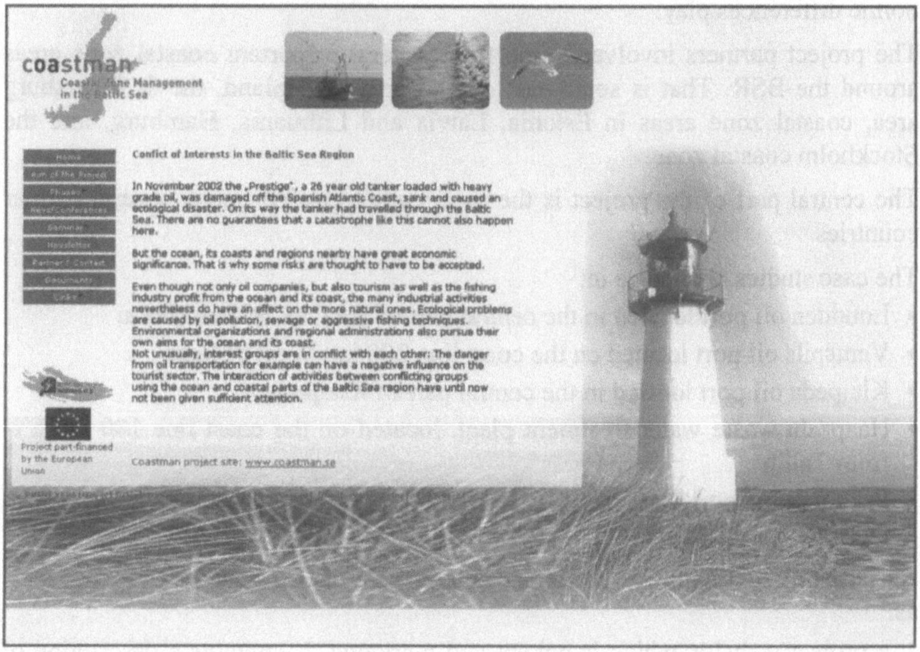

During the course of the project, a joint educational program on coastal zone management in the Baltic Sea Region with focus on conflict resolution in spatial planning is being developed. One pilot course will be soon be realized during the project, and it is intended to be given at regular intervals after the project.

Another feature which deserves mentioning is the set of cross-sectorial training schemes, focusing on practical methods for conflict resolution. These are being designed and documented. Courses will be available on the web after the project has been completed, to be used by people working in government agencies, NGOs, universities and companies.

Moreover, a "Handbook" on conflict resolution will be prepared and translated in various languages. It is also envisaged that the technical development of an information exchange platform will be completed during the project, where a "Pool of Excellence" will be created, and where the description of cases of successful conflict resolution in the Baltic will be disseminated.

The Coastman Case Study in Hamburg

Hamburg is culturally and commercially the centre of all Northern Germany region. The Hamburg Metropolitan Region consists of 3.5 million people. For all of them, Hamburg is the shopping and cultural metropolis. The municipal area with its 755

km^2 is seven times the size of Paris and 2,5 times that of London. For this very reason, Hamburg presents an exceptionally high standard of living and housing.

With 30 m^2 living space per person, Hamburg enjoys the largest average personal living space of all big cities in the world. In fact, 14% of the city is made up of green and recreation areas. Hamburg is administered by a local government (Bürgerschaft) and its main economic activities are all harbour related (e.g. shipping, transport of good, etc).

The Hamburg case study focuses on the conflict which exists in respect of the use of the Elbe river, which is essential for harbour operations. Existing regulations are focused on local impacts of the relocation of contaminated sediments and do not take the whole catchments area into account. The European Water Framework Directive demands a holistic river basin approach when considering the water resources. However, the Water Framework Directive does not explicitly mention sediments nor sediment quality and quantity and it is not even clear how sediments will be incorporated into the new regulations which have to be developed.

The case study will refer to the need to address the problem posed by the accumulation of sediments, taking at the same time environmental questions into account, i.e. the impact of dredging and natural conservation and environmental protection concerns, as well as the stakeholders, which can be seen in Figure 2.

Fig. 2: Stakeholders in the Hamburg case study

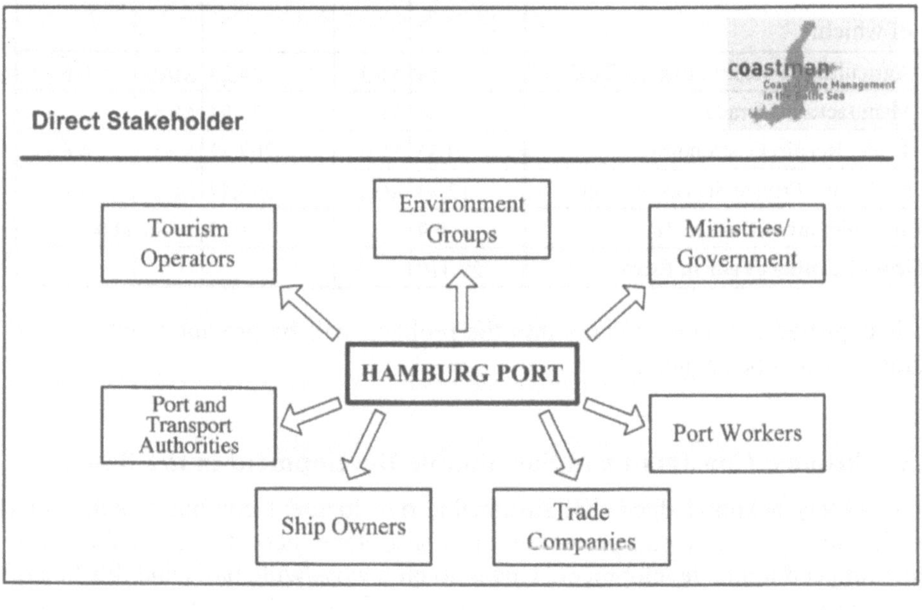

The Hamburg case study deals with a long-term conflict since the problem of sedimendation is a long-term one and which needs constant attention. The subject matter of sediments in the Elbe has been extensively analysed before (Arevalo &

Heise 2003; Behrendt & Hoffmann 2002; Frey, Borchardt, Funke, Geffers & Schleiter 2002), whilst experiences from other countries also exist (Ditschke & Markofsky 2003). However, there are many gaps which need to be fulfilled.

The case study will among other things analyse the extent to which a reduction of the contamination of dredged materials – as a contribution to the concept of a sustainable clean port and environment – can be achieved.

Tab. 1: Facts and Figures about Hamburg

	Hamburg		Germany		Germany [%]
Inhabitants	1.729.000		82.537.000		2,1
Total Area	75.532	ha	35.703.099	Ha	0,2
Built-up Area and Open Space	26.878	ha	2.308.079	Ha	1,2
Recreation Area	5.702	ha	265.853	Ha	2,1
Road Area	8.860	ha	1.711.764	Ha	0,5
Arable Land	21.000	ha	19.102.791	Ha	0,1
Forested Area	3.432	ha	10.531.415	Ha	0,0
Water Area	6.115	ha	808.462	Ha	0,8
Share of Gross Domestic Product	70.243	Mrd.	1.984.300	Mrd.	3,5
Gross Value Added	67.933	Mrd.	1.919.020	Mrd.	3,5
of which:					
Agriculture, Forestry, Fishing Trade	0,13	Mrd.	24,23	Mrd.	0,5
Manufacturing Trade	11,64	Mrd.	536,84	Mrd.	2,2
Trade, Repair, Gastronomy	9,52	Mrd.	207,72	Mrd.	4,6
Public and Private Service Industry	12,35	Mrd.	396,47	Mrd.	3,1
Unemployment Rate in %	9,9		10,7	(Stand 04/2005)	
Gross Earnings/Year in Euro	29.319				

It is expected that new insights into the problem will be provided and ways to address it will be suggested.

Conclusions: Coastman and Sustainable Development in the Baltic

It is widely acknowledged that sustainable development combines both social equity and environmental protection and these items need to be considered in pursuing economic development. This is so on a worldwide basis and this line of thinking is especially relevant to the Baltic Sea, a semi-closed system where environmental impacts and contamination are felt in a especially strong way.

In order to reach the integration between sustainability and coastal conservation, it is equally important to realise that this has to be done in a democratic process, where facts and values of different stakeholders are considered. The project *Coastman* is supporting this democratic process by applying system analysis, as well as simple but validated methods in identifying practical solutions to complex and quite serious problems.

In respect of the role of *Coastman* and its contribution towards integrative approaches towards sustainability in the Baltic, a matter extensively discussed by Leal Filho and Ubelis (2004), it can be said that the project considers not only the problems seen in coastal zones, but also pays a great emphasis on the action needed to solve them. In addition, *Coastman* is establishing an educational platform where stakeholders from different countries and sectors can meet to discuss actual problems, thus catering for the involvement of the various segments of society. Thanks to its holistic approach, flexible structure and problem-oriented approach, it is expected that Coastman may provide a small contribution towards making the goal of sustainable management of coastal areas in the Baltic, a reality.

References

Arevalo, E. & Heise, S. (2003), Sustainable Solutions to the Sediment Management on the River Basin Scale: The Elbe River Case under specific concideration of risk management and communication aspects, 2nd SedNet-Conference, September 2003.

Bach, H. (2004), "Methodology and process for indicator development" 30 August, 2004; www.nea.gov.vn/EIR/english/Workshop_Training/Documents/ Workshop_Indicator/ TA/ Methodology-process-indicator-development_ver1-10.htm

Behrendt, H. & Hoffmann, J. (2002), The Elbe Catchment and Related Coastal Area: German Bight and Wadden Sea. Part A: Catchment Profile, Institute für Gewässerökologie und Binnenfischerei im Forschungsverbund Berlin e.V, May 2002.

Ditschke, D. & Markofsky, M. (2003), Optimization of a Structure to Reduce Sedimentation in an Estuarine Harbour, COPEDEC VI, 2003, Colombo, Sri Lanka.

Frey, M., Borchardt, D., Funke, M., Geffers, K. & Schleiter, I. (2002), "Heavily Modified Waters in Europe – Case Study on the River Elbe" www.wasser-blick.net/ servlet/ is/ 13045/ Elbe.pdf?command=downloadContent&filename =Elbe.pdf

Langaas, S., Aliakseyeva, N., Gooch, G., Lopman, E., Nilsson, S. & Timmerman, J. (2002), Working Paper on "Environmental Information in Transboundary River Basin Policy-Making and Management: Selected European Case Studies", October, 2002. www.lwr.kth.se/ publikationer/ PDF_Files/ MAN-TRA_East_WP. pdf.

Leal Filho, *W*. (ed.) (2004), Ecological Agriculture and Rural Development in CEE Countries. Kluwer Academic Publishing, Doordrecht.

Leal Filho, *W. & Ubelis*, *A*. (eds.) (2004), Integrative approaches towards sustainability in the Baltic Sea Region. Peter Lang Scientific Publishers, Frankfurt a.M.

Second Plenary

Sustainable Agriculture, Agroforestry, linked Resources and Developments

The Use of Environmental Information in Decision Making Promoting a Sustainable Food Chain

C. Lagerberg-Fogelberg[1], F. Fogelberg[2], G. Grankvist[3]

[1] Swedish University of Agricultural Sciences,
Centre for Sustainable Agriculture,
POB 7047, SE – 75007 Uppsala, Sweden

[2] Hedmark University College,
Fac. Agriculture and Natural Sciences,
Blaestad, N – 2322 Ridabu, Norway

[3] University College of Trollhättan/Uddevalla,
Dept. Social and Behavioural Studies,
POB 1236, SE – 46228 Vänersborg, Sweden

Abstract

Sustainable consumption behaviour is an essential and integrated part of a future sustainable society. The food chain from production to consumption involves many different actors interacting in complex patterns. The different actors require different information packaged in accordance with their needs, skills and decision situations.

The wholesale and food service sectors control food delivered to a large number of consumers. About 5.6 million meals are served each day in Swedish institutional households. The purchasing managers of food service institutions and the wholesale sector have a key position in the food chain controlling large flows of foods, thereby allowing them to practice potential influence on the food supply chain. Consequently these sectors have a great impact on the overall environmental effect of food consumption patterns. In this ongoing project we focus on how different factors interact in food purchase and how environmental information of foods affect the decision of food purchasing managers of institutional households and wholesalers. One preliminary finding is that price is by far dominating the preference of a product although environmental parameters could vary ± 90%.

Introduction

Six years ago the Swedish parliament established 15 national environmental targets (Figure 1) to promote an ecologically sustainable society. Seventy-one subtargets have been identified which include details and timetables in order to reach the main environmental targets. The overall aim is to hand over to the next generation a society in which the major environmental problems have been solved. A target on biodiversity is in progress, which will be the sixteenth target

(Naturvårdsverket 2003). Consumption of food can be associated with most of these objectives.

Fig. 1: The Swedish national environmental targets (Miljömålsrådet 2003; 2004).

- Good-quality groundwater
- A balanced marine environment, flourishing coastal areas and archipelagos
- Thriving wetlands
- Sustainable forests
- A varied agricultiral landscape
- A magnificent mountain landscape
- A good built environment
- Reduced climate impact
- Clean air
- Natural acidification only
- A non-toxic environment
- A protective ozone layer
- A safe radiation environment
- Zero eutrophication
- Flourishing lakes and streams

The political framework for a more sustainable society was set by the 1992 UN conference in Rio de Janeiro, thereby initiating the Agenda 21 work. The world summit follow up in Johannesburg, 2002, established agreements for further work on measures to i) promote the inclusion of social aspects and environmental costs into the prices of goods and services, ii) develop programmes to promote increased awareness in areas such as education, public information, consumer information, advertising and media, iii) develop and introduce efficient tools for consumer information on issues relating to human health and safety, and iv) promote public procurement that develops and distributes environmentally sound goods and services.

Recently a governmental investigation on sustainable consumption of foodstuff, housing and transport (Edman 2004; 2005) was presented with a wide range of suggestions to make Sweden a more "green" society.

Fig. 2: Economic and material welfare nolonger automatically leads to social welfare (Edman 2004).

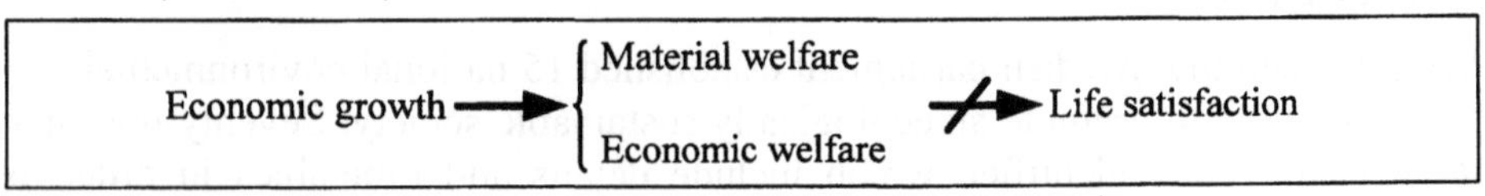

Some opinions of the governmental investigator were to consider our values, attitudes and habits in favour of a smarter and more caring lifestyle. Our modern society displays that the former relationships between economic growth and life satisfaction are no longer valid (Figure 2). The economic growth and material welfare we enjoy also cause considerable stress on people. He stressed that policy must be used more forcefully to speed up development of greener technology and change price relations in favour of greater social consumption. Among others, six main strategies on food consumption were presented (Edman 2005): i) strengthen the market for staple foods produced in Sweden, ii) invest in extensive Swedish meat production, iii) increase the proportion of organic foodstuff, iv) increase the proportion of fair trade products, v) strengthen retailer responsibility through green driving licences for shop employees, and vi) switch to Smart food.

Moreover, strategies for enhanced public procurement was presented:

> "Central government, municipalities and county councils must carry out their award procedures in the best quartile of the product assortment, taking balanced sustainability perspectives into account. Within the framework of the best quartile strategy, the Swedish Environmental Management Council will be given increased resources to act as a national body for knowledge management, analysis and the training of public procurement officers." (Edman 2005)

The Swedish Environmental Management Council is during 2005 developing an easy-to-use instrument for purchasing organisations caring about environmental issues for all types of simple non-composed foods and food services. The suggestions will be ready in 2005 and implementation will start in 2006.

In the ongoing project "Designing and evaluating the impacts of environmental information in food service institutions and the food wholesale sector" we have investigated the behaviour of food purchasing managers in general and in particular regarding their use of environmental information. Thereby we add knowledge to the governmental strategies presented by Edman (2004; 2005). In our work we carry out environmental profiles of foods and construct scenarios for the future. About 5.6 million meals are served every day in Swedish commercial and public households. The purchasing managers of food service institutions potentially impact the overall environmental performance of food consumption patterns through their key positions in controlling large flows of foods, that is through their power to exert potential influence on the food supply chain.

Public Procurement as an Instrument to empower Sustainable Consumption

Environmental profiles of a number of food products were estimated. The products were chosen so as to represent both animal and vegetable products as well as fresh and preserved products. The objective was to address differences in environmental profiles of products of different origin and processing level. Furthermore, addressing protein sources of plant origin as well as from animal sources allow for further discussions on dietary choices. Thus apples, broccoli, carrots, onions, tomatoes, peas, chick peas, chicken and beef were selected for this study. Product cases from different countries of origin were assessed.

Fig. 3 displays a generic system of the carrot supply chain. The inventories of the investigated foods include a selection of parameters such as use of water, chemicals, plant nutrients, packaging materials, and land. From this data for instance greenhouse gas emissions and then the global warming potential resulting from these emissions are estimated.

Fig. 3: The supply chain of fresh carrots including the key positions of purchasing managers. Black arrows between compartments represent transportation.

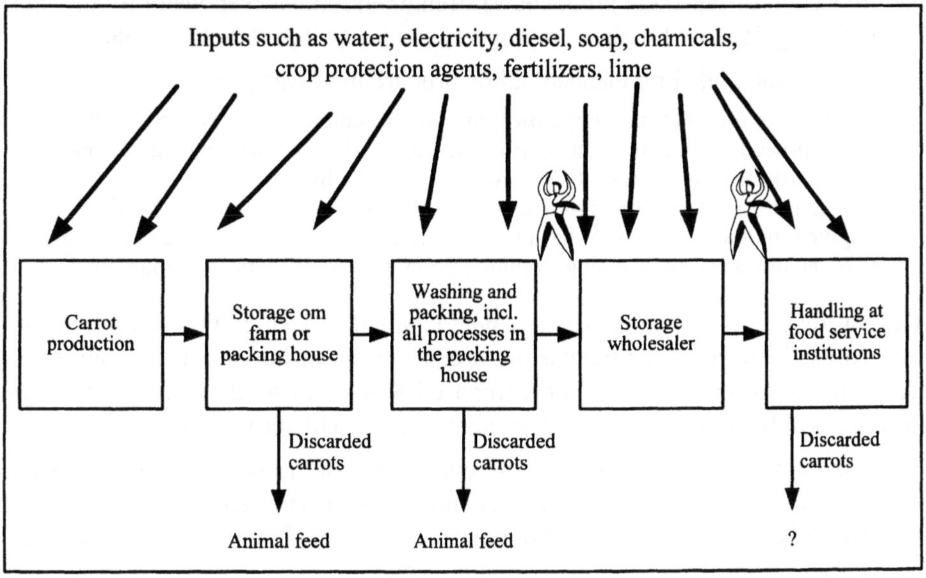

An important issue is to study actual behaviour of the purchasing managers. Little is known about how this group reacts and responds to environmental information in a purchasing situation. In order to test effects of environmental information Dr Gunne Granqvist of the College of Trollhättan/Uddevalla developed an internet based questionnaire. The purchasers were tested using minced beef meat and fresh apples as examples. In the test they were required to choose between two alternatives (Figure 4), each associated with varying levels of “Energy use”, “Greenhouse gas emissions”, “Toxic substances”, “Land use” and “Price”. The values of the environmental parameters could vary ±90% of the mean and the price varied ±7% of the mean (average price in Sweden by the time for testing).

Fig. 4: Testing environmental information – basic version.

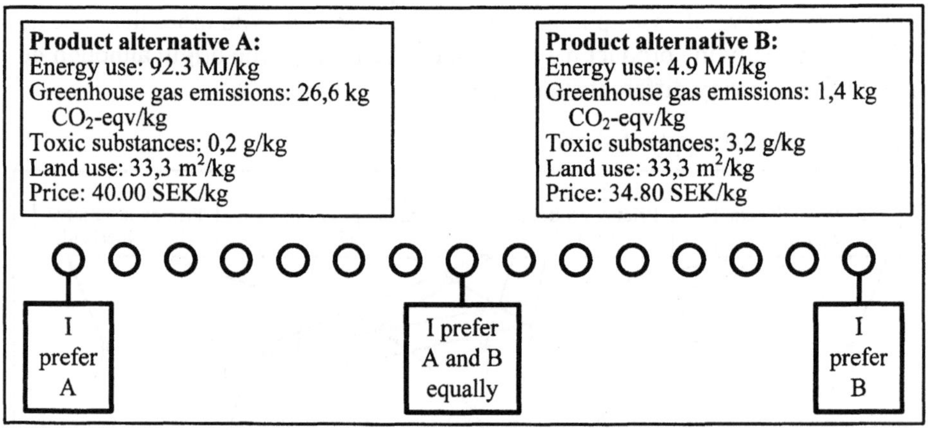

In an extended version of the test, we supplied the purchasers with a simple guiding system by adding + or – signs. Hence, the test person obtained somewhat more information about the products. The + sign indicated that the product in the case of a specific parameter was environmentally better than average while the – sign indicated poorer performance than the average (Figure 5).

Fig. 5: Testing environmental information – extended version.

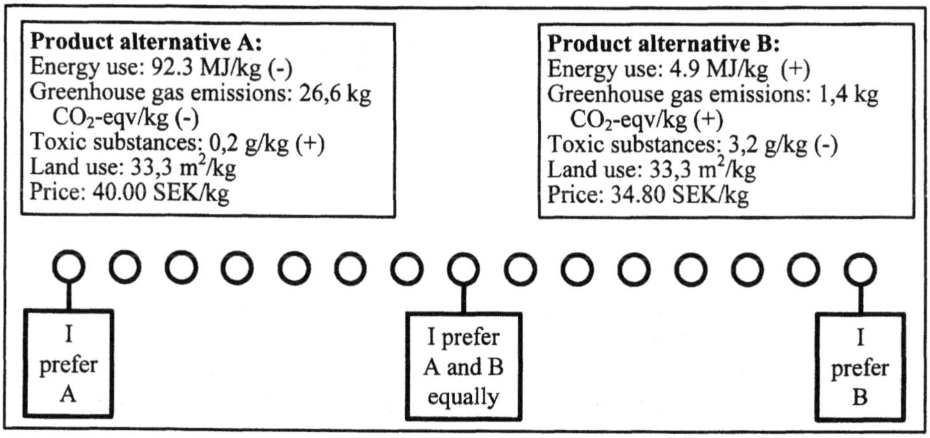

Results

Data for the environmental profiles is still coming in and is being processed. The profiles will be presented in the forthcoming project report by the end of the year (2005) and in scientific journals. However, the results suggest complex patterns of food supply (Figure 6). The companies addressed buy foods from many different companies. The complexity is not only displayed by the large number of actors at each level of the chain but also by the cross flow of products, i. e. companies in several cases purchase a specified product from more than one actor. Consequently the product may eventually be of the same origin although delivered through very different supply chains.

Fig. 6: Part of the supply chain of fresh tomatoes.

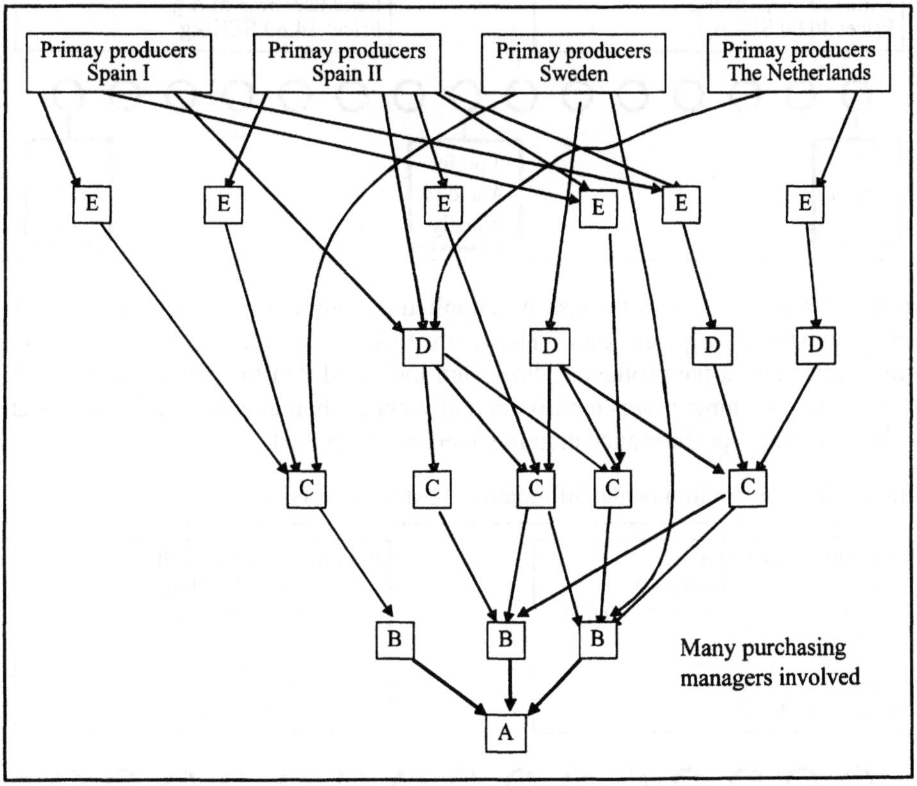

The study showed that price by far dominated the product preference. The dominating environmental parameter was "Toxic substances". However, different environmental parameters were significant for different products, i.e. in the case of beef meat "Energy use" was considered to be more important than in the case of apples where "Greenhouse gas" had a stronger influence. The use of additional environmental information (+ and -) also influenced product choice. Environmental aspects had a stronger influence on the managers' choice if additional information was presented. Another interesting finding was that the purchasers tended to avoid products that were considered less environmentally friendly than average. In contradiction, products better than average were not always chosen. A possible explanation for this is that the purchasing managers may avoid products less environmentally friendly than average instead of choosing products better than the average.

References

Edman, S. (2004), Hållbara laster – Konsumtion för en ljusare framtid. Statens offentliga utredningar SOU 2004: 19 [Sustainable burdens – consumption for a brighter future. Governmental Investigation SOU 2004: 19]. (In Swedish)

– (2005), Bilen, biffen, bostaden. Hållbara laster – smartare konsumtion. Slutbetänkande av Utredningen om en handlingsplan för hållbar konsumtion – för hushållen. Statens offentliga utredningar SOU 2005: 51 [The car, the steak and the dwelling. Sustainable burdens – smarter consumption. Investigation for a plan of action for sustainable consumption for the households. Governmental Investigation SOU 2005: 51]. (In Swedish)

Miljö- och samhällsbyggnadsdepartementet (1998), "Svenska miljömål". sammanfattning av regeringens proposition 1997/ 98: 145. ["Swedish environmental targets". Summary of the Government bill 1997/98: 145]. (In Swedish)

Naturvårdsverket (2003), Ett rikt växt- och djurliv. Förslag till miljökvalitetsmål för biologisk mångfald. Rapport 5301, aug 2003. [A rich plant and animal nature life. Suggestions of environmental targets of biological diversity]. (In Swedish)

Regeringen (2001), Svenska miljömål – delmål och åtgärdsstrategier. Regeringens proposition 2000/01: 130. [Swedish environmental targets – subtargets and fulfilments strategies, Government bill 2000/01: 130]. (In Swedish)

Green Biobased Chemistry Platform for Sustainability

Janis Gravitis

Latvian State Institute of Wood Chemistry,
Dzerbenes St. 27, LV – 1006 Riga, Latvia

Summary

The current state and future trends in green chemistry are considered. Sustainable chemistry is supposed to be founded on symbiosis of green chemistry and industrial (white) biotechnology for biorefineries based on renewable resources. The situation is analyzed on examples of case studies.

Introduction

Climate change, human health problems, food production, water quality, energy supply, and other problems are interconnected. Implementation of biorefineries based on principles of green chemistry would facilitate solution of global interconnected problems and construct a new – biobased chemical platform. Green chemistry applies technologies that reduce or eliminate the use of hazardous substances. Industrial biotechnology opens the road toward sustainable biology and chemistry based on renewable biomass to reduce dependence of industries on fossil feedstock and fuels. Implementation of new technologies is technically and economically feasible and environmentally safe.

Historical Background and Current Issues

Historically the green chemistry emerged in the USA in early nineties with implementation of the 1990 Pollution Prevention Act. Initially green chemistry concerns chemicals and production processes that reduce or eliminate the use of substances and waste hazardous to human health and the environment. Further formulation of green chemistry is associated with names of Anastas from the Environmental Protection Agency and J. C. Warner from the Boston University, when they came up with the twelve principles of green chemistry:

1. Waste prevention;
2. Atom economy (incorporation of all materials in the process into the final product);
3. Safety (low toxicity, accident prevention);
4. Use of renewable raw materials or feedstock;
5. Increase of energy efficiency;
6. Use of safer solvents and reaction conditions;

7. Design of bio-degradable chemicals and materials;
8. Design of safer chemicals;
9. Use of less hazardous synthesis;
10. Reduce derivatives;
11. Applying of selective catalytic reagents;
12. Real-Time analysis for pollution prevention.

Therefore, the green chemistry is not a strongly scientific discipline; it is more like a concept. However, the green chemistry is a science-based, nonregulatory but economically driven approach. So, the green chemistry is recently known as sustainable chemistry and is the main general concept for a whole chemical industry. In chemical synthesis, the maximum sustainability means a combination of a number of environmental, health, safety, and economic targets, the so called "ideal synthesis" (Figure 1).

Fig. 1: The ideal synthesis [1]

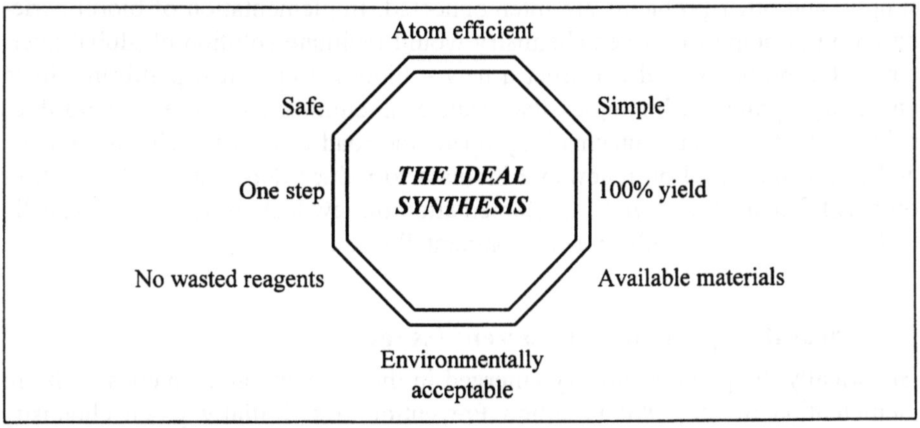

Ideal atom economy would be 100% main product yield without generation of by-products or necessity to use protecting groups. In many cases thermodynamic or kinetics factors make it impossible. However, the Diels-Alder reaction for C-C bonds forming, hydroformylation and catalytic hydrogenation present ideal atom-economy reactions (Figure 2). The elimination and substitution reactions are less sustainable in the sense of atom-economy.

European Sustainable Chemistry

To meet green chemistry challenge and maintain its global leadership, the European chemical industry (CEFIC) and the European biotech industry (EuropaBio) launched SusChem [3] on 4 March 2005 – a technological platform for EU

R&D of sustainable chemistry. The platform identifies three key sectors considered vital for innovation-led growth in the chemical sector:

- *Industrial biotechnology* (so called "white biotechnology") with the goal to use biological raw materials (glucose, vegetable oils, enzymes, etc.) to develop products such as pharmaceuticals, bio-colorants, solvents, bio-degradable plastics, vitamins, food additives, bio-pesticides and liquid bio-fuels for use in cars (Environmentally clean industrial biotechnology in Europe is labelled as white, while green biotechnology relates to agriculture and foods, red biotechnology – to health care, and black biotechnology – to bio-terrorism and bio-weapons);
- *Materials technology* with the goal to develop new materials with better performance including processing and recycling ability;
- *Reaction and process design* for faster, cheaper and cleaner production processes of existing chemicals.

Fig. 2: Examples of atom-economical reactions [2]. Application of ionic liquids, catalysts, and supercritical fluids in sustainable chemistry.

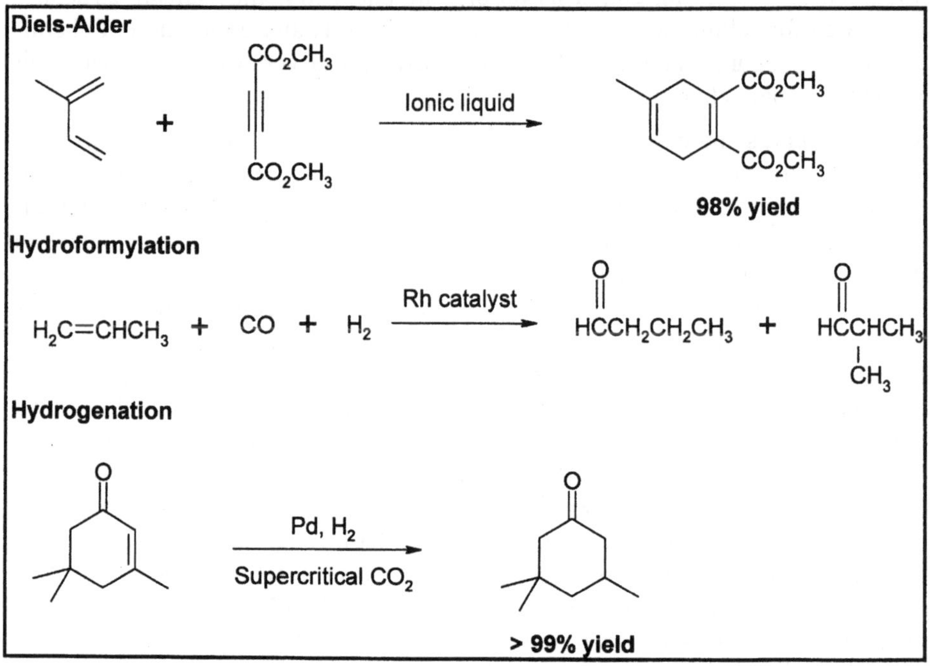

In addition, a *horizontal group* was set up to look into strategic approaches to eco-efficient innovation other than just cost reduction and environmental performance. This includes economic or financial barriers, regulation and societal acceptance of chemicals, including promotion of alternatives to animal testing.

Biomass-based Sustainable Chemistry and Biorefineries

According to the 12 principles of green chemistry the main chemistry industry raw materials and feedstock should change from fossil resources to renewable biomass. Such change is closely interrelated with the development of future biorefineries. A biorefinery is an integrated system of technologies integrating biomass conversion into chemicals, bio-fuels, power, and bio-degradable materials. By a combination of chemistry, biotechnology, engineering, and systems approach biorefineries could produce food, feed, fertilizers as well as industrial chemicals, fuels and power from biomass. Crucial for development of the biorefinery concept is the view that our planet cannot act as an infinite reservoir of fossil resources or an infinite sink for waste products. There are some current and future trends in the concept of biorefineries. At present, biorefinery is dominated by production of a few market chemicals (furfural, lactic acid, levulinic acid, etc.), bio-ethanol, and bio-diesel. In future commodity and special chemicals will increase. The share of biodegradable polymers could be increased. The problems of economic and chemical barriers for biorefinery have been understood. There is a strong emphasis on organic chemistry, mechanism, structure, and search for submissions describing novel synthesis and production of chemicals, polymers, and materials. One of the powerful drivers of biorefineries will be the white biotechnology.

A shift towards increased use of clean fuel, co-firing in power stations, and IGCC (Integrated gasification combined-cycle) technologies is expected. A transition to biomass can be exploited in the chemicals sector, gas sector, the agro-forestry-industrial cluster, intensive agriculture and silviculture, and changes in land use for production of non-food materials. In the coming decade biomass will stay an increasingly strategic raw material both for energy generation and as transport fuels, materials and chemicals. To achieve a large-scale introduction of biomass an international market will be required with cross-border trading. All biomass will be extensively utilized by integrated technologies to make value-added products, reusable and convertable to energy and materials in well developed cascade. Biorefinery concept developed by the Latvian State Institute of Wood Chemistry (LS IWC) is illustrated in Figure 3. It is interesting to note that the US National Renewable Energy Laboratory (NREL) also developed a biorefinery concept [4] based on two platforms: Sugar Platform (Biochemical) and Syngas Platform (Thermochemical). Our Biochemical Platform includes not only sugars but also phenols from lignin. The same is for Thermochemical Platform, which includes not only syngas but also bio-fuels, etc. Recently NREL corrected biorefinery interpretation and now the laboratory is one of the main players in conversion of lignified biomass into ethanol and lignin. The main backbone of the LSIWC concept is integration of technologies into clusters to reach near-Zero Emissions [5-7]. The concept emphasizes (see Figure 3) waste stream recycling and recovery (waste biorefinery) using the same integrated technologies facilities. Biorefineries generate CO_2 but this greenhouse gas is used up by the new growing biomass. So, biorefineries are not in contradiction with the Kyoto Protocol demands. CO_2 neutrality and renewability are the principal differences between biorefineries and fossil oil-based refineries. Of course, nowadays oil-based refineries are more powerful than biorefineries. However, in future the share of biorefineries will increase. Both refineries will develop parallel to reach the equilibrium then contribution from biorefineries will exceed the contribution from oil refineries.

Case Studies

Selection of building blocks from plant sugars. The US DOE Biomass Program made efforts to identify the top-tier building block chemicals as targets for biorefineries [8]. Firstly the research involved more than 300 potential compounds. Finally 12 top-tier compounds were chosen (Figure 4). Mentioned Program's 12 identified compounds can be produced from plant sugars but in future screening will cover aromatics, polysaccharides, and oils.

Fig. 3: The LSIWC biorefinery concept based on two selected platforms. The main principle – clusters of integrated technologies towards near Zero-Emissions

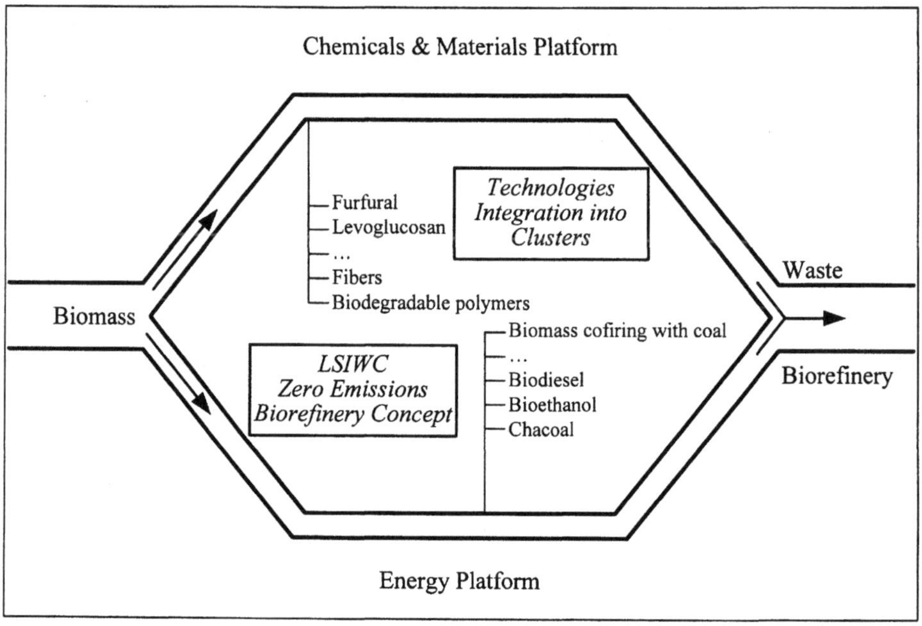

Selection was based on compatibility of compounds with the existing petrochemical processing, technical complexity of the synthesis from biomass, known market potential, etc.

Force of white biotechnology. The rocket-fuel precursor 1,2,4-butanetriol commercial syntheses involve sodium borohydride reduction of esterified malic acid, but this route generates tons of borate salts by-product. Another route needs high pressure and generates several by-products, which are difficult to separate from the main product. W. Frost and coworkers [8, 9] prepared butanetriol by microbial oxidation of D-xylose to D-xylonic acid, followed by a multistep conversion to D-1, 2, 4-butanetriol using a different engineered bacterium (Figure 5).

Fig. 4: DOE's Biomass Program selected 12 compounds, which can be produced from plant sugars and could be used as main building blocks for future biorefineries

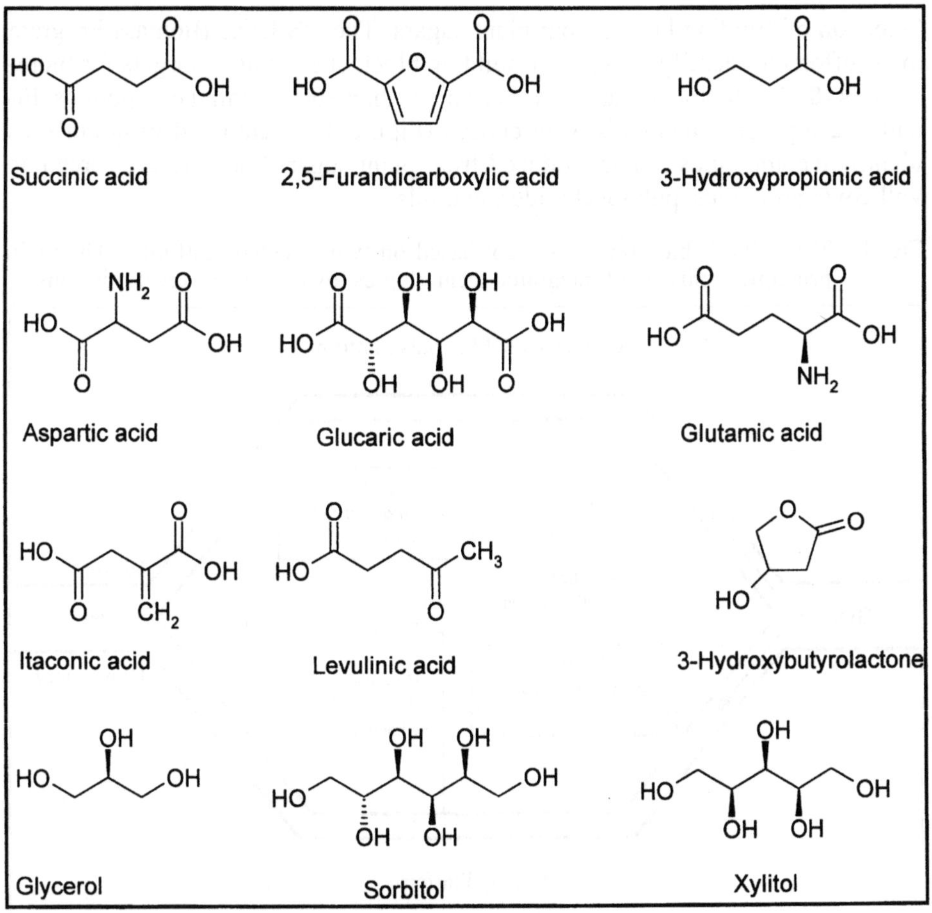

Steam explosion technology integration with enzymatic conversion of cellulose into bioethanol. Canadian industrial and food-grade enzyme producer Iogen [10] uses cellulose from wheat straw and corn stover – the stalks, leaves, and cobs to produce ethanol (EcoEthanol). The first step is raw material pretreatment with steam explosion to separate cellulose from lignin and hemicelluloses, allowing cellulose enzymes more efficiently hydrolyze the cellulose. In the current processing, mechanical and acid treatments are also used to hydrolyze cellulose. Enzymes production, steam explosion, enzymatic hydrolysis of cellulose into glucose, then fermentation to ethanol and distillation is fully integrated. Iogen operates the world's first pre-commercial demonstration facility. The $45-million plant is the final step before the construction of full-scale $250 million commercial plant.

Fig. 5: Replacement of current commercial route of 1,2,4-butanetriol synthesis with enzymatic route

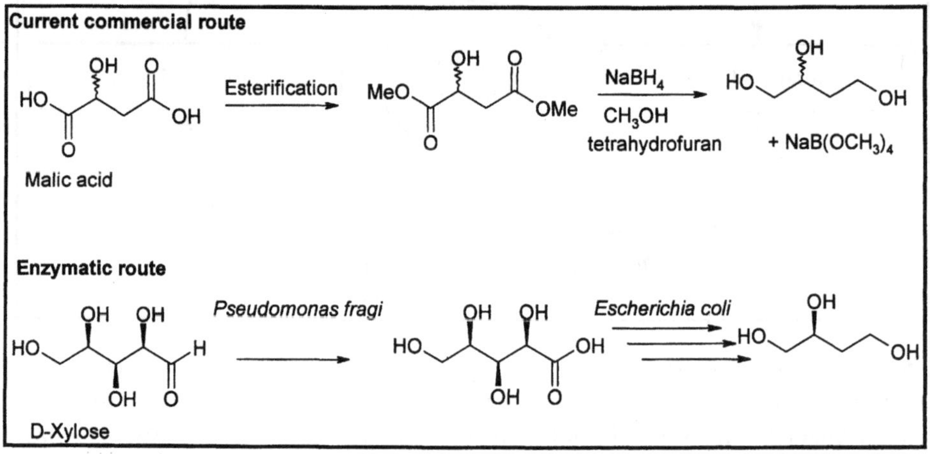

Iogen produces around 600 kg sugars from 1 ton of wheat straw, then converts sugars into 85 gal of ethanol. 200 kg of lignin is burned to produce electricity. According to Iogen, its ethanol reduces greenhouse gases emissions by 90% compared to gasoline, and is a good way to reduce dependence on imported oil. Even 5% ethanol-blended gasoline cuts CO_2 by 5 grams per kilometer driven, from European average of 163 g/km. The EU wants to cut the average to 140 g/km by 2008. There is an established commercial market for ethanol in the US and Canada. Currently about 12% of US and 5% of Canadian gasoline is blended with ethanol.

Furfural, steam explosion and carbonization technologies integration. A new eco-efficient biomass conversion into value-added chemicals and materials has been demonstrated [11, 12] by integration of the furfural, steam explosion and carbonization technologies. The integration includes production of a commodity chemical – furfural from pentoses rich plants, as well as opens the way for obtaining new value-added materials, particularly of carbon. The above-mentioned integration allows utilization of nearly 100% of the starting waste material from wood industry or agriculture (Figure 6). The biomass refineries integrated cluster offer not only economic but also environmental benefits. Real integration of furfural processing and subsequent treatment of residual lignocellulose is shown in Figure 7. Figure 8 demonstrates integration of steam explosion with carbonization technology and processing of new carbon materials.

Fig. 6: Furfural, steam explosion, and carbonisation technologies integrated in a near Zero-emissions biorefinery [11]

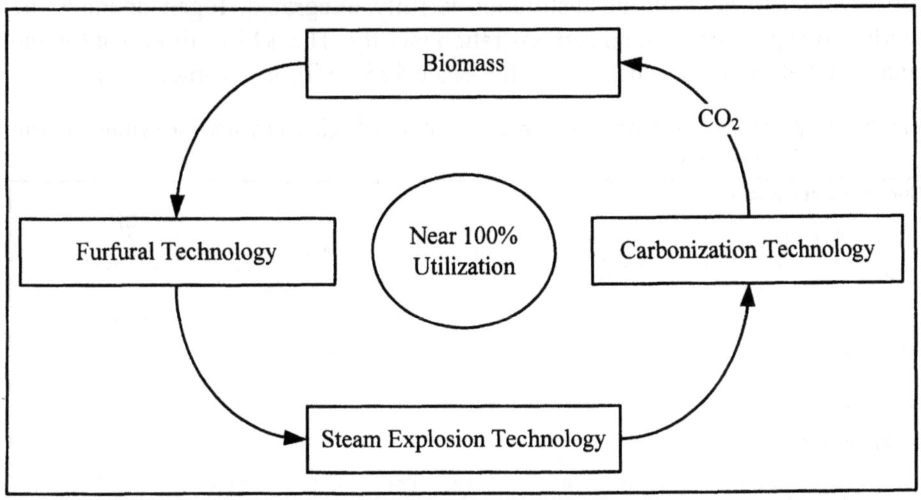

Enzymatic treatment process of cotton fibers. Novozymes – the leader in production of industrial enzymes with a market share of more than 40% of global sales replaced conventional cotton fibers treatment chemical process by enzymatic process [13]. Today, highly alkaline chemicals such as sodium hydroxide are used for complete removal of the non-cellulosic compounds (scouring). These chemicals not only remove the impurities but also attack the cellulose leading to a reduction in strength and loss of weight of the fabric. Furthermore, the resulting wastewater has a high COD (chemical oxygen demand), BOD (biological oxygen demand) and salt content.

Fig. 7: Integration of furfural and steam explosion technologies [11]

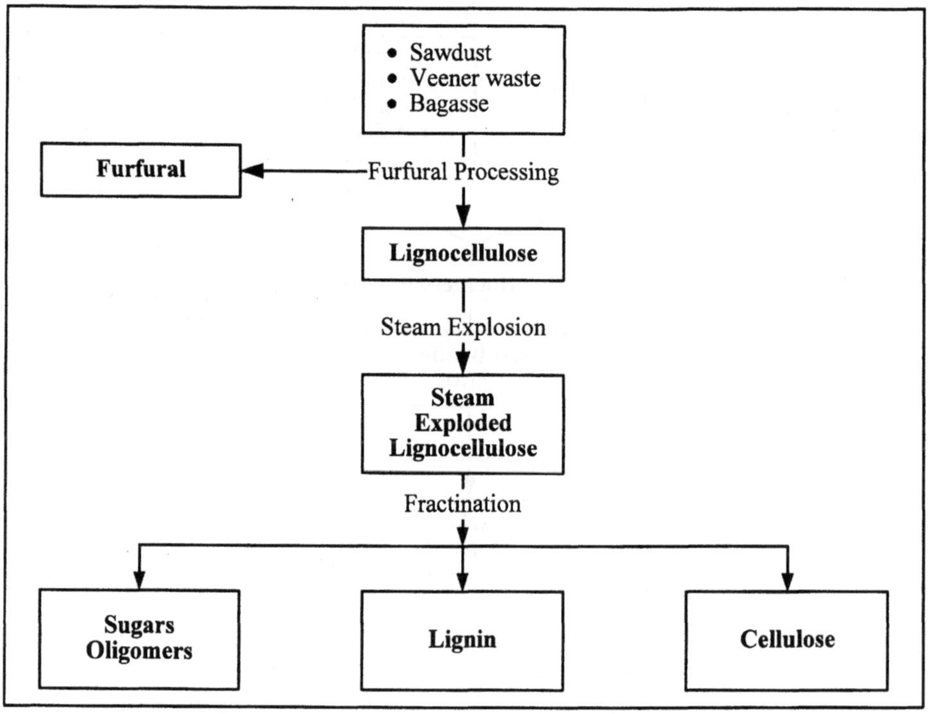

The pH optimum of new Novozymes BioPrep is 9.2 and the temperature optimum is around 60°C. At dosages as low as just 0.3 g/kg of cotton, the wetting ability obtained after less than ten minutes followed by an emulsification step is regarded as sufficient for good dyeing. COD, BOD and TDS (total dissolved solids – a measure of the salt content) are much lower than in the alkaline process partly due to a much higher cotton yield and partly due to a significantly reduced use of salts.

Fig. 8: New materials as a result of integration of steam explosion and carbonisation technologies [11]

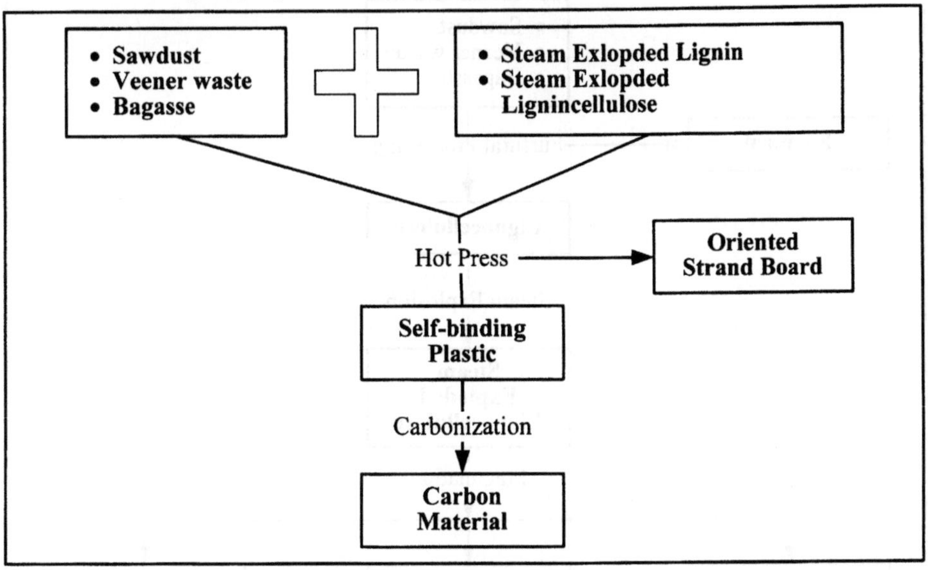

BioPrep contains a pectate lyase enzyme to which customers may add textile auxiliaries such as surfactants, emulsifiers, and chelating agents to release cotton waxes and other constituents. Pectate lease attacks and degrades pectin within a complex polymer network that makes up the outer part of cotton fibers (Figure 9).

Fig. 9: The cotton fiber outer layer. Novozymes' BioPrep pectate lyase degrades the outer polymer matrix pectin and promotes the release of cotton waxes and other constituents, leaving cleaner cotton fibers [14]

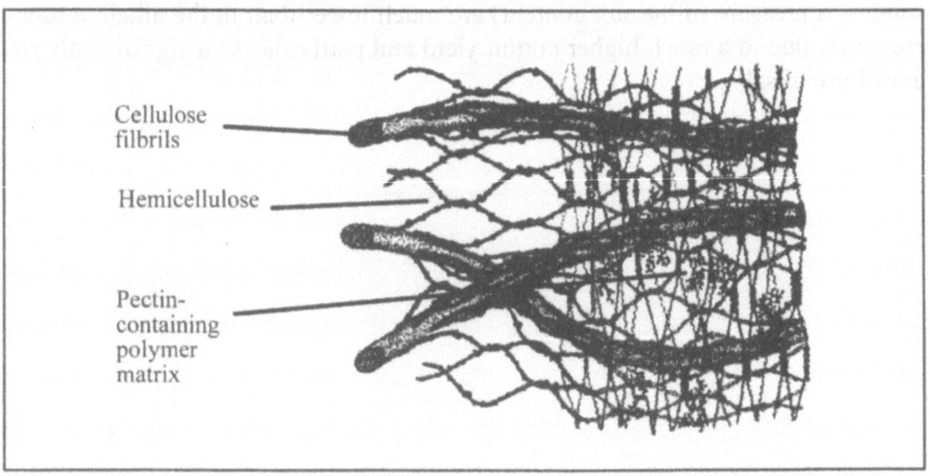

New solvents. Cellulose use has been limited by insolubility in water and most common organic solvents. Robin D. Rogers [15] demonstrated the possibility of dissolving cellulose in ionic liquids for processing cellulose into advanced functional materials for use in textiles, sensors, and plastics. Ionic room-temperature nonvolatile, stable solvents typically contain inorganic anions and organic cations. Cellulose readily dissolves in 1-butyl-3-methylimidazolium chloride (Figure 10) with gentle heating. By adding water, ethanol, or acetone cellulose can be precipitated. Ionic liquid can be recovered for reuse. Other polymers, nano-particles, metal-complexing agents, etc. can be added to dissolved cellulose to modify the end material properties. In general, the use of ionic liquids as industrial solvents for polymer processing, extraction, separation processes and organic synthesis has started.

Fig. 10: Ionic solvent for cellulose

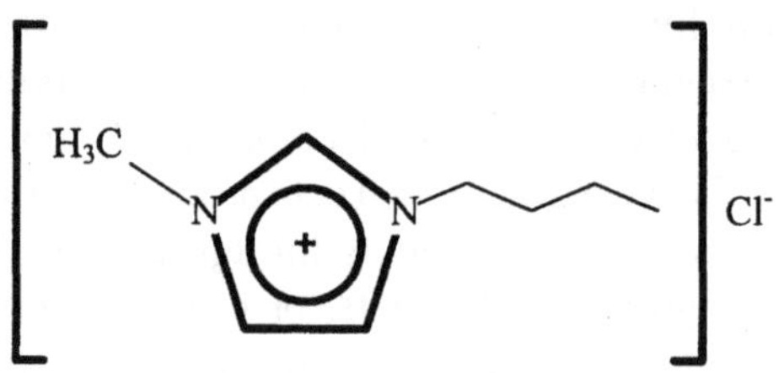

1-Butyl-3-methylimidazolium chloride

Chemical reactions in supercritical fluids. At supercritical conditions, a fluid does not meet the definition of a liquid because it cannot be made to boil by decreasing the pressure at a constant temperature. Also it is not a vapour because cooling at constant pressure wouldn't cause it to condense. Supercritical fluids can have liquid-like densities, gas-like diffusivities, and compressibility that deviates greatly from ideal gas behavior. Under supercritical conditions, solid solubility often is enhanced greatly with respect to solubility in the gas or liquid solvent. At first the supercritical CO_2 was applied in extraction of different natural plant substances, for instance, taxol. Supercritical extraction avoids thermal denaturing or distillation. Separation is achieved at low temperatures. In comparison with convential methods such as cold press, steam distillation and hexane extraction, in the supercritical extraction the yield is higher, extraction is rapid, the extract has no solvent residues, and chemical tailoring can be realized. Supercritical extraction is very prospective in future bio-based biorefineries. From the chemistry viewpoint [16] chemical reactions may be advantageously carried out in a supercritical fluid for a variety of reasons: phase behavior allowing either homogenization or ease of product separation; more rapid diffusion giving faster reactions when they are diffusion controlled, which can be homogeneous or heterogeneous reactions; control of reaction pathways and products by pressure and temperature; and environmental advantages. Mostly work is on the academic research level, but there are industrial applications, for instance, a plant in Japan producing 40,000 tons of methyl ethyl ketone per annum. Recently attempts have been made to use supercritical fluid treatment for biomass conversion [17]. Organic compounds, including solid biomass will readily dissolve in supercritical water. Once dissolved, supercritical water will efficiently break cellulose and lignin bonds. Usually, the reactions generally are not selective, resulting in rapid formation of gaseous products. So, such type of biomass treatment can be used as a technological case of biomass gasification or to produce hydrogen for use in a fuel cell. Fortunately, in this case, the hydrogen yield may be much higher than the hydrogen content of the biomass due to steam reforming where water is a hydrogen-providing participant in the overall reaction.

Polymeric products from lignin. If the polysaccharides industrial use is going as mentioned more or less advanced, lignin as an extremely abundant renewable recourse, second to cellulose, non-toxic is still only on the level of potential commercial application for the chemical industry [18-20]. Some companies' success [21] is more exception than rule. Potential reactivity of lignin has not been fully realized. However, Borregaard commercial application of lignin is impressive. The use of Borregaard LignoTech products for concrete structures, applications as diverse as dispersing agents for agrochemical formulations, performance enhancers for animal feed pellets, or even as a critical component in lead acid batteries. One of advanced possibilities may be application of lignin as a source for dendritic nano-structures, e.g., use of lignin fragments as a macromonomer core for branched dendrimers (Figure 11). Dendrimers or dendritic polymers are hyper-branched nano-particle polymers.

Fig. 11: Lignin macromonomer as a branching core for dendrimeric nano-particle structures [20]

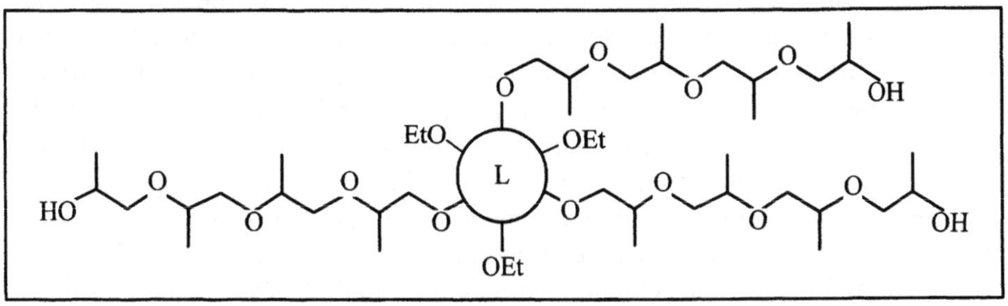

Learning from Nature

Plants are most advanced, usually non-toxic, and clean of green house gases biodegradable biorefineries. So, we should learn from the nature. Studies of natural plant walls nano-cell structure and dynamics would diverse opportunities for preparation of natural nano-composites and natural nano-technologies. Such approach will help to obtain synthetic intermediates and substrates for use in chemical process imitating nature (in biomimetic manner). The forest is not only the source wood and polymers but also a source for particularly value-added substances. Important example is shown in Figure 12 – the cancer drug ingredient paclitaxel (taxol). Paclitaxel's antitumor activity was discovered in the 1960s during a large scale 35,000 plants-screening program sponsored by the National Cancer Institute (NCI) in the US. Early research using paclitaxel was limited by a restricted supply due to several difficulties in obtaining the drug from the *Taxus brevifolia* (Pacific yew). The concentration of the compound in yew bark is low, and paclitaxel extraction is complex and expensive. The *Taxus brevifolia* (Pacific yew) is a limited resource; it grows very slow and is located in old-growth forests that are the habitat of the endangered spotted owl. As demand for paclitaxel increased, researchers have been exploring new resources to increase the availability of paclitaxel. There have been many research reports on the constitutions of other yew tree species worldwide. Scientists in US, China, Japan, Taiwan and other countries have done detailed research on *Taxus chinensisi* (Chinese yew) as the new resource of paclitaxel. There are four Chinese yew species growing in the territory of China. Scientists proved that all of them contain paclitaxel at a similar level as the *Taxus brevifolia*. Scientists have isolated and identified 110 taxanes from Chinese yews, and 36 of them were new taxanes. *In vitro* and *in vivo* studies proved that paclitaxel has the strongest anticancer activity [22]. Recently there are attempts to combine biotechnological and chemical synthesis in paclitaxel production.

Metabolic Engineering

Yeast fermentation of ethanol is a technology thousands of years old. *E. coli* breaks down five- and six-carbon sugars. However, from five-carbon sugars (obtained from hemicelluloses) there is intermediate – pyruvic acid, which turns into many by-products and very little ethanol. Inserting genes from the bacterium *Zymomonas mobilis* into *E. coli* the bacteria produce mostly ethanol as a bio-fuel (Figure 13). The next example is production of natural plastics-polyhydroxyalkanoate (PHA). Scientists incorporated a series of genes from various PHA producing bacteria into *E. coli*. Sugars and oil then convert into HPA [15]. Genetically engineered plants producing PHA, for instance, a native prairie grass, are being developed.

Fig. 12: Paclitaxel (taxol) untitumor drug

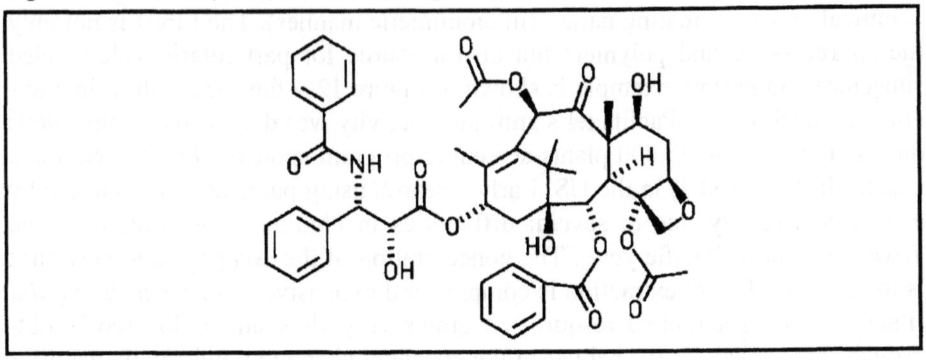

Goals of Future sustainable Chemistry

Together with social, scientific and teaching problems development of sustainable chemistry and biotechnology is also significant. Industrial biotechnology accounts for about 5% of global chemical sales, primarily ethanol, pharmaceuticals, citric acid, lactic acid, enzymes, and amino acids.

Fig. 13: Metabolic engineering of *E. coli* for biofuel production

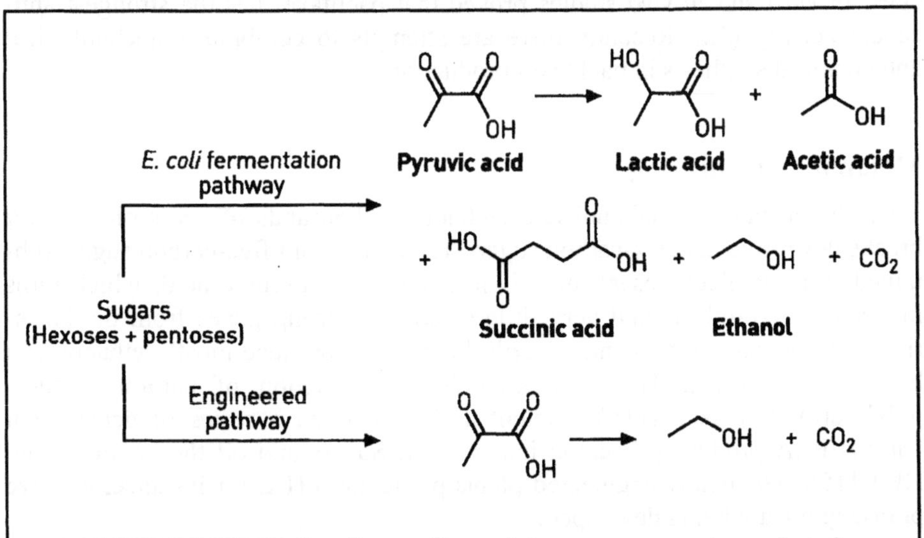

ALCOHOL FOCUSED *E. coli* normally produice numerous waste products from pyruvate, but with the help of genes from *Z. mobilis*, the bacteria produce mostly ethanol.

Experts believe that two thirds of the global $1.5 trillion chemical industry can eventually be based on bio-based feedstock [8]. Such a level of development demands integration of new recombinant technologies, bioinformatics, and other advanced interdisciplinary and transdisciplinary approaches. The last but not least question concerns bioethics, which, apart from-medical ethics, includes additional issues of research ethics related to new scientific techniques such as cloning, embryonic stem cell research, genetically modified food, environmental policy, etc. No doubt bioethics as well as environmental protection is a driving force for sustainable biotechnology. The futuristic goal of science is to implement industrial activity without using machines. If the information included in a molecule is I, the material and energy consumption for the technology is T, then, according to Frenkel's "pseudoformula" [6], an approximately correct equation is: $I \times T = const.$ Small values of I and high values of T are typical for our industrial technology. On the contrary, in natural technologies the value of I is high and value of T is low. Small values of T mean that in future technologies energy, raw materials and industrial machinery will form one entity to minimize wastes toward Zero Emissions.

Conclusions

Production of chemicals, fuels, and materials should be based on sustainable chemistry with deep implementation of the industrial biotechnology using principles of near zero emissions biorefineries. Industrial biotechnology opens the road toward sustainable biology and chemistry, based on renewable biomass, reducing dependence of industries on fossil feedstock and fuels. Implementation of new technologies is technically and economically feasible, environmentally safe and should not violate principles of the bioethics.

References

1. *Clark, James H.* (1999), Green chemistry: challenges and opportunities. Green Chemistry, vol. 1, no. 1, 1-8.
2. *Ritter, S.K.* (2002), Green chemistry progress report. Chem. & Eng. News, vol. 80, no. 47, 19-23.
3. www.suschem.org/.
4. www.nrel.gov/biomass/biorefinery.html.
5. *Gravitis, J.* (1999), Biorefinery and Lignocellulosics Economy Towards Zero Emissions. In: Targeting Zero Emissions for the Utilisation of Renewable Resources (Biorefinery, Chemical Risk Reduction, Lignocellulosic Economy), eds. K. Iijama, J. Gravitis & A. Sakoda, Tokyo, Japan, Published by UNU/IAS, ANESC/UT and IIS/UT, 2-11.
6. *Gravitis, J. & Della, Senta T.* (2001), Global Prospects of Substituting Oil by Biomass In: World Forests, Markets and Policies, eds. Matti Palo, Jussi Uusivuori and Gerardo Mery, Kluwer Academic Publishers, Chapter 2, 23-39.

7. *Gravitis, J., Zandersons, J., Vedernikovs, N., Kruma, I. & Ozols-Kalnins, V.* (2004), Clustering of bio-products technologies for zero emissions and eco-efficiency. Industrial Crops & Products, vol. 20, Issue 2, 169-180.
8. *Ritter, S.K.* (2004), Biomass or bust, Chem. & Eng. News, vol. 82, no. 22, 31-34.
9. www.pharmacy.wisc.edu/grad/seminars/pdf/Frost_2_4_05.pdf.
10. www.iogen.ca/
11. *Gravitis, J., Zandersons, J., Vedernikov, N., Kruma, I. & Bikovens, O.* (2002), Chemicals and materials from integrated furfural, steam explosion and pyrolysis technologies. Proc. 7th European Workshop on Lignocellulosics and Pulp. Turku/Abo, Finland, August 26-29, 503-506.
12. *Gravitis, J., Zandersons, J., Vedernikov, N., Kruma, I. & Ozols-Kalnins, V.* (2004), Clustering of bio-products technologies for zero emissions and eco-efficiency. Industrial Crops & Products, vol. 20, issue 2, pp. 169-180.
13. www.novozymes.com/.
14. *Ritter, S.K.* (2001), Accepting the green challenge. Chem. & Eng. News, vol. 79, no. 27, 24-28.
15. *Ritter, S.K.* (2005), Green success. Chem. & Eng. News, vol. 83, no. 26, 40-43.
16. www.criticalprocesses.com/Reaction.html.
17. www.energy.iastate.edu/renewable/biomass/cs-water.html.
18. *Feldman, D., Lacasse, M. & Beznaczuk, L.M.* (1986), Lignin-Polymer Systems and Some Applications. Prog. Polym. Sci., vol. 12, 271-299.
19. *Gandini, A. & Belgacem, N.M.* (1998), Recent advances in the elaboration of polymeric materials derived from biomass components. Polymer International, vol. 47, 267-276.
20. *Gandini, A.* (1992), Polymers from renewable resources. Comprehensive Polymer Science, First Supplement, Pergamon Press, 527-573.
21. Borregaard LignnoTech Brochure. 1-11 (the source kindly sent by Borregaard).
22. www.21cep.com/sft/pxsft.htm.
23. *Wilson, E.K.* (2005), Engineering cell-based factories. Chem. & eng. News, vol. 83, no. 12, 41-44.

Bioenergy Development in Ukraine

–

State of the Art and Perspectives

G. Geletukha, T. Zhelyezna, Y. Matveev, N. Zhovmir

The Institute of Engineering Thermophysics of National Academy of Sciences of Ukraine, Scientific Engineering Center "Biomass"
P/O Box 66, Kyiv, 03067, Ukraine

Abstract

The paper presents data on biomass potential in Ukraine. Analysed is the present state of bioenergy development and future targets. Examples of successful implementation of biomass-to-energy projects and their economic indicators are given. The conception of bioenergy development in Ukraine is proposed. Total energy potential of biomass in Ukraine is estimated at 12.3 mtoe (9% of the total primary energy consumption). The most promising technologies are direct combustion of biomass for heat production, production of biogas from manure on large farms, extraction and utilization of landfill gas, co-combustion of biomass and fossil fuel for power production.

1. Introduction

Ukraine is a country with a lack of domestic energy sources and rather high dependence on imported energy carriers. According to official data, by the end of 2000 extraction of fossil fuels in the country was 56 mill toe while primary energy consumption (PEC) amounted to 140 mill toe. That is why successful development of bioenergy is of vital importance for Ukraine.

Historically oriented towards Russia as a source of raw material for industry and energy production, the structure of PEC in Ukraine is based on the preferential consumption of natural gas and coal, and exclusive use of Russian nuclear fuel (Table 1). Ukraine's self-sufficiency in primary energy is less than 50%, including oil 10-15%, natural gas 20-25%, coal 80-85%, uranium 100%. Ukraine does not have closed fuel-nuclear cycle despite that uranium ore is available in a big amount.

Annual production of electricity in Ukraine comes to 169.1 TWh. Of them the share of green electricity (excluding large hydro) is very little, only 5.6 GWh (0.003% of the total production). The main part of electricity is produced from nuclear energy, coal and natural gas (Table 2).

Tab. 1: Structure of primary energy consumption of Ukraine (2000)

Prime energy carriers	**mill toe**	**%**
Oil	14.3	10.5
Natural gas	62	45.5
Coal	36.8	27.0
Nuclear energy	19.3	14.2
Hydro energy	3.22	2.3
Biomass*)	**0.58**	**0.5**
Total	**136.2**	**100**

*) mainly fire wood

Tab. 2: Energy mix in electricity production in Ukraine (2001)

Coal	25.5%
Oil	1.4%
Hydro	7%
Natural gas	23%
Nuclear	42.6%
Other renewables	0.003%
Total	**100%**

2. Present State of Bioenergy Development

Today the total use of renewable energy (RE) in Ukraine (including large hydropower) is 3.9 mtoe that is 2.8% of the total PEC. Of all renewables the share of biomass in energy mix is the biggest after large hydro and amounts to about 18% (Table 3). In 2001 29 PJ (only heat) was produced from biomass that corresponded to 0.5% of the total PEC. Almost all energy produced from biomass comes from the combustion of wood residues.

Tab. 3: Energy mix for renewables in Ukraine (2001)

Hydro large	78.8%
Bioenergy	**17.79%**
Hydro small	3.1%
Wind energy	0.2%
Geothermal energy	0.07%
Solar heat	0.04%
Total	**100%**

Now the ways of biomass utilization in Ukraine are as follows:

- A lot of wood processing and timber enterprises in Ukraine have converted their boilers originally designed for oil and coal into wood fired boilers and use them for heat production from own wood residues. According to data of State Committee of Forestry the total number of wood fired boilers is 1006 that makes up 75% of all boilers operating at Ukrainian timber enterprises.
- A lot of householders, especially in rural area have domestic wood fired boilers and stoves. Exact number is unknown.
- A few farmers have individual small-scale units for the production of biogas from manure. Total number does not exceed ten.

There is no biomass plant/unit for power production in Ukraine at the moment. Ukraine does not import/export biomass for energy purposes.

At the moment there are a few modern biomass plants and units operating in Ukraine. They are:

- 5 MW steam wood fired boiler at plywood factory Odek Ukraine (the town of Orzhiv, Rivno oblast). The boiler has been operating successfully since 2000.
- 1.5 MW steam wood fired boiler at Malin State Timber Enterprise (the town of Malin, Zhitomir oblast). The boiler has been operating successfully since 2000.
- 980 kW straw fired boiler at agricultural enterprise "DiM" (the village of Drozdy, Kyiv oblast). The boiler has been operating successfully since 2000.
- 160 kW_e+300 kW_{th} CHP biogas plant at agricultural enterprise Agro Oven (the village of Olenivka, Dnipropetrivsk oblast). The plant is under construction now.
- Demonstration system for the extraction and collection of landfill gas (LFG) at Lugansk landfill (the town of Lugansk). Three demonstration wells made at the landfill are giving 90 m^3 of biogas per hour. Extracted LFG is being burnt in flare now but in the near future it is planned that it will be used for energy production. Total volume of the landfill allows to drill 30 wells, and amount of biogas from them will be enough to install 1.5 MW_e engine for power production.

All the above mentioned equipment was installed within the frame of international projects of technical assistance and other international programs. The main economic indicators of the projects are presented in Table 4. Experts from SEC "Biomass" performed technical accompany of all the projects.

Tab. 4: Economic indicators of bioenergy projects implemented in Ukraine

Place of implementation	Type and capacity of equipment	Investment outlay [thous. $]	Payback period [years]	Reduction of CO_2 emission [t/yr]	Replacement of natural gas [mill m^3/yr]
Plywood factory Odek Ukraine	steam wood fired boiler 5 MW	430	1.8	15000	4.3
Malin State Timber Enterprise	steam wood fired boiler 1.5 MW	215	3.2	4000	1.2
Agricultural enterprise "DiM"	hot water straw fired boiler 980 kW	100	4.7	1050	0.385
Agricultural enterprise Agro Oven (15 thousand pigs)	CHP biogas plant 160 kW_e+300 kW_{th}	413	8.1	7500	0.51
The Lugansk landfill (calculation for a full-scale project)	LFG power plant 1.5 MW_e	1395	3.9	61700	3.36

3. Potential of Biomass

Ukraine has rather big potential for renewable energy sources (RES) on the whole and biomass in particular. Table 5 presents the structure of energy potential of biomass in Ukraine based on optimistic and moderate approaches. Both have been performed by the experts from the Institute of Engineering Thermophysics. The main part of the potential is made up by straw. Optimistic estimation gives total potential of 12.3 mill toe/year, and moderate assessment gives 7.4 mill toe/year. The main difference between them comes from the evaluation of straw potential. Moderate approach supposes that only 20% of the total amount of straw can be used for energy production.

Tab. 5: Energy potential of biomass in Ukraine (2001)

Type of biomass	Energy potential [mill toe/year] (moderate)	Energy potential [mill toe/year] (optimistic)
Cereal crops/straw (without maize):		
– Wheat	0.97	3.9
– Barley	0.79	
– Oat	0.10	
– Rye	0.15	
– Other	0.20	
Total on cereal crops	2.54	

Type of biomass	Energy potential [mill toe/year] (moderate)	Energy potential [mill toe/year] (optimistic)
Maize for grain/stems, ears	0.83	1.70
Sunflower/stems, husk	1.62	1.60
Manure/biogas	1.1	1.13
Sewage gas	-	0.16
Landfill gas	0.21	1.12
Wood wastes – Wood remained at felling areas (felling residues), W 50-60%*). – Residues generated at timber enterprises during the sawing of round timber, W 40-45%. – Residues generated at woodworking enterprises during the production of finished products, W 25-30%. – Firewood taken out of felling areas, W 40-45%. Total on wood wastes	0.314 0.114 0.180 0.969 1.1	1.4
Fuel from municipal solid waste	-	1.3
TOTAL	**7.42**	**12.3**

*) W – mass moisture content

4. State Regulation and Targets

Now Ukraine is showing increasing interest towards the development of renewable energy sources. In 1996 the Parliament of Ukraine (Verkhovna Rada) approved a national energy program for the period till 2010. According to the program non-traditional and RE will cover 10% of the total energy demand of the national economy in 2010. In 2000 topicality of this item of the program was confirmed in Recommendations of the Parliament session called "Energy policy of Ukraine".

In 1997 the State Committee of Ukraine for Energy Conservation and National Academy of Sciences of Ukraine elaborated "The program of state support for development of non-traditional and renewable energy and small hydro- and heat-power engineering as a component part of national energy program of Ukraine". According to the program 10% of the total energy and fuel consumption of Ukraine will be covered at the expense of non-traditional and renewable energy in 2010. The program consists of three stages: 1 stage – 1998-2000, 2 stage – 2001-2005, 3 stage – 2006-2010. The main task of the second stage is starting production of the equipment for the use of non-traditional and renewable energy and its implementation in Ukrainian regions. Biomass and combustible gases (biogas, producer gas), which can be produced from biomass, are considered in the chapter "Non-traditional fuel". Production of liquid fuels from biomass (fuel ethanol and bio-diesel fuel) is considered there too.

Governmental organizations responsible for RE issues in Ukraine are State Committee for Energy Conservation and Ministry of Fuel and Energy. Lately profile energy committee of Ukrainian Parliament has undertaken coordination in preparation of energy strategy of Ukraine for the period till 2030 including the section on RE. According to draft version targeted utilization of RES is 4.88 mtoe (3.4% of PEC) in 2010 and 15.1 mtoe (10.7% of PEC) in 2030 (Table 6). Corresponding figures for biomass are 1.16 mtoe and 7.08 mtoe. Detailed information on the present and targeted utilization of biomass is presented in Tables 7 and 8.

Tab. 6: The use of RES in 2001-2030 in Ukraine according to "Energy Strategy of Ukraine till 2030"*

	Heat and power production from RES in 2001-2030							
	2001		**2010**		**2020**		**2030**	
Indices	**mtoe**	**%**	**mtoe**	**%**	**mtoe**	**%**	**mtoe**	**%**
Wind energy	0.008	0.2	0.15	3.15	0.7	6.97	1.5	9.95
Solar power			0.001	0.02	0.007	0.07	0.02	0.14
Hydro small	0.1	3.1	0.1	2.16	0.33	3.36	0.45	3.01
Hydro large	3.1	78.69	3.35	68.7	3.91	39.06	4.56	30.3
Solar heat	0.001	0.04	0.08	1.72	0.49	4.88	0.9	5.93
Bioenergy	0.7	17.9	1.16	23.8	4.4	43.9	7.08	46.9
Geothermal energy	0.003	0.07	0.02	0.49	0.17	1.73	0.58	3.84
Total	3.9	100	4.88	100	10.02	100	15.1	100
Share of total consumption of primary energy in Ukraine**	2.8		3.5		7.1		10.8	

* The Strategy has been elaborated by a group of Ukrainian energy experts on the decree of President of Ukraine and is under consideration/approval at the moment

** By the end of 2000 total primary energy consumption in Ukraine was 140 mtoe

Tab. 7: Utilization of biomass in 2001-2030

		Replacement of fossil fuel by years			
Directions of the use of biomass	**Units**	**2001**	**2010**	**2020**	**2030**
Energy utilization of biomass including:	thous. toe	692	1863	4423	6451
Power production	MW installed	0	151	772	1158
	GWh	0	926	4623	6974
	thous. toe	0	227	1133	1708
Heat production	MW installed	91	1301	4455	6520
	thous. toe	49	491	1652	2422
End use of biomass as stove fuel	thous. toe	646	963	1001	1047
Production of ethanol as a motor fuel	thous. toe	11.3	183	637	1274

Tab. 8: Total indicators of biomass utilization in 2000-2030

Types of biomass and direction of their energy utilization	Units	Annual biomass utilization for years			
		2000*	2010	2020	2030
Wood totally, including for:	th cu m	2,720	3,970	5,490	60,605
– firewood utilization in all sector of economy	th cu m	2,400	3,113	3,787	4,076
– collection and energy usage of forestry woodwaste	th cu m	0	252	659	781
– wood processing waste for energy	th cu m	320	605	1,043	1,207
Straw and stalks utilization for energy needs, including:	th t	0.4	909	3,853	6,015
– in boiler houses of agricultural enterprises	th t	0.4	153	297	359
– in farmer's boilers	th t	0	676	2006	2956
– in boiler houses of energy enterprises	th t	0	80	1200	2000
– for power generation at TPS and CHP	th t	0	0	350	700
Biogas production in animal and poultry keeping	mill cu m	0.4	18	478	938
Sewage gas production	mill cu m	2.1	35.9	172	295
Refuse derived fuel utilization at power plants and incinerators	th t	350	680	1709	2121
Landfill gas extraction	mill cu m	0	173	460	529
Peat utilization for energy needs	th t	408	1110	1258	1605
Fuel ethanol and biodiesel production	th t	17.8	286	1000	2000
Total biomass utilization for energy needs	mill toe	0.69	1.86	4.42	6.45

* – estimation of current state

5. Economic Aspects

Bioenergy equipment of foreign manufacture is rather expensive, and most Ukrainian enterprises cannot afford it. This fact forms one of serious barriers to the development of bioenergy technologies. The most reasonable seems to begin production of such equipment in Ukraine, the share of foreign components would come to 25-40%. At that cost of bioenergy equipment of Ukrainian manufacture would decrease considerably. For example, a 5 MW boiler with step moving grate of Dutch firm KARA costs 750 thous. $. A similar Ukrainian boiler with 30% of foreign components is expected to cost about 1330 thous. UAH (about 240 thous. $) that is more than three times cheaper. The same can be said about under screw wood fired boilers and straw fired boilers. Boilers of Ukrainian manufacture would have rather short payback period – 1-2.5 years (Table 9).

Positive example of Ukrainian manufacturers of bioenergy equipment can be Close Corporation "Zhitomirrempischemash" (the town of Zhitomir). The enterprise produces hot water wood fired boilers in the range of 40-820 kW. The cost of

500 kW and 820 kW wood fired boilers with under feed stoker is 12.5 thous. $ and 17 thous. $ Correspondingly (including VAT). It is approximately six times cheaper than similar boilers of foreign production.

Technoeconomic parameters of a biogas plant of Ukrainian manufacture with 40% of foreign components are presented in Table 10. Capital outlays on construction of the plant amount to about 280 thous. $ that is nearly 1.4 times less than for a foreign plant. Payback period of Ukrainian biogas plant is estimated at 5.6 years. Financial benefit caused by the implementation of the plant includes saved electricity, heat and mineral fertilizers.

As an example of technoeconomic assessment of LFG extraction and utilisation technologies, Table 11 presents calculation results for Lugansk landfill containing 1.6 mill t of municipal solid waste (MSW). There were considered two variants of equipment for a small-scale power plant operating on LFG: engine of Ukrainian manufacture (public corporation "Pervomayskdiselmash") and engine of American manufacture (Caterpillar). In both cases prime cost of electricity is lower than current market price of electricity in Ukraine – 0.021 $/kWh.

Tab. 9: Technoeconomic parameters of biomass fired boilers of Ukrainian manufacture

Parameters	**5 MW wood-fired boiler with step moving grate**			**2 MW under screw wood fired boiler**			**800 kW straw fired boiler**	**990 kW straw fired boiler**
Capital outlays, thous. UAH	1330	1330	1330	520	520	520	289	303
Share of foreign components, %	30	30	30	30	30	30	30	25
Fuel moisture,%	60	60	60	40	40	40	15	15
Operation time, h/year	4400	8000	8000	4400	8000	8000	4400	4400
Operating costs, thous. UAH								
Amortisation	89.11	89.11	89.11	34.84	34.84	34.84	19.36	20.18
Profit (annual interest is 10%)	133	133	133	52	52	52	28.9	30.3
Labour	3.6	7.2	7.2	3.6	7.2	7.2	3.6	3.6
Fuel	367	0	750	134	0	274	17.9	40.1
Maintenance	10	10	10	5	5	5	2.5	0.8
TOTAL	603	239	989	230	99	373	72	98
Fuel consumption, t/year	14685	30000	30000	3357	6857	6857	895	1114
Heating value, MJ/kg	6	6	6	10.5	10.5	10.5	14	14
Fuel cost, UAH/t	25	0	25	40	0	40	20	36
Heat production, TJ/year	70.49	144	144	28.2	57.6	57.6	11.28	12.52
Prime cost of heat, UAH/GJ	8.55	1.66	6.87	8.15	1.72	6.48	6.41	8
Saved natural gas, mill m^3/year	2.01	4.11	4.11	0.81	1.65	1.65	0.32	0.39
Saved natural gas, thous. UAH/year	889	1817	1817	356	727	727	142	167
Payback period, year	2.6	0.7	1.3	2.4	0.7	1.2	2.4	2.2

Tab. 10: Results of feasibility study for biogas plant of 2000 m^3 digester volume

Parameters	Share of foreign components	
	100%	**40%**
Capital outlays, thous. $	413	280
Operating costs, thous. $/year	21	22
Savings, thous. $/year		
Electricity	41.4	41.4
Heat	5.6	5.6
Minerals (N, P, K)	25.0	25.0
Total	72	72
Payback period, years	8.1	5.6

Tab. 11: Results of technoeconomic assessment for 2 MW power plant at Lugansk landfill

Parameters	**Engine of public corporation Pervomaiskdieselmash (Ukraine)**	**Engine Caterpillar (USA)**
Volume of landfill, mill m^3	2	2
Landfill gas make, mill m^3/year	8	8
Heating value of LFG, MJ/m^3	18	18
Installed capacity of power plant, kW	2×1000	2×1000
Electricity production GWh/year	14.4	14.4
Capital outlays, thous. US$	571	1346
Operating costs, thous. US$/year		
Amortisation	38.1	89.7
Profit	57.1	134.6
Labour	3.6	3.6
Maintenance	5	5
TOTAL	103.8	232.9
Prime cost of electricity, US$/kWh	0.007	0.016
IRR,%	39	17
Payback period, year	2.0	4.6

Given that the above economic calculations do not take into consideration potential opportunity to sell quotas of reduction of CO_2 and other greenhouse gases achieved from the implementation of bioenergy projects. However now exists a real opportunity to sell mentioned quotas, for instance, within the programme Eru-PT supported by Dutch Ministry of Economy. The quotas will be bought at a price of 4.54-9.08 EUR/t on condition that during 2008-2012 total reduction will be not less than 500 thous. t of CO_2 or CO_2-equivalent. Taking into account the opportunity of such a sale, economic parameters of bioenergy projects may be improved much. For example, reduction of methane emission achieved from the implementation of above described project at Lugansk landfill amounts to about

62 thous. t/yr of CO_2-equivalent or 248 thous. t of CO_2-equivalent for four years (2008-2012). Selling the quotas at a price of 5 EUR/t one can obtain about 1.24 mill EUR. The sum is several times more than capital outlays required for the construction of a 2 MW LFG power plant equipped with Ukrainian gas engines.

6. Analysis of Barriers to Bioenergy Development in Ukraine

There are a number of technological barriers in the way of bioenergy development in Ukraine. At present there is no serial production of wood fired boilers, straw fired boilers as well as biogas plants in the country. Foreign equipment is also practically absent in the market of Ukraine. One can find only 15-50 kW domestic wood-fired boilers mainly of foreign manufacture, and also up to 200 kW wood fired boilers chiefly equipped with gasification furnace extension. Maximal capacity of steam boilers produced in Ukraine is 10 tons of steam per hour (6.5 MW), and the pressure is up to 13 bars. It forms technological barrier for starting production of equipment for power plants and CHP plants operating on wood or straw.

Today the most widespread straw handling system in Ukraine is harvesting chaffed straw and its storage in fields in the form of stacks. Baling technology has not become widespread yet. Due to the fact that long distance transportation of straw is not profitable, and it puts obstacles in the way of construction of large-scale straw fired plants and spreading of boilers for baled straw. However it should be marked that more and more farms are turning toward the use of balers for harvesting straw regardless of their plans concerning installation of straw-fired boilers. In themselves purchasing balers and harvesting straw in the form of bales are profitable actions. They are repaid at the expense of reduced grain loss during harvesting and also at the expense of reduced expenditures on storage and transportation of straw. According to expert evaluation, up to 50% of Ukrainian farms will have balers by 2010.

A great amount of large-scale pig-breeding farms in Ukraine are equipped with water wash systems for manure. In this case moisture of manure (about 99%) is much higher than moisture suitable for anaerobic digestion (88-90%) with the obtaining of positive energy effect. To implement biogas plants on such farms, it is necessary to reconstruct existing water wash systems.

Economic barriers to bioenergy development in Ukraine lie in the absence of any economic stimuli in the form of subsidising, soft loans, setting free of taxes etc for the development and implementation of biomass-to-energy technologies. Even the most profitable technologies (such as industrial wood fired boilers or LFG extraction and utilisation plants) need financial stimuli for producers as well as for consumers of the equipment, at least at the beginning of their developing and launching production.

A legislative barrier is connected with the absence of state policy in the field of biomass combustion, biogas production and other technologies. A positive example of changing the point of view of the state in this field is adoption of the law "On alternative types of liquid and gaseous fuel" (N 1391-XIV of 14/01/2000). According to the law, the projects on utilisation of landfill gas, biogas and liquid fuel produced from biomass must receive some financial support. Unfortunately there are no laws encouraging the use of wood and straw as a fuel.

An ecological barrier is connected with existing rather soft ecological legislation that does not let to include ecological constituent in economic effect from the use of bioenergy technologies. In developed countries the main effect from the implementation of biogas plant is ecological. In Ukraine biogas technologies have to compete with other technologies only at the expense of energy effect and effect from production of high-grade fertilisers.

In addition an almost complete informational vacuum on bioenergy technologies is typical for Ukraine. This forms an informational barrier in the way of bioenergy development. In libraries there are not relevant profile periodicals issued in developed countries, Ukrainian specialists rarely participate in specialized foreign conferences. Insufficient knowledge about advantages of energy utilization of biomass provokes a skeptical attitude towards intentions and plans to use biomass as a fuel. As a positive change in recent years one can mark an increased interest of Ukrainian mass media toward renewable energy including bioenergy. A number of journals include publications on these technologies in practically each issue. Among them "Installation market", "Green energy", "Ecotechnologies and Energy Saving" should be marked. Besides, the First International Ukrainian Conference on Biomass for Energy held in September 2002 in Kyiv gave a powerful incentive for the further development of bioenergy in Ukraine and attracted close attention of national and local authorities and officials to this problem. Such a conference is expected to be a regular event in Ukraine (every other year). The next conference is planned on September 2004. Corresponding information about the forum will be placed on web site of SEC "Biomass" www.biomass.kiev.ua.

An organizational barrier is formed by the absence of a co-coordinated policy in the area of bioenergy in Ukraine and absence of a state structure responsible for the policy. With regard to administrative barriers, it should be mentioned that independent producers of power from RES (excluding wind plants) are confronted with serious difficulties when selling electricity to the grid. The last concerns first of all LFG utilization systems. As a change for the better in state policy on bioenergy we can mark development of "Energy Strategy of Ukraine till 2030". The Strategy has been elaborated by a group of Ukrainian energy experts on the decree of President of Ukraine and is under consideration/approval at the moment. The Strategy contains a section on RE including bioenergy.

7. Conception of Bioenergy Development in Ukraine

Based on estimations of biomass potential in Ukraine, experts from SEC "Biomass" and the Institute of Engineering Thermophysics have elaborated the conception of bioenergy development in Ukraine. The following technologies can be considered as the most promising for commercial use in Ukraine:

- 0.1-5 MW industrial wood fired boilers for installation at timber enterprises and wood processing enterprises;
- 1-10 MW wood fired district heating (DH) plants;
- 0.1-1 MW farm and neighbor heating straw fired boilers;
- 1-10 MW straw fired district heating plants;
- Biogas plants for large-scale cattle farms, pig-breeding farms, poultry farms and food industry enterprises;
- Systems for LFG extraction and utilization at small-scale power plants of 0.5-5 MW capacity;
- Co-combustion of biomass with fossil fuels for power production.

Technologies of direct combustion of wood, first of all for heat and process steam production, require prior development in Ukraine. It is due to rather low electricity prices existing in Ukraine (0.03 $/kWh) and at the same time quite high heat and fuel prices. Implementation of small-scale power plants and CHP plants operating on solid biomass (wood, straw, husk) will be profitable if electricity price rises significantly or in the case of subsidies. Heat production from biomass is profitable now even with the use of foreign equipment. Besides, Ukraine has opportunity to launch its own production of wood and straw fired boilers.

Technologies of straw combustion are very promising for Ukraine too. To implement them widely it is necessary to solve a number of problems connected with arrangements on collection, baling, transportation and storage of straw. First of all 0.1-1 MW farm and neighbor heating boilers for implementation at agricultural enterprises have the best prospects. After demonstration of their advantages large-scale district heating plants also have good opportunities for commercialization. As for 1-10 MW_e biomass-based CHP plants, we restrict their position in the conception of bioenergy development in Ukraine only by two demonstrational plants (one wood-based and one straw-based) until electricity price rises greatly.

Large-scale biogas plants also play an important role in the conception. Their widespread implementation is possible on pig-breeding farms (with a livestock more than 5 thousand heads), cattle farms (with livestock more than 600 heads), poultry farms, food industry enterprises and waste water treatment plants. According to our estimations, it is possible to implement 2903 biogas plants in Ukraine with average digester volume of 1000 m^3 including 295 plants on pig-breeding farms, 130 plants at poultry factories and 2478 plants on cattle farms, food industry enterprises and waste water treatment plants.

The use of LFG is the most profitable at industrial enterprises located near landfills. If it is impossible to utilize LFG in nearby industry, it can be used at small-scale power plants or CHP plants equipped with gas internal-combustion engines.

Liquid fuels are unlikely to be produced from biomass in Ukraine in the near future because their prime cost is much higher than cost of traditional liquid fuels. The main efforts in this area should be focused upon research and demonstrational projects. The same can be said about fast pyrolysis and gasification of biomass.

Table 18 contains data on the equipment that may be installed in Ukraine under the realization of developed conception.

Tab. 12: Bioenergy equipment that can be installed within the implementation of proposed conception

Type of equipment	Approximate capacity of Ukrainian market, units	Installed capacity,		Operation time	Replacement of fossil fuels,	Reduction of CO2 emission,	Total capital investments,
		MW_{th}	MW_e	h/year	mtoe/year	mill t/year	mill US $
Wood-fired DH plants, 1-10 MW_{th}	250	500	–	4,400	0.21	0.49	38
Industrial wood-fired boilers, 0.1-5 MW_{th}	250	250	–	8,000	0.19	0.45	25
Wood-fired CHP plants, 1-10 MW_e	1	10	5	8,000	0.014	0.05	5
Domestic wood-fired boilers, 10-50 kW_{th}	53,000	1590	–	4,400	0.67	1.57	80
Farm straw-fired boilers, 0.1-1 MW_{th}	15,900	3180	–	4,400	1.34	3.14	254
Straw-fired DH plants, 1-10 MW_{th}	1,400	2800	–	4,400	1.18	2.76	280
Straw-fired CHP plants, 1-10 MW_e	1	10	5	8,000	0.014	0.05	8
Large-scale biogas plants	2,903*	711	325	8,000	0.93	22.36	290
Small-scale LFG power plants	90	20	80	8,000	0.17	3.26	48
TOTAL	73,795	9,071	415		4.7	34.13	1,027

* including 2478 plants on cattle farms, 295 plants on pig-breeding farms, 130 plants at poultry factories.

7. Conclusions

Ukraine has a rather big potential of biomass available for energy production. Biomass (excluding the share that is used by other sectors of economy) can cover up to 9% of the total primary energy demand. Technologies of biomass utilization are mostly at the beginning of their development in Ukraine, but they have good prospects for commercialization in the near future.

Results of performed feasibility study indicate that heat production from biomass is competitive right now even on application of foreign equipment. With regard to electricity production from biomass, the most profit-making plants now may be small-scale power plants operating on LFG. Production of electricity from wood and straw is not profitable under existing tariffs on power. The following technologies are the most promising for Ukraine:

- 1-10 MW wood fired district heating plants and 0.1-5 MW industrial wood-fired boilers for installation at timber enterprises and wood processing enterprises;
- 0.1-1 MW farm and neighbor heating straw fired boilers; 1-10 MW straw fired district heating plants;
- biogas plants for large-scale cattle farms, pig-breeding farms, poultry factories and food industry enterprises;
- plants for biogas extraction and utilisation at large-scale landfills (0.5-5 MW power plants);
- Co-combustion of biomass with fossil fuels for power production.

Production of corresponding licence European equipment at industrial enterprises of Ukraine seems to be the most promising strategy for the development of bioenergy technologies in Ukraine, at least at the first phase.

If elaborated conception for bioenergy development in Ukraine is realized, total installed capacity will come to 9071 MW_{th} and 415 MW_e. It would result in the replacement of 4.7 mill toe/year and reduction of CO_2 emission at rate of 34 mill t/year. Development of bioenergy technologies would decrease Ukraine's dependence on imported energy carriers, enhance its energy security at the expense of organizing energy supply based on local renewable sources, create a lot of new jobs (mostly in rural area), contribute greatly to the improvement of ecological situation.

Third Plenary

Education for Sustainable Development in the Context of University – Municipality Partnership

Urban Sustainability Strategies – Is there a Common Structure?

Lars Rydén

Baltic University Programme, Uppsala University,
POB 256, S – 75105 Uppsala, Sweden

Abstract

A survey was made of sustainability strategies in 20 cities in 9 Baltic Sea region countries taking part in the Baltic University Urban Forum project. Cases of such strategies were also collected in media and literature. The issue was divided into three areas – *urban flows*, *urban planning* and *urban development*. In each of them it is possible to define limited resources, and develop a quantitative theory of sustainability. Below a few cases of processes in the area of *urban flows* will illustrate strategies used by the cities.

The most common strategy used is *reduction* of flows. This refers to using resources more efficiently, such as using less water in households, and energy conservation. This strategy was used more or less in all cities studied.

A related strategy is *replacement*. The material used may be changed to renewable and non-toxic, or in other ways that is improving sustainability.

The next most common strategy is *rescaling*. This either refers to upscaling, such as when locally heated houses are joined to a district heating network, or downscaling, as when a district heated house is switching over to e.g. its own heat pump or solar panel, or (in the Baltic States) to local heating by gas. Both up- and downscaling have their sustainability gains. In general scaling is heavily dependent on the quality of the infrastructure.

The least common, but in the longer term most interesting, strategy is *recycling*. Recycling not only reduces resource use, but also closes biogeochemical cycles of e.g. carbon, nitrogen or phosphorus. One example is the fermentation of organic waste, which produces biogas to be used for public transport (buses). Here the carbon cycle is closed.

In recycling *integration* of different areas of municipal management occurs. Flows are interconnected when waste becomes energy, and nutrients are used for food production. Traffic is improved when biogas is used for public transport, and socio-economy if working places are created locally.

In general the *incentives* to introduce a new strategy are short-term economic gains. Price policies are thus very important. *Reduction* of course leads to lower costs. The decisions by individual households to *rescale* are dependent on quality of infrastructure and the cost, while the municipality when upscaling is acting to reduce pollution. The investments to introduce *recycling* schemes are often high, which is an obstacle.

Also in the area of *urban planning* the strategies of reduction, replacement, rescaling and recycling are found. The more efficient use of the built environment, so-called 'urban healing' or 'densification' is one example or replacement (of surface area use). The uses of buildings for new purposes are examples of recycling of the built environment. Also in the area of *urban development* these strategies are observed.

Further work focuses on quantitative evaluation and developing guides for urban management.

1. Introduction

1.1 How to make Sustainability Operational

Sustainable development (SD) is often defined according to the well known Brundtland Commission (1) statement on what constitutes sustainable development, that is, the development of intergenerational equity. Equally often we see that SD means that we simultaneously address the three dimensions of sustainability, that is, environmental, economic and social dimensions (2). None of these definitions and statements are in themselves very useful in practical work, or easy to operationalise. What is precisely intergenerational equity, or what constitutes the three so-called dimensions is not defined. This may be one reason why there are hundreds of definitions of SD (3), and that SD is not easily included explicitly in practical work.

In a co-operation between universities and cities in the Baltic Sea region a different road to work with SD was taken. The background was a three years practical work in 11 cities of towns in 8 countries in the Baltic Sea region in the project SUPERBS, Sustainable Urban Patterns Arround the Baltic Sea (4). The work has been documented in a total of 35 case studies published by the Baltic University Programme (5-8). They describe activities in many areas including administration, democracy, urban planning, environment, economic development, housing etc., as each of the cities included what was most important to them.

Based on this experience 10 areas of key importance for a local sustainable development were pinpointed. These were:

1. Energy management
2. Water management
3. Waste management
4. Traffic and transport
5. Urban green structures and culture

6. Rebuilding the city, brown fields
7. Socio-economic development
8. Urban rural cooperation
9. Information and education
10. Integration.

In an ongoing three years follow-up project called BUUF, Baltic University Urban Forum (9), an enlarged network of 20 cities and 15 universities worked with these ten issues. The intention is to develop strategies for dealing with them, and in the end define in a practical way how to work with SD on the local level. This bottom-up approach, lend itself to operationalisation of the concept of sustainability and is expected to produce manuals and handbooks.

1.2 SD as Management of Limited Resources

Each of the ten areas was thoroughly discussed at a best practice conference hosted by one of the participating cities. After this we decided to group the topics as follows:

A. *Urban flows*: Energy, water and waste management

B. *Urban planning*: Traffic and transport, urban green structures, and the built environment, rebuilding the city, brown fields.

C. *Urban development*. Socio-economic development, information and education

D. *Integration*: Creation of sustainable communities.

It is clear that each of the three integrated areas addresses an aspect of resource management. The resources are:

- For urban flows: material resources, the "kilos" or the "kilowatts".
- For urban planning: the urban space or the surface area of city, the "square meters".
- For urban development: the people living in the city, the "human resources".

It is also clear that each of these resources is limited.

Management of sustainability is thus here identified as the management of limited resources.

It should be noted that the three types of resources are not interchangeable. Kilos cannot be turned into square meters, nor can it be made into a human resource. This is different from money, which can be used to buy either of these resources. Money thus has a different character than the three resources defined here as key concerns for local SD.

It is clear that when going into detail we may identify hundreds of items in a city, which are limited. It may be that each of these can be included in the three categories above. This has not been further investigated.

1.3 Integrating the Sustainability Areas

Integration is needed *within* all the three areas as well as *between* the three areas. Special conferences were organised to discuss integration and strategies for urban sustainable development. Three integration conferences were organised to discuss how to deal with the items in the three areas in an integrated way, urban flows, urban planning (urban space) and urban development.

In each of them two aspects were treated. The first was how to solve problems in a very concrete way, to find *technical solutions*. The second was how to develop *management solutions*, especially integration of management. At each conference background presentations and cases, site visits, group discussions and presentations and reports from the cities were included.

Based on these meetings a number of strategies actually used by the cities were identified, as were ways to deal with the management of the resources.

In addition to the survey of sustainability strategies made in the 20 cities taking part in the Baltic University Urban Forum project, cases of such strategies were also collected in media and literature for other cities in the region. Especially valuable were the reports published from Swedish LIP, Local Investment Programme (10). The programme with a total budget of 6,2 billion (600 MEuro) run during 1998-2002, but in many places is still ongoing. 1800 projects were conducted in 161 municipalities.

2. Urban flows Strategies

2.1 What is Included in Urban Flows

Urban flows are dealing with energy, waste and water. This corresponds to all material flows in a city. Even the most dear objects and materials, sooner or later become worn out, and end up as waste, which the city needs to deal with.

The urban flows may also be called urban metabolism, and compared to the metabolism in a body. The food we eat contains energy, the water we drink will become wastewater. It is clear that energy, waste and water are all connected: Energy, water and matter goes in; Waste goes out; Energy is carried by matter; It is one system.

There are a number of technical solutions to address the integration of energy, water and waste. The technical solutions are answers to the questions: how to take care of energy in wastewater or energy in air from ventilation; how to take care of the resources in wastewater flows, such as nitrogen and phosphorus, or organic waste; and how to take care of resources in gaseous emissions, such as sulphur. The solutions discussed included: heat pumps in waste water; heat exchangers in ventilation systems; how to deal with sludge from wastewater treatment; production of biogas from organic waste; solid waste incineration; district heating systems; and various forms of heating based on biomass (11).

2.2 The Management Solutions

Management turned out to be a key concern to deal with the urban flows in a good way. There were several issues here.

The first is the *public/private choice*. Many cities have formed companies taking care of special tasks, such as water supply, wastewater treatment, energy supply, waste management etc. Those with experience from companies underlined the importance that the city keeps control of how the companies are managed. Sometimes several neighbouring municipalities have a common company for technical management. It is clear that in most cases privatisation of urban flows were not favouring sustainability. There were however other cases when privatisation worked very well. An example is Stockholm Water, where a sizable and increasing part of the budget comes from selling energy produced from the wastewater, as biogas and heat from heat pumps (12).

In small municipalities the management of urban flows are easily integrated as a single department or few individuals work with them. In larger municipalities different departments in the municipality are dealing with different parts of the urban flow. It is then important that the *activities should be coordinated.* This is not always the case.

Some municipalities have introduced and implemented a *management system*, such as EMS or IMS. In Nacka in Sweden a large part of the administration is certified according to both ISO 14001 and ISO 9000 (13). The IMS include 6 different departments in the city and a total of 90 employees, including all environmental administration. This is so far the most efficient and advanced version of integrated management of resources.

Several *legal requirements* have to be addressed in this sector. Today waste management is the greatest concern for legal compliance, but in the near future the EU water directive and energy certification will need to be dealt with. The European Union Directives will push the municipalities to integrate resource management.

2.3 The Strategies

The urban flows strategies observed in the BUUF project may be summarised as Reducing, Replacing, Rescaling, and Recycling.

Reducing refers to reducing the flows of energy (energy saving), water and waste. These are well known options for better resource management. In many municipalities they could be much better utilised than they are at present. Reductions in resource use are possible in the longer term in our countries by a factor of about five (14). Reducing energy for heating by better buildings is a major future sustainability strategy.

Replacing refers to replacing fossils with renewable energy sources, such as biomass, or replacing toxic material, e.g. mercury, with non-toxic. Replacing oil or coal is fairly frequent, but often the new choice is gas, another non-renewable. This is an improvement but still not sustainable. In the longer term we have to replace all fossils with renewable, which is a major challenge for our societies.

Rescaling refers both to downscaling and upscaling. Upscaling means that single operations are replaced with a larger common operation. The most typical is when a municipality is introducing district heating to replace the individual heating of houses. If the individual households were very polluting (e.g. if they used coke or wood) this strategy is very good for environment. Other cases of upscaling refer to district cooling, common composting instead of individual, and common wastewater treatment instead of individual.

Downscaling refers to when a common operation is replaced with an individual one. Typically when a household introduces a heat pump and buys less energy from district heating, or when a building is provided with solar panel to heat water for that house only. Which, upscaling or downscaling, is the best strategy depends on the situation. A certain degree of self-sufficiency is probably required to approach sustainability. In Hamburg very interesting neighbourhoods with essentially their own supply of energy (passive energy houses, with solar panes and photovoltaic cells) gives an interesting example (15).

Recycling refers to the fact that a material stream is made useful several times. Best know is perhaps recycling of waste flows, which leads to product reuse, material (e.g. paper) recycling, or simply incineration of waste to recover energy. But there are also recovery of nutrient flows in e.g. composting, production of biogas, and return of sludge to agricultural fields.

2.4 Comments on Urban Flows Strategies

Which sustainability strategy for urban flows a city adopts depends on the incentives and possibilities. The most common incentives are the economic ones, that is, to make the resource flows more profitable or less costly. Secondly there are the environmental incentives, e.g. when a municipality improves wastewater treatment to improve water quality to the recipient, or improve air quality. Thirdly, the legal incentives become more important especially as the EU Directives are introduced.

When it comes to the strategies of the individual inhabitants they are often in the east short sighted and economic. Even if an investment in the longer term is good, the time for return of investment is a key issue. If there is no money to invest, or no credits exist it will not happen at all. In the west individuals can afford to invest to get the return of investment in a longer perspective (typically 5-10 years), e.g. when buying a heat pump. Companies often require economic return in shorter time than several years. A municipality should be able to think in terms of decades, for example when building a district heating system, since the city is sure to stay that long.

By far the most common strategy is rescaling, probably because of economic concerns. The most sustainable but less common strategy is clearly recycling. This is the only one that is clearly looking into integration of the different parts of the flows. Recycling not only reduces resource use, but also closes biogeochemical cycles of e.g. carbon, nitrogen or phosphorus. One example is the fermentation of organic waste, which produces biogas to be used for public transport (buses) (16). This is used in several Swedish cities. Here the carbon cycle is closed. Another case is the use of mussel cultivation on the Swedish west coast to take care of nitrogen-rich effluents (17). As mussels are canned, sold and eaten the nitrogen cycle is closed. The most common area, recycling of solid waste, is well established in the Nordic countries, but often poorly developed on the eastern side of the Baltic.

3. Urban Planning Strategies

3.1 What is Included in Urban Space

Urban planning or urban space is dealing with green structures, traffic and transport structures and the built environment. This corresponds to all surface area in a city. Any piece of a city is either:

- green (blue, that is water surfaces, is most often included);
- a street or road, or similar areas such as a square or other public space;
- resides a building.

Here urban space includes all area for which a city has a planning responsibility and monopoly for both a master plan and detailed plans. It includes both properties owned by the city, and privately owned properties.

Urban planning is obviously a single concern, as for each part of a city the three alternatives mentioned are all present – it may become a (green) park, a (traffic) street or give place for a building. Still the management of urban planning is not always integrated. Departments dealing with parks and green areas, are not necessarily well coordinated with city departments dealing with traffic, or with those authorising building permits.

The surface area is a limited and expensive resource in a city. It needs to be carefully taken care of. Green structures are for this reason often threatened, as traffic structures are expanding, and built structures are added as investors are pushing the city for building permits. The master plan and detailed plan should, however, be respected and not changed easily. Fighting commercial interests and investors – the city need to be strong – is one important part. We have several examples of so-called corrupt cityscapes, in which money got the upper hand.

Traffic structures turn out to be very formative for the city plan. Traffic structures should be discussed at an early phase of planning. Still it is often the other way around. Improved coordination was asked for in many cities.

Many experienced urban planners in the BUUF meetings argued that it is important to start with what you have. Urban planning is urban development (18). Another point is that one always should look for all types of functions in an area. They may be both green and communicative. A building may also be green. Zoning is not a sustainability strategy and is not needed as much now as in the past.

3.2 The Management Perspective

The most often underlined aspect of urban planning, is that it has to include those living in the area, which is to be developed; *participation* is a key concern. This is *governance* of urban planning. A number of management tools exist for this purpose. With these tools one can manage a common process of planning, and deal with conflict of interest and organise a process of conflict resolution. A developed process for stakeholder participation has been developed at Chalmers University of Technology (19).

Cooperation between city departments is essential. It is as mentioned not always strong, although it is easier in smaller cities. Common meetings and information are obvious tools for such coordination.

In studies of successful cities two success factors are repeatedly mentioned. One is competence development of the city staff. The *learning municipality* is a successful municipality. Secondly *commitment from the leadership* is crucial. It seems to be of special importance to stage a process of common planning. These conclusions are underlined in a report from the DISCUS project on Governing Sustainable Cities (20).

3.3 Urban Planning Strategies

The Urban planning strategies observed in the BUUF project are summarised below.

In order to take care of the resource efficiently cities may *grow inwards*. This means that each part of the city needs to be well managed. Thus polluted areas, the brown fields, should be remediated and used properly. Less well-used areas, although not necessarily polluted, here called grey fields, need to be developed. This is sometimes called urban healing (the positive expression) sometimes densification (the negative expression). Grow inwards means that the city is not expanding outwards, it is not building on green fields, that is agricultural ground. Suburbanisation is avoided in this way.

In order to be as efficient as possible the individual parts of the cities also become *multifunctional*. It may be illustrated in many ways. Thus green areas may also be used for biking and thus combine greenery with transport. Neighbourhoods may include residential functions, work places, educational institutions, and social functions such as cafés, restaurants, sport arenas etc. This becomes more possible as work places are less polluting, as part of post industrialism. Zoning is thus not a sustainability strategy. One should be able to work, live and recreate in the same neighbourhood. This should decrease the pressure on the traffic system, and increase the social value of a neighbourhood.

Traffic in cities is a major dilemma in urban planning. The private car is a problem in cities for many reasons. The emissions, the noise, the space requirements, the accidents are parts of the problem as is, not the least, the fact that many planners feel that they are forced to build cities for cars rather than for people. *Reduction of private car traffic* is thus a key strategy. Technically it may be pursued through the development of good alternatives to make the personal car less needed. Important strategies include development of public transport, increased costs for private cars (e.g. through costs for going into the city centre) or strategic arrangement of parking places.

Finally the relationship between the city and its rural surrounding is a key issue for developing urban sustainability. Thus *improve and increase the built vs blue/ green interface* is an important strategy. Here built is the built environment, green is rural and blue is water. There are many examples. Building at the waterfronts is pursued in many cities as a major development. Costs for living close to water often become high, but waterfronts remain very attractive. An increased urban/rural interface reduces the ecological footprint of the city (21). One way to maximise this is to build finger cities, rather than pursuing urban sprawl or a build dense cities, the major alternatives. Finger cities in particular let the inhabitants have a good access to the countryside and at the same time easy access to the city centre if good transport is available in the "fingers".

3.4 Comments on Urban Planning Strategies

The strategies observed exist in many cities, from the largest, e.g. Hamburg, Germany, where they are part of the official policy, or in smaller cities, such as Uppsala, Sweden. The largest obstacle, when developing these strategies is the economy.

Urban healing is in many countries difficult due to private ownerships of "grey field" properties. It may stop the city administration from doing anything, even if a property in the centre of the city is ugly and problematic. The owner may keep the property as it is, for economic reasons. The law does not easily allow the city to expropriate a property, or the city cannot afford to buy it, or to pay the legal procedure needed to able to buy it. It is thus another case of a corrupt cityscape.

Also when it comes to remediation of brown fields economic concerns are often an obstacle. It is not always that the price of land is enough to motivate investment in often very costly remediation. Then the cityscape stays polluted. In the long term it will be enormously expensive to take care of all the pollutants that previous generations have left to us. In Hamburg, where the price of land is high, the city has a very extensive activity for remediation. This is not the case in the countries in transition, where this just has started.

Also development of good public transport alternatives is very expensive. Thus often car traffic is expanding much faster – it is not so expensive to buy a car – and the city is challenged with traffic jams and polluted air. Car traffic is today a major threat to decent city life in most cities in the region. All strategies possible to "reclaim the city" to people would be welcome. The classical, globally admired, projects in Bogota or Curitiba are characterised by good public transport and well-designed public space (22).

4. Urban Development Strategies

4.1 The Challenge of the Inhabitants

Municipalities today have very large responsibilities for the inhabitants living in the city. This goes for the whole life of the citizens. In the Nordic countries this typically includes care and education from infant care, kinder gardens, schooling, primary health care, elderly care, and social care, unemployment care and care of disabilities. The budget for this sector, so called "social costs", constitutes today in Sweden some 80% of the city budgets and is increasing in most municipalities.

At the same time municipality income is mainly the taxation on salaries of its inhabitants. It is thus crucial that the inhabitants are in good health, are working, and have a good income. Sustainability requires that the municipality can promote and support a good and productive life for its inhabitants.

We should not forget that the inhabitants are the reason that the municipality is there at all. It is important that the inhabitants, especially those that are productive, do not leave the municipality. They should find that the city offers a good and attractive life. It is critical that the city can manage their human resources well.

4.2 Management of Human Resources

Management of human resources consists of several components, care, education, work and free time. Work is not only paid work, or employment. It is equally work in the household and in the so-called co-operative sector, e.g. within an NGO. Some (west) European surveys indicate that we spend per day on an average about 6 hours of paid work, 4 hours of work in household and the co-operative sector, 4 hours free time and 10 hours for sleep and meals (23).

The usual administrative structure looks at the money, that is, manage financial resources. Here we will rather take the approach of dealing with individuals, management of human resources. Economic management may be seen as secondary in all areas – urban flows, urban planning and urban development.

The inhabitants of a municipality are very different. A few general rules exist however. In both short and long term, all cities in the Baltic Sea region will all have more elderly than before. The population is not increasing, and we will have fewer children. One conclusion is that it is important to find ways to let the elderly contribute to society.

An estimated 8% of the inhabitants will need support in some way due to physical, mental or social handicaps. The latter group includes drug addicts and criminals. Cities need groups, which manage with handicaps to make society accessible to them, and try to find jobs that they can manage. Clearly many of these will not contribute at all. It is then a challenge to organise a support in an economic and human way.

4.3 Urban Development Strategies

The Urban development strategies observed in the BUUF project are summarised below.

A main strategy for cities is to support *education*. This includes first of all the ordinary schools, where children spend much of their childhood and adolescence. Education is also essential for development of professional competence, a training that often needs regular updating to be on the best level. A special concern are the immigrants especially those from far away. Education together with encouragement, empowerment, is a key way to make them contribute to society. Youngsters with a weak social identity also may be brought to a good and productive life with proper education (24). Cities in the BUUF network have many different ways to support schools and other forms of education. Education makes the inhabitants more competent in all parts of their lives.

A strong economy is important for all municipalities. Most people look for work and employment in an existing company, or public sector. However companies also need to be created. Much effort is therefore made in many cities to stimulate and support *entrepreneurship*. The many ways to do this include education (how to start a business), offering a place to start (so-called incubators), and possibly offering loans on favourable conditions (this is done e.g. by Russian municipalities). Most companies are very small, 1-2 persons, but some get larger and a few much larger. All need to be supported. Considering the costs connected with individuals in unemployment, support of companies is a good strategy.

The work done in the public sector (as employed by the municipality) and private sector constitutes the white economy. But equally much work is done in households and the so-called third sector, NGOs. An important strategy for the city is to *support and stimulate the third sector*. Many organisations contribute very much to municipal life. This include schooling – e.g. in nature schools and summer camps – social care – e.g. in churches and welfare associations – and management of common resources – e.g. birdwatcher associations and neighbourhood associations. The work done here is all unpaid and the support from the city is normally only to cover direct expenses. Many individuals active in an association see this as a main part of their lives and it gives their life a meaning. Work in such associations thus has an important social function in addition to the economic contribution to the society.

Life in the city is also life in a neighbourhood. It is crucial that those living in a neighbourhood like to live there, and are proud of their neighbourhood, contribute to its management, e.g. take care of their houses, meet each other and are good neighbours. To *support a good neighbourhood life* is another important strategy in many cities. Well functioning neighbourhoods may be very important social support structures for young people with problems, for unemployed and for children and elderly.

In the work on European Common indicators for sustainable city development indicator number one was "Do you like to live in your municipality?" (25). It is clear that the society as whole need to be attractive to live in. The municipality administration of course has a main role to play here. The strategy pursued is to *support culture* in the city. It is done in many ways, e.g. by making the cityscape beautiful with art, flowers and beautiful buildings (26). They may also support art schools, e.g. the municipal music school, or, in other places art, music and dance school, which even may be compulsory.

4.4 Comments on Urban Development Strategies

The role of the municipalities in the region is becoming increasingly more comprehensive. Before it was limited to legal demand and control, that is, being an authority. Today it includes a number of activities that are connected to the maintenance and development of the local society as a whole. In order to be able to carry out this task the city administration need to have the proper budget and legal authority. Municipal reforms to make this possible have been ongoing in the Nordic countries for more than a hundred years and are still continuing. In the countries in transition this process is still at a much earlier stage. A number of administrative reforms in the last ten years have strengthened the local level, and the local budget is increasing. The social responsibilities are however more typical for the Nordic municipalities. Here the costs are skyrocketing since the costs for elderly care is increasing fast (27).

But also in the Nordic municipalities the resources are not enough to implement sustainability strategies. They need to recruit resources from many corners of its local world to manage its community. These may be business, education, NGOs and research. This is a new role for the city. It may be called the cooperative culture or negotiation culture (28).

Many societies are challenged by out- and in-migration; out-migration when large workplaces close, so that the inhabitants cannot support themselves; immigration since migrants from other countries settle. It is very important here that it is attractive to live in the municipality – so inhabitants do not leave – and that new economic activities are created continuously and build on local resources. Large working places based on import of resources and with outside owners are high risks from a sustainability point of view. A more diverse economic life, which at least partly builds on local resources would be important for the city's economic stability.

5. Connecting the Strategies, the Systems Perspective

5.1 Do the three Sets of Strategies have a Common Structure?

There are some similarities between the strategies of management of the different kinds of resources. The parallels between urban flows and urban planning are not difficult to see.

The *reduction* strategy is obvious in the resource flow. In urban planning efforts to reduce the number of cars, part of the development of public transport, may also be recognised as a reduction strategy.

The *replacement* strategy might be recognised in the urban healing as part of the "grow inwards" strategy. Here polluted and less used areas are replaced by well used areas.

The *rescaling* strategy may be identified in the strategy of creating multifunctional, partly self-reliant neighbourhoods. This may be seen as creating a functioning city in the smaller scale, downscaling. Upscaling is about creating common infrastructure, e.g. a public transport system instead of individual cars.

The *recycling* strategy is there to use a resource in many ways. One may see the strategy of multi-functionality in the urban space as a recycling approach.

It is not impossible, but more far fetched to make a corresponding comparison between he management of human resources and the management of urban space and urban flows.

5.2 Integrating the Strategies

To connect the management of the three resources is the challenge of integration. This is done on a *systems level*, on which we may deal with urban development in general.

On this systems level resource management is integrated, and the connections between the socio-economic situation, the urban space and the urban flows are made clear. Just as we referred to the bodily metabolism when introducing the urban flows, we may refer to the whole body when introducing the systems level. Anyone knows that to be healthy we need not only to have a functioning metabolism, but also be mentally in good shape (refer to socio-economy) and in good physical shape (refer to urban space). It is clear that all systems should work to make a city sustainable.

Dealing with the system in this way is a question of processes. It was not the intention with this work to study processes on the systems level. However it is useful to point out in which context the strategies specified will be useful.

The management on the systems level may be seen as a process which starts with finding out *why* a certain action is wanted, in which way it contributes to a good city life. Here the socio-economy, the inhabitants, is in focus. The process then goes on to see what changes may be made in other sectors to improve the situation. For example if the inhabitants find that the air in the city is not healthy it may be addressed by changing the traffic structure or the energy management.

5.3 Strategies as Tools for the Middle Level

Management of general resources, that is urban flows, urban space and human resources described here, constitutes a middle level in the cities administration. It is connected to the levels below and above.

The level below refers to the components of the resources. These include the ten key areas listed in the beginning – energy, water, waste, urban green structures etc. The components are typically managed in the city by the use of indicators.

In the BUUF project, nine of the components have been carefully discussed and assigned indicators (29). As an example water management may use indicators, which follows the amount of water, the purity of the water, the quality of surface water in the municipality etc. It is partly traditional management, but the indicators allow a city to better know to what extent it is sustainable. BUUF has six groups of indicators for each of nine components.

The level above is the management of the whole city, the systems level, which is steered as a process as mentioned above.

We may see the strategies as tools to be used for the middle level, management of resources. The corresponding tools on the other levels are then indicators (the level below) and processes (the level above) respectively. In the longer terms the strategies identified here may be used in a general management plan for how a municipality may improve its sustainability.

Acknowledgements

This work has been made possible through the creative and generous long-term contribution of many individuals from the cities and universities which have been part of the SUPERBS and BUUF projects. Only a few of them have been mentioned in the list of references, while many more have contributed importantly. This is here gratefully acknowledged.

The SUPERBS and BUUF projects have been financed through contributions of the European Commissions Interreg IIC and IIIB programmes, the Swedish Institute, Sida, The Swedish KK Foundation, the C-Framåt programme of Uppsala County, Uppsala University and all of the participating cities and universities through co-financing mechanisms.

References

Berg, Per G. (2005), The factor five city. Swedish University of Agricultural Sciences.

Curitiba: The Dream of a Better City is Always in the Heads of Its Residents, in The Donella Meadows Archive; Voice of a Global Citizen. The Sustainability Institute; www.sustainabilityinstitute.org; Bogota: How Bogota beats cars, by Tooker Gomberg, in Greenspiration! Articles; www.greenspiration.org/Article/BogotaBeat.html.

Ekonomi för en ny tid (1997), Lars Ingelstam Carlssons Bokförlag, Stockholm.

Engström, Carl-Johan (2005), Uppsala City. Report at the Best Practice Conference Lodz, June 8, 2005.

Governing Sustainable Cities (2004), Report from the DISCUS project. ICLEI 2004. www.governingsustainablecities.org.

Granvik, Madeleine & Christoferson, Inger (2002), Cultures of municipal administration, in Basic Patterns of Sustainability, SUPERBS Report II (L. Rydén, ed.), Baltic University Press, Uppsala.

Guidebooks series of BUUF, to be published on www.balticuniv.uu.se/buuf.

Kain, Jaan-Henrik (2005), Urban Development, governance and participation, Chalmers University of Technology, Report at the Best Practice Conference Lodz, June 8, 2005.

Kliucininkas, Linas (2003), Local Sustainability Indicators, in "Urban environmental management", SUPERBS Report III (L. Rydén, ed.), Baltic University Press, Uppsala.

Kustmiljöns framtid (2005), Erfarenheter från forskningsprogrammet SUCO-ZOMA, Sustainbel Coastal Management, 2005. Länsstyrelsen Västra Götalands län Rapport 2005: 10.

Nacka, a case of an integrated municipal management system. In: "Environmental Management Systems and Certification", P. Weiss & J. Bentlage, Baltic University Press, Uppsala, in print 2005. See also http:// infobanken.nacka.se/www/miljo_halsa/arbete/certifierat_ledningssystem.htm.

Regenerative Energies and Sustainable Urban development (2004), Hamburg.

Rydén, Lars (1996), Faces of sustainability, in "Challenges of Sustainable Development", Amsterdam, 22-25 August 1996.

– (ed.) (2002), Basic Patterns of Sustainability, SUPERBS Report I, Baltic University Press, Uppsala.

– (ed.) (2003), Building and rebuilding sustainable communities, SUPERBS Report II, Baltic University Press, Uppsala.

– (ed.) (2003), Public Participation and Democracy, SUPERBS Report III, Baltic University Press, Uppsala.

– (ed.) (2003), Urban environmental management, SUPERBS Report IV, Baltic University Press, Uppsala.

Rydén, L. & Sundström, T. (2003), The Prospect of Sustainable Development, in "Environmental Science". Baltic University Press, Uppsala.

Stockholm water; report at the BUUF Best Practice Conference Sept 3, 2003.

Sundström, Björn (2005), Swedish Association of Municipalities and Counties, report at the BUUF Best Practice conference, Norrtälje Sept 2, 2005.

The BUUF indicator book, report from the BUUF project. www.balticuniv.uu.se/buuf.

UNCED (1992), United Nations Conference on Environment and Development; the United Nations Publications, New York.

WCED (1987), World Commission on Environment and Development, Our Common Future; Oxford University Press, Oxford.

Weiss, P. & Bentlage, J. (eds.) (2005), Uppsala Bus, using biogas for public transport, in "Environmental Management Systems and Certification", Baltic University Press, Uppsala, in print.

www.balticuniv.uu.se/buuf.

www.dzivesprieks.lv/default.htm.

www.hellefors.se/ kommun/ boab/. A series of photos from several richly decorated residential areas in Hällefors, Sweden.

www.ieh.se/lip/utvardering.cfm; Some ten reports have been published from the LIP project by the Swedish EPA (in Swedish).

www.pg.gda.pl/cerso/superbs/.

Teaching for Sustainable Development and University-Municipality Partnerships (Project "Inversed Brain Drain")

Arnolds Ūbelis

Institute of Atomic Physics and Spectroscopy
Faculty of Physics and Mathematics, University of Latvia,
Raina Blvd. 19, LV – 1586 Riga, Latvia

Introduction

Education for sustainable development is one of the challenges of this century demanding innovative approaches, relevant instruments and fast solutions. It is evident, that traditional education systems are slow to react to the challenge, but alternatively there is little chance to find a better solution in the short-term perspective.

Therefore, successful and easily repeatable examples inside the system are very valuable. It is well-known and it is expected that municipalities and universities are the main actors when it comes to the identification, research and development of solutions for sustainable development related problems, but it is not easy to create partnerships, where the interests of both sides are addressed. Without a doubt there is already much project expertise in place and formal recognition of the need for partnerships can be found in various political statements from academia and municipalities. At the EU level the European Commission COM (2001) 549 final "The Regional Dimension of the European Research Area"[1] emphasizes the importance of regions as a bridge between the European and local level, in research and innovation efforts, including strengthening of cooperation by mobilizing the potential of local universities together with regional and local authorities.

The objective of this report is to share experience and expertise developed during the implementation of two INTERREG projects [2, 3] where municipalities and universities worked in partnership forming teams of experts from university staff and municipal employees and decision-makers to evaluate available resources and to design visions and actions towards local sustainable development. The key problem in actions for sustainable development is ensuring broad involvement and continuity in the next generations. Universities are institutions where the intellectual capital of society is concentrated and consequently attract the most capable youth from any society. On the other hand municipalities are distinct administrative units of society in the country and the quality of decisions on the municipal level forms an economic, social and political framework that shapes processes (speeds-up or slows-down) towards sustainable development on a local, as well as regional or even global scale. It is likely that only significant, well-argued comprehensive bottom-up actions at universities and inside municipalities can speed the processes towards sustainability. Such initiatives need strong human potential and a "critical" mass of relevant knowledge in society. Unfortunately, traditional education systems are out-sourcing more talented young professionals to the universities and rarely do they return back to their native municipalities after graduation. This process has especially negative consequences for remote municipalities and the New Member states of the EU, including Latvia. Developing university-municipality partnerships and strengthening information technologies as an instrumental part of the emerging knowledge society creates new opportunities to tackle the "brain drain" problem and convert it to a "brain gain" or, at least, "brain circulation" [4].

The Problem, Proposed new Approaches, Target Groups and Instruments

Experience from the previously mentioned two INTERREG projects created the necessary motivation for new project proposals to the Co-operation Programme in the Baltic Sea Region, PHARE 2001 and 2003 (Latvia) [5, 6], where innovative solutions to sustainable development related education and human resources problems of municipalities were sought. The first of the projects [5] was implemented in 2003-2004 and the second /6/will follow in the period 2005-2006.

The Co-operation Programme in the Baltic Sea Region is intended to respond to perceived needs and constraints in the targeted countries. Analysis shows, that the labour market and employers in both the public and private parts of the economy of Latvia encounter a noticeable lack of skilled labour with knowledge levels consistent with Western standards, which manifests itself in the form of low indices of economic and social development in Latvia compared to the EU member countries. There are a number of problems demanding immediate solution in the entire country and which are complementary also to municipalities:

- A lack of qualified and well-educated employees in the public administration who are capable of implementing new, innovative social and economic development ideas, particularly at the regional and local level, to promote development of regional territories;
- A lack of professionals in industry, especially pronounced in small and medium-size enterprises, who are knowledgeable in the high-tech fields, innovation transfer in Europe and the world, and have skills to implement them in their enterprises;
- A permanent "brain drain" to the capital area (Riga) from all parts of the country, which has a negative impact on regional development.

Fig. 1: Students, as the main target group, are able to ensure sustainable knowledge transfer between universities and municipalities and across generations

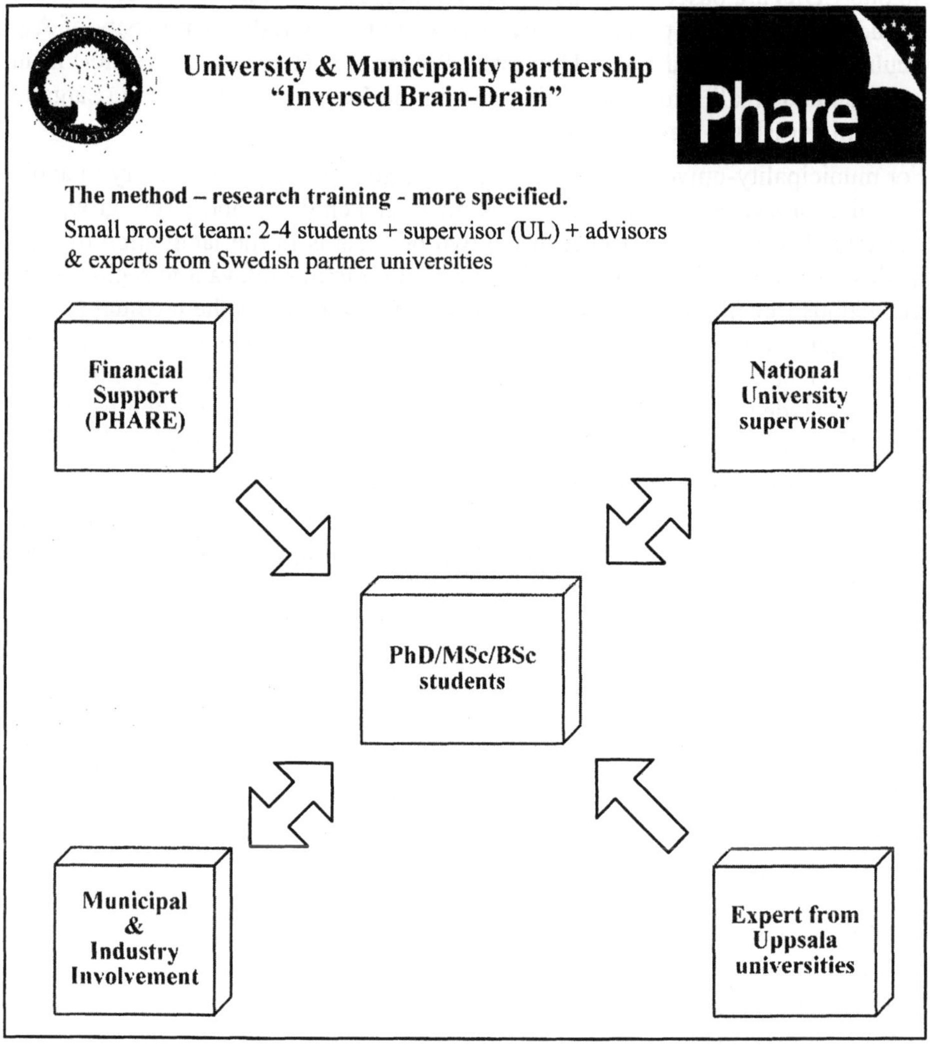

Present day problems are the result of the historical heritage of education and ideology related to 50-years under the Soviet regime and by the difficulties of transition and reconstruction of the educational system. Particularly important is the final stage of university studies – the quality of Master's theses up to now demonstrates an absence of or weak research training in the real environment of RTD or the economy. The problems especially pertain to the social science disciplines. At the same time there is an immediate and urgent need at the municipal level for university graduates trained in solving real life problems related to the economy and public administration of the country.

Both projects at a pilot scale (each project with a budget of approximately 60,000 EUR) are based on transnational municipality-university cooperation and include the following partners: University of Latvia, Swedish University of Agricultural Sciences and Baltic University Programme (Uppsala University) from one side and small remote municipalities in the Eastern part of Latvia (Latgale) – Livani, Kraslava, Dagda, Ludza.

For municipality-university partnerships to be successful it is necessary to avoid formal approaches and to influence as many stakeholders and target groups as possible. The innovative aspect of the two projects is in the facilitation of high quality research and the transfer of innovation on issues relevant to region using indigenous (locally born) PhD, M.Sc., and B.Sc. students as the medium (workforce) and one of the main target groups for local development capacity-building. Under the supervision of experienced researchers from the partner universities students undertake research projects in cooperation with the local municipality and in the process receive hands-on research training that contributes directly to their thesis work. This partnership satisfies the needs of the local student and municipality, as well as the university (Figure 1).

Therefore the projects address all the main target groups of the partners (in municipalities and universities) directly or indirectly:

Direct beneficiaries

- B.Sc., M.Sc. and Ph.D. students coming from small municipalities in Latgale;
- Local authorities – decision-makers, administration, municipal institutions;
- Local business community in the private and public sectors;
- University staff from the University of Latvia and universities from Uppsala – Swedish University of Agricultural Sciences and Baltic University Programme (Uppsala University).

Indirect Beneficiaries

- All municipalities of Latgale as well as Latvia;
- The entire student and academic community of Latvia;
- National business community.

Due to the innovative approach of the projects the academic community of Latvia receives valuable expertise concerning the fruitfulness of partnerships with municipalities, and other municipalities can learn from and replicate the success of this pilot initiative.

The first project targeted the small municipality Livani with the following objectives:

General objective

To demonstrate the effectiveness of local and international university-municipality, university-industry partnerships as an imperative precondition for:

1. Effective transfer of knowledge from academia to public administrations and private businesses and the advancement of a knowledge-based economy in Livani as well as in Latgale.
2. Emergence of an attractive intellectual environment in the regions which among other benefits should *stop or even reverse the "brain drain"* of the most talented students from the region to capital city Riga and abroad, consequently increasing social and economic welfare and competitiveness.

Specific objectives

The project aims to:

- Establish sustainable partnerships and cooperation between Livani and surrounding municipalities, local industry on the one hand and the University of Latvia, Swedish University of Agricultural Sciences and the Baltic University Programme (Uppsala University) on the other hand to facilitate high quality research and the transfer of innovation on relevant issues of the region basing on
- To ensure high level research training of students, exchange of know-how and experience at various levels and disciplines with the involvement of national and Baltic Sea Region universities as well as expertise of other municipalities and universities involved in INTERREG III B Baltic Sea Region complementary project "Baltic University Urban Forum – no. 025 in BSR IR3B.
- To commence the development of a Virtual Regional Sustainable Development and Innovation Relay Centre (VRS-INNO) at the University of Latvia, to increase capacity and promote advanced and wider use of existing and upcoming information and communication technologies in business enhancement, regional sustainable development and public administration.
- To motivate people to work together in virtual institutions.

Main outcomes

- 40 Livani born students were involved in 8 project research teams for research training. In total approximately 160 students were associated with the project in various ways.
- 15 Livani born students received research training in Sweden for a duration of 0.5 to 1.5 months;
- First steps were taken in the creation of a Virtual Regional Sustainable Development and Innovation Relay Centre.

A project team approach was used. In total eight Project Research Teams (PRT) consisting of 2-4 mostly M.Sc. students (40 in total and several B.Sc. and two PhD students), one university academic staff member acting as an advisor, a local expert or advisor from the municipality of Livani and an associated advisor from one of the partner universities in Sweden. Students in the PRT acted as the main work force and in return received excellent research training on selected local development issues under double direct and virtual guidance from local and foreign tutors. It is also evident, that the knowledge base and training of students was enriched by the expertise of the other project partners in the PRT. Selection of M.Sc. students to take part in the PRT was undertaken on a competitive basis. Experts from Swedish universities were important as they contributed in covering still existing knowledge gaps of academic staff in Latvia regarding socio-economic and sustainable development issues pertaining to market-driven knowledge-based economies. Table 1 contains a list of the urgent needs of Latvian municipalities, in particular Livani in advice and RTD efforts.

Tab. 1: Proposed sectors, themes and problem areas for research and research training.

Sectors or themes	Identified problem areas with urgently needed intervention and solutions
High-tech sector: Optoelectronics, Renewable Energy	Research on options to support innovation and development of high-tech industry sector in Livani. Creation of relevant human potential A. Light guides in optoelectronics, health care, environmental protection. B. Development of new methods of production, processing and usage of wood pellets. C. Problems of development and usage of solar panels.
Space-borne data and research on agriculture, forestry resources	D. IT methods of processing of space-borne or air-borne data on agriculture and forestry resources and their sustainable usage.
Business environment; Processing of wood and agricultural products of SMEs; Sustainable Development and Regional planning	E. Development of the business environment for diversification of small and medium-size entrepreneurship in the area; F. Project perspectives for the use of EU agriculture support schemes and EU Structural Funds in municipalities, in particular to support sustainable agriculture, forestry and fishery G. Perspectives in the development of new enterprises and modernization of existing ones for the processing of wood and agricultural products enhanced by the production of higher value added products. H. Business and society – joint social responsibility: what can be done on municipality level. I. Sustainable use of natural resources and preservation of biodiversity in municipalities. J. Sustainable development and regional planning with a bottom-up aproach at the municipality level: • The long term goals and visions for the year 2025; • The medium term strategic objectives for 2010; • The short-term priority actions for the next 5-6 years (2006).

Sectors or themes	Identified problem areas with urgently needed intervention and solutions
Mobilization of resources and attractive social and intellectual environment	K. Local community development approach, intellectual capacity and brain-drain problems. L. Rich natural and cultural environment as a basis for a good and healthy social life in the municipality. M.Family training for new employment generation, promotion of ethnic participation and integration.

Finalisation of research training efforts of all the PRT revealed unexpected practical outcomes in terms of raising professional skills and the knowledge-base for students, contributions to curricula being developed by involved university staff and experts, and new promising network activities across the Baltic Sea Region. The work completed to date will result in new co-operation arrangements, projects and will contribute to the emerging European Research and Innovation Area in the Baltic Sea Region. In general one can say that such outcomes are logical and more or less foreseen concerning the academic community, but outcomes for the public and private sectors of Livani are very surprising and encouraging. The synergy with the previous INTERREG SUPERBS and the still running INTERREG BUUF project has created needed preconditions for investments of 1-2 millions EUR coming to Livani in nearest years and possibly transforming Livani into a "high-tech" industry centre in Latgale.

The project sent a serious message to involved municipal stakeholders about significant changes taking place in the intellectual environment of the municipality and the perspective emergence of well-trained top-level human capacity in the municipality. The project identified approximately 100 university level students that have their roots in Livani. About 40 expressed a willingness to co-operate with the project and in various ways contributed to the project. 15 students used the opportunity and benefited from research training visits organised and supported by the project's Swedish partners. Eleven university teachers and researchers, three Ph.D. students among them, acquired new experience, training, an improved curricula, and were involved in studies of the problems of Livani. Additionally, about 15 Swedish experts, professors, researchers and five municipal experts acquired knowledge regarding Livani and some have expressed a readiness to support development in Livani in the future.

The efforts of four PRT resulted in comprehensive project outlines for the public and private sectors in the municipality: Development of Bicycle-friendly Livani, Development Towards Solar Panel Production in Livani, Cultural Links between Livani and Hellefors. The work of the other two PRT could contribute substantially to the development of the renewable energy sector in Livani and bring new large-scale investments in optoelectronics.

Additionally, the project resulted in the creation of a knowledge repository, a VSD-INNO final structure design and start-up of several Virtual Institute Departments – future project generators.

The final conference of the project in September 20, 2004 in Livani was a great success. Conference participants constituted a very strong multidisciplinary team of 10 doctors of sciences, seven leading Livani municipality experts, a leading innovation expert from Riga, as well as more advanced Livani born students. The merit of the "INVERSED BRAIN-DRAIN" project as well as of the project final conference is a clear demonstration of how impressive intellectual capacity can be mobilized to foster the development of a small municipality if appropriate instruments are used, motivation is created and an attractive and creative environment is designed. Apart from various excellent contributions and presentations during the conference plenary sessions, the final round-table discussions require special mention. Panelists of the round-table were confronted with an audience searching for ways to raise the human capacity of Livani and surrounding territories above the threshold of "critical mass" needed to open opportunities for the intensified emergence of "high-tech" industry in the town and to ensure continuity in the process of human resource development, which would contribute to a rapid increase in the welfare of residents and the development of lifestyles compatible with sustainable development. Several important conclusions from the round-table discussion should be mentioned:

- Strong municipal support for studies and research training of locally-born students is required;
- The municipality should provide support to advanced students as well as their local teachers;
- It is essential to ensure good starting conditions for young talents together with continuous monitoring of their progress;
- Innovations are always created by top-level specialists and the municipality should make serious efforts to retain such people, to support them and to use physical and virtual space to maintain contact with locally-born talents both in-country and internationally;
- The intellectual environment of the municipality is crucial for future development. Alumni Clubs should be created and traditions to target long-term goals developed;
- Targeted and aggressive municipal policy towards human resource development should be implemented based not only on keeping local talents in mind, but also in making efforts to ensure "brain circulation" to Livani from other parts of Latvia;
- Emergence of the Joint European Research and Innovation Area and a common labour market in the field of RTD development should be considered an

ideal opportunity to move Livani towards capturing the leading position in the "high-tech" sector in Latgale region;
- High-tech industry oriented technical college together with relevant support institutions should be opened in next two years in Livani.

The experience gained to date will be put to use in the next project /6/, which will cover several Latgale region municipalities.

References

1. COM (2001), 549 final "The Regional Dimension of the European Research Area".
2. EUROPEAN REGIONAL DEVELOPMENT FUND, INTERREG II C, Community Initiative concerning Transnational, Co-operation on Spatial Planning 1997 – 1999, SUPERBS – Sustainable Urban Patterns around the Baltic Sea, Model project 3 – Livani, LATVIA.
3. Baltic University Urban Forum – Implementing Sustainable Development in Urban Areas, INTERREG IIIB, no. 025 in BSR IR3B.
4. Conclusions and recommendations of the EU-Conference: "Brain Gain – the Instruments". The Hague, 29-30 September 2004.
5. Co-operation Programme in the Baltic Sea Region, PHARE 2001, grant contract LE01.07/SPF/0023 (LU 2000-2033). Human Resources for the Economic Development of Livani District and Surrounding Territories Basing on University Municipality Partnership. Acronym: "Inversed Brain-Drain".
6. Co-operation Programme in the Baltic Sea Region, PHARE 2003, grant contract no. MPF14 (LV2003/005-876/01) Human Resources for Know-ledge Society in South Latgale Based on Transnational Municipality University Partnership. Acronym: "Knowledge Transfer".

The Local Agenda 21 Change Agent

–

Why, What, Who, How?

Bjarne Rasmussen

International Director & Local Agenda 21 Coordinator,
Storstroem County, Department of Technology and Environment,
Parkvej 37, DK – 4800 Nykoebing F., Denmark

1. The importance of Local Agenda 21 is not understood

Local authorities are pivotal in creating sustainable societies. This fact is stressed in Agenda 21, chapter 28, defining the Local Agenda 21 as a new approach. In several contexts even this key function of the local authorities are completely underestimated, or even forgotten. This latter is for instance in the case of the Agenda 21 for the Baltic Sea Region, Baltic 21, from 1998, and also in the revised Baltic 21 Strategy from 2004 there is no special emphasis on Local Agenda 21.

Most of the national Agenda 21s or Strategies for Sustainable Development in the BSR do mention Local Agenda 21, but it is difficult to detect the commitment and actions of the governments that agreed on the importance of Local Agenda 21 in the Joburg Plan of Implementation, and thereby promised to support the local implementation of Agenda 21.

What is a "real" Local Agenda 21?

- an action plan has been developed and adopted,
- it is an result of a multi-stakeholder consultation process, and
- awareness on SD has been raised

2. The Local Agenda 21 Change Agent is the Key

The function of the Local Agenda 21 coordinator is one of the most crucial success factors. Also this fact seems to be heavily underestimated, even though everybody working on the ground – in the real world – with the development and implementation of local and regional action, processes and projects, are aware of this.

This article deals with the function of Local Agenda 21 coordinators and other key actors that may play the role as change agents, be it coordination, innovation or facilitation of processes – in the wider community, in businesses, in the educational systems, or even within the local governmental organizations.

Based on cases and observations a number of recommendations will be described, with regard to:

- The organizational set up of the Local Agenda 21 structures and processes,
- The role of the Local Agenda 21 Coordinator and other Change Agents,
- The competences needed by Local Agenda 21 Change Agents, and
- The type of education and training needed in order to strengthen the competences of the Change Agents and build the capacity of the implementation structures, in which the LA21 Coordinators and other Change Agents are embedded.

3. Roles and relations

3.1 The Role of the local authority

- Overall responsible of the Local Agenda 21-processes
- Ensuring citizen involvement /public participation
- Cooperation with local actors /stakeholders
- LA21 Strategies and Action Plans

3.2 The Relations of the local authority

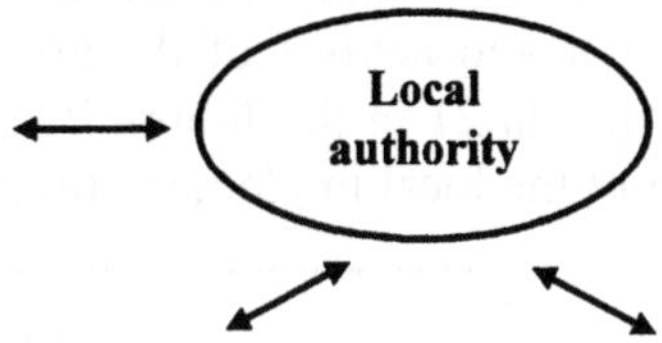

- All societal actors in the community and the region are to be involved:
  - citizens and citizen groups
  - companies
  - institutions
  - associations and organisations
  - other local authorities
- In specific processes and projects: special target groups /segments should be involved in a targeted way.
- In addition to this: cooperation and networking with actors from other parts of the country and in other countries can play an important role as inspiration and resource base.

3.3 The internal culture of the local authority

- The necessary culture is characterized by project culture and cross-departmental interplay.
- The traditional hierarchic structure and culture supplemented by networking.

3.4 Roles of the Local Agenda 21 coordinator

- Coordinator
- Ensuring citizen involvement
- Facilitator (development and decision-making processes)

- Policy-maker (decision-making preparation and alternatives)
- Interpreter /Adviser /Counsellor
- Idea developer /shaper
- Project leader, team leader
- Campaign leader, event manager (awareness raising + action)
- [Trainer /coach] (competence development in the organisation and in the community)

4. Competences

4.1 Competences of the Local Authority – Which?

Key preconditions for LA21 processes:

- A LA21 organisation has been established
- Focus on citizen involvement and interplay with local actors
- LA21 competence development is taking place throughout the entire organisation

- The principles of sustainable development is systematically integrated in all policies, plans, programmes, processes and projects
- Management for sustainable development is to be developed (as an umbrella for other management concepts)

4.2 Local Agenda 21 competences – Which?

- Understanding and overview (complexity, different cultures, thinking across sectors and disciplines, sense importance)
- Think future (understand trends and attitudes, imagine alternative futures)
- Political touch (policy-making, choose "good" themes, timing)
- Coordination & Organisation (processes, involvement of many actors, balancing interests)
- Creativity & Innovation (development of concepts, unique products og processes)

- Communication & Dialogue (targeted information, listening, networking, knowledge sharing)
- Facilitation /process management (motivate, create confidence, vitalize processes, shape resultants)
- Action /realization (convert ideas into concrete projects and activities)
- Public Communication (create interest, sense for events and the good story)
- Project Management
- Work in – and perhaps lead – teams
- Cross-sectoral thinking and action (environment – culture – health – education – business etc.)

- Understanding organisation culture (and knowledge on management concepts and processes)

4.3 Local Agenda 21 competences – How?

Important starting points:
- WHO AM I /ARE WE?
- WHAT CAN I /WE?
- WHAT DO I /WE WANT?

A focus on personal competences and skills, by the means of
- Tests (e.g. personality, team)
- Exercises (e.g. creativity, empathy, dialogue, interplay, techniques)
- Training (e.g. courses, on-the job-training, in nature, in other organizational cultures)
- Workshops (future, theatre, practical tasks etc.)
- Projects (learning by doing)
- Study tours /Excursions
- Knowledge sharing
- and much more!

and next, these personal skills and competences should be combined with social, communicative and technical competences and skills.

Environmental Management (EM) and Sustainable Development (SD) in University's Strategic Management
–
Motivation and Experience from the Hochschule Zittau/Goerlitz – University of Applied Sciences (F.R.G.)

Bernd Delakowitz

University's EMS, Hochschule Zittau/Goerlitz (FH)
– University of Applied Sciences,
Theodor-Körner-Allee 16, D – 02763 Zittau, Germany

Abstract

Educational institutions in Germany currently suffer several problems such as limited budgets and, in some parts of Germany, the consequences of limited demographic growth. Eastern Germany also experiences a considerable out-migration of young people to the west. The serious threats posed by reduced student numbers after 2007/08 and increasing competition means that universities need to pursue strategies to make themselves more attractive.

The University of Applied Sciences Zittau/Goerlitz has chosen to do so by developing an individual university profile focusing on a policy of sustainability and by introducing an environmental management system.

This paper demonstrates the experiences and practical examples from the University of Applied Sciences Zittau/Goerlitz, which looks back on more then seven years of successful and intensive work as the first institution of higher education in Germany with a validated and registered environmental management system (EMS). The EMS and the SD profile are considered to be suitable instruments of competitiveness and safeguards for the university's profiling.

Introduction

The University of Applied Sciences (Fachhochschule) Zittau/ Goerlitz, was founded in 1992 based on a former well renowned University School of Engineering for Electronics and Information Processing. It is located in the eastern part of the Federal State of Saxony right where the Polish, Czech and German borders meet. The University is a resident at two locations – Zittau and Goerlitz – with a distance of 35 km between them (Figure 1). The university has some 400 personnel and close to 4 000 students. The Zittau campus has a focus on engineering and natural sciences, while the Goerlitz campus is dealing with social and computer sciences (see Box 1). At present (2004) 5,000 m^2 of new laboratories and lecture halls are being built. The external funding for R&D amounts to ca. 4.2 Mio € p.a; of this some 42% is environmentally related.

Fig. 1: The University of Applied Sciences Zittau/Goerlitz is located in the southeast of the Federal State of Saxony at the Polish and Czech borders

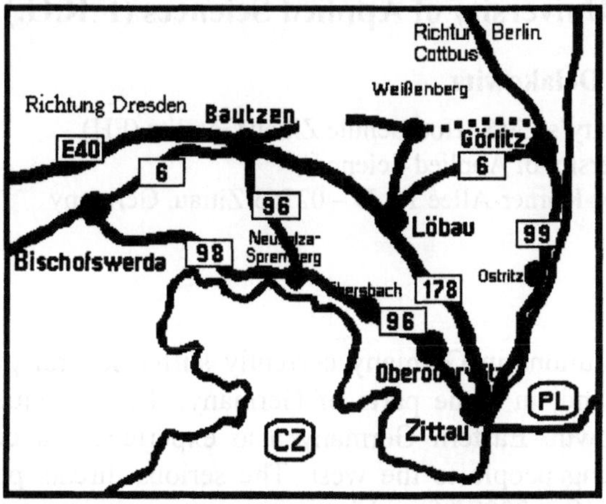

The area has undergone already one big change in 1989 due to the reunification of Germany and had been facing another major change on May 1st 2004 when the EU was enlarged by 10 more mostly eastern European states. The most challenging consequence so far is the negative demographic development caused by a dramatic move of jobs to western Germany after the break-down of the former German Democratic Republic and large portions of its former industry (textile, power generation, heavy machinery, mining). Consequently a huge and still increasing number of mostly young, well educated and skilled people have been and still are moving away leaving the region with an unemployment rate of currently more than 22%. About two thirds of graduates of the University of Applied Sciences Zittau/Goerlitz are migrating to work either in western Germany or other parts in Europe which – in other words – means that the Hochschule Zittau/Goerlitz is investing all its strengths, efforts and resources for the education and training of students for the benefits of far more wealthier parts within Germany.

Innovation and strategies to maintain and to improve both the positive image based on very well established R&D as well as high quality education and the attraction of the university are needed. The particular region in the heart of Europe may, therefore, serve as a good small-scale case study to highlight the positive effects of future oriented management strategies such as Environmental Management and Sustainable Development.

The Environmental Management System (EMS)

The Environmental Management System of the University of Applied Sciences Zittau/Goerlitz was introduced in three steps: (1) The University senate decided on an all over environmental policy in 1995. (2) In 1997 an Environmental Management Representative was appointed by the rector and the EMS team started the implementation process. The first check of the university's environmental performance was conducted in 1998 based on internal and external validation and certification audits. (3) The validation procedure according to the EMAS[1] requirements was concluded in March 1999 when the university's approved EMS was registered by the Industrial Chamber of Commerce (Industrie- und Handelskammer, IHK) at Dresden. The Hochschule Zittau/Goerlitz then became the first university in Germany and one of the first in the world to be fully certified and registered.

In the following years a considerable amount of work has been done to maintain the EMS. A second audit and certification was made in 2001, now according to the amended EMAS II[2]. External Supervisory Audits were carried out in 2002, 2003 and 2004. In spring 2005 the university achieved its third EMAS-registration. From the very beginning students' work has been an important part of the environmental management at the university (see Table 1).

Tab. 1: Time schedule and individual steps (milestones) of implementing the EMS at the University of Applied Sciences Zittau/Goerlitz (F.R.G.)

1995	1st Senate decision		
1997	Founding of the Environmental Management Team		
1998	First environmental check 1998 – Formation and implementation of the EMS		
3/1999	**First validation** and registration according to EMAS I. Audits at relevant areas	assisted by	Student's project works
2000	Audits at relevant areas and buildings inspection	assisted by	Student's project works
2001	Audits in relevant areas **External supervisory audit**	assisted by	Students' project works Audits by students
2/2002	**Second validation** according the new EMAS II requirements (EG 761/2001)	assisted by	Student project works Audits by students
2003	Audits in relevant areas **External supervisory audit**	assisted by	Students' project works Audits by students
2004	Audits in relevant areas **External supervisory audit**	assisted by	Students' project works Audits by students
3/2005	**Third validation** according the EMAS II	assisted by	Student project works Audits by students

1 Council Regulation (EEC) 1836/93 of 29 June 1993.

2 Regulation (EC) 761/2001 of the European Parliament and the Council of 19 March 2001.

The annual monitoring of specific energy, material and waste flows as well as the determination of specific emissions resulting from individual university processes provided the EMS team in the beginning with fix starting points (reference values). These values have later allowed them to determine and measure continuous progress, and where needed to improve and correct actions or conduct research.

The success of the environmental management system was determined by the inclusion of as many active participants as possible. This has been made possible by a variety of actions in the course of introducing the environmental guidelines, and has strengthened and stabilised these guidelines. Periodic audits, inspections and students project work have been conducted to test and realise the environmental strategy since the first EMS run in 1998.

EM and Health & Safety

A board for *Environmental protection and Health & Safety* (Figure 2) was established in 2002 as part of the university administration. It consists of the representative for the EMS, the safety inspectors, and the radiation protection representative (because the University of Applied Sciences Zittau/Goerlitz is operating a small nuclear reactor for education and research with an engine power of 10 Watt). Also included is the staff council.

The internal documentation, required according to the EMAS regulation, has been integrated in the exiting university administrative manual. This has proved to enable an effective handling of the EMS and to use synergies between older relevant documents and the new environmental process instructions and other EMS documents. The external documentation is summarised and integrated in the annually published environmental declaration according to the EMAS requirements.

Fig. 2: Organization of the Environmental Management and Health and Safety Management at the University of Applied Sciences Zittau/Goerlitz

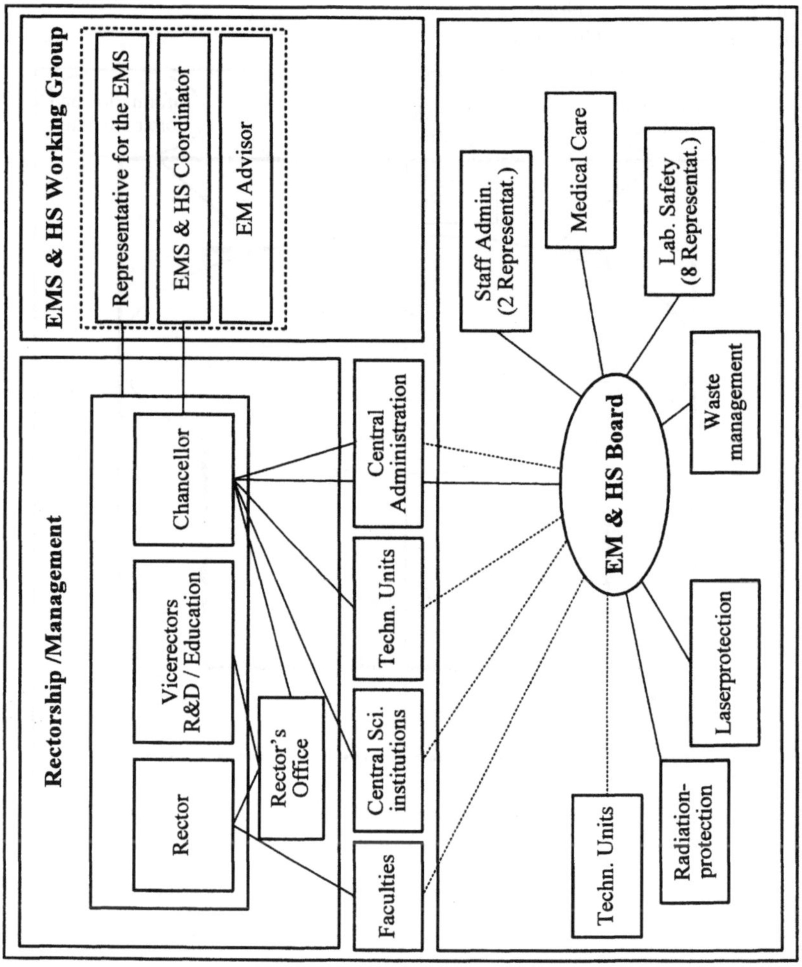

The Sustainable Development (SD) Policy of the University

In July 2002 the senate of the Hochschule Zittau/Goerlitz passed the "Strategic Profiling Concept" officially declaring "Sustainability" as a dominating guideline in its education and research policy. As a result the university is involved in a variety of initiatives in both social-economic and cultural areas of communal SD. The sustainability profile is based on three sub-profiles (economy, ecology, society) and 6 core competences (which are more or less reflecting the individual faculties) and is completed by centres of excellences. All SD-linked elements are supported by the central management ofhe university (see Figure 3).

Fig. 3: The strategic SD profile of the University of Applied Sciences Zittau/Goerlitz

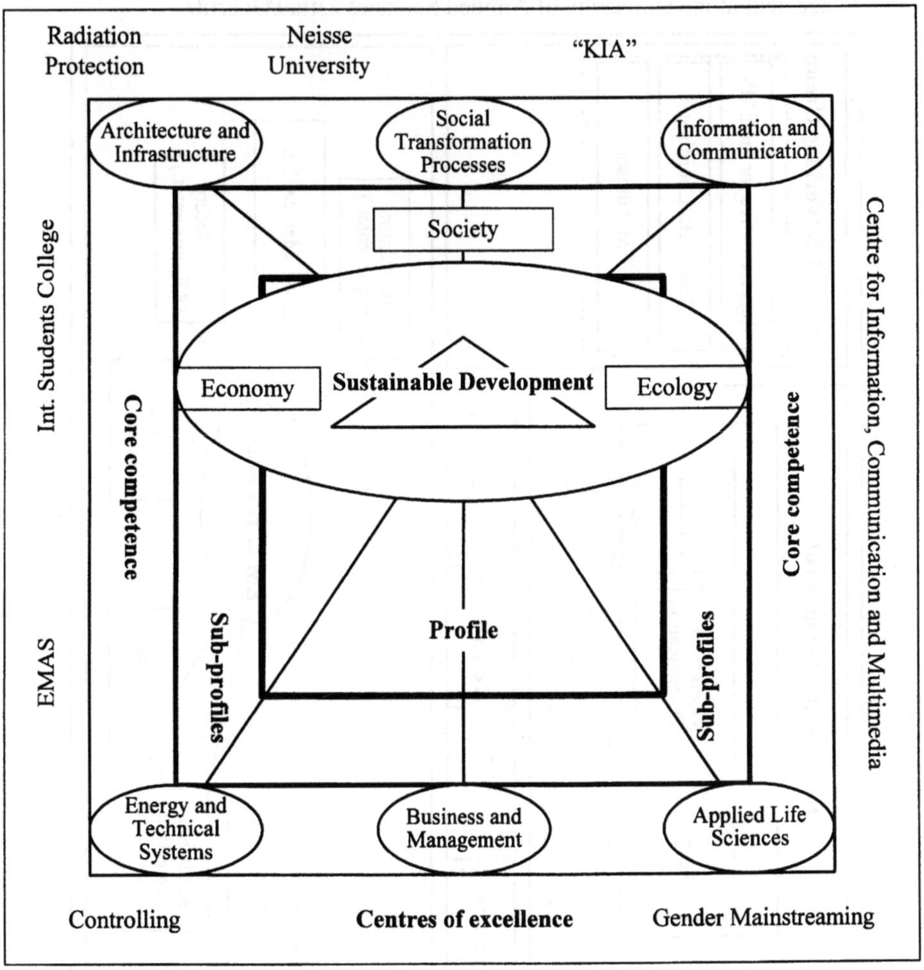

Introducing EM and SD into the University's Everyday World

Information and Communication

Staff and students of the university are continuously invited to become active in the environmental management process. A number of events are used as a platform for information and motivation, e.g. meetings of the university senate, assemblies of personnel and staff and faculty colloquia. On these occasions information is given about the state-of-the-art of the EMS and everyone is invited to exchange views on the environmental performance within the university.

The university also communicates with comparable external institutions. An important objective is to continuously optimize and adapt the management system to varying external conditions such as legal requirements.

Information on the environmental management of the university is made through electronic publications (Internet), conferences and seminars open to the public, newspaper articles and publications in scientific journals and books. An important channel is the annually published EM declaration of the university.

A special and today well-known event is the annual Environmental Protection Day linked to the United Nations annual Day of the Environment (June 5). This was started in 2001 focusing on (global) energy demand and supply including nuclear energy as well as the disposal of radioactive waste, an issue of highly political importance, but too often discussed emotionally rather than by considering scientific, economic and social facts. The 2001 seminar also featured well-known lecturers, e.g. Prof. Stefan Rahmstorf, an international renowned researcher in the field of global climate change. The 2002 seminar focused on biotechnology and bioethics (the university has started a new Bachelor course in biotechnology in 2001) and in 2003 we organized an elective short course for students on sustainable development in the framework of the Baltic University Program, and a colloquium on the prevention of flooding, in the aftermath of the so called "Century Flood" of August 2002. This caused enormous damage particularly in the State of Saxony where the Hochschule Zittau/Goerlitz is located. The 2004 Environmental Protection Day focused on waste management.

Education and Research/Greening the Curricula

In the education at the university environmental issues are introduced both in special programs and as parts in all programs. According to a decision of the university senate in 1993 all students, no matter what their specific studies are, have to participate in a one semester (2 classroom hours per week) lecture on fundamentals of ecology, environmental protection and sustainability as an integral part of their individual courses curricula. This lecture course, which is quite unique amongst German universities, addresses biological, ecological, technical, social and economical concerns and dynamically refers to current events. During an academic year an average of 550 students of various faculties and courses attend the lectures. The students become more aware of socio-economic and ecological issues and understand better complex interrelations between natural and anthropogenic processes. Some of those participating in the lectures may serve as multipliers in their future professions and private lives and may, thus, constructively contribute to environmental protection and sustainable development.

Fig. 4: Environmental oriented lectures, courses and other areas of study in various faculties of the Hochschule Zittau/Goerlitz. The lecture course on Fundamentals of Ecology, Environmental Protection and Sustainability, an integral part of the curricula of all faculties, is marked in the middle column

	Areas of Study				
Faculty	Course of Study	Focal Point Education	Individual Lectures	Dissertations	Projects
Architecture, Construction Engin		x	x	x	x
Electrical Engineering		x	x		
Computer Sciences			x		x
Mechanical Engineering	x		x	x	x
Mathematics/Natural Sciences	x	x	x	x	x
Social Sciences			x		x
Languages			x		
Economics	x		x	x	x

A four years study program on "Ecology and Environment Protection" leads to an academic degree Diploma in Engineering. The program includes (i) technical environmental protection, such as waste management and recycling technique, (ii) natural conservation and spatial planning, and (iii) environmental and sustainable oriented business management with particular consideration of health and safety risk management.

Finally all faculties offer projects, which include environmental subjects within the scope of their individual study programs (Figure 4) and projects for which interdisciplinary work is called for.

The research profile of the university also mirrors an interest for environmentally relevant fields of teaching and research. These include building construction and the environment, energy efficiency and the environment, nuclear safety, and protection of the natural environment and natural resources.

Student's Participation in EM and SD Processes

Tell them, and they will forget,
Demonstrate, and they will remember,
Involve them, and they will understand.[3]

This guiding idea has been a key principle for education at the University of Applied Sciences Zittau/Goerlitz and the implementation and maintenance of the university EMS. The principle of practical learning and taking over individual responsibility by being actively involved has been part of the university environmental policy since 1998. This shall be achieved by

> „qualifying and enabling students to play an active role in a continuous process of improving the environmental performance of the Hochschule Zittau/Goerlitz, thus, to promote their potential of creativity and innovation"[4].

Between 1997 and 2003 an annual average of 30 to 40 students, mainly from the "Ecology and Environmental Protection" course, have been participating in all steps of the university EMS by either assisting in internal audits or performing projects focusing on the improvement of the environmental performance. The student projects have become both an integral part of the curriculum (courses on Life Cycle Assessment and Environmental Management) and part of the annual environmental checks within the EMS of the university. Students have, under supervision of lecturers, been involved with tasks such as

- Acquisition and determination of the university energy and water consumption
- Managing waste production and emissions using defined environmental measures and codes
- Participation in preparation, maintenance and application of an administrative disposal guideline
- Studies on technical measures on specific facilities in order to improve energy efficiency.

Students from the "Ecology and Environmental Protection" program have since 2003 formed small teams (10 students yearly) carrying out project studies on EM or related issues on both internal university and external industrial (in particular SMEs) related tasks. Thus one team worked in 2003 on a two semester project in cooperation with a regional market garden (SME) to reduce its consumption of heating energy. The students developed an ecological sound and economical favourable solution in designing a new type of wooden chips fuelled heating installation. The company installed this system in the summer of 2004.

---

3 Confucius/ Epigraph to the 2nd European Conference of Sustainable Cities and Towns, Lisbon, 1996.

4 The Environmental Policy of the Hochschule Zittau/ Goerlitz (FH), 1st guideline (www.hs-zigr.de/ verwaltung/ ag-um).

Students always have been and still are playing an important role in preparing and assisting internal audits of the university environmental and safety management system. Those exercises aim to enable students to apply theoretical knowledge, from lectures or in the literature, in solving practical problems. Results are presented and discussed in seminars and are integrated into the EM documentation and program.

Some Results of the EM and SD Work

Cost Savings and Expenses Related to the EMS

So far the economic consequences of the environmental management has not been systematically monitored. Cost reductions have so far not been the predominant motivation for universities to implement an EMS. Of course, this will be and in parts has already been changed.

We may however estimate savings due to improved energy management. The costs (for both electricity and heating) have up to 2004 stayed on the 1998 level although the number of students have increased since then by approximately 20%, new laboratories and lecture halls with an equivalent of about 15% of the beoverall university area have started to operate since, and the energy price has increased by nearly 15%. The relative cost saving linked to the EMS is then approximately 25%.

The re-organization and re-contracting of waste disposal as a result of the environmental checks has also lead to substantial cost savings. Higher safety and risk reduction on the handling of hazardous substances and biological materials have also led to lower costs.

Other Benefits Related to EM and SD

Organizations are repeatedly reporting benefits other than economic ones from implementing an EMS and introducing the principles of SD. These include improved understanding of internal processes, improved information and communication, motivation and quality of teaching and research. An organization, not only a university, which properly applies EM and related management systems such as QM and health & safety management may achieve other advantages. These include the ability for quicker and more flexible reactions towards modified market situations and better conditions to meet stakeholder expectations.

In the case of institutions of higher education competing in a situation of negative demography factors which force, too, into the race for competition Environmental Management and Sustainable Development may also help to improve the university's attraction.

References

Delakowitz, B. & Hoffmann, A. (2000), The Hochschule Zittau/Goerlitz – Germany's First Registered Environmental Management System (EMAS) at an Institution of Higher Education. Int. J of Sustainability at Higher Education, vol. 1 no. 1, 35-47.

Delakowitz, B., Hoffmann, A. & Will, M. (2005); Vom operativen Umweltmanagement zum Leitmotiv Nachhaltige Entwicklung. Das Beispiel der Hochschule Zittau/Goerlitz (FH). In: Leal Filho, W. & Delakowitz, B. (Hrsg.): Umweltmanagement und Nachhaltigkeit an Hochschulen, Peter Lang Verlag.

Delakowitz, B., Leal Filho, W., Hoffmann, A. & Will, M. (2005), Sustainability as a tool towards an improved university profile – Experiences, practice and expectations at the Hochschule Zittau/ Goerlitz (FH). ENCOS 2004 Conference READER (www.encos2004.net), Berlin.

Leal Filho, W. (ed.) (1999): Sustainability and University Life. Environmental Education, Communication and Sustainability, vol. 5, Peter Lang.

– (2000), Communicating Sustainability. Environmental Education, Communication and Sustainability, vol. 8, Peter Lang.

– (2002), International Experiences on Sustainability. Environmental Education, Communication and Sustainability, vol. 12, Peter Lang.

Leal Filho, W. & Delakowitz, B. (eds.): Umweltmanagement und Nachhaltigkeit an Hochschulen, vol. 18, Peter Lang.

Leal Filho, W. & Hoffmann, A. (2004), Flyer for the student course "Principles and Practices of Environmental Quality Management and Sustainable Development", November 2004.

Leal Filho, W., Macdermott, F. & Padgham, J. (eds.) (1996): Implementing Sustainable Development at University Level. CRE-COPERNICUS, European Research and Training Centre on Environmental Education, University of Bradford, UK.

Leal Filho, W. & Sauer, P. (eds.) (2002): Towards International Environmental Cooperation, The German-Czech Environmental Information Programme Seminars, Hamburg.

City as a Vehicle of Education

Krzysztof Nawratek

Dr. eng. arch., PhD at The Silesian Technical University
Katowice – Poland/Riga – Latvia

> "The city is the original, the first, archetypical medium of interaction. The new media are simply an extension, a completion, of the city as the locus of interactive life. In the village, each person has a place; in the city each has instead an address, an address that permits options, random access, choice"[9].
>
> "… social capital, connections, social networks are much more correlated with human happiness than is financial capital" [12].

In my opinion there are two main, secular institutions on which European culture and civilization is based. One is the University, the other is the City.

There are several features which are characterizing both the city and the university.

They are: concentration/ density of people and human interrelationships, high density of information, sharing of information and experiences, mixture of official, hierarchical organization and unofficial, net-organized groups.

However, the most important and amazing feature is educational aspect of these two institutions. This is obvious when one is talking about the university. Is it worthwhile to consider educational aspects of the city?

Definitely yes.

Before we start the proper analysis we should explain the dual nature of the city.

As we know, the city has both a material/physical and a social body.

> "How does society produce forms and the forms reproduce society? What roles do history and social institutions play in generating the built environment? What is the relationship between space and power" [7].

These are crucial questions not only for anthropologists, but also for everyone who is dealing with society and space.

The process of producing the material city by the people and reproducing society structures by the material city is the most fascinating aspect in studying urban problems.

This process – which is tightly connected to some kind of "memory of the city" we can consider as educational.

There are three main educational aspects of the city.

The city is obviously also an area where institutional education could take place. Huge amount of people and artificial "containers" of information somehow provoke and support educational processes. Sometimes, authorities realize the power of the city and advantage of education and organize schools, universities, museums and other culture institutions in so called "education zone".

The city is the space of interaction.

The "social capital" theory enables us to explain the economic power of cities rather as a result of nonmarket interaction than vice versa.

Some urban economists neglect this aspect of city,

> "...a brilliant theorist can explain cities without addressing nonmarket interactions. But it is less obvious to me why one would want to do so. The flow of ideas and values that occurs through face-to-face interaction may be the most interesting feature of cities. Furthermore, the market for space – that is, the real-estate market – also appears to be driven to a large extent by the relative presence or absence of particular forms of nonmarket interactions in different areas" [4].

This fact is an important argument pro high density areas. However, we should describe this kind of "density" not only as a density of population, but also as the area of rich and diverse interrelationships. This fact is, by the way, one of the most important arguments against car-oriented cities and pro pedestrian areas.

> „Walking to an appointment in a European capital (indeed, in most of the world's cities) can be more pleasant than a drive to achieve the same end in a North American metropolitan area. In the former case, one sees other people, some of whom one might wish to talk to; observing others may provide clues on social currents and interactions; window displays provide information on available products and services, etc.(...). Drivers are subjected to unwanted information from billboards, while they choose aural information from music, news/talk radio, or chatting on their cellphone. Again, the primary exchange is loaded with secondary (wanted or unwanted) information" [2].

Yes – urban environment is "broadcasting" huge amounts of information.

However, the existence of information is not enough. If we use a metaphor and describe "material city" as a huge library, or more contemporary – as an "analog-internet", it would be clear that users of the city should have ability to learn and understand and should have any kind of "interface".

This is the crucial problem.

Traditional society had some common ideological foundations. For this kind of society, city's symbolic language was quite easy to learn and understand. It was not coincidence that gothic cathedrals were called "stone catechisms". This fact shows that the teaching role of architectural artifacts was obvious at that time.

Nowadays, the situation is different.

Contemporary society does not share common symbolic language. Urban environment became to some extent mute.

However, city is not voiceless. It is constantly "broadcasting" information. The problem is that understanding of this information is extremely subjective.

As Kevin Lynch explored by “mental maps” research, urban space is slightly different for each person. Differences between members of the same community are not so big but “mental maps” of members of different communities, despite living in the same territory, could be extremely varied. In some cases the result is so different that, in fact, communities live somehow separated from each other; living “next door” to each other but apart. It is kind of a mental (or in more drastic situation – real) ghetto.

But sometimes the same area is important for different communities, and various people try to possess – in real or symbolic way – the same territory, the same spatial artifact. For example in the case of Jerusalem, which shows how dangerous such situation can be.

The process of “teaching” has one more important aspect. People who grew up in a specific urban structure tend to repeat a model of interpersonal relationships which they already know.

The example of Riga is significant, although not unique.

Majority (almost 60%) of Riga population live in post-soviet panel housing. This type of settlement, despite “The socio-political idea of a homogeneous “middle-class society” in the East and in the West, as was the case after World War II, was supposed to be realized on the spatial level through uniform apartments for nuclear families” [5]. This reduces the number of interpersonal, non-market relationships.

In this context, panel-block housing is a very un-urban environment.

Because of a lack of defined semi-private and semi-public spaces this type of urban structure somehow “is teaching” residents how to live apart from each other. Definitely, it is neither “teaching” nor “nurturing”. It is not creating an environment in which people can learn how to be part of a healthy society, how to cooperate with each other. There are just no places *ergo* no occasions for this type of interactions.

The result is that nowadays Riga people prefer to repeat this “egoistic” style of living. They would still like to live in high-rise buildings (of course – luxurious versions of the Soviet ones) or in one-family housing.

Exactly the same trend exists in every city where panel-housing is a significant percentage of residential areas – “The shrinking of East German cities was further exacerbated by increasing suburbanization – anyone who had money built a little house in the countryside on cheap land in the outskirts” [5].

However, because the percentage of panel-block residents in Riga is the highest in Europe – the problem is very dramatic in this city.

This anti-urban approach seems to be one of the most important factors which could destroy Riga as a city.

This is not only a theoretic and abstract problem. As we mentioned at the beginning of this text – "dense nonmarket relationships" which can only appear in "dense urban environment" are a real stable background for economic grow of the city.

The conclusion is very simple – no urban environment means no interpersonal relationships. This would cause a decline of "social capital" of the city and the decline of the city itself.

The city is learning.

It has some kind of memory, or even two types of memories. One is the memory of the society; the other – even more fascinating – is the memory of the city itself. The second type is saved in the urban structure and buildings.

The process of how spatial artifact becomes part of city memory is extremely interesting.

Some buildings or places are important from the very beginning – however, it is a very rare situation. Such a "building-sign" often has national or religious meaning: monuments, parliamentary buildings, cathedrals etc. However, the majority of buildings became important for people because of the history. Not necessarily because of great historical events; very often just because passing time "covers" spaces and buildings by meanings and stories.

It is important to stress that some buildings and places are "meaning-friendly" and some are not.

Definitely "traditional" aesthetics and traditional urban structure are more supporting of this process than modernist ones.

However, new, progressive architecture and urban design could also be suitable "containers" for stories and meanings. There is only one condition under which it can be successful: new urban structures should sustain interpersonal relationships and a level of information broadcast should be high enough.

In my opinion only under this condition is spatial artifact "prepared" to learn.

It is possible to analyze separately how memory – or even both types of memory work. After the Second World War some cities almost completely changed their population (e.g. Wroclaw/Breslau, Vilnius), and some cities lost almost completely their "physical body" (e.g. Warsaw, Dresden).

The process of rebuilding these cities shows perfectly a tight connection between manipulation of urban structure and architecture and a manipulation of history.

The example of rebuilding of the Old City in Warsaw is very significant.

Communist government rebuilt Old Warsaw not as it looked before the 2nd WW, but its urban structure from the 18th Century. Almost all "capitalist" urban tissue from 19th and beginning of 20th century was not rebuilt [6].

Actually, it is only my hypothesis, which should be proved in deep sociological and anthropological research, but I am deeply convinced that the process of "exchanging" citizens could show how deep urban/architectural structures can influence the society.

The examples of Wroclaw/Breslau and Vilnius show that rich and dense existing urban structure has power to "shape" new citizens, new society in a form which would be quite similar to the society which created it.

The educational aspect of the city and understanding the city structure as a form of memory provoke rethinking urban planning process.

From this point of view, each planning decision and piece of information coded in buildings and urban structure should be considered as an appropriate or inappropriate element of the process of educating the society. In fact, it has to be considered as the element of education or the element of manipulation.

We are talking exactly about some kind of "social engineering". No building and no planning decision is neutral. Each of them just pushes or pulls society somewhere. People who create an image of the city should be aware of long term results of each their decisions.

If we are talking about sustainable development we must not neglect these educational aspects of cities.

The city as a space of education should be supported – it means that all educational and cultural institutions should be organized as a system, tightly connected to existing city structure – to its history, society and spatial structure.

The city should "learn" what the healthy society is. Only buildings and other spatial artifacts which are "meaning-friendly" should be supported. Spaces able to support nonmarket interactions should be developed.

After that, we can hope that the City will be teaching us and our children how to be better people and a better society.

References

Benevolo, Leonardo (1995), The European City, Cambridge.

Brookings-Wharton Papers on Urban Affairs (2000).

Coward, L. Andrew & Salingaros, A. Nikos (2004), The Information Architecture of Cities, Journal of Information Science, vol. 30 no. 1.

Glaeser, L. Edward (2000), The Future of Urban Research: Nonmarket Interactions.

Hannemann, *Christine* (2004), Architecture as Ideology: Industrialization of Housing in the GDR, Humbolt-Universitat zu Berlin, Working Paper.

Krzysztof, *Nawratek* (2002), PhD Dissertation, Spatial Ideology: Public Spaces and Buildings as Manifestations of Power. Silesian Technical University, Poland.

Lawrence, *Denise L. & Setha*, *M. Low* (1990), The Built Environment and Spatial Form. Annual Reviews In Anthropology 19: 453-505.

Lynch, *K.* (1960), The Image of the City. Cambridge: The MIT Press.

McClintock, *Robert*, Cities, Youth and Technology: Toward a Pedagogy of Autonomy. Institute for Learning technologies Teachers College, Columbia University.

Norberg-Schulz, *Ch.* (1975), Meaning in Western Architecture, International Thomson Publishing.

Przestrzeń, filozofia i architektura (1999) (red) Ewa Rewers. Poznań, Poland.

Putnam, *Robert D.* (1999), "The Marshall Lecture." Mimeographed.

Fourth Plenary

Education for Sustainable Development in the Context of University Municipality Partnership

A Challenge for the Design
–
The Systems Design Thinking

Luigi Bistagnino, Pier Paolo Peruccio, Carlotta Montanera[1]
Politecnico di Torino, Dipartimento di Progettazione Architettonica e Disegno Industriale,
Via Pastrengo 19, I – 10128 Torino, Italy

Introduction

The term "design" regularly enters our vocabulary to describe some of the most common aspects of our everyday life [1-6]. Nowadays we usually hear terms such as fashion design, transportation design, food design, management design, product design, urban design, etc, considering design as a value added to most of our activities and experiences. As a consequence, these specializations of design have been growing up in many industrial design courses at universities; the borderline between architecture, design, art, fashion, at one time rigid, now disappears. It is time to debate the real and fundamental principles of industrial design that address the careful ethics of the quality of life, that makes aware the people involved in the design and the production of manufactured articles. We have been able to create a vast number of artefacts but we have to go beyond what we know today. The time has come to bring design to the next level of performance.

The first challenge is to educate young people, to help them to understand how things are interconnected and give them much more knowledge about natural science then we had in the past. And, at the same time, it is necessary to take the environment and climate crisis seriously, and try to use these problems as an instrument for change: the transformation of our production and consumption systems should be the top objective for our next strategies.

> „The global GDP (Gross Domestic Product) of the world is today more than 8 times bigger than in 1945 and 15 times more than in 1900. Many people expect this kind of development to continue to bring us more welfare, products and services, but what needed in future is some kind of development of what we have today"

stated Anders Wijkman, member of the European Parliament, during a meeting in Turin, Italy [7]. Besides in the contemporary globalized economy we have new very strong actors like China or India, and we have no other way to compete except investing in knowledge and new technologies, information and communication technologies, biotech, nanotech etc.

---

1 Author of Case study: the slaughtering.

The introduction of a system perspective in the business community then becomes more and more urgent, because today most business models are very much looking for core business.

We need to rethink a lot in order to come up with something that is less "vertical", more "horizontal" and more allied with systems thinking. We need bridge building in society, a better understanding between science and society, business and politics, politics and science, business and science, because at the moment the interface between these different sectors is not developed enough.

There are many individual disciplines at university and in the scientific system, but very few efforts to understand how things are interconnected. We are used to dealing with only one problem at a time, as confirmed by Wijkman: "most politicians are specialized on one or two issues and they are very good at it, but we simply lack people who have a more broad outlook of life and who can see how things are interrelated".

This is the way we have to start looking at, innovating science technology, seeing connections among different businesses and considering things we have not considered before [8].

This requires a narrow and deep program. Its development asks us not to be worried about the answers to the questions, but to pose the right questions. Unfortunately our educational system is too much based, at all levels, on knowing the right answers to questions that have been posed decades before and answered in the same way. This is not the base for innovation, it will not bring creativity and leadership: "we have to be prepared to have exams where the students ask the questions. Questions to which professors have no answers. Then we will be on the path of innovation and creativity" Gunter Pauli said [9].

The other challenge is the increasing problems that we are facing with regard to the negative impact on the environment, on climate, on our present production and consumption systems. And indeed the increasing competition for scarce resources.

The environment challenge is profound, we have to present things with a positive angle even if the total effects of increasing demand on energy and materials all over the world are serious. Historically our activities as a human society were very limited if compared to nature. Global stocks of what we call "natural capital" as forests, soils, fresh water, marine resources, etc., were abandoned and very little exploited, and whatever we did it had limited consequences. There were of course examples in history of societies who mismanaged the environment resources.

The challenge is really to make sure that civilization as a whole can continue to prosper if nature is becoming more and more scarce. Not in the sense that we will run out of materials like minerals – there is plenty of those in the crust – but rather that the living systems are being put under pressure. And of course that the stability of the atmosphere is at risk: there has been a number of scientific reports emphasizing that it may be more urgent than we thought before to start reducing total emissions of gases. And still in Europe today we are not really doing much to reduce emissions. So there is really a need to rethink our industrial systems and the only way to tackle this problem in an intelligent way is trying to do more with less and to make sure that the services and products that are being offered to consumers in the whole world are as clean as possible. Besides, if we take a look at all the materials that have been used in society, it is only a minor part that has been used and having an impact during production. Most of the impact is through consumption: despite having a lot of initiatives in terms of environment legislation since 1990 in Europe, the overall of volume of waste materials has been increasing over the last 15 years by almost 20%.

The key is the education: without education, without exposing people, children in kindergartens or students at universities, to this interconnectivity – that is quite obvious once seen it – we will not be able to design systems as an innovative and competitive business strategy [10].

The objective is to be able to respond to basic needs of all, in co-evolution with each other with what is available. It is possible to implement this aim by designing competitive strategies which are based on innovative science and technology.

The educational system needs to bridge *idea* with *reality*. It is not enough to have a bright idea, implementation is the difference. Implementation needs to have respect for local culture and it needs to be based on solid scientific insights.

In order to enrich the designers' culture the next design challenge is to create open systems design. It represents a shift in the concept of industry, away from linear models in which wastes are considered the norm, to integrated systems in which everything has its use because converted into value-added input for other processes [11]. This will lead to a new model of business, a kind of new industrial revolution, in which industry mimics nature's sustainable cycles.

As we know, a competitive and sustainable industry, which is capable of looking at the future, responding to the needs of our society and the requirements of the market, depends on some factors such as the ability to innovate, introduce new technologies, operate in an ever-changing environment.

To make it possible it calls for the capacity to operate multiple agendas simultaneously, maintaining focus on profitability and sustainability.

Therefore the necessity to answer to diverse needs, while operating in complex markets, leads to new teaching models and new schools to prepare the future competent generation of designers.

At a time when science education is facing reduced interest and that educational standards are lowered, the challenge is to convert every student into a person who masters the arts and sciences, while imagining innovative systems of production and consumption, within an interconnected world, and who has the capacity to implement breakthroughs [12].

During the Zeri 10th anniversary world congress [13] in Japan, Gunter Pauli remembered: "implementation is the key, but we need to start looking for a wisdom in the natural systems around us. A lotus flower cleans itself every day and it does not use soaps or detergents. How can it do that? Some experiments have been made to mimic the lotus flower system: there is a new industrial product designed by a German company, a new type of paint, that whenever it rains it cleans the facade of the building in the same way that the lotus flower does.

Besides, 80% of all the colours that birds make on earth are not made with colour pigments. It regards the reflection of the light and this is what happens to the peacock with no toxic heavy metal based colour pigments.

How can we fly without making a noise? An owl flies at 150 km/hour, producing no noise: its feathers can do what not other plane is able to do. Actually the owl knows something that we do not yet know; the particular feathers on its wings take the noise away.

As many people know, the law of gravity tells how the apple falls down from a tree, but no one explains first how the apple got up in the tree, defying the law of gravity before it could subject itself to that law of gravity. It is an amazing hiatus, an amazing gap in our educational system. So we are starting already in elementary school with a way of thinking that does not make sense for most of the kids. And most of the people, if asked to write the formula of the law of gravity, are not able to write it down. If we only teach what we know, the next generation can only do things as badly as we do. Education needs to create space for doing things different and more creative.

In fact, 2005 will mark the commencement of the initial class meetings of the Master of Systems Design at the Politecnico di Torino in collaboration with Zeri and a team of universities: Technical University-Hamburg Harburg (Germany), Royal Technical University of Stockholm (Sweden), Ecole Nationale des Arts et Metiers, Paris (France), Universidad de Mondragon (Spain), University of Latvia, Riga (Latvia), University of Wageningen, (the Netherlands).

Fritjof Capra will be the inaugural instructor for this innovative post-graduate program designed to prepare students for a new career genre premised in systems thinking while combining scientific breadth and an innovative spirit unparalleled in modern education[2].

---

2 The professors of the Master of Systems Design are: *Fritjof Capra*, director of Center for Ecoliteracy, Berkeley, USA, author of books: "The Tao of Physics" and "Hidden Connections"; *Gunter Pauli*, founder and director of ZERI (Zero Emission Research and Initiatives); *Janine Benyus*, researcher and author of the book "Biomimicry"; *Anders Wijkman*, member of the Royal Academy of Sciences of Sweden and Member of European Parlia-

The purpose of the master in Systems Design is to prepare a diverse group of students, with a wide variety of backgrounds to become that new generation of leaders capable of navigating through science and technology, psychology and history, while at the same time have the aesthetics and artistic expression to combine functionality, effectiveness with beauty and joy.

After three years of trial courses at the Politecnico di Torino, Industrial Design Degree, it has become clear that this educational approach is not only feasible, it is much welcomed by the students who enjoy being challenged.

The vision is the development of a school that educates about 50 students – in a first phase – at both the master and doctoral level, and builds on a network of professors and mentors. In parallel other initiatives will reach out to elementary, middle and high school in order to create an innovative framework for learning for life.

Just like an executive training program, the master is based on courses in which the professors engage the students in one full week sessions at the time. It is a remarkable system whereby there will be one teacher for one week teaching one course. It is 40 hours socratic dialog. This permits to develop an intensive learning program where the student shifts insights, creates linkages and builds up capacity to envision innovative approaches solidly based on the best of science, in a close relation with professor. The graduates will discover hundreds of linkages during these one week intensive and interactive sessions.

The courses are a combination of theory and practice, organized in a horizontal manner cutting across traditional subjects.

At the outset there is an intensive program on Systems Theory (Autopoietic Systems, also known as self-organizing systems) and afterwards a focus on the connections which exist in natural systems in order to understand how open systems work.

---

ment; *Ralf Otterpohl*, Director Institute of Municipal and Industrial Waste-Water Management at the Technische Universitat Hamburg-Harburg; *Luigi Bistagnino*, Dean of Industrial design Degree, Politecnico di Torino; *Carlo Olmo*, Dean of the 1 Faculty of Architecture, Politecnico di Torino; *Amory Lovins*, President of the Rocky Mountain Institute, author of the book Natural Capitalism and Small is Profitable; *Margrit Kennedy*, author of the book on Local Currencies; *Lynn Margulis*, professor at University of Massachusetts, USA; *Ashok Khosla* President of Development Alternatives, PhD Massachusetts Institute of Technology; *Paolo Lugari*, founder of Las Gaviotas, Colombia; *Lucio Brusch*, founder of ZERI Brasil, Porto Alegre, Rio Grande do Sul; *Anders Nyquist*, architect; *Carl-Goran Heden*, member of the Royal Swedish Academy of Sciences, director of Microbiology Department of the Karolinksa Institute; *Thebe Medupe*, professor at Cape Town University; *Jorge Reynolds*, the inventor of the pacemaker, professor at the National University of Colombia; *William Ury*, professor at the Law School, Harvard University; *Paul Schatz Jr*, scientist of water; Davide Scabin, chef of Combal.Zero, Rivoli, Torino.

Graduate students will conclude the program with acquired ability to work in multidisciplinary groups, to manage the vastness of scientific knowledge with emotional intelligence, artistic expression and the capability to implement projects while having a hands-on experience.

Master prerogative is to reach the scientific advisors network connected to ZERI and to the database of already started worldwide projects. Therefore study trips are planned in two different periods of the year to discover how ecosystems function: from coastal deserts in Namibia to Andean highland tropics in Latin America, from the tundra in Scandinavia to integrated biosystems in Fiji, from converted cement plants in Japan to native American water treatments systems in the USA.

Case Study: the Slaughtering (by Carlotta Montanera)

This project was carried out during the semester of Systems design of the Laurea Magistrale in Ecodesign at The Politecnico di Torino [14]. The topic of this project was the "water", like a resource, like a source of life, like something to preserve undamaged for the future.

The first month of work was dedicated to some important investigations about the water theme, like the studies effectuated by M. Emoto about water's crystals or the theories of F. Capra.

The analysis of water's uses has brought to this project, which has in the centre the water flow of the slaughtering activities, with all his properties.

At the present time the slaughtering has a water input made up of cleaned water and, after the slaughtering activities, an output of water polluted by organic substances, which must be purified and the sludge eliminated in a dump. In this way the properties of the water are not exploited and what we do is to change the water's properties with our activities and than we search to rectify the damages in the slowest way. We don't care about the best thing to do and we destroy the ecosystem without any worry. But, what has been learnt in this project session is to think about an output like an input, like a resource for other activities.

The water output of the slaughtering is polluted by organic substances, blood and little pieces of meat, a source of possible bacterical infections. So a purification of this water flow is dictated by the European law.

All around the slaughtering activity there are frequently some other satellite activities, like cow breeding, and consequently fodder culture and fertilizer's and integrator's industries.

The objective of the study is to change the whole of this system to create a better new system, more ecological and one that can preserve the ecosystem, increasing also profit and production.

Analysing the output, without changing the input, because the slaughtering operations are all necessary, the project changes first of all the depuration: from a mechanical or chemical process to a biological depuration, as effective, but less expensive and more ecological. In this way we can preserve an effective purification and have no risk of bacterical infections. We increase the purification process using an interesting machine, projected by Paul Shatz, the oloid. The oloid enables this process to be faster agitating the mud and airing it. Also, if the process is faster, we can save energy and time, estimated around 30%.

The sludge that we get with this first process, instead of being eliminated, is used to produce biogas, a volume of about six million cubic metres per year. The biogas is a clean and effective resource, it provides good energy, so the system can be independent with regards to heating.

The mud obtained from this production is finally inert and can be used in agriculture as fertilizer. It contains a lot of important minerals, like potassium and nitrogen, essential for every culture and included in chemical fertilizer. The only difference is that chemical products change the pH of the ground and consequently also the pH of water and in superficial tension, destroying an ecosystem based on these parameters. The mud is as effective as a fertilizer, but it doesn't damage the equilibrium of the ecosystem. This mud is especially good for the idroponic culture, because this special type of agriculture can better exploit the good properties of the sludge. With idroponic technique we can grow a lot of species of plants, like little flowers, and algae. With this technique we can have a continuous harvest throughout the years; in fact it gives a good water temperature and a good quality of subsoil.

We decided on a culture of three different species of algae: the Spirulina algae, an excellent proteinic and mineral integrator for human use, with a good commercial value; the Chlorella algae, used like integrator for the cow's diet; the blue-green algae, a good and totally natural fertilizer.

The spirulina is an algae really exploited in the parapharmacy and in biological diets. With this we can make product tablets as good as the best integrator of minerals and amminoacids that are available. The commercial value of Spirulina is really high, so we can increase the profit of our system's project.

Also Chlorella and Blue-green algae are two species which are really interesting because of their nutritional properties.

Using these products instead of chemical fertlizers and chemical integrators, the biological value of our system increases and this means that also the quality of human life is increasing.

Thanks to the products of these activities it is possible to observe the increase in cow-breeding, the principal source of slaughtering. We produce not only a bigger quantity of meat, but we also search for an increase in the quality.

But, with this system we can see how this process means an increase so great, that meat can be certificated like a biological product, and its value increases by 30%.

Besides, with the water used in the idroponic culture coming from the slaughtering, it is possible to introduce fish-breeding of three species of fish.

This water has a perfect pH, a good superficial tension and includes the best organic properties: algae purify the water and enrich it with minerals and little pieces of leaves, that are a good food for fish.

Cow breeding, in this way, doesn't need any chemical input, a fact that means a better meat quality, and consequently a human advantage.

Instead of chemical pesticides for the fodder culture, a major source of food for cows, we can use a natural pesticide. From the past we have learnt that a plant, the Azadirachta indica, called Neem in the used language, mixed with the cow's urine, can be an excellent pesticide, very effective and biological.

With solid waste from the slaughtering we can develop fleshfly-breeding, that provides nourishment to some quails, that we can sell. Two advantages of this are: quail-breeding is really economic and makes a lot of profit, with eggs and also with meat, but, doing fly-breeding also, we can get rid of the flies from the cows and from the purification systems. So it is possible to solve two problems with one solution and we also get money from an activity run on waste.

Therefore the water output of slaughtering, before considered a waste, can create a new system. Now we can obtain from this "waste" a biogas production, a good production of algae, fish and quails. Together we produce a better quality of meat, that we can call biological, because with this system we eliminate fertilizers and antiparasitics, replacing them with blue-green algae and Azadirachta indica, a natural antiparasitic used 5.000 years ago.

We are also independent with regards the heating, we created new jobs, we created an innovative system and we diversified the production and the ecosystem.

This study is of a little region of production, with medium slaughtering activity.

Our ambition is to apply a system like this in every economic system.

Like this we don't have any waste, but the output is utilized as input by another activity.

Looking at the economical aspects, no less important that the other points, we can observe how this system is excellent for saving money and to increasing the production: in one year it produces a profit of about 1.400.000 Euro, excepting the increasing of the value of the meat, about 30%.

References

1. *Forty, Adrian* (1986), Objects of Desire. Design and society since 1750, Thames and Hudson, London.
2. *Gregotti, Vittorio* (1986), Il Disegno del prodotto industriale. Italia 1860-1980, Electa, Milano
3. *Bistagnino, Luigi* (2003), Design con un futuro, Time and Mind.
4. *Lanzavecchia, Carla* (2000), Il fare ecologico. Il prodotto industriale e i suoi requisiti ambientali, Paravia, Torino.
5. *Maldonado, Tomas* (1976), Disegno industriale: un riesame, Feltrinelli, Milano 2003 (I ed.).
6. *Manzini, Ezio & Jegou, Francois* (2003), Quotidiano sostenibile. Scenari di vita urbana, Edizioni Ambiente, Milano.
7. *Wijkman, Anders* (2005), speech at the presentation of the Master of Systems Design, Skillab-Unione Industriale, Turin (Italy), 08/04/2005.
8. *Capra, Fritjof* (2002), La scienza della vita. Le connessioni nascoste fra la natura e gli esseri viventi, Bur, Milano 2004, Orig. Tit. The Hidden Connections.
9. *Pauli, Gunter* (2005), speech at the presentation of the Master of Systems Design, Skillab-Unione Industriale, Turin (Italy), 08/04/2005.
10. *Pauli, Gunter* (2004), From Fairy Tales to Reality, Casa Editorial San Pablo, Bogotà, Mexico, Madrid 2005; *Gunter Pauli*, Out of the Box: 21 inspiring fairy tales to create and innovate at work, Future Managers, Cape Town, South Africa.
11. *Peruccio, Pier Paolo* (2003), La ricetta Pauli: bambù invece di PVC, interview with Gunter Pauli, in "Il Giornale dell'Architettura", no. 3, january 2003, pp. 1 e 25.
12. *Pauli, Gunter* (1996), Svolte epocali. Il Business per un futuro migliore, Baldini&Castaldi, Varese 1997. Orig Tit.: Breakthrough: what business can offer society, Tokyo.
13. Zeri (Zero Emissions Research and Initiatives) 10th Anniversary World Congress, United Nations University, Tokyo October 2004
14. *Caiati, Michele & Montanera, Carlotta* (2004-2005), Fabiola Salman research work at Politecnico di Torino, Laurea Magistrale in Ecodesign, A.A. 2004-2005 (Professors: Luigi Bistagnino and Gunter Pauli).

Adriatic Sea and Coastal Zone of Serbia and Montenegro – Characteristics of Biodiversity and Possibilities for their Sustainable Exploitation

Vladimir Stevanovic[1], Ivica Radovic[1], Slobodan Regner[2], Sreten Mandic[2], Zlatko Bulic[3], Aleksandar Joksimovic[2], Dejan Radovic[1]

[1] Faculty of Biology, University of Belgrade, 11000 Belgrade, Serbia and Montenegro

[2] Institute of Marine Biology, POB 69, 85330 Kotor, Serbia and Montenegro

[3] Institute for the Protection of Nature of Montenegro, Podgorica, Serbia and Montenegro

Abstract

The availability of biological resources, their diversity and quantity, structure and geographical distribution, as well as related questions, were always of prime economic, social and political importance to humans. Regions of high biodiversity with all their applicative potentials are analytically incorporated in relevant geostrategic and geopolitical studies and regional sustainable developmental plans.

It is pertinent to ask ourselves which biological resources and how many of them do we possess? What are their current and potential values? How are they distributed? and what are their long term perspectives of conservation, restoration and rational exploitation?

According to the IUCN-WCMC criteria, the territory of Serbia and Montenegro, together with the mountainous part of Greece, Albania and Bulgaria, represents one of the six European or one of the 153 world's centers of biological diversity. Approximately 70% of European mammals, 75% of European birds, 52% of European freshwater fishes, and 39% of European vascular plants inhibit the territory of Serbia and Montenegro. Biodiversity of Serbia and Montenegro includes more than 1,600 internationally significant (endemic and endemo-relict) animal and plant species.

Considering the biodiversity of Adriatic Sea and Coastal Zone (Mediterranean part) of Montenegro and Serbia it is important to emphasize that, at the European level, they are distinguished with a great ecosystem and species diversity. The marine aquatory of Montenegro and Serbia is characterized with a high level of biological diversity on the species level (79% of Mediterranean fish species, 68% of Mediterranean echinoderms).In the territory from Adriatic Sea to Prokletije Mt, Durmitor Mt. and Bjelasica Mt. along the aerial distance not longer than 100 km, nearly all zonobiomes of Europe with more different ecosystems are present: marine ecosystem, freshwater ecosystems (eutrophic lakes, rivers), terrestrial ecosystems of coastal zone (sand and pebble coastlines, salt marshes, ecosystem of peat-bogs, eu Mediterranean evergreen sclerophyllous and sub Mediterranean deciduous forest, scrub and pasture), ecosystem of montane and sub alpine beech and white-bark pine coniferous forests, ecosystem of scrub and grassland above timber line and arctic and alpine tundra-type vegetation.

Both the territory and aquatory of the Mediterranean part of Serbia and Montenegro have high potentials for exploitation of some components of biodiversity. However, it is necessary to emphasize that the majority of natural ecosystems have been exploited for centuries and that their recent state is different from climatogenic, since they represent progradational-degradational derivates of climazonal ecosystems still under anthropogenic pressure and constant stress.

Concerning these facts, it is clear that in the present and future, exploitation of the components of biodiversity has to be based on the principles of sustainability. Practically, this means that the components of biodiversity cannot be exploited without control, especially without a previous assessment of their quantity, model of use, conservation status, resources fragility and renewable of resources.

In this paper the threat and degree of human impact, possibilities and regime of biodiversity exploitation were presented, for each natural ecosystem of Mediterranean part of Serbia and Montenegro in general.

Introduction

Biodiversity

Biodiversity is the variety of all living things and their interactions. The term and concept of "BIODIVERSITY" today represents a commonly used expression in biological sciences, and it has become a customary word. At the Conference for Sustainable Development, held in Rio de Janeiro in 1992, biodiversity has moved to the centre stage as one of the focal issues, as well as one of the major scientific and political concerns worldwide (Alonso et al. 2001).

Scientists often speak of three levels of diversity: genetic, species, and ecosystem diversity. In effect, these levels cannot be separated. Each one is equally important, interacting with and influencing the others. A change at one level can cause changes at other levels (Wilson 1992; Heywood & Watson 1995; Pimm et al. 1995).

Benefits of Biodiversity: Biodiversity provides valuable ecosystem services free of charge, provides food, fuel and fiber, provides shelter and building materials, purifies air and water, detoxifies and decomposes wastes, stabilizes and moderates Earth's climate, moderates floods, droughts, wind and temperature extremes, pollinates plants, provides traditional medicines and high-tech pharmaceutical products for control of pests and diseases, maintains genetic resources as key inputs to crop varieties, livestock breeds and medicines, provides cultural and aesthetic benefits, provides us the means to adopt to change. Life, as we know it, will not be the same if our rich heritage is dramatically altered (Peyton et al. 1995; Stevanovic 1996).

The major causes of Biodiversity Loss and Species Extinction: Scientists have coined an acronym *HIPPO to* denote the five primary causes of species decline and extinction: Habitat alteration, Invasive species, Pollution, Population growth of human, Overexploitation (over harvesting of species from the wild and over consumption of resources).

Sustainable Development

The United Nation defines sustainable development as: "development that meets the needs of the present without compromising the ability of future generations to meet their own needs." U.N. Commission (called the Brundtland Commission) *Our Common Future*, 1987.

> "From the African continent, the cradle of humankind, we solemnly pledge to the peoples of the world and the generations that will surely inherit this Earth that we are determined to ensure that our collective hope for sustainable developments is realized." (Declaration signed by 193 nations represented at the World Summit, Johannesburg 2002)

Sustainability depends, in large part, on ability of the current human population to limit its environmental impact. For the successful future of both human civilization and nature, it is necessary that sustainability is maintained though the balanced contribution of their three integral elements (Environment, Society and Economy), achieved by adequate global policy. Environmental scientists recognize that human civilization depends on an intact and functional natural environment; in fact, civilization cannot exist without it.

Biodiversity and Sustainability

Looking at these facts, it is clear that in the present and future, exploitation of the components of biodiversity (biological resources) has to be based on the principles of sustainability. Practically, this means that the components of biodiversity cannot be exploited without control, especially without a previous assessment of their quantity, model of use, conservation status, resources fragility and renew ability of resources.

Biological Resources

Biological resources – genes, species and ecosystems that have actual or potential value to people are the physical manifestations of the global biological diversity. The availability of biological resources, their quantity and diversity, structure and geographical distribution, as well as other related questions, were always of prime economic and political importance for each country. Moreover, regions of high biodiversity with all their applied potentials, as well as their geographical position and human resources, are analytically incorporated both in the relevant geostrategic and geopolitical studies and regional developmental plans.

Montenegro as an Ecological State

In proclaiming Montenegro as an Ecological State in September 1991 the Montenegrin Parliament set out for the future its relationship with nature. Promotion of this Declaration was done in UN Conference on Environment and Development "Earth Summit" held in Rio de Janeiro in 1992.

The Main Natural and Social Characteristics of South Adriatic and Coastal Zone of Serbia and Montenegro

The Adriatic is one of the cleanest and most transparent warm seas, linked to the central Mediterranean by the Strait of Otranto. It is widest between Montenegro and southern Italy (about 200 km). This is also the deepest part of the Adriatic (1,330 m) at a point 120 km southwest of Boka Kotorska Bay.

The coast is very rugged but with hardly an island off it. The coastline indentation ratio is about three: the full length of the coastline is 293.5 km, but only 90 km of aerial line. Throughout the length of the coast there are 52 km of sand beaches. The 13 km. long beach in Ulcinj, in the southernmost part of Montenegro, has the finest sand imaginable. That in Becici was declared Europe's most beautiful beach in 1930.

There are many bays (Budva, Bar) and coves (Petrovac, Lucice, Buljarica, Canj, Ulcinj), the most beautiful of them being Boka Kotorska Bay. Boka Kotorska Bay (total costal-line of 106 km) consisting of four smaller basins (Herceg Novi, Tivat, Risan and Kotor), crisscrossed by inlets and cutting deep into the mainland, it is one of Europe's most beautiful bays and the world's southernmost fjord.

The littoral is very narrow. It is separated from the hinterland by the steep karst mountains of Rumija (1,595 m), Sutorman (1,180 m), Orjen (1,895 m) and Lovcen (1,749 m). It is widest in the region of Boka Kotorska Bay (about 10 km) and narrowest at Pastrovici (less than 2 km).

Ulcinj, Bar, Petrovac, Budva, Tivat, Kotor and Herceg Novi have remained typical coastal towns.

Modern urbanization has not destroyed their Mediterranean soul, enriched by a singular mixture of religions and cultures. The old well-preserved town centres are precious monuments of bygone days, but they are as much a part of everyday life as are the modern-day buildings. The ancient ramparts, towers, castles, monasteries, summerhouses and royal residencies are not just museums but places where people still live and work.

The Kotor city and inner basin of the Boka Kotorska Bay at the Conference held in Kairo-Luxor in 1979 are included on UNESCO World Natural and Cultural Heritage List.

Main Characteristics of the Biodiversity of Serbia and Montenegro

The most recent analyses of flora and fauna in Serbia and Montenegro exhibited that they are amongst the richest in Europe. According to the IUCN-WCMS criteria, the territory of Serbia & Montenegro, together with the mountain parts of Albania, Bulgaria and Greece represents one of the six European or one of the 153 world's centres of biological diversity (biodiversity hotspot).

The abundance and diversity of taxa, ecosystems and landscapes may be explained by various climatic, geomorphologic, petrographical, orographical and hydrographical features, as well as by a great number of refugial habitats throughout the land, where a significant number of relict and endemo-relict species and communities – dating back as far as the Tertiary and Ice Age – have been preserved.

This great biodiversity of the territory of Serbia and Montenegro is characterized not only with rather heterogeneous florogenetic and faunogenetic composition and mosaic distribution, and with a great spatial density of its components – species, ecosystems and landscapes – but also with more than 1,600 internationally significant species, including a great number of local endemics – in accordance with the CORINE classification.

Biogeographically, almost all basic bioms of Europe are present within the territory of Serbia & Montenegro: Steppe zonobiome, Zonobiome of deciduous forests, Zonobiome of Mediterranean forests and shrubs, Orobiome of coniferous boreal forests, Orobiome of high-montane "tundra". Beside these biomes on the territory of Serbia and Montenegro we also recognize: different kind of fresh – water ecosystems in the Adriatic Sea as a part of the complex marine biome of the World (Fukarek & Jovanovic 1983; Stevanovic & Vasic 1995; Radovic & Mandic 1998).

Approximately: 70% of European mammals, 75% of European birds, 52% of European fresh-water fishes, 79% of Mediterranean marine fishes and 39% of European vascular plants inhabit the territory of Serbia & Montenegro (Stevanovic & Vasic 1995).

Main Characteristics of the Biodiversity of Montenegro

The main characteristic of the biodiversity of Montenegro is a high concentration of different species and ecosystems in limited area. Almost all European zonobiomes, from boreal coniferous and mideuropean deciduous forests to sub Mediterranean deciduous and eu Mediterranean evergreen sclerophyllous forests and scrub formations can be found along the line from the Prokletije Mountain (Durmitor Mountain, Bjelasica Mountain) to the Adriatic coast not exceeding 100 kilometers. High concentrations of different species and ecosystems in a limited area, sometimes represented as "Europe in small", indicates that most of the ecosystems distributed in limited areas are enclaves distant from the borders of main biomes in Europe. Besides, all basic types of ecosystems and landscapes in Montenegro have an adequate biome equivalent in Europe, but are specific in numerous endemic, particularly locally endemic species (the globally important component of biodiversity), special physiognomy and structure, thus emphasizing uniqueness and complexity of the biodiversity of Montenegro (Stevanovic & Radovic 2001).

Certain smaller geographical areas have an extremely high biological diversity of particular groups of animals and plants (Table 1). For example, 54 species (including many endemic ones) of amphibians and reptiles live in a relatively small area of Skadar lake region (southeast Montenegro). Most of the larger Mediterranean countries cannot boast such diversity of herpetofauna on their territories (Italy 62, Greece 68 and Spain 66 species of herpetofauna) (Dzukic & Kalezic 2004). As a rule, such special areas are elsewhere proclaimed as biogenetic reserves (Stevanovic & Vasic 1995; Stevanovic et al. 1996; Bulic & Hadziablahovic 2001).

Tab. 1: Numerical synopsis of species diversity of selected groups of biota in Serbia and Montenegro

TAXA	**Number of species**				**Percentage of Europe species**		
	Europe	**Ser&Mg**	**Mg**	**Med Mg**	**Ser&Mg**	**Mg**	**Med Mg**
Vascular plants	11,000	4,182	3,136	1,725	38.93	28.50	15.68
Insecta	app. 70,000 (estimate)	app. 40,000 (estimated)	app. 30,000 (estimated)	app. 21,000 (estimated)	57.14	42.85	30.00

TAXA	Number of species				Percentage of Europe species		
	Europe	Ser&Mg	Mg	Med Mg	Ser&Mg	Mg	Med Mg
Vascular plants	11,000	4,182	3,136	1,725	38.93	28.50	15.68
Freshwater Osteichthyes	215	110	72	53	51.16	33.48	24.65
Marine Fishes	540 (Mediteranean)	407 (Adriatic)	402 (Adriatic)	402 (Adriatic)	75.37	74,44	74,44
Amphibia + Reptilia	277	70	53	51	25.27	19.13	18.14
Aves	516	260 (nesting)	246 (nesting)	185 (nesting)	50.38	47.67	35.85
Mammalia	146	104	67	31	71,20	45.90	21.23

Ser&Mg-Serbia and Montenegro, Mg-Montenegro, Med Mg-Mediterranean part of Montenegro

Review of the Basic Types of Biomes of the Mediterranean Part of Montenegro: Causes and Levels of Biodiversity Threats and Possibilities and Regimes of their Exploitation

Both the territory and the aquatory of the Mediterranean part of Montenegro have high potentials for exploitation of some components of biodiversity. However, it is necessary to emphasize that the majority of natural ecosystems have been exploited for centuries and that their recent state is different from the climatogenic, since they represent progradational-degradational derivatives of climazonal ecosystems still under anthropogenic pressure and constant stress. Concerning these facts, the exploitation of components of biological diversity of some ecosystems has to be harmonized with their state and sustainable capacities, where the establishing of critical points of sustainable capacities of each ecosystem is the prerequisite for rational and long-term exploitation of its biological components.

The Threat and Degree of Human Impact *(TDHI)*, Possibilities of Biodiversity Exploitation (*PBE*) and Regime of Biodiversity Exploitation (*RBE*) can be presented for each natural ecosystem in general (Stevanovic 1997).

The threat and degree of human impact on ecosystems could be recognized as: High *(H)*, Moderate (*M*), Low (*L*), Insignificant (*I*);

The possibilities of biodiversity exploitation as: High *(H)*, Moderate (*M*), Low (*L*), Insignificant (*I*);

The regime of biodiversity exploitation as: Liberal (*L*), Restrictive (*R*), Highly Restrictive (*HR*); Exclusion of Using (*ExU*).

These qualifications are general and based on the recent state, historical dimension of anthropogenic impact, and distribution of some basic types of ecosystems in the Mediterranean part of Montenegro. In regard to their biological components, the possibilities and regimes of exploitation can be determined.

The main characteristics of biodiversity of the South Adriatic

The South Adriatic could be considered as the centre of the Adriatic Sea biodiversity (74,4% of Mediterranean and 98,8% of Adriatic fish species). All existing data indicate that almost all plant and animal species found so far in the Adriatic live in its southern part as well. Even more so, many of the species, especially those belonging to the fauna of the profound are solely to be found in the South Adriatic. The total biodiversity index decreases if going from the South to North Adriatic. Sometimes also the organic productivity and with it the biomass and the number of specimens are growing from the South to the North Adriatic.

1. Marine Ecosystems

About 407 fish species have been observed in the Adriatic (540 in the Mediterranean Sea), either permanently living or occasionally entering (353 of Osteichthyes and 54 species of Chondrichthyes) (Stevanovic et al. 1997; Radujkovic 2001).

1.1 Populations of pelagic fishes permanently or periodically inhabiting open waters of Southern Adriatic

Pelagic fish populations that permanently or periodically inhabit the open waters of the southern Adriatic, such as sardine, anchova, Spanish mackerel, mackerel, and tunny, are of great economic significance for the development of the sea fishery in Montenegro. This component of the biodiversity is far from being completely utilized in regard to its potentials. Estimated MSY per year (maximal sustainable yield) of Sardina and Anchova is 30.000 t; (Sardina/Anchova =1/3).
Negative human impacts: Polution; Exploitation of oil, Changes in nutrient regime

TDHI: M-L; PBE: H-M; RBE: L-R

1.2 Populations of Benthic Organisms (Fishes, Crustaceans and Mollusks) Inhabiting Lower Parts of Littoral and Sub Littoral Zones

It is necessary to emphasize that, contrary to the case of pelagic fish, the intensive research of the demersal (trawl) resources was realized during the period 1997-2004. Estimated MSY per year 600 t. The analysis of the mean structure of weight of the trawl catch according to species of organisms (Figure 1) indicates that the percentage of the cartilaginous fish (Elasmobranchia) decreased from 32% (1948) to 20% (2000). The decrease of these fish by more than 10% is regarded as a sign of over fishing. (Regner 2001).

Fig. 1: Structure of weight of the trawl catch in the period 1998-2000

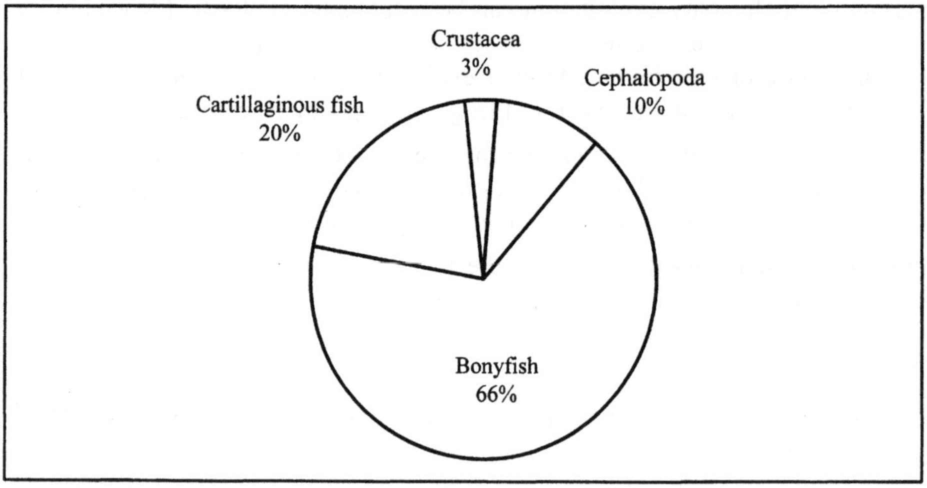

Negative human impact: Overfishering and harvesting of bentic communities by trawl catch, particularly between 50 and 100 m depth; Pollution by urban and industrial waste waters.

Benthos of litoral zone: TDHI: H-M; PBU: M-L; RBE: R-HR; Benthos of sublittoral zone: TDHI: H-M; PBE: H-M; RBE: R

2. Freshwater ecosystems

2.1 Eutrophic lakes

Eutrophic lakes and salt and brackish marshes in Montenegro are among the most important spots ("hot spots") of waterfowl diversity in southeastern Europe, both during the nesting and wintering seasons (e.g. "Lake Skadar" National Park is included in the Ramsar Convention list of protected areas, clearly showing the value and importance of this region, yet also creating a serious commitment). The winter identification of water birds (IWC 1991-2000) counted 43 species, with numbers of individuals ranging from 100,000 to 280,000. Nine of the mentioned species surpass the 1% threshold that defines the international importance, which means that the population size of these birds during winter stay is significant for the population *in toto*. In addition to this, Skadar Lake is one of the diversity centers for freshwater snails, crabs, frogs,klizards, freshwater fishes and other animals. At the same time, waterfowl is favored for hunting, which represents a significant part of the Montenegro tourism, and one of the many examples where hunting and conservation are opposed and incompatible. However, waterfowl hunting can be approved when organized and controlled, especially of those species with numerous populations and if not affecting their age structure and abundance. Waterfowl represents an exceptionally important element of the biodiversity of wetland areas of Montenegro as well as of stability of the wetland and water ecosystems. The regime of biodiversity exploitation can be recognized as: R-HR (depending on the game bird species and hunting closed seasons). The majority of waterfowl is permanently protected by law. Similar exploitation policy (in Lake Skadar) can be applied on some sibling species of frogs, among which some are endemic.

Negative human impacts: Overfishering during whole season; Overhunting of waterfowl, particularly during winter season; Trade of rare birds; Polution by urban and industrial wastewaters, Rapid eutrophication.

Populations of freshwater fishes in the lakes: *TDHI*: *H*; *PBE*: *H-M*; *RBE*: *R*; Birds of eutrophic lakes and salt marshes: *TDHI*: *H*; *PBE*: *M*; *RBE*: *R-HR*.

2.2 Rivers

In regard to the amount of fish caught and the type of fishing, the Bojana River, as a migration route (Adriatic Sea – Lake Skadar – inland rivers of the Adriatic Sea basin and vice versa) for a large number of anadromous and catadromous commercially important fish species, has to be more strictly controlled.

Negative human impact: Fishering without control during whole season, Disturbance of bentic and litoral habitats by pebble exploitation; Polution by urban and industrial waste waters.

Fish populations of oligotrophic rivers: *TDHI*: *H*; *PBE*: *M*; *RBE*: *R – HR*; Populations of anadromus and catadromus fishes: *TDHI*: *H*; *PBE*: *M*; *RBE*: *R – HR*.

3. Terrestrial Ecosystems

3.1 Coastal Zone (Sand and/or Pebble Coastlines)

The threat and degree of human impact on this type of biome can be recognized as high (tourism); the possibilities of biodiversity exploitation as low to insignificant, and the regime of biodiversity exploitation as highly restrictive.

Negative human impact: tourism including various ways of habitat disturbance

TDHI: H; PBE: L-I; RBE: R – HR.

3.2 Salt Marshes, Swards and Mud Flats of Coastal Zones

The high tide river mouths, low muddy highly saline coasts influenced by high tides and overgrown by communities of *Thero-Salicornion* and *Arthrocnemion fruticosi*, as well as coastal salty meadows, salty pseudosteppes, and pastures of *Limonion dalmaticum* (river mouth of the Bojana, bay Solila in Grbaljsko polje, salt basins in Ulcinj, Igalo), are mainly used as salt pans, active (Ulcinj), or abandoned (salt pans of Tivat). Ulcinj salt pans represent an internationally important bird area (IBA YU 045), which consists of circa 1,350 ha of salt basins, mud flats, reedbeds, pools, canals, mud banks, freshwater and brackish pools and lakes (former Zagajsko Blato). Being the last nesting places of the shelduck (*Tadorna tadorna*) and one of the last habitats of oystercatcher (*Haematopus ostralegus*), they are of special importance for the biodiversity of Serbia and Montenegro, but also internationally important as nesting places for several species of birds (*Himantopus himantopus*, *Glareola pratincola*, *Charadrius alexandrinus* and *Sterna albifrons*) as well as an internationally important concentration site for a great number of water birds during migration (especially *Platalea leucorodia* and *Egretta alba*). At the same time, the salt pans are the only habitats in Serbia and Montenegro for some species such as the brine shrimp (*Artemia salina*), which is very important for aquaculture.

Negative human impacts: bird hunting, trade of rare species

TDHI: M; PBE: M-L; RBE: R-HR

3.3 Forest and Scrub Ecosystems

3.3.1 Sclerophyllous Evergreen Mediterranean Forests and Scrubs

The most common type of evergreen Mediterranean vegetation in Montenegro is maquie and pseudomaquie, in their initial forms and stable coenotic structure restricted to small areas, mainly of conservation and decorative value. For this type of habitat, Mediterranean and sub Mediterranean laurel forests of *Laurion nobilis* type, are characteristic along the coastline, although they can be found in the area of Lake Skadar as well. The tradition of collecting laurel leaves is long and is still maintained today, since the laurel represents a valuable spice in the food industry, and its oils are used in pharmaceutical and cosmetics industry.

Negative human impacts: fire, clear-cutting, goat grazing, and building of human settlements

TDHI: H; PBE: M; RBE: R

3.3.2 Forest Ecosystems of Evergreen Oak Species

Forest ecosystems of evergreen oak species Quercus ilex, Querqus coccifera are almost completely absent from the Montenegro coast. These forests consisted of holm oak populations not so long ago, but today only single trees or small groups of this species can be found wether as relic associations deeper in mainland (Cijevna canyon, northern side of Rumia Mountain above Lake Skadar) or as small groups of old trees in the Mediterranean part of Montenegrin Coast (Lustica).

Negative human impact: Selective cuting of holm oak (Quercus ilex)

TDHI: H; PBE: L-I; RBE: HR

3.3.3 Mediterranean Forests of Aleppo Pine and Umbrella Pine

Autochthonous forests of Aleppo pine have restricted distribution in the Mediterranean part of the Montenegro coast, but this species is commonly planted whether in groups, as part of coastal landscapes, or as single old specimens in urban areas.

TDHI: H-M; PBU: L-I; RBU: HR

3.3.4 Submediterranean Deciduous Forest Ecosystems

They represent very important ecosystems spontaneously developed on eroded carst, with lower economic value, but with an indispensable anti-erosion protective function. These ecosystems are economically important as firewood for local communities concentrated in carst depressions of the mainland. This type of ecosystem on the carst includes the Macedonian oak and chestnut forests.

Negative human impacts: three cutting for fuel; fire, goat browsing, overgrazing.

TDHI: H; PBE: L-I; RBE: R-HR

3.3.5 Secondary Ecosystems of Submediterranean Scrubs and Limestone Grounds

These ecosystems originate during the degradation of mediterranean and submediterranean types of forest vegetation. With protective anti-erosive function, they represent a successive phase of progradation towards spontaneous renewal of the forest vegetation.

Negative human impacts: three cutting for fuel; fire, goat browsing, overgrazing.

TDHI: H-M; PBE: M-L; RBE: L-R

3.3.6 Upper Timber Line of Inner (Maritime) Chain of Dinaric Alps-Oromediterranean Mountains

The upper timberline in Montenegro consists of subalpine beech, along with associations of endemic and relic species of munika and molika pines (white bark pine), scrub and grassland above the timber line. Coniferous ecosystems of these species represent one of the most important vegetation belts of the Montenegro mountains.

Negative human impacts: forest cutting of old stands and fire.

TDHI: H-M; PBE: L; RBE: ExU

4. Ecosystems of Peat-Bogs

Peat-bog ecosystems are present in the mediterranean and submediterranean part of Montenegro (flooded area of Lake Skadar near Bojana River).

TDHI: L; PBE: L-I; RBE: ExU

Genetic Resources of Wild Flora and Fauna

This component of biodiversity of the Mediterranean part of Montenegro has not drawn too much attention recently, but considering the fact that genetic resources represent the focus of interests of scientists worldwide, it is very likely that this interest will increase in the future as a result of the constant demand for new biotechnologies. High potentials of wild flora and fauna of Montenegro can be summarized as follows:

a) Endemic and relic flora and fauna, especially of international (global) significance
b) Relic or ancestral populations of widely distributed species of trees in mediterranean and submediterranean canyons and gorges
c) Wild related species of cultivated plants and domesticated animals
d) Well-preserved populations with stable age structure of widely used wild plants and animals.

Possibilities for the exploitation of this level of biodiversity components may be recognized as high to moderate and the regime of exploitation as restrictive.

Establishing and Maintaining Mariculture in the Coastal Area of Montenegro

The Institute of Marine Biology in Kotor during the last five years, together with the Norwegian Institute for Aquaculture "Aquaforsk", is actively engaged in the programme of estimation and establishing of mariculture in Montenegro. The chosen species was the rainbow trout (*Onchorinchus mykiss*), and later the Atlantic salmon (*Salmo salar*) was also included. A joint project with the Institute of Marine Biology in Trieste on breeding of the Mediterranean mussel (*Mytilus galloprovincialis)* and oyster (*Ostrea edulis*) is underway, as well as a joint programme with the Chinese Academy of Science on possibilities of the lagoon breeding of the prawn (*Penaeus vanamei*). Significant attention is paid to the condition of the fish stock in the region of Ulcinj and the river mouth of the Bojana River, particularly of the eel (*Anguilla anguilla*), bass (*Dicentrarchus labrax*), and Grey mullets (Mugilidae), as well as to the determination of the condition of populations of the brine shrimp *Artemia salina* (Mandic, 2001; Mandic and Joksimovic 2001).

Conclusions

Scientific arguments indicate that both the territory and the aquatory of Montenegro have a high species, ecosystem and landscape diversity (biodiversity hotspot). The biodiversity of Montenegro and its Mediterranean part is unique and of global significance, especially concerning the internationally important species with majority of Montenegrin endemics.

However, it is necessary to emphasize that nearly all natural ecosystems in Montenegro are more or less sensitive (fragile), depending on their size, duration and kind of athropogenic influence. Ecosystems with high species diversity, as major ecosystems in Montenegro, are more fragile in comparison with those that have lower species diversity and simpler structure.

Today, it is clear that the exploitation of biological resources of the Mediterranean part of Montenegro has to be based on the principles of sustainability.

As basic criteria of conservation and exploitation of these biological resources, including relict, endemo-relict and endemic species, we recognize the following categories: quantity of resource, modes of use, conservation status, resource fragility and resource renewability.

References

Alonso, A., Dallmeier, F., Granek, E. & Raven, P. (2001), Biodiversity: Connecting with the tapestry of life. Smithsonian Institution/Monitoring and assessment of Biodiversity Program and President's Committee of Advisors on Science and Technology. Washington, DC.

Bulic, Z. & Hadziablahovic, S. (2001), The importance and protection of floristic rarities in Montenegro. Prirodni potencijali kopna, kontinentalnih voda I mora Crne Gore I njihova zastita. Plenarni referati I izvodi iz saopstenja, 56. Zabljak.

Dzukic, G. & Kalezic, M. (2004), The biodiversity of amphibians and reptiles in the Balkan Penisula (167-192). In: Griffiths, H., Krystufek, B. & Reed, J. (eds.), Balkan Biodiversity: Pattern and Process in the European Hotspot. Kluwer Acad, Publ., London (336 pp).

Fukarek, P. & Jovanovic, B. (1983), Karta prirodne potencijalne vegetacije SFR Jugoslavije. – Nau~no vece vegetacijske karte Jugoslavije.

Heywood, V.H. & Watson, R.T. (1995), Global biodiversity assessment. – UNEP, University Press, Cambridge.

Mandic, S. (2001), Institute of marine Biology, Kotor, Montenegro, Yugoslavia.Prirodni potencijali kopna, kontinentalnih voda I mora Crne Gore I njihova zastita. Plenarni referati I izvodi iz saopstenja, 11-17. Zabljak.

Mandic, S. & Joksimovic, A. (2001), Mogucnosti razvoja marikulture u priobalnom podru~ju Crnogorskog primorja. In: Istra~ivanje, koriscenje I zastita litoralnog podru~ja Juznog Jadrana. Studija. Projekat OSI-267. Institut za biologiju mora Kotor.

Peyton, B. et al. (1995), Biological diversity. Environmental education module. UNESCO – UNEP 1-152.

Pimm, S.G.J., Russell, S.L. & Brooks, T.M. (1995), The future of biodiversity. Science, 269: 347-350.

Radovic, I. & Mandic, R. (1998), Zasticena prirodna dobra i zastita i unapre|ivanje biodiverziteta – osnova prirodne bastine Srbije. Zastita prirode, 50; 13-32. Beograd.

Radujkovic, B. (2001), Check-list of the South Adriatic fishes. Prirodni potencijali kopna, kontinentalnih voda I mora Crne Gore I njihova zastita. Plenarni referati I izvodi iz saopstenja 101. Zabljak.

Regner, S. (2001), Razvoj morskog ribarstva (Bentos I pelagijal). In Istrazivanje, koriscenje I zastita litoralnog podru~ja Juznog Jadrana. Studija. Projekat OSI-267. Institut za biologiju mora Kotor.

Stevanovic, V. (1996), Biodiverzitet i zastita zivotne sredine. V kongres ekologa Jugoslavije. Zbornik plenarnih referata, pp. 21-34, Beograd.

– (1997), Biogeni potencijali (biodiverzitet) razvoja Crne Gore. In: Pravci razvoja Crne Gore ekoloske drzave (finalni izvestaj). Evropski centar za mir i razvoj, Beograd.

Stevanovic, V. & Radovic, I. (2001), The term, concept and significance of the biodiversity conservation. Prirodni potencijali kopna, kontinentalnih voda I mora Crne Gore I njihova zastita. Plenarni referati i Izvodi iz saopstenja, 31.-49. Zabljak.

Stevanovic, V. & Vasic, V. (eds.) (1995), Biodiverzitet Jugoslavije sa pregledom vrsta od medjunarodnog znacaja. Bioloski fakultet i Ecolibri, Beograd.

Stevanovic, V., Vasic, V. & Radovic, I. (1996), Biodiversity in Yugoslavia – A review of landscapes, ecosystems, and biota, with an outline of the conservation status and activities (including the list of internationally significant species). International Senckenberg Conference "Global Biodiversity Research in Europe", 9-13. Dec. 1996. Frankfurt a.M. Germany, Abstract vol., p. 75.

Stevanovic, V., Vasic, V. & Regner, S. (1997), Biological diversity of FR Yugoslavia: assesment, threats and polices. Ecolibri-Bionet, Beograd i Ministarstvo za razvoj, nauku i zastitu zivotne sredine SR Jugoslavije, Beograd (reprint: 2000).

Wilson, E.O. (1992), The diversity of life. Harvard University Press. 424 pp.

Towards an International e-Learning Didactic for Sustainable Development
–
The European Virtual Seminar on Sustainable Development (EVS)

Hans-Peter Winkelmann

Executive Chairman of COPERNICUS-CAMPUS,
the European University Network for Sustainability
Dortmund, Germany

It is a long way from Nicolaus Copernicus' heliocentric system to the modern COPERNICUS-CAMPUS of European universities. But the fact that the name Copernicus was chosen is certainly no mistake: Nicolaus Copernicus (1473-1543) changed the world by his academic observations resulting in the notion, that man is not the centre of the universe which was a revolutionary idea at that time. Changing the perspective on a more and more globalized and knowledge-based world to guarantee sustainable development is at the heart of the COPERNICUS-CAMPUS.

Higher education occupies a central position in shaping the way in which future generations learn to cope with the complexities of globalization In Europe. The educational targets set out by the European Union in the Lisbon Strategy can not be realised without paying attention to ecological and social/ cultural aspects. Therefore, in addition to the economic function of universities also the cultural function of universities remains important. Social and cultural diversity could also be seen as a potential resource. To achieve sustainable development and to bring Europe to the leading edge in the knowledge economy and society will present universities with a number of challenges and opportunities. The emergence of a global knowledge market, more competition, new competitors, institutions that are expanding beyond traditional territories, all these effects of globalization force universities to improve their educational and training offers. Universities need to react to new market requirements with more flexibility, high quality, institutional branding and ranking. Changing demographics, changing student population create an increasing need for lifelong learning and permanent up-grading of knowledge. No longer can universities see themselves as only part of a national system. In their strive for excellence, in a much more competitive world, universities must more and more support each other with the emergence of "university strategic alliances". The development of new technologies makes E-Learning an excellent instrument for fulfilling all these requirements.

Through E-Learning universities can take up theses challenges and will present a pro-active approach on what they could do to realise the Bologna reforms in order to face globalization, to serve the learners of the future and to contribute to the "Europe of knowledge". E-learning offers a new dimension in the European Higher Education Area to contribute to the "Europe of knowledge", fit for a society in which new media and virtual space are part of the standard procedures of universities. Using the potential of Information and Communication Technologies (ICT) for sharing and developing knowledge in a time- and place independent way, networking among universities and university students is of vital importance for learning how to live with and draw from this European diversity. Networking among European universities and explorations of the usefulness of ICT in this respect is very important and an important value in addition to sustainable development.

The benefits E-Learning will offer can be summarized by the following aspects: There is a widened access to higher education for traditional and new groups of learners. Students are offered more flexible, high-quality education opportunities integrating different teaching and student learning styles, offered at convenient (self determined) times. The programmes meet the needs of students on the one hand, on the other hand learners gain access to many different types of educational resources, they can choose out of a variety of offers – which enhances quality. Furthermore E-Learning allows them to overcome limitations of traditional learning settings by modelling, simulation and visualizing of complex processes. E-Learning is an effective educational training for lifelong learning. The tool is learner-controlled and learner-centred, encourages active participation, interaction, group collaboration and cooperative learning, as well as deals with problem solving strategies and discourse. High quality E-Learning can empower students, encourage critical thinking and lead to the involvement in lifelong learning opportunities which is in particular important for education for sustainable development.

The Lisbon Strategy is a commitment to bring about economic, social and environmental renewal in the EU seeking to improve the economic and social performance of the Union and guarantee sustainable development. To achieve this ambitious goal, Europe should be the world leader in terms of the quality of its education and training systems[1]. Making this happen, COPERNICUS-CAMPUS follows the institutional model of a consortium of partners constituted to develop and offer E-Learning for sustainable development in higher education.

As the major 'visiting card' in this context, COPERNICUS-CAMPUS has developed and coordinated projects in e-learning for sustainable development, such as the European Virtual Seminar for Sustainable Development (EVS). Initiated by COPERNICUS-CAMPUS and the Open University of the Netherlands with a

1 Communication from the Commission „Education & Training 2010, the success of the lisbon strategy hinges on urgent reforms", (com(2003) 685 final).

support of the EU, the EVS started in 2000 with 6 Universities. In the academic year 2003/2004 the EVS was going to its third round with now 15 Universities from 10 European countries throughout Europe. In 2004/2005 already 20 partners from 13 European countries participated in the EVS. Via learning by doing, continuous monitoring and feedback the aim is to develop a didactic model for virtual collaborative learning in the field of sustainable development.

Fig. 1: Number of institutions participating in the EVS

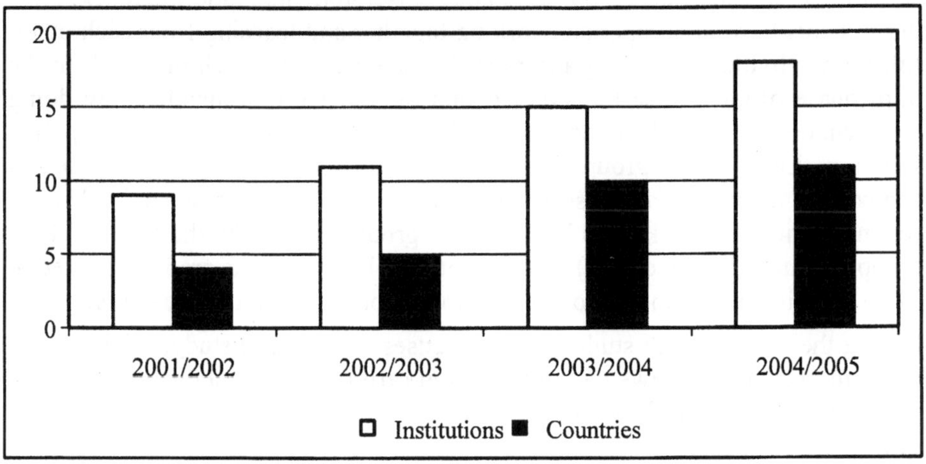

The aim of the EVS is to promote an international and multidisciplinary dialogue on sustainable development issues between students from all over Europe by using modern information and communication technology (ICT). The European Virtual Seminar on Sustainable Development focuses on sustainable development issues in an enlarging Europe. After a general introduction on sustainable development, the European Union and the enlargement process, students have to deal with case studies in the field of sustainable development. In the EVS students are directly confronted with different views and opinions from other students on societal processes of change towards sustainable development in Europe. Views and opinions that are highly related to different social, economic, political, cultural and environmental backgrounds the students are living in. To make this idea work the student groups in EVS are international and multidisciplinary compound, using modern ICT and the internet to overcome the constraints of place and time.

The didactic concept of EVS is that of virtual collaborative learning based on ideas of constructive learning theories. The student groups are international and multidisciplinary compound. A student group consist of five members spread over Europe. All communication between group members (and staff members) is facilitated by modern ICT. The virtual cooperation between group members is organised by a using a common timetable, clear activities and phases, and stimulated by tutors (responsible for the monitoring and coaching of the group process) and experts (authors of the case studies and responsible for answering content related questions of the students and they give comments on the group products). In the first phase of the course (starting in October in each academic year) students fulfil first individual activities as introducing themselves to other students, and become familiar to the electronic learning environment. In the second phase of the course students become enrolled into a heterogeneous student group (about the size of 4-6 students) based on nationality, discipline, institution and sex. Now the students have to co-operate, working together and learning from each other, fulfilling activities as writing and rewriting the group research proposal. In the third phase of the course the students and the other group members are doing research, consulting stakeholders, and deliver a common group report and policy summary. Each student group is coached and monitored by a tutor, and students can consult an expert for questions about the content of the case study. The assessment and final marking is based on the group process (by the tutor) and the group products (by the expert). To evaluate the EVS students are being asked to fill in questionnaires, and a group reflection report at the end of the course.

During the seminar each student group focuses on one case study, discusses the problem, exchanges views and opinions, and tries to find sustainable solutions. Before the EVS start all students are asked about their case study preference. The case studies are:

1. *Agriculture* – The future of agriculture in Poland.
2. *Biodiversity* – Nature and biodiversity conservation: Romanian forests.
3. *Communication* – Sustainability Communications. Strategies for Communicating the Idea of Sustainable Development.
4. *Spatial Planning* – European spatial planning and policy. Case: the regions with the Mediterranean forest Dehesa.
5. *Water* – Integrated water management. Implementation of the European Water Framework Directive in an international perspective: the Danube.

Formal Issues

The EVS is offered in the period between October and February. The study load is 120 hours, approximately 8-10 hours a week. Students that pass for the EVS receive a 'certificate of completion' from the EVS coordinator, and – if arranged in advance – credit points from their institutional contact person. In the EVS almost 20 institutions from 13 different European countries are participating. The Open University of the Netherlands is the coordinator of the EVS.

Requirements to Participate

Students that like to participate in the EVS should:

- be interested in sustainable development issues
- be interested in group work
- be able to read and write English
- be able to study 8-10 hours a week on this course
- have access to a computer and internet connection
- have an open mind to E-learning.

Project Objectives

- develop a didactic concept for virtual collaborative learning
- develop a model for co-operative development and exploitation of electronic courses via Internet
- applicable in other knowledge domains (content) and at different scales (international, national, local).

Fig. 2: The EVS web site: http://blackboard.ou.nl/evs

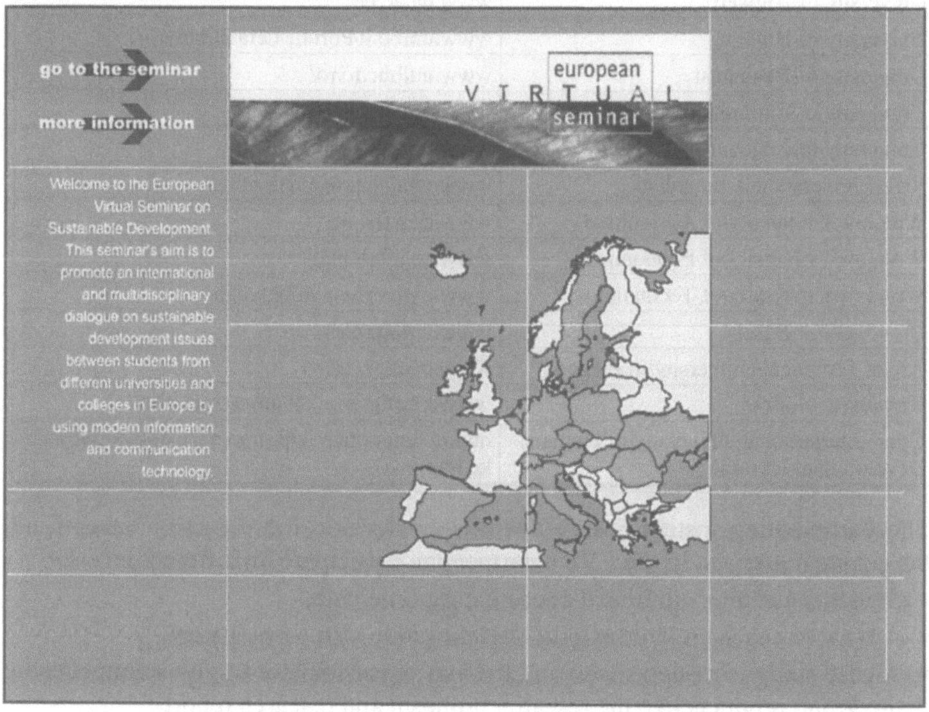

Learning Objectives Students

- link regional, national and European policy-related issues to sustainable development and enlargement
- collaborate in an international and multidisciplinary team
- become a member of a virtual community of learning.

Tab. 1: Institutions that participated in the EVS (2004/2005) are:

Institutions	**URL**
Charles University in Prague	www.cuni.cz/
COPERNICUS-CAMPUS	www.copernicus-campus.org/
Czech University of Agriculture in Prague	www.czu.cz/
Eszterházy Károly College	www.ektf.hu/index.php
Karkonosze College	www.kk.jgora.pl/
Karlstad University	www.kau.se
Open Universiteit Nederland	www.ou.nl
Technical University of Liberec	www.vslib.cz/
University of Amsterdam	www.uva.nl
University of Antwerp	www.ua.ac.be
University of Bologna	www.unibo.it/Portale/default.htm
University of Bucharest	www.unibuc.ro/ro/
University of Extremadura	www.unex.es/
University of Lüneburg	www.uni-lueneburg.de/
Vrije Universiteit Amsterdam	www.vu.nl/home/index.cfm
Wraclow University of Agriculture	www.ar.wroc.pl
Wraclow University of Economics	www.ae.jgora.pl/
Wraclow University of Technology	www.pwr.wroc.pl/index.php
University of Basel	www.unibas.ch/
Czech Technical University in Prague	www.cvut.cz/cz/
University of Graz	www.kfunigraz.ac.at/homepage.html
Open International University of Human Development Ukraine	http://samoilenko.iatp.org.ua/educorp/vmurol-ke.htm

The participating institutions co-operatively develop the didactic concept and the course materials. In the EVS a partner can participate on different levels:

- Students and an institutional coordinator participate;
- A tutor to coach an international student group also participates;
- A case study is made as well and the expert participates to give comments on the group products like the research proposal and research report.

Internationalisation and Virtual Collaborative Learning in the EVS: Lessons Learned

The EVS is being organized by sharing expertise and investment of staff time and resources by all partners. That means that there is no joint or extra budget for exploitation or further development/improvement of EVS. The total costs of the EVS as a strategic university alliance for e-learning for the co-ordination of the activities of all involved universities and of their representation outward will be lower therefore than the sum of the costs, which must be spent otherwise on single activities of the universities. The benefits of this partnership are clear: Students from any member university enjoy access to a European network no single university could build up on its own. The students benefit from coordinated virtual courses, easy access to coordinated learning options from one university to another, mutually recognized credits, and a cooperative European learning and working environment in an international, intercultural and interdisciplinary approach. The EVS ultimately results in both guaranteed quality and increased mobility. The most important aspect of this strategy is building up mutual trust between universities. This will open the door to increased cooperation between the institutions, and as a result, better opportunities for mobility. The EVS is one such initiative aimed at reinforcing mutual trust. Therefore the EVS pays off for all partners.

Formal Issues and Quality Guarantee

In view of the more formal issues of the EVS it can be concluded that international cooperation takes a lot of time, because of differences in culture, different educational paradigm, etc. Especially the students, but also the EVS staff, must learn to work together with people from different cultural backgrounds which is quite time-consuming. However, both students and staff regard this challenge as very rewarding.

The EVS staff members and most students are enthusiastic about the EVS concept of international and multidisciplinary collaborative learning which is understood as a necessary skill for learning for sustainable development.

At the start of each EVS round most students also lack specific computer skills for virtual collaborative learning, and often they don't have experience in group work. The reason for these basic lack of skills might be seen in structural problems in today's university teaching procedures and didactical methods.

Fig. 3: Institutions that participated in the EVS (2004/2005)

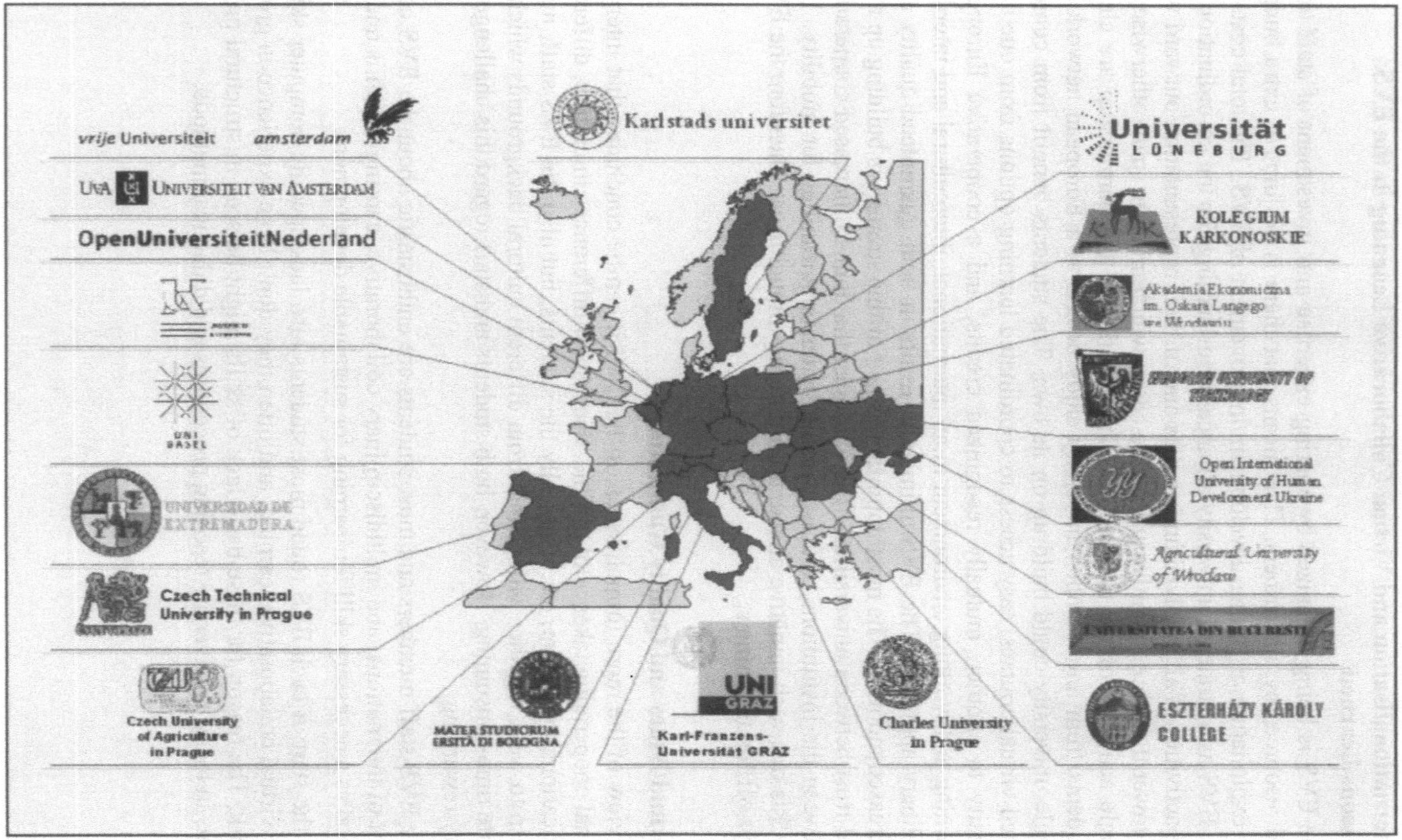

The team building process in the EVS requires a lot of attention by the course coordinators, because students don't know each other at the start and all social processes are dependant on virtual communication and interaction. One basic principle of the EVS is that there are no face-to-face meetings organized for students, mainly because of budget constraints and for organizational reasons. The EVS must be understood as a sole virtual course. However, after each EVS round students very often suggest as part of the evaluation to organise a summer school after EVS because they miss the physical contact or they are keen to become acquainted with their colleagues after a period of virtual collaboration over about 4 months.

The role of the tutor is crucial for coaching and monitoring of the student groups, especially when problems arise in a group. For that reason students ask for more steering and feedback from the tutor in order to get more guidance. In this regard it must be taken into account that students feel completely different about "virtual guidance" compared to the much more common traditional tutoring in ordinary courses and seminars.

The total number of dropouts is on average about 50%. In some participating institutions it is considerably lower. The quite high drop out rate can be explained by the complexity of the EVS (international, multidisciplinary, virtual, group work, etc.). In order to lower the drop out rate the course management should be stricter (selection of students, credit points for students, formal position of EVS in curriculum, etc.).

In view of perceiving the EVS as an approach towards an international didactic it can be concluded that the EVS model of sharing expertise, staff time and resources works. Together all partner institutions in the EVS offer an international learning experience, impossible to realise by an individual institution on its own. Furthermore, modern ICT makes it possible to internationalise the content of the curriculum (ICT as delivery tool), but also to communicate and interact worldwide (ICT as learning tool). Modern ICT makes E-learning possible, but E-Learning requires a didactic concept to be successful. At the same time international E-Learning requires specific skills from staff (teaching in an intercultural context, using modern ICT, etc.) and students (language, miscommunication due to cultural differences, etc.).

An international program on sustainable development should offer students international, intercultural and multidisciplinary knowledge and skills, to prepare them on their professional and societal functioning in an international and multicultural context. Online courses as the EVS can be offered as building blocks in international programs on sustainable development

The Future of the EVS as Building Block of the Virtual COPERNICUS-CAMPUS

One basic principle of COPERNICUS-CAMPUS is that universities must learn from each other and need not re-invent the wheel time and time again. For that reason COPERNICUS-CAMPUS is developing a European Virtual Campus on Sustainable Development, hosted by COPERNICUS-CAMPUS and based on a consortium of five European universities that are frontrunners in the field of E-learning on/for sustainable development. The five universities are:

1. University of Graz (Austria)
2. University of Lüneburg (Germany)
3. University of Macedonia (Greece)
4. Open University of the Netherlands (The Netherlands)
5. Charles University Prague (Czech Republic).

The Virtual COPERNICUS-CAMPUS (VCC) will offer new learning possibilities for the students by bundling existing E-Learning opportunities and the knowledge and experiences of the university teachers from the participating universities and will thereby support the virtual mobility of students between the five participating universities and beyond.

The Virtual COPERNICUS-CAMPUS will establish a strategic alliance of leading European universities ready to share experiences and develop together innovative approaches in E-Learning. By using ICT for sharing and developing knowledge in a time- and place independent way, the Virtual COPERNICUS-CAMPUS addresses three areas which are absolutely central to the future of most higher education institutions in Europe:

- Cost-effectiveness;
- Improving the quality of higher education and making it more attractive and future-oriented;
- Internationalization.

The local activities, projects, expertise and experience of the five universities are "building blocks" to create this European Virtual Campus. This will be the 'first step' in the project: sharing courses, access for all students, flexibility, etc. The 'second step' in the project will be that the participating experts learn from each other and develop together new instruments, methods and modalities for European Cooperation in E-Learning on/ for Sustainable Development (mobility schemes, ECTS), upgrading the European universities in effective ICT use. The 'third step' of the project will be that the participating universities create a European Virtual Master on Sustainable Development as a long-term perspective. The Virtual Master would be the spin-off of the Virtual Campus, and would 'formalize and institutionalize' the cooperation.

Students from any member university will enjoy access to a European network no single university could build up on its own. The students benefit from coordinated virtual courses, easy access to coordinated learning options from one university to another, mutually recognized credits, and a cooperative European learning and working environment in an international, intercultural and interdisciplinary approach. At the end of the project other universities can become a partner in the Virtual COPERNICUS-CAMPUS, and probably also the Virtual Master.

COPERNICUS-CAMPUS is a European University Network for Sustainability founded by the former European Rectors Conference (CRE) which became an independent network in 1999. To date (1 August 2004), it counts 324 member universities in 38 European countries. The focus of COPERNICUS-CAMPUS is on the activities of higher education institutions in the field of sustainable development. This implies that in curricula, institutional management and services to the local/regional society a responsible balance between economical aspects, ecological aspects and societal/cultural aspects has to be worked out.

COPERNICUS-CAMPUS supports cooperation between its members through thematic projects and offers them a communication platform on sustainable development. As the European University Network for Sustainability are responding to the need expressed by its members, COPERNICUS-CAMPUS identified the initiatives best positioned to develop the European virtual campus. These initiatives are leading European initiatives in terms of E-Learning for sustainable development. They are very different in size and scope and have a European or international dimension, which makes the project group the best able to understand and carry-out an in-depth study of the requirements to create a European Virtual Campus and to develop scenarios for its implementation.

The VCC project will bring together existing E-Learning initiatives in the field of sustainable development from the five universities in Austria, Germany, the Netherlands, Greece and the Czech Republic. The project aims directly at the creation of a virtual campus that should serve as a platform for E-Learning for sustainable development in Europe to pool the resources from the five participating universities and to offer a common E-Learning platform for the students from these institutions. The Virtual COPERNICUS-CAMPUS will bundle existing E-Learning opportunities from the participating universities and will therefore support the virtual mobility of students between the five participating universities. The project targets first on the five European universities involved which will constitute the founders of the VCC. The VCC as a European cooperation platform might also be used as a promotion tool for E-Learning for sustainable development and could therefore be opened to any other European university interested in offering E-Learning on sustainable development to its students after the successful completion of the two-year pilot project 2005-2007. This means new learning possibilities for the students from the five participating universities; bundling of the knowledge of the best university teachers in

Europe. For the students the VCC will extend the learning options offered by their university and will create in addition new possibilities to complement their courses with a European perspective.

To gather together the knowledge and experiences of the best university teachers from the individual universities is the main aim of the VCC. The VCC as a group of committed European universities from the Copernicus network will offer the suitable platform for that. The leading role of those universities engaged in the VCC in the area of E-Learning and the prominent position of these universities with regard to research on sustainability is the commitment to offer this leading teaching wherever the new technologies make this possible. Such trailblazing offers beyond each university will create the basis for co-operation with universities in other countries and their developments into virtual teaching and learning offers. The fruitful development for all involved universities will lie in a balanced reciprocal exchange of their virtual offers, not in the one-sided assumption.

Individual universities – including the larger among them – would be in most cases overtaxed with the necessary investments and infrastructures of the next few years. Only a co-ordinated bundling of all forces offers the necessary critical mass in order to be able to keep up in the global competition in virtual higher education. By bundling the individual strengths and competences of the involved institutions the excellence of these universities will become even internationally more visible and simply accessible and can create so the basis for a wider demands. The VCC will be therefore use an organizational form which offers an appropriate co-operation of the universities involved.

Learning about sustainability on a university level is due to the fact that VCC means they are no longer limited to the geographical university locations and their lecture-rooms and seminar area at firm term times limits. Teachings and highly-qualified support come to the learners, to the job, into the dwelling, at arbitrary times, at the decision of the individual learner, adapted to the preferred learning style of each individual, with a custom-made profile to contents and qualifications.

The VCC is in the first instance a thematic-oriented alliance of universities in E-Learning that will promote virtual mobility of their students within this consortium. A virtual university as a new independent university – independent from the existing – would require a long time and huge financial resources. It would also neglect to a big extent the existing entire expertise and the extensive existing infrastructures of the universities. The VCC will therefore bundle the existing experience and competences as well as the study programmes and the infrastructure at the involved universities. Thus it becomes a comprehensive virtual teaching and learning consortium of traditional universities. Experienced researchers and university teachers in prominent positions in their fields of activity are available thereby as authors, responsible persons and examiners.

The VCC does not want and cannot replace the proven forms of the university teaching. Therefore no new independent university is in competition with the existing. The existing university mechanisms will be extended with the help of the VCC through modern service structures and alternative training forms: The partner universities of the VCC guarantee by their extensive research activities the necessary scientific basis that the training offers. Through its partner universities the VCC can fall back also to all proven and recognized forms of university examinations, diplomas and certificates. It needs therefore no additional new study and examination regulations. The VCC will promote a coordinated cooperation between the participating universities around a common learning content by using the most appropriate methods of virtual teaching.

The UniAdrion Experience

–

A Virtual University for the Adriatic-Ionian Basin

Andrea Segrè[1], Luca Falasconi[2], Marina Mantini[3], Sabina Morganti[4], Elisabetta Nanni[5], Hajnalka Petrics[6], Matteo Vittuari[7]

[1] Professor of Agricultural Policy and Chairman, Department of Agricultural Economics and Engineering, University of Bologna, Italy
Scientific Coordinator of the "Environment and Sustainable Development Group" of UniAdrion – Virtual University of the Adriatic – Ionian Basin

[2] PhD in Agricultural and Food Economics, Department of Agricultural Economics and Engineering, University of Bologna, Italy

[3] PhD candidate in "International Cooperation and Sustainable Development Policies", Department of Agricultural Economics and Engineering, University of Bologna, Italy

[4] PhD candidate in Agricultural Engineering, Department of Agricultural Economics and Engineering, University of Bologna, Italy

[5] PhD candidate in Agricultural Engineering, Department of Agricultural Economics and Engineering, University of Bologna, Italy

[6] PhD candidate in "International Cooperation and Sustainable Development Policies", Department of Agricultural Economics and Engineering, University of Bologna, Italy
Tutor of the "Environment and Sustainable Development Group" of UniAdrion – Virtual University of the Adriatic – Ionian Basin

[7] PhD candidate in "International Cooperation and Sustainable Development Policies", Department of Agricultural Economics and Engineering, University of Bologna, Italy
Tutor of the "Environment and Sustainable Development Group" of UniAdrion – Virtual University of the Adriatic – Ionian Basin

Abstract

This paper is based on three years work experience within the UniAdrion (The Virtual University of the Adriatic – Ionian Basin) project, that is aimed to promote culture, cooperation and a permanent connection among Universities and Research centres from the Adriatic-Ionian Region. This "University Network" was born as a consequence of the Adriatic-Ionian Initiative established in two different conferences: "Culture as a Bridge – The Interuniversity Cooperation in the Adriatic-Ionian Basin" held on 15th-16th December 2000 in Ravenna, Italy and "Development and Security in the Adriatic-Ionian", held the 19th-20th May 2000 in Ancona, Italy[1].

The aim of this paper, on the one hand, is to present a case study on *Distance Learning* and on the use of *Information Technologies* for education in the field of *agriculture and rural development*. The case study has been prepared on the basis of the results of the Working Group 2 (Environment and Sustainable Development) of UniAdrion, which has developed the Course of Specialization on International Cooperation and Sustainable Development of Agricultural, Environmental and Rural Systems, a specialization course based both on frontal and virtual (e-learning) modules. On the other hand, the paper desires to contribute to the broadening of the perspective on developments in the use of Information Technologies. Among the results of the Project particular attention will be given to Peace Parks (Transboundary Natural Parks) as tools to promote peace, cooperation and development in the Balkans.

Members and Associated Members of Uniadrion[2]

Regular Members[3] of UniAdrion are all the Universities that have signed the Ravenna Declaration before the 31st January 2001. After that date, only the Universities that forward their request to the UniAdrion General Assembly (responsible for examining and approving the applications), can become members of UniAdrion. Each Member University is represented in the UniAdrion General Assembly by the Rector or his/her delegate. Both individuals and juridical entities may join

---

1 More information can be found on the UniAdrion website: http: //www.uniadrion.net.

2 More information can be found on the UniAdrion website: http: //www.uniadrion.net.

3 Regular Members of UniAdrion: Shkodra University "Luigj Gurakuqi" (Albania); University of Tirana (Albania); Technological University Ismail Qemal Vlora (Albania); University of Banja Luka (Bosnia-Herzegovina); Džemal Bijedić University (Bosnia-Herzegovina); Croatian University of Mostar (Bosnia-Herzegovina); University of Sarajevo (Bosnia-Herzegovina); University of Rijeka (Croatia); University of Split (Croatia); University of Zadar (Croatia); University of Ioannina (Greece); Ionian University (Greece); University of Patras (Greece); Università Politecnica delle Marche (Italy); University of Bologna (Italy); University of Lecce (Italy); University of Teramo (Italy); University of Trieste (Italy); University of Udine (Italy); University of Venice Ca' Foscari (Italy); University of Arts in Belgrade (Serbia and Montenegro); University of Belgrade (Serbia and Montenegro); University of Montenegro (Serbia and Montenegro); University of Maribor (Slovenia).11 Associate Members of UniAdrion: Inter-University Centre Dubrovnik (Croatia); Institute for International Relations (Croatia); Alternative Academic Educational Network (Serbia and Montenegro); University of Primorska (Slovenia); University of Ss. Cyril and Methodius Skopje Macedonia (FYROM).

UniAdrion as Associate Members[4]. Associate Members are entitled to take part in the Assembly but have only advisory-consultative rights.

UniAdrion at Glance

UniAdrion currently counts a total of 29 members, 24 full member Universities and 5 associated members (4 Research Centres and also the University of Skopje, not eligible for a full membership due to the fact that Macedonia does not belong to the Adriatic-Ionian Initiative). A total of eight sub-Projects (post-university courses or research projects) were devised for the implementation of the Network. They refer to the following topics: two of them refer to the commissioning of seven multimedia sub-stations/Focal Points (one for each UniAdrion member Countries) and to the transfer of the specific know-how needed to guarantee their proper functioning. Six training Courses refer to either higher education or conceptual activities, such as preservation of cultural heritage, environment conservation and sustainable development, promotion of a sustainable tourism and transport in the Adriatic-Ionian area. The major part of financial support has been guaranteed and offered by the Italian Ministry of Foreign Affairs and the Italian Ministry of Education. Moreover the University of Bologna contributed to the implementation of the project through a financial contribution (about 13% of the total amount).

The "environment and sustainable development" Working Group

Sustainability in agricultural, rural and coastal areas has at least four interrelated dimensions: the ecological, the social, the economic and the international dimension. Just referring to the economic dimension, World Trade Organisation rules do not allow any more projection policies to intensify production, while sustainability coupled with multi-functionality may represent an occasion to multiply additional sources of income (i.e. organic and natural farming, agro-tourism, recreational use of parks, and the like) promoting at the same time employment in rural areas. Present societies, both in the West and in the East, are looking for safe food, high quality products, wide range of variety, better protection of the environment, preservation of rural and coastal landscapes, and serious concern is mounting over animal welfare. For high quality, it is clear that one should accept paying the higher price and also to compensate farmers for their active role in preserving public goods.

4 Associate Members of UniAdrion: Inter-University Centre Dubrovnik (Croatia); Institute for International Relations (Croatia); Alternative Academic Educational Network (Serbia and Montenegro); University of Primorska (Slovenia); University of Ss. Cyril and Methodius Skopje Macedonia (FYROM).

The Course of Specialization on International Cooperation and Sustainable Development of Agricultural, Environmental and Rural Systems seeks to offer a deep analysis of the theoretical, legal and operational aspects related to the planning and management of agro-environmental systems, and, on this basis, develop sound methodologies for an active policy of endogenous sustainable development in agricultural, rural and coastal areas of the Adriatic-Ionian Basin. The Course, entirely in English, have been articulated in frontal and virtual modules that have been developed with the aim to create an open forum and an useful tool to learn and exchange ideas. The modules have been scheduled as follow:

- Module 1, First Summer School, Cervia (Italy), 2-13 September, 2002;
- Module 2, Second Summer School, Cervia (Italy), 1-12 September, 2003;
- Module 3, On-line work with tutorship, October-November, 2003;
- Module 4, On-line video-lectures (smil modules), March-August, 2004;
- Module 5, Third Summer School, Tirana (Albania), 20 September-2 October, 2004;
- Discussion of the dissertation, November 2004.

The Course of Specialisation has been addressed to graduates with at least a three-year degree, as well as to officials already working in local, national or international agencies belonging to the countries of the Adriatic-Ionian Initiative. The Course should have provided participants with a specific set of competence to aim a career at various levels in local, national, and international governmental, as well as non-governmental, organisations. Particular relevance was given to the "virtual modules" with the aim of creating a tool to develop and promote distance education and to build a permanent "virtual platform" to improve communication among Universities and Research Centres, professors and researchers from the Adriatic-Ionian Basin. Among the main issues of the Course the following topics were included: sustainable development and international relations; environmental management systems: valuable tools to achieve sustainable development; peace parks as a tool to achieve sustainable development and stability; cross-border Balkans peace park; economic development and agrarian systems analysis; agriculture, rural development and international relations; agricultural and rural development policies and European enlargement; the agrarian system diagnosis: an analytical tool for regional studies; introduction to systems analysis: background to the Cropping Systems Simulation Models; cropping systems: definition, characterisation and simulation models; sustainable rural development policies; indigenous knowledge in agriculture and rural development.

Parks for Peace

According to the World Commission on Protected Areas (WCPA) a "Park for Peace" covers "transboundary protected areas that are formally dedicated to the protection and maintenance of biological diversity, and of natural and associated cultural resources, and to the promotion of peace and co-operation"[5]. Parks for Peace might be taken into consideration at different levels with different implications. A Park for Peace might be considered important from an ecological perspective (to protect an area from destructive large-scale development, mass tourism and external exploitation) as well as from an economical perspective trough the promotion of rural development and rural tourism, as from a social perspective trough the promotion of cooperation among different communities, a better border control (to fight smuggling ...). In this perspective a Park for Peace can be considered as a perfect example of sustainable development because they involve environmental protection as well as economic and social development.

The International World Conservation Network (IUCN) has published a substantial guide listing over 600 protected border areas worldwide[6]. The creation of some "trasboundary protected areas" in the Balkans could have particular importance and could be particularly effective. In fact, some projects have already been under implementation.

---

5 Trevor Sandwith, Clare Shine, Lawrence Hamilton and David Sheppard, Transboundary Protected Areas for Peace and Cooperation, IUCN, The World conservation Union, Cardiff University, Best Practice Protected Area Guidelines Series no. 7, p. 3.

6 Trevor Sandwith, Clare Shine, Lawrence Hamilton and David Sheppard, Transboundary Protected Areas for Peace and Cooperation, IUCN, The World conservation Union, Cardiff University, Best Practice Protected Area Guidelines Series no. 7, p. 3.

EU Targeted Plenary

Research for Sustainable Development – EU Strategy and Policy

The Contribution of Energy Policy to Sustainability

Stina Soewarta

Cabinet of Commissioner Andris Piebalgs,
European Commission
200, Rue de la loi, B – 1140 Brüssel, Belgien

It was a great honour for me to be invited to take part in the conference on sustainability.

EU energy policy is one of the keys to sustainability. Let us look at what can be achieved in the next five years. Let me first look at the priorities for the Energy policy then at the trends, and finally at the possible solutions.

EU Energy Policy Priorities

For the EU, energy policy must target three main objectives:

- to ensure the security of energy supply – we do not want more blackouts;
- to make the internal market in energy work and be competitive – which should lead to competitive prices for consumers and businesses;
- and to contribute to sustainable development – meet our Kyoto commitment notably in terms of reduction of CO_2 emissions.

To meet these objectives, Commissioner Piebalgs have identified six priorities:

- Reduce energy demand through energy efficiency;
- Ensure the proper functioning of the internal markets for gas and electricity for the benefit of all our citizens;
- Promote renewable energy sources;
- Create a better linkage between energy policies and environmental and research policies;
- Strengthen nuclear safety and security, and
- Further develop external energy policy relations.

Trends

Within the EU, domestic energy production and resources are on a clear downward trend. At the same time, demand is rising across all sectors and for all energy forms, at a steady 1-2% per year. The growth of GDP and the use of energy has been decoupled at the level of 1% already some years ago. But there is a worrying trend, this decoupling is now slowing down to around 0,5%.

The EU and other parts of the world will become increasingly dependent on external energy supplies if nothing is done. If current trends continue, the EU import share will grow from 50% today to 70% in 2030. By then, the EU will be 90% dependent on oil imports and 70% on gas imports. We are also confronted with growing energy consumption – not only in Europe (mainly lead by the transport sector), but also worldwide, in the US and especially in China and India.

Energy Efficiency Should be our Focus

It is clear that we only have a limited margin for manœuvre as far as the energy supply is concerned. Therefore we should act on the demand side.

Improving energy efficiency is the most cost-effective way to reduce our energy dependency and the only way to reverse current trends. It is also crucial to reduce our CO_2 emissions in order to meet our Kyoto commitments. Energy Efficiency is where Kyoto meets Lisbon via Gothenburg.

The EU has developed an important legislative framework for instance to improve energy performance in buildings, to promote cogeneration and to enhance energy efficiency labelling for electrical appliances. Other current initiatives include a directive on eco-design that will enable us to set efficiency requirements on a wide range of energy-using appliances and equipment. They also include a directive on energy efficiency and energy services with a proposed binding savings target of at least 1% per year.

Although much has been done in Europe, it is generally accepted that because of new energy-efficient technology, Europe can still improve its energy use by some 20% in the coming 15 years. This savings potential lies in all the end-use consumption sectors, including industry and transport. It is very cost-effective, often with zero or negative costs. We should redouble our efforts in terms of energy efficiency. This will be the message of the forthcoming Green Paper on Energy Efficiency.

Still, a lot remains to be done in the transport sector, which accounts for 32% of all energy consumption and 40% of CO_2 emissions. There is an agreement with the automobile industry to reduce CO_2 emissions for new cars to 140gCO2/km by 2008/2009. We have mandatory labelling of fuel efficiency for new cars and are now considering a proposal for a Directive on public procurement requirements for clean and energy-efficient vehicles.

Beyond energy efficiency, there are other areas we should be looking at to help a different energy future.

Other Policy Areas that can help a Different Energy Future

How can other areas contribute? Diversification of energy sources is part of the answer and in particular of renewable energy sources where, certainly, Europe is the world's leader especially in wind energy. We have set targets at EU level to reach 12% of overall energy consumption to come from renewables, and 21% of "green electricity" to be produced from renewable sources by 2010. We are also promoting the use of bio-fuels or other renewable fuels for transport. In connection with the agricultural policy, new action in the field of biomass is needed, as its potential is not yet fully exploited.

Research and technology can surely help develop a more sustainable energy future for example through clean coal technologies like carbon sequestration, but also fuel cells, hydrogen and bio-fuels.

The Road Ahead

Energy efficiency is the key priority for Commissioner Piebalgs this year. It will not only deliver important savings in terms of economy, but also create jobs, a better environment, and less dependency on imported energy sources. The solutions are there, it is now a question of having these solutions communicated and debated among the public. Habits need to be changed, and all citizens can save a substantial amount of money, and help the environment, without a significant change in lifestyle. The debate now needs to start.

Parallel & Poster sessions

A. Economic + Political Framework

Sustainable Development Concepts and Tools

–

The World Bank's Approach to Sustainable Development: Environmental Sustainability and Poverty Reduction

Hajnalka Petrics

Faculty of Agriculture, University of Bologna
Viale Fanin, 50 – 40127 Bologna, Italy

Abstract

Target 9 of the seventh Millennium Development Goal, environmental sustainability, calls for the integration of the principles of sustainable development into country policies and programmes and for reversing the losses of environmental resources. The World Bank's mission is to fight poverty and improve the living standards of people in the developing world. To achieve this goal growth is essential, but growth has to be responsible, growth that embraces both environmental sustainability and social development is needed to maintain the increases in human welfare in consumption, in health, in human skills and in social equity, which are targets of the MDGs. The aim of the paper is to review some of the instruments the World Bank is applying to realise its mission of poverty reduction, at the same time meeting the requirement of mainstreaming environment.

Introduction – The World Bank and Sustainable Development

The term sustainable development first came to prominence in the World Conservation Strategy (WCS) published by the World Conservation Union (IUCN) in 1980. It was then in 1987 when at the World Commission on Environment and Development (WCED) the concept of sustainable development was given a specific framework. Sustainable development has been defined as meeting "the [human] needs of the present without compromising the ability of future generations to meet their own needs." Moreover, the Commission saw the "possibility for a new era of economic growth, one that must be based on policies that sustain and expand the environmental resource base … such growth to be absolutely essential to relieve … poverty"[1]. Poverty thus has been seen as a major cause and effect of environmental degradation, for which the solution lies in economic growth that is equitable, and environmentally sustainable. The Commission called for, amongst others, institutional reform in terms of changing the nature of institutions and

1 Reid David, *Sustainable Development, An Introductory Guide*, Earthscan Publications Ltd, London, 1995.

laws to reflect the interconnectedness of environmental and economic problems. Since the WCED numerous different international forums have been organised, the United Nations Conference on Environment and Development in 1992, the World Summit for Social Development in 1995, the Millennium Summit in 2000, and World Summit on Sustainable Development in 2002, which all aimed at the reinforcing of actions for implementing sustainable development.

Among the different stakeholders, the leading international organisations have a particular role in realising sustainable development[2]. Considering this and the close relationship between environment, poverty and development, the mainstreaming of environment into other policy areas has become crucial. Environmental improvement should come not just from interventions directly on environmental issues but also from the incorporation of environmental sensitivities and objectives into the work of other sectors. Taking the World Bank as an example, this process of "mainstreaming" has been underway for a number of years and was strongly endorsed in the World Bank Environment Strategy (*Making Sustainable Commitments*: *An Environment Strategy for the World Bank*). The Strategy recognizes that sustainable development, which balances economic development, social cohesion, and environmental protection, is fundamental to the World Bank's core objective of lasting poverty alleviation.

The World Bank is committed to a comprehensive approach to sustainable development that builds on the three key pillars of development – enlightened public policy, a responsible and accountable private sector, and proactive civil society organizations – to achieve the UN Millennium Development Goals (MDGs)[3]. According to the Bank, sustainable development depends on the nurturing and strengthening of the five key forms of development capital:

- Financial capital: sound macroeconomic planning and prudent fiscal management.
- Physical capital: infrastructure assets such as buildings, machines, roads, power plants, and ports.
- Human capital: good health and education to maintain labour markets.

---

2 The UN Development Programme (UNDP) promotes sustainable human development; the International Monetary Fund (IMF) and the Organization for Economic Development talk of sustainable economic growth; the European Union, since Maastricht, has explored the notion of sustainable economic and social development; and the World Bank is committed to sustainable development and equitable development.

3 1. Eradicate extreme poverty and hunger. 2. Achieve universal primary education. 3. Promote gender equality and empower women. 4. Reduce child mortality. 5. Improve maternal health. 6. Combat HIV/ AIDS, malaria and other diseases. 7. Ensure environmental sustainability. 8. Develop a global partnership for development.

- Social capital: people's skills and abilities as well as the institutions, relationships, and norms that shape the quality and quantity of a society's social interactions.
- Natural capital: natural resources, both commercial and non-commercial, and ecological services which provide the requirements for life, including food, water, energy, fibres, waste assimilation, climate stabilization, and other life-support services.

Progress on the MDGs will be possible only if each form of capital is recognized, supported and enhanced. In 2003 the World Bank identified seven areas on which to focus its efforts what regards sustainable development among which we can find environmental sustainability at the first place. The other areas are water supply and sanitation; education; HIV/AIDS; maternal and child health; investment climate and finance; and trade.

Environmental Sustainability

Natural and human-made environmental resources – fresh water, clean air, forests, grasslands, marine resources, and agro-ecosystems – provide sustenance to people and are the foundation of social and economic development. The Environment Strategy of the World Bank[4] focuses on forming links among environment, poverty, and development, with particular emphasis on the health, livelihoods, and vulnerability of poor people. Protecting the environment is thus a key pillar of the Bank's mission of alleviating poverty and sustaining the quality of development.

The Word Bank uses different tools for promoting environmental sustainability; these include the Country Environmental Analysis (CEA), the Strategic Environmental Assessments (SEA), and finally the incorporation of environmental considerations – as cross-cutting issues into other sector strategies, such as those for water resources, rural development, and poverty reduction. This approach was emphasised in the document "*Linking Poverty Reduction and Environmental Management, Policy Challenges and Opportunities*"[5] too, stating that "environmental management cannot be treated separately form other development concerns. Rather, it must be integrated into poverty reduction and sustainable

---

4 The Environment Strategy makes part of the group of the Bank's Sector Strategy Papers (SSPs). The SSPs are Network products that review the Bank's operational experience within a sector or thematic area, identify priorities for the Bank at the global level, and set out and implementation plan for pursuing the priorities. They define indicators that can serve as a basis for assessing country and Bank performance, and for identifying priority country programs for Bank involvement.

5 The "*Linking Poverty Reduction and Environmental Management, Policy Challenges and Opportunities*", July 2002, was prepared as a contribution to the 2002 World Summit on Sustainable Development, Johannesburg.

development efforts in order to achieve significant and lasting results". The paper is going to concentrate on how the World Bank is promoting sustainable development incorporating environmental considerations into its Poverty Reduction Strategies (PRSs).

1. The World Bank's Approach Towards Environment and Poverty

1.1 Environmental and Poverty Issues – The Context

Environmental quality is closely linked with the quality of people's lives. Since it is the poor whose health is most endangered by air and water pollution, whose livelihoods are most affected by the loss of forests and fisheries or by soil erosion, and who are most likely to be at risk from environmental catastrophes, environment plays a particularly important role in the lives of the poor. Efforts to achieve concrete environmental improvements that make a difference to people's lives must be an integral part of economic development and poverty reduction programs.

The close relationship among poverty, environment, and development are increasingly understood. The links between poverty and environment are particularly close when poverty is viewed as a multidimensional phenomenon rather than simply a matter of income[6]. This is reflected in the UN Millennium Declaration's International Development Goals for 2015, which closely associate the goals of poverty reduction, heath improvements, and environmental protection. The Millennium Development Goals framework captures the environmental concerns in *Goal 7*: *Ensure environmental sustainability*. The targets associated with this goal refer to mainstreaming the environment in policy and programs, reversing the loss of environmental resources, and improving access to environmental services. It makes sense to try to achieve the goals together, because of the many synergies among them. Addressing environmental issues would help to achieve the other goals, and achieving the other goals would help to ensure environmental sustainability. The centrality of the environment to the Millennium Development Goals is reinforced by its strong linkages to the rest of the goals. Environmental factors influence the different dimensions of poverty – affecting people's opportunity, security, and empowerment – in many different ways.

---

6 Poverty now is viewed as encompassing both income and non-income dimensions of deprivation – including lack of income and other material means, lack of access to basic social services such as education, health, and safe water; lack of personal security; and lack of empowerment to participate in the political process and in decisions that influence someone's life. (The World Bank, *"Linking Poverty Reduction and Environmental Management, Policy Challenges and Opportunities"*, July 2002).

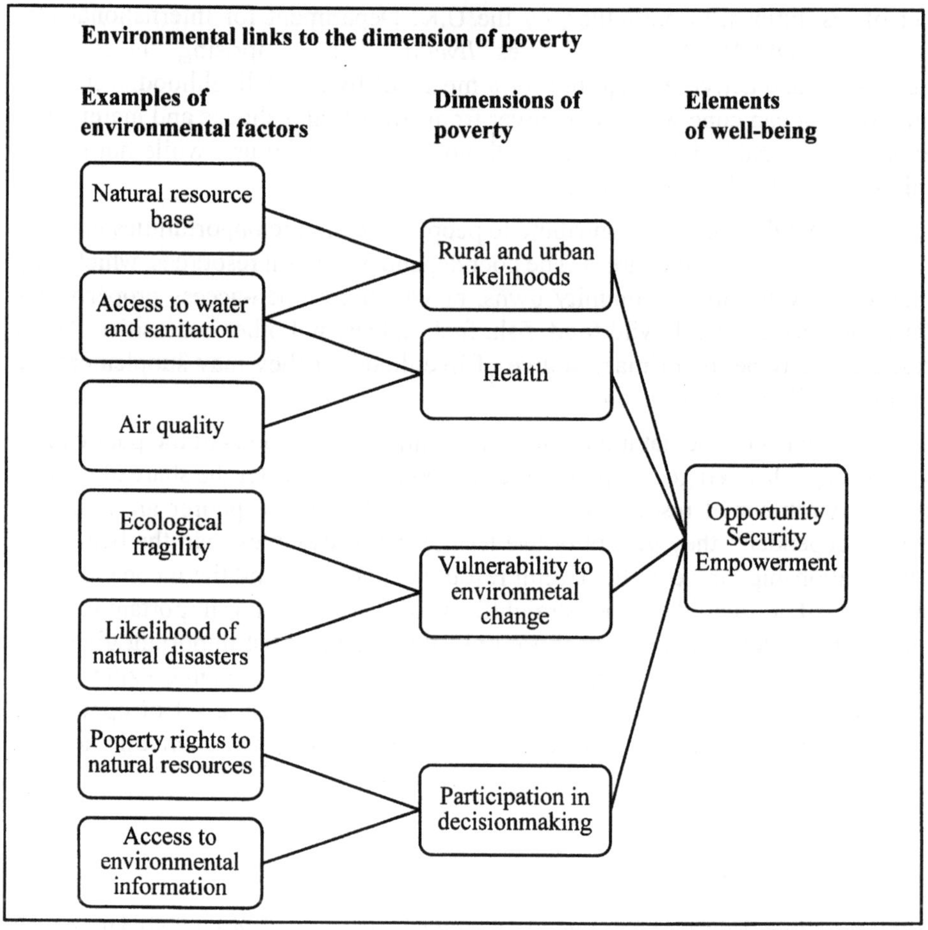

Source: *J. Bucknall, C. Kraus & P. Pillai* (2001). "Poverty and Environment." Environment Strategy Paper. World Bank, Environment Department, Washington, D.C.

The Bank's view on the links between environment and the dimensions of poverty can be seen in the following figure. Since a number of environmental factors affect the different dimensions of poverty, environmental improvements may be fundamental in strategies and actions to reduce poverty and sustain development.

The Bank describes how environmental factors influence different elements of well-beings (natural resource-based livelihoods and opportunities, environmental health and security and empowerment) in the following way.

Environment and Opportunity

The analysis of the relationships between the environment and economic opportunity is consistent with the sustainable livelihoods approach adopted by a number of institutions, in particular, by the U.K. Department for International Development (DFID). The DFID defines "*livelihood*" as comprising "the capabilities, assets and activities required for a means of living. A livelihood is sustainable when it can cope with and recover from stresses and shocks and maintain or enhance its capabilities and assets both now and in the future, while not undermining the natural resource base".[7]

Environmental conditions contribute to people's economic opportunities in many ways, particularly in rural areas. The poor depend on natural resources, which they may own, which the community owns, or which are open-access property, for farm- and grazing land, wild food, fish, fuel, fodder, and other necessities. These resources may be the primary source of livelihood, or they may supplement the family's daily needs or income.

The links between the natural environment and the livelihoods of the poor can be very strong. On average, the poorer the household, the greater the share of income from environmental resources. However, even though the poorer are more resource dependent, they generally use less of these resources than the better-off. The relationship between the natural resource conditions and the income or consumption of households are not straightforward. Therefore, it is important to build on country-specific analysis of those links that are the most important given the income level, income distribution, and nature of natural resource exploitation. Environment represents not only a set of problems but also a set of opportunities, for example, nature-based tourism can be a mainstream economic activity in many countries[8].

Environment and Health

Environmental health issues concern the linkage between improvements in human health and poverty reduction, taking into consideration a wide range of environmentally related diseases such as, vector-borne disease, diarrhoea and acute respiratory infections.[9]

---

7 The World Bank, Environment Chapter of the PRSP Sourcebook.

8 Ibid.

9 Disability-adjusted life-years (DALY) are a measure of the burden of disease. They reflect the total amount of health life lost to all causes, whether from premature mortality or from some degree of disability during a period of time. („*Linking Poverty Reduction and Environmental Management, Policy Challenges and Opportunities*", July 2002).

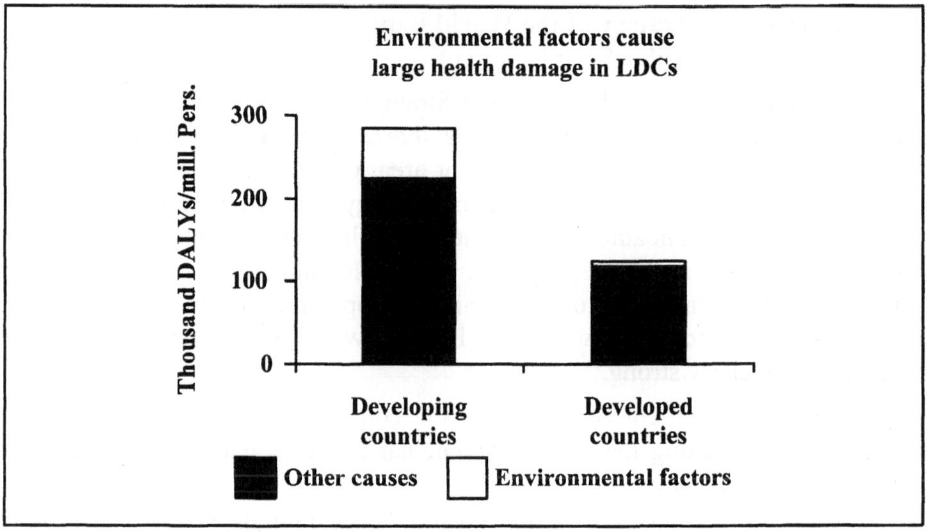

Environment and Security (Vulnerability)

While all members of a community suffer from external shocks, the poor are often disproportionately affected because they have the least ability to cope. Poor countries suffer more from environmental shocks than better-off countries. For example, in 1999 Hurricane Andrew hit the southern coast of the United States and caused 32 deaths, the same year a cyclone, a storm of similar intensity, hit Bangladesh and caused 100,000 deaths. Poor people are thus more vulnerable to micro- and macroeconomic level environmental shocks, environmental disasters, and environmental degradation.[10]

Environment and Empowerment

Environmental activities can contribute to the empowerment of poor people in at least two ways. First, they can use knowledge about environmental resources to organize themselves. Second, the ability of local communities to participate in decision-making about environmental resources can help them maintain their livelihoods, gain equitable access to resources, and use these resources in a sustainable way. The Environment Chapter of the Bank's PRSP Sourcebook suggests four factors which can make a community successful in managing common property resources, such as transparency, participation, inclusion, and ownership.[11]

---

10 Environment Chapter of the PRSP Sourcebook.

11 Ibid.

1.2 Environment Strategy of the World Bank – The Central Role of Environment

The Bank had prepared its Environment Strategy (*Making Sustainable Commitments: An Environment Strategy for the World Bank*) by 2001, to guide the World Bank's actions in the environment area, particularly over the following five years. The Goal of the Environment Strategy is to promote environmental improvements as a fundamental element of development and poverty reduction strategies and actions. In keeping with the World Bank's mission of reducing poverty within a framework of economic development, the Environment Strategy gives priority to issues where the links between poverty and the environment are particularly strong.

The Bank's Environment Strategy sets the connections between environment and poverty associating the key environmental concerns with three interrelated aspects of development:

1. Quality of life – especially with respect to the livelihoods, health, and vulnerability of the poor.

   The Strategy groups the effects of environmental conditions on poor people into three categories: livelihoods, health, and vulnerability.

Tab. 1: Quality of life – Effects of environmental conditions on poor people

1. Threats to livelihoods	**A.** Overuse, mismanagement, and contamination of freshwater resources.	**B.** Degradation of soil.	**C.** Rapid depletion of forests, fisheries, and biodiversity.	
2. Threats to health	**A.** Water-related diseases.	**B.** Exposure to indoor air pollution.	**C.** Exposure to urban air pollution.	**D.** Exposure to agricultural and industrial chemicals.
3. Vulnerability	**A.** Ecological fragility.			

2. Quality of growth – with particular regard to property rights, access to environmental information and education, institutions.
3. Quality of the regional and global environmental commons – especially with respect to climate change prevention, preservation of the ozone layer, conservation of biodiversity.

The Strategy, in relation to the connections set between environment and poverty, outlines as its objectives the followings:

1. Improving the quality of life
2. Improving the quality of growth

3. Improving the quality of the regional and global environmental commons.

The Bank foresees the achievement of the these three interrelated objectives through the realization of the following actions (Table 2):

Tab. 2: Bank actions

1. Improving the quality of life	**2. Improving the quality of growth**	**3. Improving the quality of the regional and global environmental commons.**
A. Enhance livelihoods by protecting the long-term productivity and resilience of natural resources and ecosystems.	A. Help improve the policy, regulatory, and institutional frameworks for sustainable environmental management.	A. Focus on the positive linkages between poverty reduction and environmental protection.
B. Prevent and reduce environmental health risks.	B. Support environmentally and socially sustainable private sector development.	B. Focus first on local environmental benefits, and build on overlaps with regional and global benefits.
C. Reduce people's vulnerability to environmental hazards.		C. Address the vulnerability and adaptation needs of developing countries.
		D. Facilitate transfer of financial resources to client countries to help them meet the costs of generating global environmental benefits not matched by national benefits.
		E. Stimulate markets for global environmental public goods.

Source: Environment Strategy Paper. World Bank, Environment Department, Washington, D.C.

2. Poverty Reduction Strategy Initiative

The Poverty Reduction Strategy (PRS) Initiative is a major effort by the World Bank and International Monetary Fund to make aid more effective in reducing poverty. The PRSP approach initiated by the IMF and the World Bank in 1999, and has resulted in a comprehensive country based strategy for poverty reduction. The aim of the PRSP is to provide the crucial link between national public actions, donor support, and the development outcomes needed to meet the United Nations' Millennium Development Goals, which are centred on halving poverty between 1990 and 2015. PRSPs provide the operational basis for Fund and Bank concessional lending and for debt relief under the Heavily Indebted Poor Countries (HIPC) Initiative. Governments of low-income countries thus prepare their own PRSPs in order to obtain debt relief and concessional lending from the International Development Association (IDA) or the Poverty Reduction and Growth Facility (PRGF) through a participatory process involving civil society and development partners, including the World Bank and the International Monetary Fund (IMF).

To achieve the overarching goal of poverty reduction, the key objective of the initiative enunciated by the Bretton Woods Institutions is to assist low-income countries in developing and implementing more effective strategies to fight poverty (that is, strengthening the poverty impact of public actions, including policies and spending) through supporting and sustaining a country-driven Poverty Reduction Strategy (PRS) process in low-income countries

The PRSP Initiative is characterised by two key features, the five underlying principles and several process requirements that link PRSPs to concessional BWI lending and HIPC debt relief.

The five underlying principles were drawn from the Comprehensive Development Framework and are set out in general terms for countries to follow. They indicate the PRS process should be:

- Country-driven, involving broad-based participation.
- Comprehensive in recognizing the multi-dimensional nature of poverty.
- Results-oriented and focused on outcomes that benefit the poor.
- Partnership-oriented, involving coordinated participation of development partners.
- Based on a long-term perspective for poverty reduction.

The process requirements call for preparation of PRSPs and their endorsement by the Boards of the BWIs for countries to gain access to resources. To obtain irrevocable HIPC debt relief, countries are required to produce an initial PRSP and implement it successfully for a year. Recognizing the considerable time required to formulate a full PRSP and the need for HIPCs to receive debt relief in the short term, debt relief is permitted to begin when a country completes an Interim PRSP. To obtain concessional resources from the BWIs, low income countries are required to complete PRSPs, or signal progress toward a full PRSP through an Interim PRSP or a PRSP Preparation Status Report.

Typically, the preparation of PRSPs involves three stages:

1. Developing a comprehensive understanding of poverty and its determinants.
2. Choosing the mix of public actions that has the highest impact on poverty reduction.
3. Selecting and tracking outcome indicators.

3. Mainstreaming Environment in Poverty Reduction Strategies

Because of the links between environment and poverty, and because a poverty reduction strategy must be environmentally sustainable over the long term, the Bank and the International Monetary Fund (IMF) have encouraged governments to consider environmental factors in their PRSPs. The Bank is helping to build the analytical base needed to quantify the links and prioritize potential environmental interventions compared with those in other sectors. Guidelines were prepared for the *PRSP Sourcebook.* The Environment Chapter of the Bank's PRSP Sourcebook sets out the relevance of the environment to the PRSP at three key stages:

- poverty diagnostics,
- choosing the most effective and efficient public actions to address the identified problems,
- monitoring and evaluation.

Because poverty-environment links are location specific, the Bank has taken a three-pronged approach toward helping integrate environmental issues into the PRSP process in individual countries: analytical work, training, review. The pilot program of integrating environmental considerations into the new PRSPs was started in November 2000. It includes

1. a systematic review of all interim and full PRSPs from an environmental perspective to rate the degree of mainstreaming,
2. workshops on mainstreaming for PRSP teams in client countries,
3. training for World Bank staff,
4. development of Environment Chapter in the PRSP Sourcebook, and
5. analytical work on poverty-environment links.

To date, four main studies were undertaken examining how environmental issues are mainstreamed in the PRSPs

- Jan Bojö, Rama Chandra Reddy, Poverty Reduction Strategies and Environment, A Review of 40 Interim and Full PRSPs, The World Bank Environment Department, December 2001.
- Jan Bojö, Rama Chandra Reddy, Poverty Reduction Strategies and Environment: A Review of 40 Interim and Full PRSPs, The World Bank Environment Department, June 2002.
- Jan Bojö, Rama Chandra Reddy, Status and Evolution of Environmental Priorities in the Poverty Reduction Strategies, An Assessment of Fifty Poverty Reduction Strategy Papers, Environmental Economics Series, The World Bank Environment Department, November, 2003.
- Jan Bojö, Kenneth Green, Sunanda Kishore, Sumith Pilapitiya, Rama Chandra Reddy, Environment and Poverty Reduction Strategies and Poverty Reduction Support Credits, The World Bank Environment Department, November 2004.

The latest systematic review of the available interim and full PRSPs from an environmental perspective to rate the degree of mainstreaming was prepared in 2004[12], and the World Bank and the IMF is currently undertaking a *2005 PRS Review* of implementation experience. This year marks the fifth anniversary of the PRSP initiative. In light of five years experience in helping to implement the approach, World Bank and IMF staff plan to use the 2005 review to undertake an in-depth assessment of progress, challenges, and good practice related to several key issues, such as that of environment too.

3.1 Method of Environmental Review of PRSPs

First of all, it is appropriate to define the term "mainstreaming". What the Bank means by mainstreaming was summarised in the 2003 Review in the following way: the term mainstreaming is not the existence of a stand-alone section or chapter in the PRSP, nor is it the frequent reference to the environment in the PRSP. The term mainstreaming used to denote the

- description of environmental issues and opportunities
- analysis of links between poverty and environment
- design of responses to meet the identified challenges
- inclusion of the environmental constituency. (The World Bank 2003)

The assessment framework used in the reviews is built on the first work of Jan Bojö and Rama Chandra Reddy[13] and on previous works on mainstreaming environment in the Country Assistance Strategies[14]. The review suggests a simple method that enables one to assess the environmental focus of PRSPs on a set of multiple criteria in a structured format. The scoring format provides a basis for assessing the progress on environmental variables at each stage, from the IPRSP to PRSP and from one 3-year PRSP cycle to the next (The World Bank 2001).

The framework for assessing the environment mainstreaming according to the definition of mainstreaming is across 17 variables under four categories:

1. *Issues* in focus: concerns the resource base, environmental problems, and opportunities;

---

12 "Environment in Poverty Reduction Strategies and Poverty Reduction Support Credits", November 2004, The World Bank Environment Department, Washington D.C. USA.

13 Jan Bojö, Rama Chandra Reddy, *Poverty Reduction Strategies and Environment, A Review of 40 Interim and Full PRSPs*, The World Bank Environment Department, December 2001.

14 Country Assistance Strategy (CAS) is the central tool for reviewing and guiding the Bank's country programs and the vehicle for judging the impact of its work. It summarizes the development context for the Bank's assistance in a client country and sets out the level and composition of the Bank's lending and non-lending activities. Its objective is to identify the key areas where Bank assistance can have the largest impact on poverty reduction.

2. *Causal links* assessment: poverty-environment link analysis (e.g. why is the environment degrading, how does this impact the poor);
3. *Responses* to environmental challenges: the country-level response systems to environmental challenges and opportunities are assessed based on the PRSP's proposal to improve the environment management capacity, targeted investments in promoting environmental objectives, and measuring the progress of interventions through monitoring and evaluation.
4. *Process*: participation and inclusion of environmental constituencies and voices. While these aspects are not uniquely environmental they are added because an inclusive and participatory process is an important requirement if the basic concerns of environmental health, natural resource degradation and vulnerability to natural disasters are to be heard, registered and acted upon. (The World Bank 2001).

Tab. 3: Format for scoring in the PRSP Assessment

Issues in Focus	1. Land use	Degradation deforestation, erosion, overgrazing, etc.
	2. Water	Drinking water, irrigation, fisheries and water pollution, etc.
	3. Air and climate	Air quality, solid fuel usage, emissions, climate variability
	4. Biodiversity	Threats to ecoysystems, species and genes, nature-based opportunities
Casual Link Assessment	5. Poverty and natural resource degradation	Resource dependence and inequality
	6. Environmental health	Water and air pollution related to disease
	7. Vulnerability	Impacts of natural hazards
	8. Property rights	Tenure and user rights
	9. Incentives	Pricing interventions, taxation, subsidies, exchange rate, trade, etc.
	10. Empowerment	Community-based management, decentralisation, and partnership
	11. Gender and environment	Role of women in environmental management
Response system	12. Environment management capacity	Legislation, regulation, institutional reform, data systems, cross-sectoral coordination, environmental standards
	13. Investment in natural capital	Investment in sustainable natural resource management, e.g. watershed management.
	14. Investment in human-made capital	Investment in environmental infrastructure, e.g. sewage treatment plants.
	15. Monitoring natural resource outcomes	Deforestation, afforestation, rehabilitated areas, protected areas, soil and water conservation measures, renewable energy use, etc.
	16. Monitoring human resource outcomes	Infant and child mortality, disease burden related to environmental risk factors, time spent collecting fuelwood and water.
Process	17. Description of the participatory process and inclusion of environmental constituencies, particularly with respect to the identification of environmental issues, poverty links, and actions.	

Source: *Jan Bojö, Kenneth Green, Sunanda Kishore, Sumith Pilapitiya & Rama Chandra Reddy*, Environment and Poverty Reduction Strategies and Poverty Reduction Support Credits, The World Bank Environment Department, November 2004.

The 17 variables identified and described above are scored with respect to each country's PRSP. A score range of 0 to 3 is used depending on the treatment of relevant issues:

0 = no mention,

1 = mentioned but not elaborated

2 = elaborated

3 = good practice.

The first three scores are related directly to the level of attention given, Overall the top score implies a judgment of the quality of the text.

4. Conclusions

According to the latest report which has reviewed 53 PRSPs, *Environment and Poverty Reduction Strategies and Poverty Reduction Support Credits*, there is a significant variation in the integration of environmental priorities into PRSPs, with the average country score ranging from 0.3 to 2.4. The average score shows a slight improvement – from 1.3 to 1.5 on the 0-3 point scale. All in all, PRSPs show an uneven attention to environment, but good practice exists[15], and full PRSPs tend to be much better mainstreamed than interim ones.

Tab. 4: Average country environmental mainstreaming score

S. No.	Country	Region	PRSP	Over all Score
1	Zambia	AFR	F	2.4
2	Ghana	AFR	F	2.2
3	Mozambique	AFR	F	2.2
4	Cambodia	EAP	F	2.2
5	Azerbaijan	ECA	F	2.1
6	Bosnia and Herzegovina	ECA	F	2.1
7	Honduras	LAC	F	2.1
8	Yemen	MNA	F	2.1
9	Sri Lanka	SA	F	2.1
10	Bolivia	LAC	F	2.0
11	Nicaragua	LAC	F	2.0
12	Kenya	AFR	I	1.9
13	Vietnam	EAP	F	1.9
14	Cameroon	AFR	F	1.8
15	Madagascar	AFR	F	1.8
16	Mongolia	EAP	F	1.8

15 Azerbaijan, Bolivia, Bosnia-Herzegovina, Cambodia, Ghana, Honduras, Mozambique, Nicaragua, Sri Lanka, Yemen, and Zambia have a score $\geq$ 2.0.

S. No.	Country	Region	PRSP	Over all Score
17	Djibouti	MNA	F	1.8
18	Burkina Faso	AFR	F	1.7
19	Malawi	AFR	F	1.7
20	Mali	AFR	F	1.7
21	Rwanda	AFR	F	1.7
22	Senegal	AFR	F	1.7
23	Ethiopia	AFR	F	1.6
24	Guinea	AFR	F	1.6
25	Lao PDR	EAP	I	1.6
26	Albania	ECA	F	1.6
27	Serbia and Montenegro	ECA	F	1.6
28	Benin	AFR	F	1.5
29	Chad	AFR	F	1.5
30	Niger	AFR	F	1.5
31	Guyana	LAC	F	1.5
32	Burundi	AFR	I	1.4
33	Mauritania	AFR	F	1.4
34	Armenia	ECA	F	1.4
35	Georgia	ECA	F	1.4
36	Bangladesh	SA	I	1.4
37	Pakistan	SA	F	1.4
38	Gambia	AFR	F	1.2
39	Kyrgyz Republic	ECA	F	1.2
40	Tajikistan	ECA	F	1.2
41	Uganda	AFR	F	1.1
42	Nepal	SA	F	1.1
43	Cape Verde	AFR	I	1.0
44	Tanzania	AFR	F	0.9
45	Cote d'Ivoire	AFR	I	0.8
46	Moldave	ECA	I	0.8
47	Dem. Rep. of Congo	AFR	I	0.6
48	Lesotho	AFR	I	0.6
49	Sierra Leone	AFR	I	0.6
50	Guinea-Bissau	AFR	I	0.5
51	Dominica	LAC	I	0.5
52	Central African Republic	AFR	I	0.3
53	Sao Tome Principe	AFR	I	0.3
		Average		1.5

Source: *Jan Bojö*, *Kenneth Green*, *Sunanda Kishore*, *Sumith Pilapitiya & Rama Chandra Reddy*, Environment and Poverty Reduction Strategies and Poverty Reduction Support Credits, The World Bank Environment Department, November 2004.

What regards the "issues in focus", more attention is given to land and water issues, while air pollution, energy use and biodiversity are given limited attention. As for the links between environment and the dimensions of poverty environmental health and climate vulnerability are better treated than the aspects of poverty rights, empowerment and gender. As far as the response system is concerned proposals for mainstreaming of environment are generally outlined, but, on average, information on the cost of interventions and schedule for interventions are not indicated. With regards to the participation and inclusion of environmental stakeholders, it is difficult to assess the extent to which environmental concerns of the poor are considered in developing the implementation priorities.

Mainstreaming environment into the Poverty Reduction Strategy Papers has reached a good level, however improvement is desired. Some links between environment and poverty deserve to be better addressed, such as indoor air pollution, health of women and children. Future Strategies should emphasize the establishment of monitoring systems with explicit focus on poverty-environment indices and outcomes.

Progress of the guide on the process for preparing environmental sections of a poverty reduction strategy would be desirable. Moreover, encouragement of environmentalists to take an active interest in the PRS process shall be one of the Bank's priorities.

With regards to the PRS policy it is still ambiguous that to be successful in its objective, it has to be essentially country specific and countries shall be given total freedom to formulate their PRS, or, in certain cases it may still be better if the PRS is imposed on a country, as the country's government may not be competent enough (due to for example corruption) to formulate an effective and efficient PRS. In whatever way the PRSP is prepared, it is very important that it reflects the national budget and that PRSP is regarded by the local government as an instrument for programming domestic resources. To achieve good results in mainstreaming environment, the existence of a functional link between the national budget, the priorities of national development projects and the PRSP is crucial.

References

An Environmental Strategy for the World Bank, Introduction, Chapter 1, Chapter 3, Annex B.; tp: //lnweb18.worldbank.org/ESSD/envext.nsf/41ByDocName/ Making Sustainable Commitments An Environment Strategy forthe-World Bank 2001 Full Document in Color 5MBPDF/ $FILE/ Environment Strategy2001FullDocument.pdf.

Bojö, Jan & Reddy, Rama Chandra (2001), Poverty Reduction Strategies and Environment, A Review of 40 Interim and Full PRSPs, December. www.wds.worldbank.org/ servlet/ WDSContentServer/ WDSP/ IB/ 2002/ 09/ 24/ 000094946_02090504023465/Rendered/PDF/multi0page.pdf.

– (2002), Poverty Reduction Strategies and Environment: A Review of 40 Interim and Full PRSPs, June 2002. http://www-wds.worldbank.org/servlet/WDSContentServer/WDSP/IB/2002/09/24/00009496_02090504023465/Rendered/PDF/multi0page.pdf.

– (2003), Status and Evolution of Environmental Priorities in the Poverty Reduction Strategies, An Assessment of Fifty Poverty Reduction Strategy Papers, Environmental Economics Series, The World Bank Environmental Department, November, 2003. http://lnweb18.worldbank.org/essd/envext. nsf/44ByDocName/ StatusandEvolutionof Envrionmental Priorities in the PovertyReductionStartegiesAnAssessmentofFiftyPovertyReductionStrategyPapers/$FILE/Status and Evolution of Environmental Priorities in the Poverty ReductionStrategies.pdf.

Bojö, *Jan*, *Green*, *Kenneth*, *Kishore*, *Sunanda*, *Pilapitiya Sumith & Reddy*, *Rama Chandra* (2004), Environment and Poverty Reduction Strategies and Poverty Reduction Support Credits, The World Bank Environment Department, November 2004. www.wds.worldbank.org/servlet/WDSContentServer/WDSP/IB/ 2004/ 12/ 16/ 000090341_20041216111209/ Rendered/ PDF/ 308900PAPER0EDP0102.pdf.

Carley, *Michael & Ian*, *Christie* (2000), Managing Sustainable Development, Earthscan Publications Ltd, London.

Chambers, *Robert & Conway*, *Gordon* (1992), Sustainable Rural Livelihoods: Practical Concepts for the 21st Century. IDS Discussion Paper no. 296. Institute of Development Studies, University of Sussex, Brighton, UK.

Environment Chapter of the PRSP Sourcebook (2001), http:// poverty.worldbank.org/files/4145_chap11.pdf.

Environment Strategy Note no. 4: Mainstreaming Environment In Poverty Reduction Strategies, 2002 http:// lnweb18.worldbank.org/ ESSD/ envext.nsf/41ByDocName/Environment Strategy Note No4 Mainstreaming Environment In Poverty Reduction Strategies 2002607KBPDF/ $FILE/ EnvStrategyNote 42002.Pdf.

Fact sheet, Poverty Reduction Strategy Papers (PRSP), International Monetary Fund. www.impf.org/external/np/exr/facts/prsp.htm.

La Camera, *Francesco* (2003), Sviluppo Sostenibile, Origini, Teoria e Pratica, Editori Riuniti, Roma, Italy.

Linking Poverty Reduction and Environmental Management, Policy Challenges and Opportunities, DFID, EC, UNDP, The World Bank, July 2002.

Poverty and Environment: Evidence of Links and Integration into the Country Assistance Strategy Process, 1999 http:// lnweb18.worldbank.org/ ESSD/ envext.nsf/ 44ByDocName/ Poverty and Environment Evdence of Links and Integration into the Country Assistance Strategy Process 1999/$FILE/Poverty and Environment Evidence of Links and Integration into the Country Assistance Strategy Process1999.pdf.

Poverty Reduction Strategies and the Millennium Development Goal on Environmental Sustainability: Opportunities for Alignment, 2003. http://www-wds.worldbank.org/ servlet/ WDSContentServer/ WDSP/ IB/ 2005/ 01/ 10/ 00016006_20050110165441/ Rendered/ PDF/ 312190 PovertyR1ntGoalson-Environment.pdf.

Programme of Action of the World Summit for Social Development, Chapter 2: Eradication of Poverty, World Summit for Social Development, Copenhagen, March 1995. www.un.org/esa/socdev/wssd/agreements/poach2.htm.

Reaching the Rural Poor, A Renewed Strategy for Rural Development, The World Bank Group (2003), http:// lnweb18.worldbank.org/ ESSD/ ardext.nsf/ 11ByDocName/ReachingtheRuralPoorexecutivesumay/$FILE/ReachingRural-Poor-execsummary.pdf.

Reid, David, Sustainable Development, An Introductory Guide, Earthscan Publications Ltd, London, 1995.

Rural Poverty Chapter of the PRSP's Sourcebook (2001), http://poverty.world-bank.org/files/5312_chap15.pdf.

Shyamsundar, Priya (2002), Poverty and Environment Indicators, Environ-mental Economics Series, The World Bank, January 2002. http:// lnweb18. worldbank.org/ ESSD/ envext.nsf/ 44ByDocName/ PovertyandEnvironment Indicators 2002189KPDF/$FILE/PovertyandEnvironmentIndicators2002.pdf.

The Poverty Reduction Strategy Initiative, An Independent Evaluation of the World Bank's Support Through 2003, World Bank Operations Evaluation Department, The World Bank, Washington, D.C., 2004.

The World Bank (2000), Development Committee (Joint Ministerial Committee of the Boards of Governors of the Bank and the Fund On the Transfer of Real Resources to Developing Countries) Supporting Country Development: World Bank Role and Instruments in Low- and Middle-Income Countries, DC/2000-19 September 8, 2000, http:// lnweb18.worldbank.org/ MIP/ BMMMDR.nsf/ 0/ 041008E2441272DA85256C00006E7EE5/ $File/ MMM+ Website+Aid+Effec-tiveness.pdf?OpenElement.

– (2004), Sustainable Development (ESSD) Vice Presidency, World Bank Sustainable Development Reference Guide, 2004. http:// lnweb18. world-bank.org/ essd/ sdvext.nsf/ 43ByDocName/ WorldBank Sustainable Develop-mentReferenceGuideText/$FILE/ESSDReferenceGuideText.pdf.

– (2005), World Bank Focus on Sustainability 2004, Washington, D.C. USA, January 2005. ww.wds.worldbank.org/ servlet/ WDSContentServer/ WDSP/ IB/ 2005/ 02/ 07/ 000009486_20050207160411/ Rendered/ PDF/ 315170Focus OnSustainability200401public1.pdf.

Annexes

Annex 1: A Sample of Poverty-Natural Resource Indicators that Affect Income, Security, and Vulnerability of Poor People in Poor Countries

	Poverty issue	Poverty-environment indicator	Natural resource problems that could influence this indicator
1	Income and opportunity	Percentage of rural population below poverty line	Deforestation Water scarcity Overfishing Land degradation
		Rural per capita cereal production	
2		Time spent by household members to collect water and fuel wood	
3		Distance walked by household members to collect water and fuel wood	
4		Quantity of annual household consumption derived from common lands[1]	
5		Quantity of annual household consumption that is derived from forest products and fisheries[1]	
6		Percentage of irrigated area in total cultivated area by wealth/income categories[2]	
7		Percentage of rural households with adequate water for livestock by wealth/income categories[2]	
8	Food security	Rural per capita cereal production	Land degradation Water scarcity Pest outbreak Natural disasters
9		Percentage of farmers who grow drought resistant crops by income/wealth quintiles	
10		Quantity of household consumption that is derived from forest products and fisheries[1]	Deforestation Overfishing
11		Percentage of rural children under five who are underweight	Land degradation Water scarcity Water quality
12		Percentage of rural children under five who are stunted	
13		Percentage of rural children under five who are wasted	
14	Vulnerability to natural disasters	Households rendered homeless from floods/hurricanes/ cyclones/landslides per year by income/wealth quintiles	Natural disaster Deforestation
15		Number of deaths from natural disasters by income/ wealth quintiles	
16		Percentage of farmers with land on slopes/wetlands by income/wealth quintiles	
17		Percentage of rural children under five who are wasted	

Notes:

1. Among households that are largely dependent on natural resources with few alternative income/employment opportunities.
2. Field tested by a DFID research group (DFID 2001).

Source: *Priya Shyamsundar*, Poverty – Environment Indicators, Paper no. 84, Environmental Economics Series, The World Bank Environment Department, January 2002.

Annex 2: Available Interim and full Poverty Reduction Strategy Papers as at June 30, 2005

Board Presentations pf PRSP Documents
As at December 31 2005

	Country	Region	I-PRSP	PRSP	PR 1	PR2	PR 3	PR 4	PRSP II
1	Benin	AFR	13-Jul-00	20-Mar-03	10-Jun-05				
2	Burkina Faso	AFR		30-Jun-00	06-Dec-01	05-Nov-02	16-Mar-04	05-May-05	05-May-05
3	Burundi	AFR	22-Jan-04						
4	Cameroon	AFR	10-Oct-00	31-Jul-03	26-May-05				
5	Cape Verde	AFR	09-Apr-02	25-Jan-05					
6	CAR	AFR	18-Jan-01						
7	Chad	AFR	25-Jul-00	13-Nov-03	15-Jun-05				
8	Congo, DR	AFR	11-Jun-02						
9	Congo, Republic of	AFR	07-Dec-04						
10	Cote D'Ivoire	AFR	28-Mar-02						
11	Ethiopia	AFR	20-Mar-01	17-Sep-02	17-Feb-04	26-Oct-05			
12	Gambia	AFR	14-Dec-00	16-Jul-02	25-Apr-05				
13	Ghana	AFR	24-Aug-00	08-May-03	13-Jul-04				
14	Guinea	AFR	22-Dec-00	25-Jul-02	26-Aug-04				
15	Guinea Bissau	AFR	14-Dec-00						
16	Kenya	AFR	01-Aug-00	06-May-04					
17	Lesotho	AFR	06-Mar-01	12-Sep-05					
18	Madagascar	AFR	19-Dec-00	18-Nov-03	21-Oct-04				
19	Malawi	AFR	21-Dec-00	29-Aug-02	23-Oct-03	10-Aug-05			
20	Mali	AFR	07-Sep-00	06-Mar-03	22-Jun-04	21-Dec-05			
21	Mauritania	AFR		06-Feb-01	18-Jun-02	11-Jul-03			
22	Mozambique	AFR	06-Apr-00	25-Sep-01	29-Jul-03	06-Jul-04	21-Jun-05		
23	Niger	AFR	20-Dec-00	07-Feb-02	19-Nov-03	03-Mar-05			
24	Nigeria	AFR		01-Nov-05					
25	Rwanda	AFR	21-Dec-00	08-Aug-02	08-Jun-04	14-Apr-05			
26	Sao Tome & Principe	AFR	27-Apr-00	25-Apr-05					
27	Senegal	AFR	20-Jun-00	23-Dec-02	07-Dec-04	23-Nov-05			
28	Sierra Leone	AFR	25-Sep-01	06-May-05					
29	Tanzania	AFR	04-Apr-00	30-Nov-00	27-Nov-01	27-May-03	08-Jun-04		
30	Uganda	AFR		02-May-00	31-May-01	23-Jul-02	09-Sep-03		28-Jul-05
31	Zambia	AFR	04-Aug-00	23-May-02	15-Jun-04	11-Apr-05			
32	Cambodia	EAP	18-Jan-01	20-Feb-03	09-Sep-04				
33	Indonesia	EAP	21-Apr-03						
34	Lao	EAP	24-Apr-01	30-Nov-04					
35	Mongolia	EAP	27-Sep-01	09-Sep-03	26-Sep-05				
36	Timor Leste	EAP		14-Jun-05					
37	Vietnam	EAP	12-Apr-01	02-Jul-02	19-Feb-04	08-Aug-05			
38	Albania	ECA	08-Jun-00	20-Jun-02	10-Jul-03	12-Jul-04			
39	Armenia	ECA	22-May-01	20-Nov-03					
40	Azerbaijan	ECA	05-Jul-01	27-May-03	10-Sep-04	25-Aug-05			
41	Bosnia-Herzegovina*	ECA	12-Nov-02	15-Jun-04					

Board Presentations pf PRSP Documents
As at December 31 2005

	Country	Region	I-PRSP	PRSP	PR 1	PR2	PR 3	PR 4	PRSP II
42	Georgia	ECA	11-Jan-01	06-Nov-03	13-Jun-05				
43	Krygyz Republic	ECA	04-Dec-01	27-Feb-03	29-Jun-04				
44	Macedonia, FYR	ECA	14-Dec-00						
45	Moldova	ECA	19-Dec-00	11-Nov-04					
46	Serbia & Montenegro	ECA	08-Aug-02	16-Mar-04					
47	Tajikistan	ECA	31-Oct-00	05-Dec-02	29-Jun-04	11-Nov-05			
48	Uzbekistan	ECA	24-May-05						
49	Bolivia	LAC	27-Jan-00	05-Jun-01					
50	Dominica	LAC	29-Dec-03						
51	Guyana	LAC	14-Nov-00	19-Sep-02	27-Jul-04				
52	Honduras	LAC	06-Jul-00	11-Oct-01	26-Feb-04	20-Apr-05			
53	Nicaragua	LAC	21-Dec-00	25-Sep-01	16-Dec-02	22-Jan-04			
54	Djibouti	MENA	27-Nov-01	08-Jun-04					
55	Yemen	MENA	27-Feb-01	01-Aug-02					
56	Bangladesh	SAR	19-Jun-03						
57	Bhutan	SAR		08-Feb-05					
58	Nepal	SAR		18-Nov-03	02-Dec-05				
59	Pakistan	SAR	04-Dec-01	11-Mar-04					
60	Sri Lanka	SAR		01-Apr-03					
Sub Totals	AFR	31	27	25	19	12	4	1	2
	EAP	6	5	5	3	1	0	0	0
	ECA	11	11	9	5	3	0	0	0
	LAC	5	5	4	3	2	0	0	0
	MENA	2	2	2	0	0	0	0	0
	SAR	5	2	4	1	0	0	0	0
Total		**60**	**52**	**49**	**31**	**18**	**4**	**1**	**02**

Source: Official website of The World Bank Group, http://siteresources.worldbank.org/INTPRS1/Resources/boardlist.pdf. –13/1/2006.

Local Agenda 21 Process Facilitation
–
Environmental Communication and Self-Experience Development in Latvia

Raimonds Ernsteins

Institute for Environmental Science and Management,
University of Latvia
Rainis blvd. 19, LV – 1586 Riga, Latvia

Abstract

The university-municipality partnerships proved to be the main driving force behind enhancement of *Local Agenda 21* (*LA21*) process in Latvia, particularly in terms of development and wide application of *incremental environmental communication* – four partite cycle of information, education, participation and environmentally friendly behaviour components – and *self-experience facilitation* as two basic LA21 process facilitation instruments and precondition as well. *LA21* application projects in Latvia has been designed, realized in practice (at different level of self-governance with various success and further continuity) and also studied as municipal case studies. Perceived understanding of the LA 21 process development until now and necessity to elaborate further steps ahead is leading us to formulate some *basic preconditions – principles and approaches*, *working models and instruments* – required to be fulfilled now and here in Latvia for any successful continuation.

Introduction

Successful application of sustainable development principles is determined by its realization in the public administration level closest to inhabitants – in local municipalities. Local Agenda 21 (*LA 21*) or Sustainable Development Action Programs (SDAP) for local and regional levels are to be elaborated in municipalities for the integration of sustainable development's principles into municipal every days practice and promotion of cross-sectorial cooperation, and this is to be done in direct and compulsory community consultation process.

Local Agenda 21 process in Latvia as the same elsewhere in Eastern Europe has begun later than in the Western and Northern part of Europe and occurs gradually and occasionally, notably because of the general lack of joint international and particularly state support as well as minor self-interest of municipalities – municipalities has been missing basic information and it was not in practice really clear what for and what exactly and through what kind of means can take place for

LA21. Nevertheless the growing experience of the most successful *LA21* processes in Latvia does not practically differ in quality from the rest of the world. Activities which in substance correspond to *LA21* content (int. al. elaboration of municipal environmental policy and action programs etc.), but are not respectively named as LA21 take place more widely and actively – energy management projects, national water and waste management upgrade implementation programs at the local and regional level etc. have to be noted as well. For the time being *Local Agenda 21* process has been started in relatively few municipalities in Latvia, i.e., 20 different level municipalities declare this for international inquiries, but since these are mostly town municipalities, including the capital city – Riga, then formally we can conclude that in the overall almost 50% of all inhabitants of Latvia live in these sustainable development approaching municipalities.

Municipalities of Latvia like ones from other countries participate in cooperation with different international organizations as well as successfully do realize various networking projects. Cooperation with International Council for Local Environmental Initiatives (ICLEI), Baltic Local Agenda 21 Forum (*BLA 21 F*), Union of Baltic Cities (UBC) and others, int. al. in elaborating sustainable development strategy for cities of the Baltic Sea region, shows good progress. Many municipalities are involved in the Alborg Sustainable development Charter process since its very beginning and have signed accession papers.

Experience gained from sustainable development pilot projects in Riga, Jurmala, in the Bartava and North-Kurzeme regions etc. [1, 2], allows a LA21 process evaluation to be made showing that several Latvian municipalities have practically passed through the first steps of experience and acquired the major skills needed for sustainable development planning:

- the first sustainable development concepts and strategies at the regional/subregional municipal level have been developed;
- methods for public involvement have been developed and tested, the recommendation for appropriate activities prepared, etc.;
- methods for the preparation of sustainable development indicators (bottom-up, top-down) have been developed, tested and applied;
- municipality level sustainable development strategic plans and action plans have been elaborated;
- the first models for sustainable development action programmes covering the various levels of local government have been developed and tested.

In Latvia as well as in other countries the *LA 21* process has been launched very differently, activities and number of local and regional municipalities varies as well, nevertheless, *LA 21* already puts the most active municipalities in a considerably better environmental and economical position than others. Sometimes process realization is too formalized, without substantial changes in municipal administration and public participation.

Local Agenda 21 action programs are being developed not only in municipalities, but also in schools for example, which also acts as an important catalyst for local Agenda 21 development. However, it will take a much *longer time* and, most importantly, *innovative approaches and instruments*, to begin really full scale implementation of local Agenda 21, as significant changes are required in the everyday management of municipality activities and the organization of Agenda 21 work, the identification and involvement of major target groups, and the securing of necessary resources.

Chapter 1: LA21 in Latvia: Application Principles and Approaches in Practice

LA21 application projects in Latvia have been designed, realized in practice (at different level of self-governance with various success and further continuity) and also studied as municipal case studies. There is to be recognized step-wise LA21 process development in Latvia with following general governance level characteristics:

- preparatory stage at national level – neither real top-down nor bottom-up activities developed, however national strategy elaborated and implemented under monitoring as well as national Council established (at the Ministry of Environment);
- related occasional activities at regional/district level (except full scale process launched at North-Kurzeme coastal region) – e.g. initiatives on healthy communities;
- local level – pre-dominantly a top-down approach is developing with very limited public involvement, initiated mainly by:
  - international projects (EU, Baltic region or bilateral) – "outside force" and financing facilitation,
  - active and for LA21 interested municipal employees looking for different options to initiate at least separate LA21 related activities,
  - few municipal LA 21 centers established.

Public participation and partnership development between the main interested stakeholder groups indeed characterizes LA 21 development in the country. We can list most active and process influential actors:

- Interested individuals – municipal administrators and specialists,
- Knowledge institutions (universities, professional NGO's and also new consultancies),
- International project partners (municipalities and others),
- Ministry of Environment, particularly Environmental Protection Fund,

- Groups/associations of neighbour municipalities (voluntary agreements as inter-municipal cooperation facilitation and also optional legal transitional stage within national administrative-territorial reform).

Actors in supportive roles, for the time being could be listed as follows:
- Local inhabitants – established local/regional professionals (individuals/non-formal groupings) interested and self-involved in local development activities (also searching for post-experience post-graduate interdisciplinary education);
- environmentalists, planners, governmentals, teachers, social and culture workers etc.;
- Latvian National association of municipalities (incl. all sub-associations);
- National NGO's, particularly those with regional/local chapters;
- First citizen groups established.

Perceived understanding of the LA 21 process development and the necessity to elaborate further steps ahead is leading us to formulate some *basic preconditions – some principles and approaches*, *working models and instruments* – required to be fulfilled now and here in Latvia for any successful continuation at all.

One principle appears out of a general notion of integrative and disciplinary realization of the possibility of LA21 process and its documentation. LA21's ongoing processes and the main actors involved clarifies the need for reconsidering the following *principle – mutual interlinkage of integrative and disciplinary LA21 approach models* [2] – to be done in both theoretical planning and in practical realization. Consequently, elaborations of sustainable development planning and SDAP guidelines (usually the first written material in Latvian municipalities at the beginning of LA 21) in particular, can be further developed, namely:
- elaborated as a separate document and process *(disciplinary model)* – LA 21,
- integrated in all existing plans *(integrative model)* and/or used in elaborating new_development strategy plans, projects etc.
- mutually integrated and interlinked processes and documents – integrating both approach models above (*mixed model*).

For municipality is still an evolving multifunctional system in the current conditions of general development in Latvia and taking into consideration existing LA21 cases of experiences, both sustainability implementation models shall be used as much as complementary and mutually integrative e.g. mixed model.

Also in the context of sustainable development planning in municipalities in general (in seminars with interest group participation in particular) the quality of the achieved results in LA21 elaboration was ensured by the elaborated integrated *methodology*:
- *integration* of strategic planning *and* action planning approaches

- mutual *integration* of the different, frequently separated *municipal operational sectors* in the planning process (in the seminar)
- elaboration of sustainable development *indicators* used for planning and measurement and later evaluation of the *achieved progress* in sectorial or LA 21 context (preferably visions and aims), especially realizing it in a community-initiated way.

Another principle that appears important nowadays in Latvia is related to all three LA21 process implementation models [2, 3] – the participatory process itself, the cross-sectorial and interdisciplinary content of LA21 process and also the action planning structure models. These implementation models for local sustainable development action programming are both exploratory/explanatory for training/ education purposes of municipal specialists and understanding/ awareness development of the general public, but also can be succesfully realized in practice if taken into account as a coherent whole for LA21 planning and management at local/ regional municipalities. There is following principle – complimentarity of tripartite process, content and structure LA21 implementation models.

In Latvia there are already experiences of all three traditional LA 21 application approaches starting by top-down and bottom-up cases and continuing by LA 21 centre intermediation. Also there are the first elements of new innovative cases of LA21 application approaches – instrumental integration and disciplinarisation approaches, including 3 different and interesting sub-approaches taking into account existing conditions in Latvian municipalities. The following is the list of case studies developed and explored in Latvia and four Local Agenda 21 *process approaches* that have been formulated:

- Municipalities Pledging Approach – City Council Planning (Jurmala city LA21 top-down planning case),
- *Public Involvement Approach – Region Agenda 21 Participatory Process (Bartava grouping of municipalities LA21 bottom-up involvement case)*,
- *Intermediary Facilitation Approach – Regional Agenda 21 Centre (North-Kurzeme coastal region LA21 centre case)*.

Instrumental Integration and Disciplinarisation Approaches:

- Ecotourism and Local Integrative Development and LA21 (ecotourism as a tool and municipality development sector)
- Local School Agenda 21 for Municipality LA21,
- Cultural Heritage for LA 21 – Museum Involvement Case (approach, tool and centre).

Particular interest could be taken of the very last one mentioned as a perspective of culture environment to be used as a *cornerstone for LA 21* process development in comparison to the traditional European approach via environmental protection. It seems to be thoroughly embedded in Latvian culture heritage development traditions.

Also all four process approaches are by definition complimentary and the best possible application must be relevant.

A case study designed, developed and implemented in Nort-Kurzeme coastal region (Dundaga, Roja and Kolka municipalities) – "Livonian Green Coastal region 21" was realised as the LIFE ENVIRONMENT project – has been aiming to tackle most if not all eventual approaches, to apply some of the models and to widely use communication instruments and techniques [2].The case study results permit us to conclude, that a combined version of all four LA21 process approaches has been tested successfully (however with different degrees of quality fulfilment) and proves to characterize the fifth process approach – *facilitation as structural network approach.*

Public participation in general and all stakeholders co-operation for sustainable regional development is to be facilitated via:

- separate innovative demonstration projects (particularly successful model in Latvia) or other type of activities, but shall be
- planned and participatory, implemented as a coherent networking program (also to be seen as demonstration network).

Components of this coherent whole approach were developed as a kind of regional sustainable development action program (structural network):

- conflict resolution and partnership practice as an overall framework,
- round table forum and public participation as a bottom-up process,
- council for sustainable development of region as a top-down process for collaborative and integrative decision planning,
- regional *Agenda 21* centre as an intermediary facilitation and partnership coordination,
- rural communication and information network as well as regional sustainable development implementation demonstration projects etc as an instrumental integration and sectorial development.

According to the North-Kurzeme case particularly but also taking into account other case studies it is to be concluded that university-municipality partnerships proved to be the main driving force behind enhancement of the LA21 process in Latvia, particularly in terms of *incrimental environmental communication* development – information, education, participation and environmentally friendly behaviour – and *self-experience facilitation* as two basic LA21 facilitation instruments (instrumental approach) and also preconditions.

When preparing and taking a decision on the *planning and practical implementation* (parallel processes!) of sustainable development in ones own municipality, it is vitally important to start with the *experience and ideas* that have been crystallized in different municipality development projects in Latvia. Likewise, work has to be started from scratch, as every municipality has their own forerunners even if they are called something else. However, in LA 21 work it is important to choose approaches and models corresponding to the needs of the present development level of the rural areas, to choose concrete first steps and projects for *every individual, i.e. subsequently different municipality.*

Our experience as one the initiators of LA 21 in Latvia shows that it is important to start LA 21 with *local initiatives* (int. al. those from the advisors) with situation study and evaluation, consideration of the community and interest groups opinions and their participation, as well as in creating initiative groups and project development, etc. Further we will separately deal with some of the theoretical, as well as practical approaches in local initiative creation and particularly with environmental communication and so called *self-experience* development (experience acquired by stimulated active work of the individual that at the same time is applied in the further acquiring of experience). It should be pointed out that the ultimate result is a "resonance and openness for action" and the most important conclusion drawn from our experience is to do everything that could promote creation of positive attitude towards innovations in municipalities so as to encourage the local experts and local population to accept the new ideas and opportunities, initially at least paying attention to and desirably considering them.

Chapter 2: Environmental Communication – four Partite Cycle Development

Environmental awareness being one of the main preconditions for sustainable development, maintenance and improvement of environmental quality, in practice, for the general public and for every one of us can be expressed as environmentally friendly action in any field of life; work, leisure and social activities as well as active participation in decision making processes on sustainable development. Since environmental or sustainability problem solutions are strongly correlated to the level of knowledge, understanding of a situation and sense of responsibility then not only politicians and environmental/municipal specialists, but everyone of us, when in local areas and confronted directly with those problems, becomes the decision-maker – taking action or staying aside.

Sociological researchers in Latvia often have also shown [3, 4] that the public is not well enough informed on different environmental issues and also the role of the state institutions and municipalities has been evaluated as quite low. Unfortunately information and education instruments are not always incorporated in the environmental management projects, the information process is traditionally fragmented and information is located at different institutions and organisations and the public does not know which establishment/institution should be addressed and what the options are for involvement in decision making process. Also there is insufficient coordination between non-governmental environmental organisations and lack of purposive and positive sustainable communication programs. This hinders the development of effective environmental management and environmental friendly life styles. Subsequently, the development of different representation forms for promotion of dialogue and seeking compromise among official institutions and various public target groups is no doubts essential and so already perceived at the environmental protection development stage.

Results of the assessment of LA21 activities and also public environmental awareness development in Latvia indicate the need for an *environmental communication system and related process* development with involvement of all main actors in the field – Ministry of Environment and it's institutions, other ministries and institutions, municipalities, general public and public organisations, business organisations, mass media and educational establishments et al. Pretty often the application of information/education principles is complicated as the cooperation between different target groups in the context of environmental policy implementation is just under development because of the continuing process of self-organisation of different target groups..

To encourage dialogue and development of a mutual agreement process and to ensure formal and informal cooperation and environmentally friendly behaviour of inhabitants, different target groups and institutions of public administration, not only the development of normative acts and other traditional instruments, but also innovative creation of the necessary preconditions, incl. complimentarity of communication components/ steps and effective mechanisms of implementation are required.

Taking into account all of the above and after testing the effectiveness of new approaches elaborated during LA21 facilitation processes in Latvia environmental communication could be defined more comprehensive and extensive as traditionally used to, particularly including also action oriented part, aimed and created by "information and education flow" – public response and participation. Environmental communication is then viewed as multilateral information exchange and *cooperation enhancement process* based on and including four following components:

- information and
- public education (target groups oriented),
- participation and
- environmental friendly behaviour,
- being required for successful development of identification, assessment, decision making and implementation phases of environmental management.

Hereafter we propose an innovative *model of incremental environmental communication cycle [3]*. This figure (see Table 1) demonstrates the linkage between environmental communication tasks or the cyclic basic steps of communication process and pedagogical/practical results that within the particular cycle ensure applied and concrete practical case oriented environmental awareness development. Within the multi–cycle integration – the process of repeating and inter-supplementary *self-experience* development is facilitating general environmental awareness enhancement.

Appropriate environmental communication results have been measured as knowledge and practical skills, understanding and ability to solve environmental problems, self-regulation attitudes, motivation and readiness for concrete action and obtained experience for case related target groups as well as each individual in general.

Tab. 5: Incremental environmental communication process – four partite cycle model

Tasks		**Tools/Environment**		**Applied result**
1. Environmental information		Choice depends on specific/concrete problem situation:		Knowledge and intellectual action skills, situation attitudes (I)
2. Public education	⇨	– on specific tasks,	⇨	Understanding and values (value-orientation)
3. Involvement and participation		– target groups, – thematic content;		Applied action skills, practice and self-regulation attitudes (II un III)
4. Environmental friendly behaviour		– action realization etc.		Action motivation and readiness, action self-experience
				⇩
Environmental Awareness (integrated process and product)			⇦	Applied Environmental awareness ((incremental concrete practice oriented)

The four partite incremental environmental communication cycle model demonstrates the necessity for all *four basic elements and their direct and cyclic interaction* within the *environmental communication* process. This is identified in the definition and latter development of National Environmental communication and education strategy [4] which can be mentioned as one of the nation wide applications of this theory and practice based development.

Chapter 3: Local Initiative and Self-Experience Development

Putting environmental communication theory developments into practice appears to be crucial for local population/interested individuals and local experts/specialists/decision makers to create a further self-organized application working towards local municipality development [3]. If we pose simple questions on human cognition and perception, particularly in the process of learning and take a look at our capability to perceive information and learn practical activities, we can easily draw a conclusion that one of the most effective life long learning approaches is the "Learning By Doing And Doing By Learning".

Elaboration and testing/application of this approach in practice in Latvia turned out to be further designed into a complex of LA21 process facilitation activities for local interest groups and individuals as a kind of self-experience development tool-box. So successful LA21 process start-up and local ongoing facilitation, esp. in rural areas, depends directly on following *self-experience development tool-box components*:

- Self-active development,
- Project ideas,
- Community involvement wave,
- Interest groups,
- Facilitation team,
- Local experts involvement,
- Environmental communication emphasis.

Self-active Work Development

This approach of 'learn by doing' is advisable when working with Local Agenda inception in the local municipality. Far from always having to actually be done in real life, in many cases you can illustrate in a simple way or with a model and imitate self-learning in process, i.e. 'play' it in the learning room, e.g. with an advisor or self organised self-experience seminar on LA 21. For more than five years we have consistently applied this approach in praxis in municipality training, and particularly in Sustainable development projects. It has always yielded good results, even when working with 'compulsory sent' or initially negative oriented participants of courses, seminars, co-planning and discussion meetings. Needless to say there are active people in all municipalities at different levels.

Project Ideas

Even very specific training seminars, public discussions, planning meetings etc would have to be organised at the local level. Moreover, the self-experience work would have to result in concrete *local development ideas* that the participants could come up with on the spot and immediately publicly present, i.e. discuss. This would lead to already formulated, and most importantly, practical results, namely *project forerunners*. It gives people an opportunity to see the results of their work directly and serves as a motivation to continue to work on the further development of the project together with the associates immediately after the training activity. As the experience of the municipality projects indicates, the projects based on the ideas of local experts and local population and implemented by the same people are the most successful.

Community Involvement Wave

The long term success of any municipality local development project, let alone LA 21 projects, depends on the *possession of knowledge and essentially involvement* of the community which does not require hundreds or thousands of people. It is enough in some cases to have 10-20 actively dedicated representatives of the community. Besides, sometimes it can be achieved during a one day seminar with the participation of some enthusiastic individuals or, for instance, by conducting a public survey (broad, but focused surveying of the community, i.e. families, etc.) with participation of local activists and an active feedback after it. In this way, step by step, and particularly through further work of the local activists after the seminar, a "community involvement wave" is created.

Interest Groups

Community involvement and support and encouragement of separate active individuals is of great importance. However one has to work towards involvement of the main professional and other *public interest groups*. Participation of both formal and informal groups ensures a successful unfolding of the specific seminar, work team, discussion group or public forum, practical development of forerunners to be implemented in the municipality, resonance in the community (serving as a trigger for action at the local level in the municipality), as well as change in the local people opinions on *initiation of the community involvement process*. The theory and praxis in Latvia proves that most effective planning is carried out in a team, since the result achieved is more comprehensive and of better quality. Most importantly it is more creative and thoroughly discussed, besides the many local authors behind it usually carry it out as well.

Facilitation Team

A team is a well organised *group of local activists* – motivated and active people with an initially determined and accepted common goal (work vision) – that works in a new quality. They work in close cooperation, and harmonious spirit, *complimenting and supporting each other*, consequently achieving more than if working alone. The team members highly value the results of the work and are collectively responsible for it. They achieve results of good quality applying new and often *unusual and original solutions* produced by joint efforts. When starting work with LA 21 creation in a municipality or when applying the well-balanced approach to a separate sector, problem topics or even simple projects one should try to form teams consisting of 3-7 active individuals, most importantly by encouraging and maintaining any *motivation* to achieve the planned results.

Local Experts Involvement

We should also emphasize that for the municipality development projects that necessarily require involvement of external experts, i.e. different advisors and consultants, it is crucial to involve in the *preparation and execution of the commissioned work* local experts and activists, or representatives of NGOs too. Thus facilitating acquisition of highly precise knowledge on the local situation and implementation by their own efforts, as well as *local self-experience development.*

Environmental Communication Emphasis

Irrespective of the involvement maturity level of the specific community in Latvia one would have to launch an environmental communication (information, education and involvement of the community, examples of environmentally friendly activities) and the specific communication forms. Municipalities that already have the experience have to develop it further as it is the main prerequisite for local development in general, and LA 21 development in particular. We have also to highlight a seemingly *unconventional* method for community involvement and interest creation – self-experience *seminars* – application for local community target groups self-experience and initiative development seminars in municipalities. This approach has proven to be effective not only in municipalities 'beginners', but also for project planning and development in already active municipalities where the involvement of the communities at large, individual activists and the main target groups, and/or interest groups is not widely developed – consequently all Latvian municipalities.

Some of the initiatives brought forward can be implemented rather simply, others will require a longer period of time (even several years). The latter can be an indicator of a certain degree of maturity of the society, and the seminars demonstrated that there are some realistic ways for implementation. Further implementation will chiefly depend on the cooperation between the local community initiative groups and the community target groups. In most of the cases the seminar has served as a real trigger to start initiative implementation.

In self-active work seminars in municipalities their creators and/or leaders (independent experts) only suggest and open the main topics of the seminar, as well as organize teamwork of the participants, guide and stimulate discussions, generation and formulation of proposals. Most importantly all of the *seminar participants*, representatives of the main interest groups and target groups and local activists through interactive team work during the main part of the seminar have themselves put forward, evaluated and prioritised the specific *local ideas* and will do the further developing of the ideas and projects *during the seminar and later in praxis*. The seminar output is not only information acquiring and exchange, but also involvement and development of concrete ideas and projects, comprehensive self-experience and mastering of action means, finding of cooperation partners, which is also important for further development of the projects, deeper knowledge on the local activists and leaders/organisers.

The Latvian public opinion is concerned not merely with economic development. Elements and parts of local agenda for sustainable and democratic development in local and district municipalities (possibly not well enough structured yet) that joint local economic, social and environmental resources in the development of their territory and community can be observed.

Conclusions

LA21 application projects in Latvia have been designed, realized in practice (at different level of self-governance with various success and further continuity) and also studied as municipal case studies. LA21 process development will take a much *longer time* and, most importantly, besides traditional also *innovative approaches and instruments* elaborated and applied. *Basic preconditions* (besides traditional resources necessary) are to be developed in Latvia – applied LA21 *principles and approaches* as well as emphasizing development of *incrimental environmental communication – information, education, participation and environmentally friendly behaviour – and self-experience facilitation toolbox applications*.

There are in Latvia experiences with all three traditional LA 21 application approaches starting with top-down and bottom-up cases and continuing by LA 21 centre intermediation. Also there are elements of new innovative and really impacting cases of LA21 application approaches – instrumental integration and disciplinarisation approaches as well as a combined version of all four LA21 process approaches. This has been tested and proves to be characterized as the fifth process approach – *structural network facilitation approach*.

References

Ernsteins, R. (2001), Sustainable Development Planning for Municipalities: Local Initiative and Self Experience in Practice, Proceedings, University of Latvia Research Conf, Riga.

– (2002), Local "Agenda 21" in Latvian National Report "Rio+10" to the World Summit on Sustainable Development in Johannesburg, Riga.

– (2002), Models for Sustainable Development Action Programs. In: Local Agenda 21 for Self-governments in Latvia, Riga.

National Environmental Communication and Education Strategy (2000), Ministry of Environmental Protection and Regional Development, Republic of Latvia.

Climate Change Mitigation Initiatives

–

Energy Policy in Compliance with UNFCC and Kyoto Targets

Dina Bērziņa

University of Latvia, ASI,
Rainis blvd. 19, LV – 1586 Riga, Latvia

Introduction

Energy is the very lifeblood of today's society and economy. Our work and leisure, economic, social and physical welfare depend on a sufficient, uninterrupted supply of energy. As we take it for granted the energy demand continues to grow year after year. The increased consumption is provided largely by reserves of fossil fuel that emit both green house gasses and other pollutants.

Today fossil fuel provides 90% of the world's energy consumption, but demand for energy is growing at an alarming rate and is predicted an average growth rate 1.8% per annum in 2000-2030 for primary energy worldwide: global energy consumption will rise by 60% in 2030 [8]. Fossil fuels will continue to dominate global energy use, accounting for some 85% of the increase in world primary demand by 2020. Oil will remain the largest fuel in the primary energy mix; demand for natural gas will grow most rapidly; the share of coal will fall slightly [25]. Accordingly CO_2 emission will increase from the present 24 bn t to 38 bn t [15].

Climate-destabilising carbon-dioxide emissions continue to rise, giving doubt to the sustainability of the current energy system. This calls for urgent and constructive action by governments around the world. Potential effects of climate change can become serious and irreversible. The Intergovernmental Panel of Climate Change (IPCC) concluded that global average surface temperature has increased by about 0.6° over the 20th century [3] but the mean temperature in Europe increased by more than 0.9°C [5]. Globally, the 10 warmest years on record all occurred after 1990. Greenhouse gas (GHG) concentrations are higher now than at anytime in the past 450 000 years, are predicted to keep rising and emissions from human activity and past emissions still will lead to an additional rise in the temperature during the 21st century. As a consequence, global temperatures are expected to increase by 1.4 to 5.8°C by the year 2100 (compared to 1990 temperatures) and by 2.0 to 6.3°C in Europe [24].

GHG Emissions and Climate Change

Climate change is a modern problem – complicated, involving the entire world, closely linked with poverty, economic development, and population growth.

Climate change is already happening and costing us money. Dealing with it will not be easy: a 2.5° rise in temperatures could cost 1.5-2% of worldwide GDP in the future [15]. Ignoring possible climate change is worse and the price of doing nothing will be much higher resulting in [5]:

- up to 31 billion people will find access to water more difficult,
- about 1.4 C° increase in temperature will put about 10 million people at risk of coastal flooding, but predicted rise of sea levels of up to 0.88 m by 2100 along coastline will affect about 68 million people in the EU,
- more storms and extreme weather events (the number of weather related disasters in Europe has doubled compared to previous decade),
- although agriculture in northern Europe could benefit, in the south the heat wave cut up to 30% of yields in 2003,
- regional conflicts, famines and migration, especially in the developing countries worst hit by climate change,
- ireversible effects such as melting of the Greenland ice sheet could become more likely with even modest rises in temperature,
- the impact on ecosystems can already be seen – if unabated climate change will accelerate the destruction of coastal wetlands, coral refs, and the endangered species.

The first steps have already been taken: in 1992 most countries joined the United Nations Framework Convention on Climate Change to begin considering what can be done to reduce global warming and to cope with any temperature increases. The Convention sets an ultimate objective – the stabilisation of greenhouse gas concentrations in the atmosphere at a level that would prevent dangerous anthropogenic interference with the climate system and states that "such a level should be achieved within a time-frame sufficient to allow ecosystems to adapt naturally to climate change, to ensure that food production is not threatened, and to enable economic development to proceed in a sustainable manner" [23].

In 1997 governments agreed to an addition to the treaty: the Kyoto Protocol [11], which has more powerful and legally binding measures. In accordance with the Kyoto Protocol countries individually or in a joint action with another country, should reach the level when anthropogenic GHG emissions by the years 2008-2012 are below emission level of 1990 of 8%. It defines three innovative "flexibility mechanisms": Joint Implementation Projects (JIP), Clean Development Mechanism (CDM), and Emissions Trading (ET) to lower the overall costs of achieving emission targets. These mechanisms enable Parties to access cost-effective opportunities to reduce emissions or to remove carbon from the atmosphere in other countries. While the cost of limiting emissions varies considerably from region to region, the benefit for the atmosphere is the same, wherever the action is taken. The Kyoto Protocol came into force on 16 February 2005 and by

31 August 155 countries and regional economic organisations have ratified the protocol representing 80% of the world population [11].

Both, supporters and critics, agree that the first commitment period (2008-2012) under Kyoto is not enough to sufficiently address the challenges posed by climate change – the emphasis now is on the post 2012 negotiations as the global annual mean surface temperature should not exceed 2°C above pre-industrial levels. The international negotiations for developing a medium and long-term strategy to achieve this goal – emission reductions of 15-30% by 2020 by developed countries compared to 1990 are about to start.

Concepts of transition to carbon-free economy have become understood broadly and been outlined by many actors. In G8 (Canada, France, Germany, Italy, Japan, Russia, UK, US) meeting in July 2005 it was concluded that there is a need for urgent actions to slow, peak and then decline GHG emissions moving towards a low-carbon economy and establishing a pathway to new dialogue on a replacement for the Kyoto Protocol when it expires in 2012. Without the US joining the consensus to tackle climate change it will not be possible to ensure that emerging economies (particularly China and India) whose emissions are rising rapidly are a part of the dialogue. Energy consumption per capita in China today is approximately 14 times lower than that in the US and 7 times lower than in the EU. If India and China increase consumption levels to just ½ of the EU levels, or ¼ of US levels by 2025, it would represent an increased energy demand – more than the present combined US and EU oil consumption [1].

Prospective energy studies predict that in the 1st half of 21st century the world will depend on non-renewable fossil fuels to a large extent [6]. We have only today's technology for 2025 as the renewal of existing infrastructure regarding fuel consumption switch is very slow: the change by new generation takes 10-15 years for Cars, 20-30 for Aircrafts, 25 – Wind turbines, 30 – Trains, 40 – Power plants and Electricity distribution, 70 – Houses. In relation to new technologies, the focus is on wide range of cost-effective options that prevent or minimise the formation of CO_2 emissions (accelerated development of energy-efficiency and renewable technologies) and a shift to non-polluting sources such as renewables.

Energy Consumption in EU

Today the European Union imports about 50% of its energy needs, which amounts to approximately 240 billion euros each year. If nothing is done, 70% of the EU energy needs would be covered by imports by 2030: the dependence increases each day as energy sources are becoming scarce [6]. Renewable energy has not evolved enough to replace oil and gas in a sufficient quantity. The EU has set a target of 12% of energy to be supported from renewable energy sources, including hydropower in 2010.

Figure 1 illustrates the primary energy demand in the EU between 1971-2004 and energy savings calculated on the basis of 1971 energy intensity. In view of technological developments, vehicle and component manufactures, transport providers, energy industry and even households are looking at alternative energy sources and fuels and cleaner technologies [7]. Europe has developed world leadership in a number of energy technologies. It is the pioneer in modern renewable energy technologies, such as bio-energy and wind energy. The EU is also in leading positions in power generation and distribution technologies and has a strong research capability in the area of carbon capture and sequestration.

Fig. 1: Development of the primary energy demand and of "negajoules"[1] (in Mtoe)

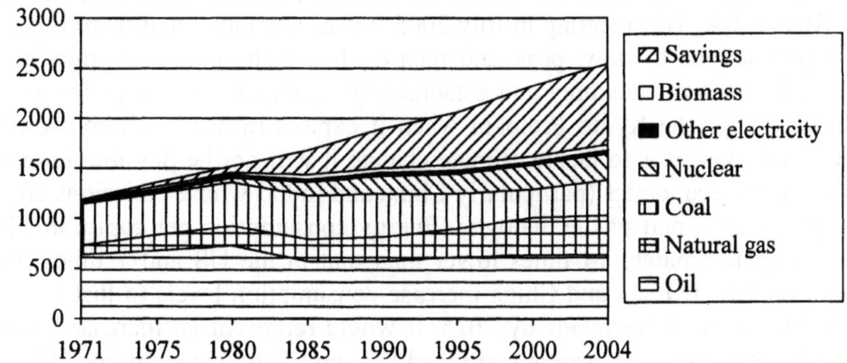

Source: Green Paper on Energy Efficiency or Doing More With Less – Brussels, 2.6.2005 COM(2005) 265 final

In 2003 5% of electricity in EU was generated on oil basis, 18% – on gas basis and 33% – on basis of coal (~⅔ – hard coal, ⅓ – lignite). Gas is the "most clean" fossil fuel: oil produces 2 times more, but coal – 3 times more CO_2 emissions as gas [15].

The EU is leading the world in power production from nuclear energy – 32%. No other region in the world covers such a large share of its electricity needs from this energy source. Within the EU the picture is very different: Lithuania and France generates up to 80% of their power from nuclear energy, but 12 countries (including Latvia) make no use at all of this option.

In 2003 EU countries generated 12% of power from renewable energies. The traditional energy source: hydropower accounted the largest share – 9%. About 3% of power is produced from wind, biomass, waste and solar energy. The front-runners in generation from renewables in absolute figures are: France, Spain and Sweden. If calculated as percentage share of renewables in national

1 „Negajoules": energy savings calculated on the basis of 1971 energy intensity.

power generation – then leaders are Latvia, Austria and Sweden (France is 11th) [15]. The benefits of renewable energy savings to greenhouse gas elimination are huge. CO_2 reduction due to the development of renewable energy sources could be 320 million tons per year in 2010 if the share of renewables reaches 12%. This represents 95% of Kyoto commitment [18]. The target for renewable energy in EU by 2020 is 20%. Predictions made by the European Renewable Energy Council based on assumptions and cumulative knowledge show that by 2040 a share of renewable energy up to 50% worldwide is possible [17]. Table 1 shows possible CO_2 emission reductions and savings for avoided fuel cost.

Tab. 2: CO_2 emissions reductions and avoided fuel costs

	2010		2020	
	CO_2, million tonnes	**Costs, billion €**	**CO_2, million tonnes**	**Costs, billion €**
Wind	99	12.9	236	63
Photovoltaic	2.2	0.2	24	4.3
Biomass	176		326	
Hydro	23	3.1	35	11.5
Geothermal	5.8	1.5	15	7.3
Solar thermal	14	2.3	92	29.7
Total	320	20	728	115.8

Source: Renewable energy: The solution to climate change – European Renewable Energy Council, 2004

Figure 2 represents Structure of primary energy demand in EU-25 and trends up to 2030 under baseline assumptions [6].

Fig. 2: Structure of primary energy demand in EU-25

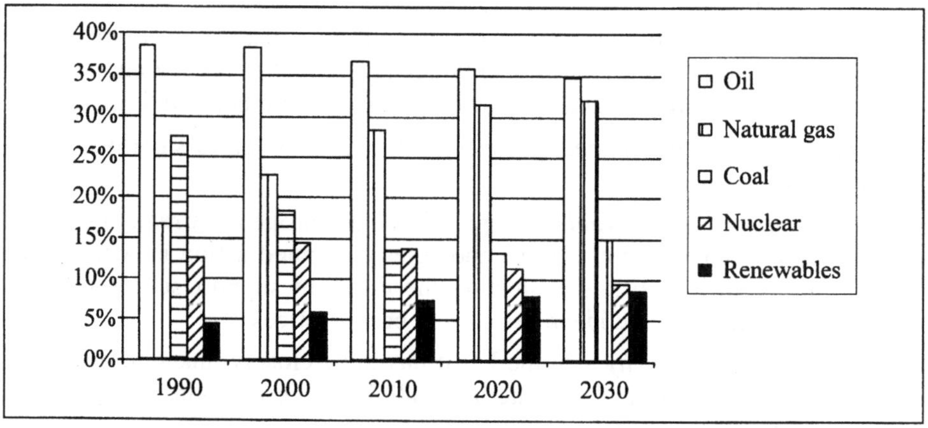

Source: European Energy and Transport Trends to 2030 – EC, Luxembourg 2003

EU enlargement means new challenges for the energy market and the demand increases every year more rapidly due to growing economies of new Member States. The energy consumption of 'the 25' per capita over time will rise to the average of the 'former 15' per capita [24], so – CO_2 emissions cannot fail to rise, quite contrary to the goals of preventing climate change. In face of this dilemma, the crucial areas are key innovations in modern power generation technologies on the basis of fossil energy sources. In the next few decades the energy sector will need replacement of old and the construction of new power plants.

EU Energy Policy and Initiatives

Setting out energy policy for the EU is a complex task: it is an area with a large number of players – governments, national regulators, large enterprises, local authorities, etc. Therefore mobilisation of all players and transformation of energy efficiency policy over long-term is essential. No energy source, whether carbon-based, petroleum-based or nuclear has ever been developed without support of the public authorities. The EC has concluded that 50% of future reductions of GHG emissions could be achieved by improved energy efficiency for reaching greenhouse gas emission levels, which are compatible with maintaining the climate [24].

EU Commissioner for Energy Andris Piebalgs has put forward three goals for EU energy policy: competitiveness, sustainability and security and outlined six energy priorities, which include a closer linkage between energy, environment and research policies.

> „Europe should put a concerted effort to the Lisbon and Kyoto objectives of making Europe the most competitive economy in the world and cutting GHG emissions by developing the new technologies that will have huge commercial potential in the coming decades" [1].

The six priorities are:

Energy efficiency – the cheapest, most competitive, most secure form of new energy for the European Union is energy savings by better regulation, by introduction of new technology, and by providing citizens with better information.

The internal market in gas and electricity – has already led to significant price reductions.

Continued support of renewable energy – although Europe is currently the world leader in many renewable energy technologies continued action is necessary to follow this trend. The EC is evaluating the different support schemes implemented around Europe and identifies current best practices.

Nuclear safety and security – in the EU it has to be closely linked to a strategy dealing with security of nuclear materials, radiological protection, nuclear safety and the reliable disposal of nuclear waste.

Further development of EU external energy policy relations – is a response to the security of energy supply presented by rapidly increasing worldwide demand, high and volatile oil prices: EU-Russia energy dialogue; Energy Community of South-East Europe.

Better linkage of energy, environment and research policy – combing these two objectives will turn Europe's environmental commitments into a competitive advantage and lead to the development and introduction of new technologies.

Catherine Day, Director General for DG Environment, EC draws consequences that Climate Change is the biggest challenge the world is facing today. It is obvious already now that the accepted 8% emission reduction is too small and it is necessary to decrease emissions more. Figure 3 presents today's output and level laid out in the Kyoto Protocol.

> „If we are serious about addressing climate change, the urgent need is to reduce GHG emissions more. Power plants account for around 30% of total CO_2 emissions in the EU and limiting their emissions is bound a part of any climate change policy" [2].

Fig. 3: Kyoto Protocol – implementation

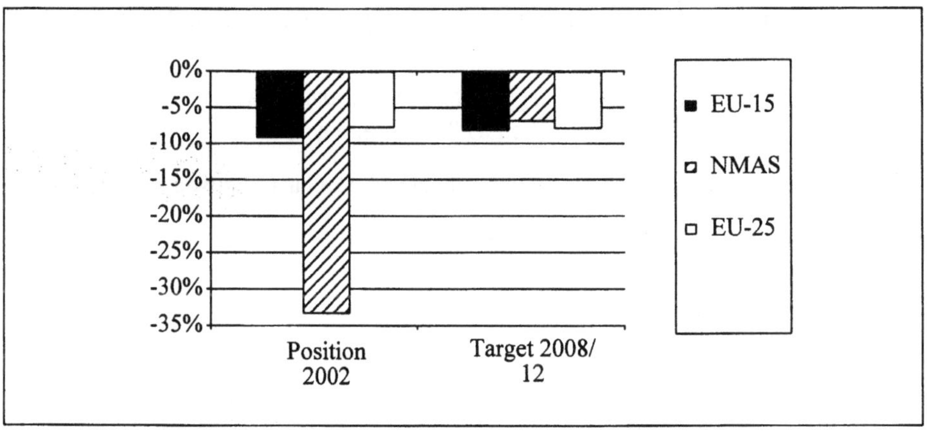

Source: Power: perspectives 2005, Innovations to prevent climate change in fossil-fired power plant engineering – RWE Power Aktiengesellschaft, Essen

Kyoto has led Europe to embark on the European Emissions Trading Scheme (EU ETS) in 2005, which is the world's first international market for carbon emissions. It is a historic milestone in Europe's fight against global climate change, and an unprecedented instrument in EU environment policy where it is becoming a key driver for the development of climate friendly technologies while minimising costs. It covers some 12 000 industrial plants across 25 Member states responsible for almost half of EU's output of CO_2 [5] including 91 objects in Latvia [14].

Green Paper on Energy Efficiency or Doing More With Less – is a European Strategy for the Security of Energy Supply which emphasizes secure energy supply at the same time reducing green house gas emissions that are associated with climate change. It calls for immediate actions to promote renewable energy sources, alternative fuels for transport, to increase energy efficiency and sets out three initiatives for energy savings [7]:

Competitiveness and the Lisbon agenda – the EU could save at least 20% of its present energy consumption in a cost-effective manner. Although considerable investment is needed to harness these potential savings in terms of new energy efficient equipment and energy services, they are largely local in character.

Environmental protection and the EU's Kyoto obligations – energy saving is the quickest, most effective and most cost-effective manner for reducing greenhouse gas emissions, as well as improving air quality.

Security of supply – by 2030 the EU will be 90% dependent on imports of oil and 80% of gas (on the basis of present consumption trends). It is impossible to predict the price of oil and gas in 2020, particularly if demand from the developing world continues to increase as rapidly as today: energy efficiency is one of the key methods to deal with this challenge. The most important barriers to increased energy efficiency are lack of: information on costs and availability of new technology and costs of our own energy consumption; technicians on proper maintenance.

Figure 4 illustrates primary energy intensity adjusted at current purchasing power parities in 2002 assuming the EU 25 average as 100.

Fig. 4: Primary energy intensity adjusted at current purchasing power parities (2002): EU-25=100

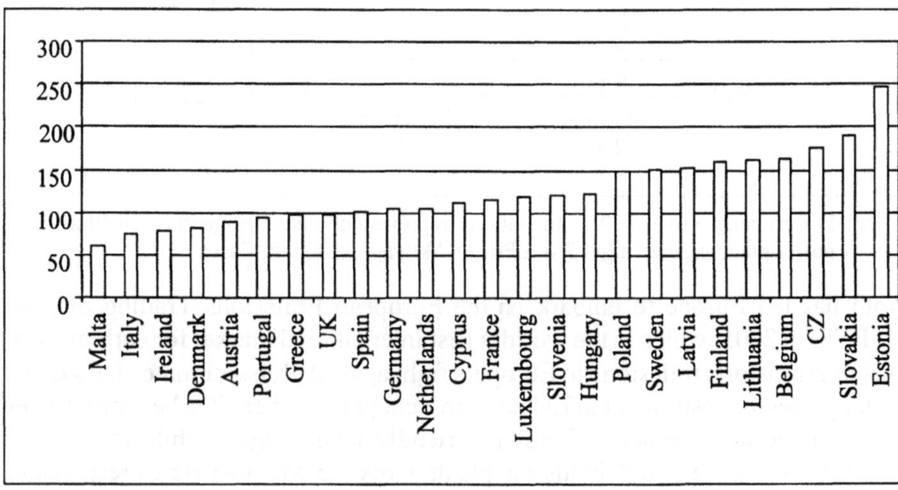

Source: Green Paper on Energy Efficiency or Doing More With Less – Brussels, 22.6.2005 COM (2005) 265 final

The Sustainable Energy – Europe 2005-2008 campaign is an EC initiative aiming to raise public awareness and promoting sustainable energy production and use among individuals and organisations, private companies and public authorities, professional and energy agencies, industry associations and NGOs across Europe. It is structured in four fields: improvement of energy efficiency and rational use of energy (in building and industry sectors); promotion of new and renewable energy sources for centralised and decentralised production of electricity and heat (integration into the local energy systems); support for initiatives relating to all energy aspects of transport (the diversification of fuels) and promotion of renewable fuels (biofuels) and energy efficiency in transport; support for initiatives promoting renewable energy sources and energy efficiency in the developing countries, in the framework of the Community cooperation with these countries.

Research and Development in the European Union

The urgency to develop adequate and timely solutions is justified by the need to limit emissions of GHG to mitigate the destructive effects of climate change. Research and demonstration can provide the most environmentally and cost-effective technologies and measures enabling the EU to meet its targets under the Kyoto Protocol (and beyond) and to implement its energy policy commitments.

Carbon capture and geological storage is a priority in EU research and development programmes in the closest future. In the short term the EU should achieve higher energy efficiency and increased supply from European energy sources, in particular renewables [4]. In the long term a hydrogen-based economy will have an impact [8]. Radical transforming of the energy system requires new technologies – all European efforts and resources should be combined in a coherent and more effective manner.

The importance of developing new energy technologies is recognised also by the ERA Working group of the Advisory Group on Energy, which has set 8 recommendations-imperatives for the seventh EC Framework Programme for research 2007-2013 (FP7) [22]:

- to accelerate the coordination of regional and national programmes,
- to accelerate progress in strategically crucial fields of R&D,
- to develop world-class facilities for European energy research,
- to establish effective information and communication systems,
- to develop human resources,
- to ensure adequate international participation,
- to build societal backing,
- to ensure the necessary scientific basis in other research fields (materials research, information and communication technologies).

FP7 has set an objective of "transforming the current fossil-fuel based energy system into a more sustainable one based on a diverse portfolio of energy sources and carriers combined with enhanced energy efficiency to address the pressing challenges of security, supply and climate change, whilst increasing the competitiveness of Europe's energy industries" and nine priority research fields in Energy sector – technologies, integrated actions, new concepts and tools for [16]:

- hydrogen and fuel cells – to provide a strong technological foundation for competitive fuel cell and hydrogen industries, for stationary, portable and transport applications,
- renewable electricity generation – to increase overall conversion efficiency, the development and the demonstration of technologies suited to different regional conditions,
- renewable fuel production – to develop fuels produced from renewable energy sources, aiming at the cost-effective production and use of carbon-neutral fuels, in particular liquid biofuels for transport,
- renewables for heating and cooling – to increase efficiencies of heating and cooling from renewable energy sources, ensuring their use in different regional conditions,
- CO_2 capture and storage technologies for zero emission power generation – to reduce the environmental impact of fossil fuel use aiming at highly efficient power generation plants with near zero emissions, based on CO_2 capture and storage technologies,
- clean coal technologies – to improve plant efficiency, reliability and cost through development and demonstration of clean coal conversion technologies,
- smart energy networks – to increase the efficiency, safety and reliability of the European electricity and gas systems and networks by transforming the current electricity grids into a service network including integration of renewable energy sources,
- energy efficiency and savings – to improve energy efficiency and savings for buildings, services and industry,
- knowledge for energy policy making – to assess the main economic and social issues related to energy technologies and to provide quantifiable targets and scenarios for medium and long term horizons.

Energy Production and Consumption in Latvia

Achieving the EU goals for energy efficiency and increasing the share from renewables is a common activity of all governments, organisations and individuals.

In 2001 the total of primary energy resources in Latvia was made up of: coal – 2%, oil products – 29%, natural and liquid gas – 31%, other solid fuels – 30%, hydroelectric power station electricity (including import of electricity) – 9% [20]

(Figure 5). Both local energy resources (wood, peat, hydro resources) and imported resources (oil products, natural gas, coal) are used by the energy sector.

The dependence of the Latvian economy on imported energy resources has evolved over a long period of time but the structure of heating fuels has been changed [6]. The share of fossil fuel in the consumption of energy resources in Latvia in comparison with other European countries is small: wood, hydro energy and importation of electrical energy are used widely. In comparison to 1995 oil products and coal have decreased, but the proportion of renewable energy resources and gas has increased [21] (Figure 6). In recent years in the primary energy resource balance, local energy resources comprise 30-35% of total energy consumption and have increased along with the proportion of wood and hydro resources in energy production [19]. It is predicted that the proportion of gas as a less-harmful-to-the-environment fuel in the total energy resource structure will continue to grow, further reducing the share of oil. The main issues in the development of energy sector in Latvia are: improvement of legislation, the privatisation and restructuring of energy industry enterprises, increasing energy efficiency – security of supply and the establishment of oil products reserve.

Fig. 5: Proportion of fuel types in the primary energy resource balance

Fig. 6: Consumption of primary energy resources

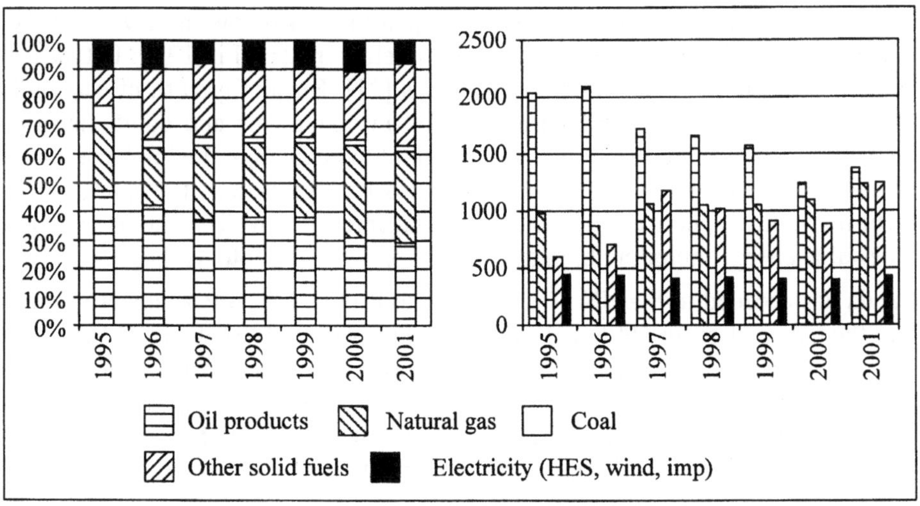

Source: Sustainable Development Indicators in Latvia 2003 – Latvian Environment agency, 2003

Source: Third National Communication of the Republic of Latvia under UN FCC Ministry of Environment

Energy and Climate Change Mitigation Policy in Latvia

Emissions of greenhouse gases in Latvia are among the lowest in Europe (equivalent of emissions per capita is a small). The main task in this respect is to ensure that the emissions of greenhouse effect gases are limited in the course of economic development of the country [19].

Transition to a market economy in Latvia after 1991 initiated essential changes in all sectors of national economy and resulted in the decrease of GHG emissions. The total emissions in 2001 in comparison to 1990 have decreased by 61% and were related to the restructuring of the economy and decrease in the production volume. GHG emissions represented by sectors demonstrate that in 1995-2001 emissions generated by the energy and agriculture sector have significantly decreased (by 35% and 12%), but the transport sector has increased by 17%, however, without reaching the 1990 emissions level.

The most important source of GHG in 2003 was the sector of energetics – 72.2%, composed of: transport – 35.3%, energy – 32.1%, manufacturing industry and construction – 12.7%, other sectors – 18.2%, and losses in transport and distribution – 1.7% [10].

The forests and agricultural lands of Latvia absorb roughly 60% of Latvia's anthropogenically emitted carbon dioxide. So, the GHG emission mitigation in Latvia is not the primary policy goal but rather a side effect. The projections testify that GHG emissions in 2010 will be equal to 45% below the level in 1990, but in the scenario "with measures"[2] – 51% below the level in 1990: Latvia will be able to fulfil the international obligations [21]. But the growth of economic development may be faster than anticipated in documents and could cause an increase in CO_2. Therefore, it is very important to encourage the implementation of measures in all energy-intensive industries (energy sector, transport, manufacturing industry, construction, house management) as well as the use of local renewable energy including biomass and the process of photosynthesis.

Latvia has acceded to the UN FCC and Kyoto Protocol. The Parliament has ratified the Convention in 1995 and the Kyoto Protocol in 2002. In 1997-98 Latvia elaborated on its first climate change mitigation policy plan, which incorporated the Kyoto Protocol objectives and requirements but in 2002 approved the Joint Implementation Strategy for 2002-2012. The objective of the Strategy is to contribute to the mitigation of the global climate change by means of attraction the

---

2 Wider use of fuel wood for centralised district heat generation; renewal of small HPS; use of wind energy; bio diesel fuel as a internal combustion fuel in small scale co-generation plants (and/ or transport); wider use of co-generation; rehabilitation of centralised district heating; increase of energy efficiency in dairy companies of Latvia; reduction of heat losses in buildings; economically cost-effective use of wood-waste in local governments' heating systems; efficient lighting program.

investments to carry out GHG mitigation projects [9]. Latvia has prepared a National Allocation Plan including the principles of allocating allowances for greenhouse gas emissions, questions on harmonisation of legislation as well as the potential for reduction of GHG emissions [14]. The prognosis of the Third National Report to the UN FCC predicts that Latvia will be able to fulfil its obligations under the Kyoto Protocol maintaining its estimated rate of economic development [21]. However, similar to other CEE countries, energy resources are not used effectively in Latvia and the emissions resulting from power generation per inhabitant are higher than for most OECD countries [12].

In 2005 the Cabinet of Ministers adopted the Climate Change mitigation programme for 2005-2010 stating the goals for Latvia and setting policy for achieving it. The priorities in the energy sector related to climate change mitigation are: raise of energy efficiency, development of environmentally sound transport system, and implementation of best available and cleaner technologies. The following measures have been accepted: increase of renewable resources in energy sector, increase of energy use efficiency, development of environmentally sound transport system, promotion of best available and cleaner technologies, promotion of environmentally safe agricultural technologies, increase of CO_2 attraction in forestry, development of waste management system by capture of biogas in landfills, participation in trading of emission quotas and other Kyoto mechanisms, promotion of environmental management systems [10].

By way of illustration a project on energy saving and efficiency rising in the street-lightening system in the Latvian town Tukums, been recognised as EU good practice. It improved the quality of life of residents and led to significant reductions of energy use. By implementing better technology the network's energy use has been cut by 630 000 kWh per year leading to savings of 37 000 € per annum [13]. The project is a model for small and medium-sized municipalities and there is potential to repeat such activities in more than 500 municipal areas in Latvia.

References

1. *Piebalgs, Andris* (2005), EU Commissioner for Energy – The European CO_2 Capture and Storage Conference "Towards Zero Emission Power Plants", Brussels, 13 April 2005.
2. *Day, Catherine* (2005), Director-General DG ENV "Climate Change in the EU and Worldwide: The Commission's view" – The European CO_2 Capture and Storage Conference "Towards Zero Emission Power Plants", Brussels, 13 April 2005.
3. Climate Change 2001: Impacts, Adaptation and Vulnerability (Eds: James McCarthy, Osvaldo Canziani, Neil Leary, David Dokken, Kasey White) – Cambridge University Press, 2001, 1032p, ISBN 0 521 80768 9.

4. Electricity from renewable energy sources: Encouraging green electricity in Europe – EC, Luxembourg 2004, 16p, ISBN 92-894-6666-9.
5. Environment for Europeans – EC, no. 19, March 2005 (Special issue on Climate Change).
6. European Energy and Transport Trends to 2030 – EC, Luxembourg 2003, 220p, ISBN 92-894-4444-4.
7. Green Paper on Energy Efficiency or Doing More With Less – Brussels, 22.6.2005 COM (2005) 265 final.
8. Hydrogen Energy and Fuel Cells, A vision of our future – EC, 2003, 33p, ISBN 92-894-5589-6.
9. Joint Implementation Strategy (2003-2012) – "Vides projekti", Rīga 2003, 32p, ISBN 9984-9655-1-1.
10. Klimata pārmaiņu samazināšanas programma 2005.-2010.gadam – Ministru kabineta 2005.gada 6. aprīļa rīkojums no. 220.
11. Kyoto Protocol to the UN FCCC (1997) – http:// unfccc.int/essential_background/kyoto_protocol.
12. Latvian National Report "Rio + 10" to the World Summit on Sustainable Development in Johannesburg – 2002, The Ministry of Environment of the Republic of Latvia.
13. Local energy action: EU good practices – EC, Luxembourg 2004, 32 p., ISBN 92-894-8218-4.
14. National Allocation Plan 2005-2007 – Order no. 270 of the Cabinet of Ministers of the RL on 27 April 2004.
15. Power: perspectives 2005. Innovations to prevent climate change in fossil-fired power plant engineering – RWE Power Aktiengesellschaft, Essen, 2005, 79p.
16. Proposal for the Seventh framework programme of the European Community for research, technological development and demonstration activities (2007-2013) – Brussels, 6.4.2005 COM (2005) 119 final.
17. Renewable energy scenario to 2040: Half of the global energy supply from renewables in 2040 – European Renewable Energy Council, 2004, 16p.
18. Renewable energy: The solution to climate change – European Renewable Energy Council, 2004, 16p.
19. Strategy for Sustainable Development of Latvia – "Mantojums", Rīga, 32p, 2002, ISBN 9984-9575-4-3.
20. Sustainable Development Indicators in Latvia 2003 – Ministry of Environment of Latvia, 2003, 164 pages, ISBN 9984-9557-2-9.
21. Third National Communication of the Republic of Latvia under United Nations Framework Convention on Climate Change – "Vides projekti", Rīga 2001, 172p, ISBN 9984-9388-9-1.

22. Towards the European Energy Research Area: Recommendations by the ERA Working group of the Advisory Group on Energy – EC, Luxembourg 2005, 85p, ISBN 92-894-8744-5.
23. United Nations Framework Convention on Climate Change – http://unfccc.int.
24. Winning the Battle Against Global Climate Change – Brussels, 9.2.2005, COM (2005) 35 final.
25. World Energy Outlook – 2004 Edition, 550 pages, ISBN 92-64-10817-3.

Foresight and Integrated Management Systems in Small and Medium sized Enterprises (SME)

–

Preliminary Considerations for a Tool-kit towards Corporate Sustainability

Markus Will

Freelance Engineer, Environmental Manager (TÜV)
in close cooperation with Hochschule Zittau/Goerlitz (FH),
University of Applied Sciences,
Theodor-Körner-Allee 16, D – 02763 Zittau
Brückenstraße, D – 02826 Görlitz, Germany

Abstract

When it comes to *Sustainable Development*, we have to recognize several premises: (1) *Sustainability* needs a long-term way of thinking, we will not achieve it over night – it is a time consuming and dynamical process, (2) *innovation* and *knowledge* are key factors and (3) most of the *stakeholders* have to be included and (4) we need massive *reorientations* in economy.

When we discuss the contribution of entrepreneurs to the Sustainability goals, indeed, we concentrate on management tools supporting *Corporate Sustainability* and reducing (economical) risks. These tools are *separate management systems* for environment (EMAS II or ISO 14001), for quality (ISO 9000), safety and social standards (BS 8800, SA 8000, etc.). However, there are also several integrated approaches, so called *Integrated Management Systems* (IMS) including environment, quality and safety management systems.

Nowadays, society is in a transition from industrial toward knowledge and information society. Thus, it becomes more and more important for SME to manage also *corporate intellectual capital* in order to be innovative. That's why also *Innovation- and Knowledge-Management* have to be included.

But, with management systems, only procedures, organizational structures and maybe processes are considered. A management system itself will not touch the product portfolio. For this reason, a systematized process of corporate strategy setting, a sort of *Visioning* and *Foresight* of products and their (potential) impacts is needed. In order to have a sustainability sound product portfolio, companies need a methodological tool kit of *Life Cycle Assessment*, *Technology Assessment* and *Technology Foresight*. The poster will present a first approach on such a tool-kit.

Introduction

This discussion paper aims in giving some explanations to the poster presentation on the Sharing Conference on "Sharing Integrated Approches towards Sustainable Development" in Jūrmala in May 2005. The poster does not represent results of a R&D project but it reveals preliminary considerations on how SME could support Sustainable Development with a methodological tool-kit, composed of *Foresight* and *Management Systems* integrating quality-, safety- and ecological aspects added by innovation and knowledge management.

In order to keep the text readable and also giving non-experts in the field of businesses and *Sustainability* the chance of comprehensive understanding, the text is structured in questions to be answered, rounded up by a short summary (see abstract) and open questions for further research and practical work. The author would be grateful for constructive critics and comments. He is also interested in cooperation with interested parties, especially from the New Member States.

The poster is comprised of six key elements: three different coloured ellipses in the centre; additionally one can see the two different kinds of arrows representing a time scale (the one below) and the Life Cycle Management approach. The other (sixth) element is the PDCA-Cycle (Deming, 1986) fundamental to all management approaches. In the first ellipse an Integrated Management System (IMS) including Quality-Management (QMS), Environmental Management (EMS) and Safety- and Health-Management (S+H-MS) is emphasized. This is the recent understanding of an Integrated Management System that could be added by Management Procedures for Knowledge and Innovation as illustrated (so called IMS++). This is consistent with the time arrow showing that IMS is situated somewhere between "yesterday" and "today" whereas it is an inherent characteristic of IMS++ to have a strategic outlook to the future. The third ellipse in the centre of the poster deals with Foresight as systematic view into future prospects on a time scale up to 30 years.

What is Sustainability?

Individual, social and economical behaviour is called sustainable when it ensures that a whole society develops in a way that "*...meet[s] the needs of the present without compromising the ability of future generations to meet their own needs*" (Brundtland Commission, 1987). The term Sustainable Development was introduced into the public discussion, most recently with the UNCED (UN Conference on Environment and Development) in Rio de Janeiro (Brazil) in 1992.

Fig.1: Poster "Foresight and Integrated Management Systems in SME. A toolkit towards Corporate Sustainability"

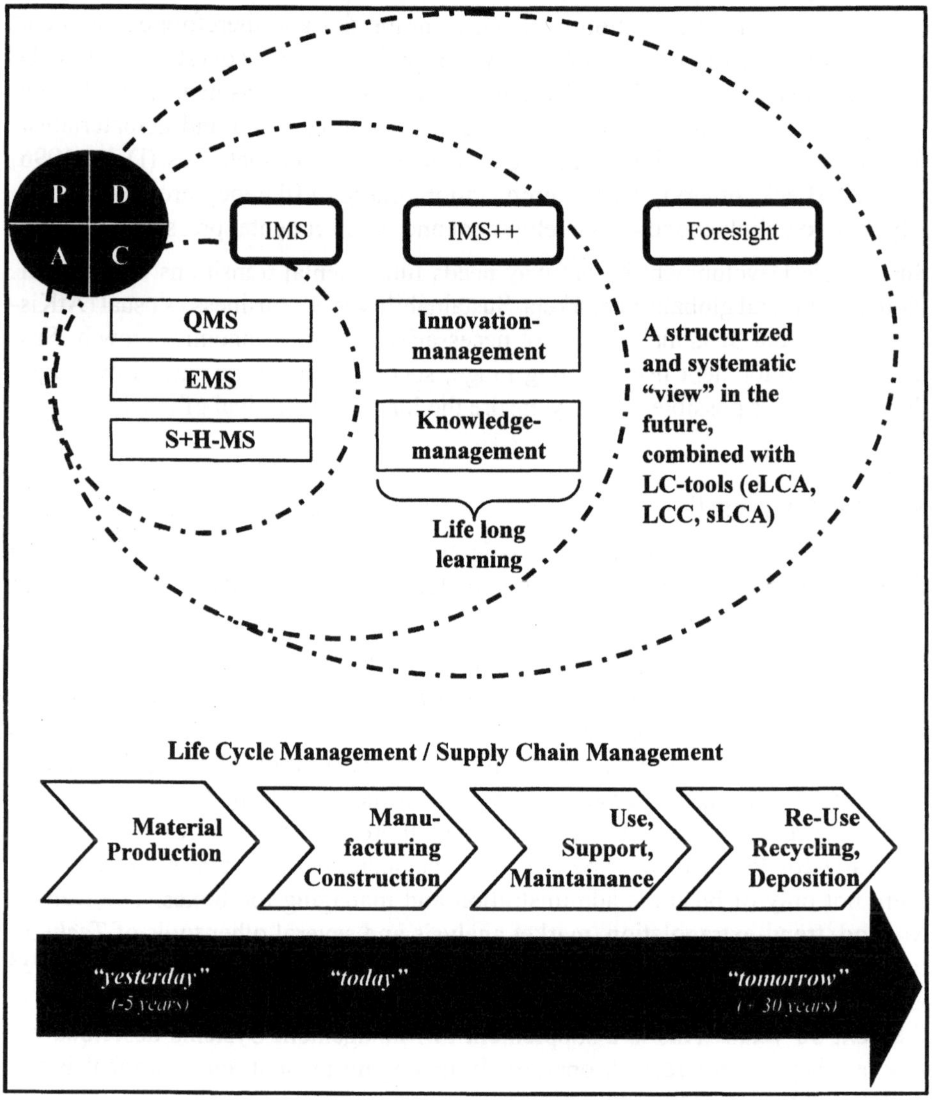

Sustainability is widely understood, but obviously it is much more than environmentalism. It is clear that *Sustainability* is a complex matter that includes at least economical, social and environmental issues. In this anthropocentric 3D-understanding of *Sustainability* maintaining the *economical* system means to sustain man-made capital, human capital as well as renewable and non-renewable resources and other forms of nature capital for adding values in the production process (similar in Spangenberg (2002), and others). Evidently, this is linked with the social and ecological dimension: Maintaining *social* or human capital stands for intra- and intergenerational justice – or fairness to speak with John Rawls- this includes education, participation in social communities and therefore employment and income. For the *ecological* dimension, resilience[1] of ecosystems has to be considered (Folke et al. 2002; Millenium Ecosystem Assessment 2005; Döring 2004). Basically important issues are resource conservation and dematerialisation in terms of reduction of throughput of energy and materials (Daly, 1996; Schmidt-Bleek, others), waste minimization, energy efficiency are more generally seen to delink economic development and environmental pressures.

Sustainable Development in that way needs fundamental transitions in both individual minds and globalised markets. Sustainability is not business as usual (AtKisson 1999). And Sustainability is not necessarily against (economic) growth. Sustainable Development needs a long range; success will not be earned overnight. Sustainability is possible and may ensure the long-term survival of mankind.

What is Foresight?

Foresight is a structured and systematic "view" into future. Generally it aims to detect direct and indirect as well as intended and non-intended consequences of today's actions; otherwise it looks for future prospects, chances and risks on a time scale up to thirty years. To get a little insight, let's switch to history.

Roots of *Foresight* can be found in the 1940's in the US: The Stanford Research Institute (SRI) and the famous RAND Corporation were founded by managers and scientists in Menlo Park, California. Indeed, SRI and RAND are the nucleus of Silicon Valley and belonged to the most influential Think Tanks after World War II. Scenario Method was developed there by Herman Kahn and Peter Schwartz for Royal Dutch/Shell. Today, the scenario approach is a central element not only of business administration and managing toolboxes. The Delphi-Method, trend-extrapolation, market analysis and several other tools of *Technology Assessment* are supplements to the scenario approach and allow a structured and systematic outlook to societal needs, market potentials, chances and risks in the next 30 years. This is a supplement to Management Systems described below, as they concentrate on operatively managing present and historical issues with only little regard to future needs.

With its science based methodology *Foresight* is far removed from a dubious and unserious look into the "glass ball". As *Foresight* is a central element in *Technology Assessment (TA)* it can be argued that research for *Sustainability* and TA and therefore *Foresight* often have analogous goals (Will 2003; also others).

---

1 "Resilience [of ecosystems; MW] provides the capacity to absorb shocks while maintaining function." (Folke et al. 2002).

What is Corporate Foresight?

"*To futurize the company*" is a central claim concerning *Corporate Foresight* (Burmeister et al. 2004). Peter Schwartz, another opinion leader of the Futurologists, called it "*The Art of the long view*" (Schwartz 1991). This is really not different from the aims of *Sustainability*.

Corporate Foresight enables companies to detect future challenges and give them – maybe in combination with Knowledge and Innovation Management – the potential to put economical challenges and *Sustainability*-needs into action, into innovation, into products.

It must be clear that *Foresight* and its application in companies and businesses (*Corporate Foresight*) represent more than a simple toolbox. Indeed, it is a process of communication to build a mid-term to long-term vision on future markets, costumer needs and societal challenges. *Corporate Foresight* helps to switch the company on the right track and supports potentials, competitiveness and innovation capability (see Burmeister et al. 2004).

What are Integrated Management Systems (IMS) and what is IMS++?

A *system* can be understood as an interconnected matter, perhaps including organisational and administrative aspects, as well as documents, policies and practices but also people.

Management can be seen as the high level structure and executive process to run a socio-technical system, as a business enterprise is. In this understanding to manage means to set goals *(plan)*, to shape and drive actions in order to fulfil these goals *(do)* and to control to which degree the aims could be reached *(check)*. In case those goals were not accomplished completely, new measures have to be taken. And in turn: in case that all aims are completed new goals have to be set up *(act)* as continual improvement. Basically this is a typical structure for a Management Cycle as introduced by Deming (Deming 1986).

A *Management System (MS)* indeed usually follows the Deming-Cycle and can be regarded as organisational basement ensuring optimal executive functioning of the company. *Management Systems* contain, corresponding to the Deming Cycle, a policy, goals, an action plan, audits and a management review process. An inherent aspect of all *MS* is the continuous process of improvement over the time. A *MS* demonstrate companies' efforts on risk management and legal compliance, also with regard to conditions of BASEL II assessments.

Ideally, an *MS* is the inherent management tool that commits both: top management and staff. The overall aim of an *MS* is, of course, to maintain companies' economic activities and to support its purpose and mission.

There are several separate MS for (i) Quality, (ii) Environment and (iii) Safety and Health, etc.

ad (i) Quality Management Systems

Quality Management Systems (QMS) aiming in supreme product quality by maximum consistency in the supply chain, by strengthening product consistency and minimising failures and thereby reducing claims and rework. A QMS ensures also meets customers satisfaction and compliance with legal obligations (in Germany: law on product liability). There is an international standard for QMS developed by ISO (ISO 9000: 2000) that is strictly formalised and gives the possibility for external certification. There are also other approaches on assuring product quality, like the Total Quality Management (TQM), Kaizen, EFQM and others.

ad (ii) Environmental Management Systems (EMS)

The overall goal for an EMS is to reduce environmental pressure while strengthening the business with a view to continuous improvement of a sound environmental performance. For a company an EMS is necessary because it offers potentials to reduce costs (esp. electricity and water consumption, waste, ...) by raising efficiency. Additionally legal compliance can be guaranteed by an EMS. Nevertheless EMS plays a key role in fulfilling customers' demands and aid marketing issues and image. Not forgetting, an EMS is an outstanding contribution to risk management.

ad (iii) Safety and Health-Risk Management

A company needs a sort of Safety and Health Management to reduce risks and to prevent accidents. Normally, there are national legal regulations forcing a company to take actions to ensure workplace and process security and implement hygiene standards and to intend competences and responsible employees. Because of this strict legal regulations there exist quite a lot of different Management Systems for Safety and Health-Risks, as the BS 8800, OHRIS or the ISA 2000 (International Safety Audit 2000) and OHSAS 18001 (Occupational Health and Safety Assessment Series) that are certifiable and that provide at least to a certain degree links to ISO 9000 standards on Quality Management.

An *Integrated Management System* (IMS) is often understood as a synergy between Quality Management Systems (QMS), Environmental Management Systems (EMS) and Safety and Health Management Systems (S+H-MS).

To overcome organisational boundaries – an *IMS* only makes sense, when it is more than combining books of documentation, or putting separated quality, environmental or safety guys in one room. More motivation, less documentation perhaps is a good claim – the goal is a living *IMS*.

Actually, an *IMS* could integrate anything that influences business. However, an *IMS* currently integrates mostly quality, environmental and health and safety aspects. But, as said, more issues could be added to the *IMS*.

As globalised societies are nowadays in a transition from industrial towards knowledge and information society, it's straightforward also to include Innovation- and Knowledge-Management (here described as *IMS++*). And this although Innovation- and Knowledge-Management approaches are not that developed, compared f.i. to EMAS II.

Beside named issues to be included in an *IMS*, also human resources, finance or security perhaps could be added. But let's do a step-by-step approach. As there are no national or international standards for implementing *IMS* in companies, there are several approaches (also EFQM, SBC, etc.). Actually drivers to implement an *IMS* originate within the company and not so much from customer demands. But they are perhaps in line with other stakeholder aspects.

To guarantee that a *MS* backs up *Sustainable Development*, some preconditions have to be fulfilled. Namely *Life Cycle Thinking* has to be considered as a key element of sustainability oriented mind-set. Although insinuated here, that a CEO concentrates mainly on tasks inside the company (gate-to-gate) any economic, social and ecological aspect over the whole (product) life cycle, beginning with resource extraction, production, usage, recycling and end-of-life-disposal has to be taken into account (Klöppfer 2003, also others). That's also the inherent meaning of systems thinking and Life Cycle Management.

Why Should SME Deal with Foresight & IMS++ ?

If there were to be a science and knowledge based toolkit of *Corporate Foresight* in addition to available knowledge on IMS++ approaches combined with a consulting product, it will cost money. That's quite clear, some might say: What costs nothing is nothing! But SME, especially micro-companies with up to 50 employees are typically underfinanced to a certain degree, human and financial resources are restricted. Management and CEO are busy with day-to-day tasks; there is no time for *future outlooks*, serious *environmental performance* or *Sustainability thinking*. The economic situation is hard today, not only in Germany, and so it is with "thinking out of the box " (Pauli, G. 2004)! But, no reason for getting pessimistic, the "*future is unwritten, but we can shape it*" (Burmeister et al. 2004)!

However, globalised markets, accelerated product cycles and the transition from industrial to knowledge society characterise the economic and societal true. Also restricted resources and the environmental situation, calls for a sustainable way of living and a *sustainable economy*. That's why there is a need for *IMS++* and *Corporate Foresight*! Here, this is not a question of choice; it's a question of taking chances!

Moreover, what Global Players are already doing (Royal Dutch/Shell, BASF, SIEMENS, DaimlerChrysler, PHILLIPS, Infineon, Deutsche Telekom) should also be attractive for SME, shouldn't it?

So, the key question is how a tool kit must look, if SME should use it to implement processes of *Corporate Foresight* and *IMS++*. Obviously there is a growing need for down-scaling the requirements of standardised Management Systems like ISO 9000, EMAS II and ISO 14000 – especially with regard to high documentation requests. A variety of small-scale EMS were developed (i.e. BS 8555, Ekoscan, Ecomapping, Ecoprofit, Eco+, Environmental Diploma, EcoLighthouse – for more information see: www.ems-for-sme.org).

Possibly, these approaches deliver a starting point for further research on developing a tool kit towards *Corporate Sustainability* that is tailored to the needs and potentials of SME and that includes the management approaches described here.

As a first consideration there are two ways of thinking: (1) a downscaled tool kit in a public-private-partnership (PPP) that includes consulting, group auditing, with shared templates and routines (as f.i. EcoMapping; www.ecomapping.org; Engel 2004)). In contradiction to a PPP-approach (2) an independent self-build-tool, preferably computer aided, can offer a SME the possibility to develop their "own" *IMS++* step-by-step independently by their own in-house-experts. Naturally, not everything can be done by in-house-experts, esp. with regard to the *Corporate Foresight* process. That is why such a self-built-tool must be added to with education and training modules following a blended-learning concept and perhaps including external training and consulting.

Foresight and IMS++ in SME: Reason why – in Short

- Reduction of risks and increasing profitability (risks of bad investment or compensation claims)
- Reduction of duplications and therefore costs, by combined approach and more efficiency
- Harmonisation and optimisation of practices – creating consistency
- Elimination of conflicting responsibilities and relationships
- Balancing conflicting objectives
- Improved communication
- Approximation of future needs and markets (consumer oriented)
- Approximation of future impacts of products (society oriented)
- Analysis of interactions of product and society (a first step of "getting out of the box")
- Exploration of possible alternatives
- Innovative product design, combined with Impact Assessment (prospective LCA)
- Stakeholder communication.

Open Questions

- How to distinguish Innovation Management and Corporate Foresight?
- How to distinguish it from usual strategic management?
- How to explain the "return-of-investment" to a CEO of a SME?
- What in detail is the need of an SME regarding to IMS++, Foresight and *Corporate Sustainability*?
- How a consultation product could be designed?
- Are there options and potentials for Multi-Client or Open-Source approaches?
- Are there potentials for creating an interdisciplinary research project? Are there interested partners? How could it be founded?

Acknowledgements

The author wants to thank all members of the organisation-team for inviting young researchers to present ideas and results of their own work at an international high level conference. Furthermore the author is grateful for interesting discussions with conference participants and young researches, especially from Eastern Europe. Special thanks to Dr. Arnolds Ūbelis, Prof. Jānis Grāvītis, MSc. Dina Bērziņa, Dr. Jānis Āboliņš, MSc. Andis Zīlāns, Prof. Dr. Walter Leal Filho, Prof. Dr. rer. nat. Bernd Delakowitz, Prof. Lars Rydén and Peter Fagerström, Christian Zwatz and not forgetting Dr. Mikkael 'Mikka' Planasch. For critical discussion and review while preparing the poster presentation and the paper the author wants to thank also Dipl.-Kff. Anke Hoffmann and again Prof. Dr. rer. nat. Bernd Delakowitz.

References

Albertshauser/Malanowski (2003), Innovations- und Technikanalyse im Management. [Innovation- and Technology Assessment in Corporate Management] Campus Verlag.

AtKisson, Alan (1999), Believing Cassandra. An optimist looks at a pessimist world. ISBN 1-890132-16-0. Chelsea Green.

brundtland commission (1986).

daly (1996), Beyond Growth. The Economics of Sustainable Development. Beacon Press, Boston.

Deming, W. Edwards (1986), Out of Crisis. MIT, Cambridge, Massachusetts, USA. (cited in Löbler 2001).

Döring, Ralf (2004), Wie stark ist schwache, wie schwach starke Nachhaltigkeit? [How strong is weak, how weak is strong sustainability?] Wirtschaftswissenschaftliche Diskussionspapiere 08/2004. Ernst-Moritz-Arndt-Universität

Greifswald, Lehrstuhl für Landschaftsökonomie. www.rsf. uni-greifswald.de/bwl/paper.html.

Engel, *Heinz Werner* (2005), EMAS and ISO 14001 wasy: 10 people, 10 pages, 10 days. Small Management Tools for Micro-Enterprises. www.inem.org.

Folke et al. (2002), Resilience and Sustainable Development. Building adaptive capacity in a world of transformations. A report for the Swedish Environmental Advisory Council 2002: 1. Stockholm, Sweden.

Klöppfer, *Walter* (2003), Life Cycle Based Methods for Sustainable Product Development. Editorial in the International Journal of Life Cycle Assessment, 8 (3), 2003. Ecomed publishers, Landsberg, Germany.

Löbel, *J.* (2001), Nachhaltige Managementsysteme: sustainable development durch ganzheitliche Führungs- und Organisationssysteme. [Sustaibability Management: Sustainable Development by Integrated Management Systems] Vorgehensmodell und Prüflisten von den Verfassern. Jürgen Löbel, Heinz-Albrecht Schröger, Heiko Clohsen. Erich Schmidt Verlag. Berlin. ISBN 3-503-05983-0.

Ludwig, *Bjørn* (2003), Analyse von konzeptionellen und organisatorischen Schnittstellen in Unternehmen zur Integration des Konzeptes der Innovations- und Technikanalysen (ITA). [Analysis of potentials towards integrating Technology Assessment in Business Management] Techknowledgement Consulting, Postfach 3521; D-37025 Göttingen.

Millennium Ecosystem Assessment (2005), Ecosystems and Human Well-being: Synthesis. Island Press, Washington, DC. World Resources Institute.

Pauli, *Gunther* (2005), Presentation on the "Sharing" Conference in Jurmala, Latvia.

Schmidt-Bleek, *Friedrich* (2000), Das MIPS-Konzept. Weniger Naturverbrauch – mehr Lebensqualität durch Faktor 10. [MIPS concept and Factor 10] Droemersche Verlagsanstalt Th. Knaur Nachf., München.

Schwartz, *Peter* (1991), The Art of the Long View: Paths to Strategic Insight for Yourself and Your Company.

Spangenberg, *Joachim* (2002), Nachhaltiges Wirtschaften und Wachstum. [Sustainable Development and Growth], in: UTOPIE kreativ, H 136, Februar 2002, S.145-155.

Will, Markus (2003), Nachhaltigkeit, Technikfolgen-Abschätzung und die Grüne Bioraffinerie. Exposé zum Praxissemester. [Sustainability, Technology Assessment and the Green Biorefinery Technology] Institut für Technologiefolgen-Abschätzung, Österreichische Akademie der Wissenschaften, Wien und Hochschule Zittau/Görlitz (not published).

Zero Emissions Research in Austria
–
A Case Study

H. Schnitzer, Professor, DI **G. Gwehenberger**, DI **Ch. Zwatz**

Institute for Resource Efficient and Sustainable Systems,
Technical University Graz,
Inffeldgasse 21 b, A – Graz, Austria

Introduction

One of the most important cornerstones of a sustainable society is an industry that uses zero waste, zero emissions processes. International research activities concentrate on realizing zero emissions objectives predominantly in developing countries. In Austria, several projects are realized within the "Factory of Tomorrow" subprogram called ZERIA – Zero Emissions in Austria. Our target groups are primary local companies.

In our first project, ZERIA 1, we focused on defining fundamental concepts and analysing possible practical implementations of waste free, zero emissions processes and technologies in Austrian industries. We discovered that *zero* must not be interpreted as zero in a strictly chemical analytical sense. If the concentration and the total mass flow of a substance in an emission are smaller than natural fluctuations in geogenic flows, we will assume that there is no impact on the environment. In these circumstances we will refer to "Zero Emissions".

Figure 1 shows a simplified model of our industrial system. To achieve Zero Emissions in this system, there are a number of proven technologies, such as:

- Cleaner Production
- Eco Design
- Upsizing
- Recycling
- Industrial Ecology
- Integrated Bio-Systems
- Renewable Raw Materials.

However, it is impossible to achieve Zero Emissions by implementing only one of these technologies. It is the aim of the ZERIA projects to combine different models and technologies.

The project ZERIA 2 builds upon the results of the first fundamental works and aims at developing an integrated strategy, combining various methods in a targeted manner and adapting them to the special needs of Austria's industries.

Fig. 1: Simplified model of our industrial system

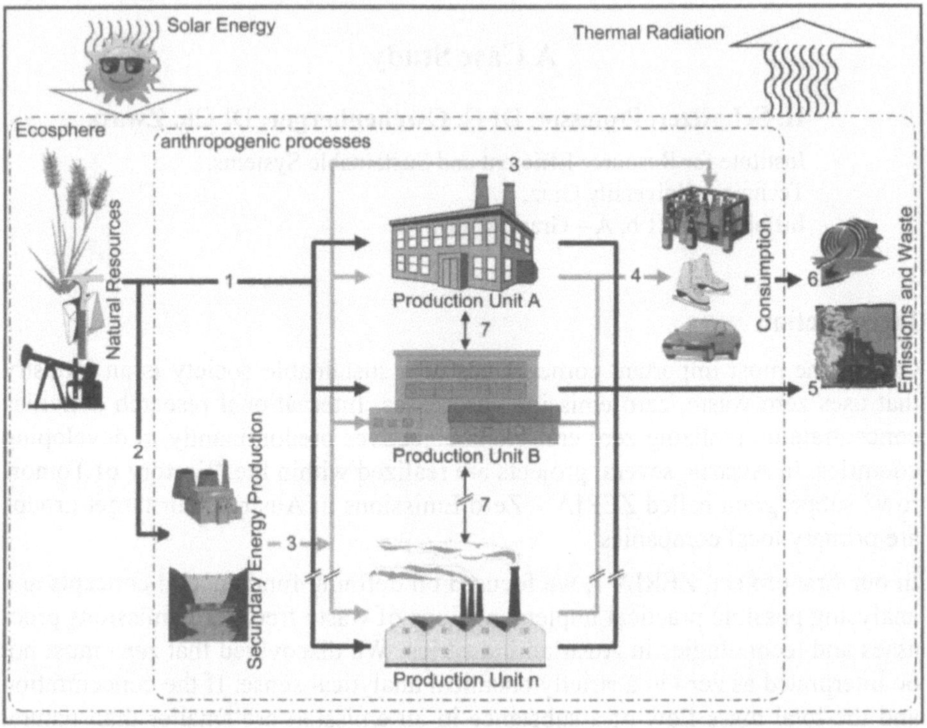

Our work is based on mass and energy flow analyses. We compare input and output streams of every single process step and follow the energy and mass streams through the production. The quality of our research depends directly on the completeness and accuracy of the data. Therefore, a great amount of time has been spent on gathering and cross-checking all available information. Gathering the data has proved to be rather difficult: very often, the data is not readily available in the companies, especially if they have little to do with product quality, purchasing, sold products or waste disposal. It is even more difficult, if third party companies are involved.

The gathered data allows us to carry out thorough mass and energy flow analyses. Based on this information we are able to determine where, when and why certain substances become an unwanted emission. With the help of the companies we analyse this information and try to determine at what point of the process the implementation of cleaner production measures can reduce the usage of raw materials and/or energy.

The following case study is an example for our approaches and possible solutions.

Case Study – Medical Products

For this case study, we analyzed the production process of a company in the north of Austria. The company produces medical infusion solutions, the main component of which is desalinated and distilled water. As cleanness and sterility are essential for the production, water and energy consumption is high. The largest amount of water is used in the cooling process. Figures 2 and 3 show simplified Block and Sankey Diagrams of the processes.

Fig. 2: Block diagram of water flows

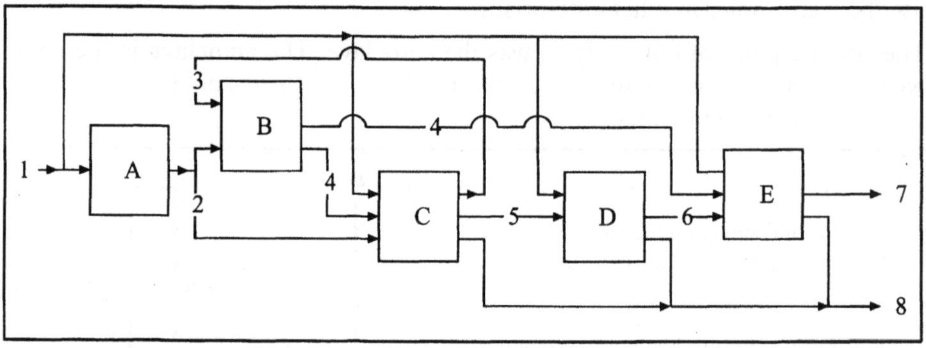

1 Fresh Water
2 Demineralised Water
3 Recirculated Steam
4 Steam
5 Distilled Water
6 Product
7 Finished Product
8 Waste Water

A Demineralisation
B Steam Boiler
C Distiller
D Base of Product, Filling, Cooling of Product
E Autoclave

Fig. 3: Simplified Sankey Diagram

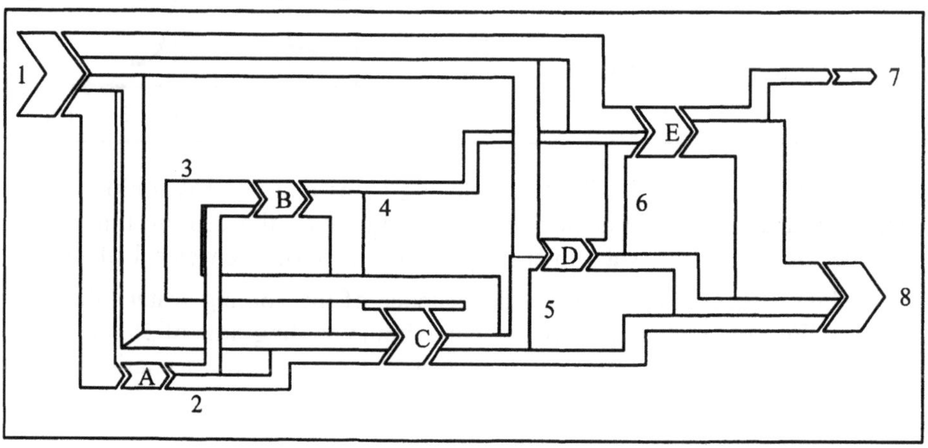

Why is this Case Study Interesting for ZERIA?

The main emission of this enterprise is waste water with the quality of drinking water. This water has a temperature of about 45°C. Without further measures this temperature is too hot to be discharged into the environment, it is too hot to be used in a closed water cycle and the usable thermal energy is too low to be used in industrial processes.

This is a very common problem in today's industries. Therefore, we were looking for a way to either reuse the water or the thermal energy in this plant or to use the warm water in other enterprises.

The starting point of our analysis was the autoclave. The autoclave in the company is used for the distillation of the product. It is a discontinuous process, which is divided into 3 phases.

Phase 1:

Charging and emptying the autoclave, about 10 minutes: – no waste water

Phase 2:

Distillation of the product for 50 minutes with a mixture of steam and high-temperature water

Less waste water (~1 m^3/h) at high temperature (90° C)

Phase 3:

Cooling down: high water flow rate (~10m^3/h), temperature decreasing from 90 C to 20°C

Time of "Cooling Phase" about 20 minutes.

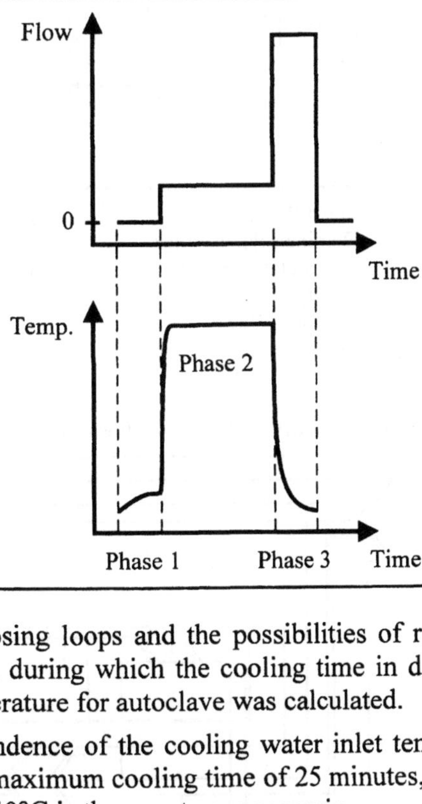

To verify the technical feasibility of closing loops and the possibilities of re-cooling, a data analysis was carried out, during which the cooling time in dependence of the cooling water inlet temperature for autoclave was calculated.

Fig. 4 shows the cooling time t in dependence of the cooling water inlet temperature. We can see that at an assumed maximum cooling time of 25 minutes, a cooling water inlet temperature of about 40°C is the worst-case scenario.

Fig. 4: "Cooling time t" in dependence of the "cooling water inlet temperature" (product and cooling water outlet temperature 60°C)

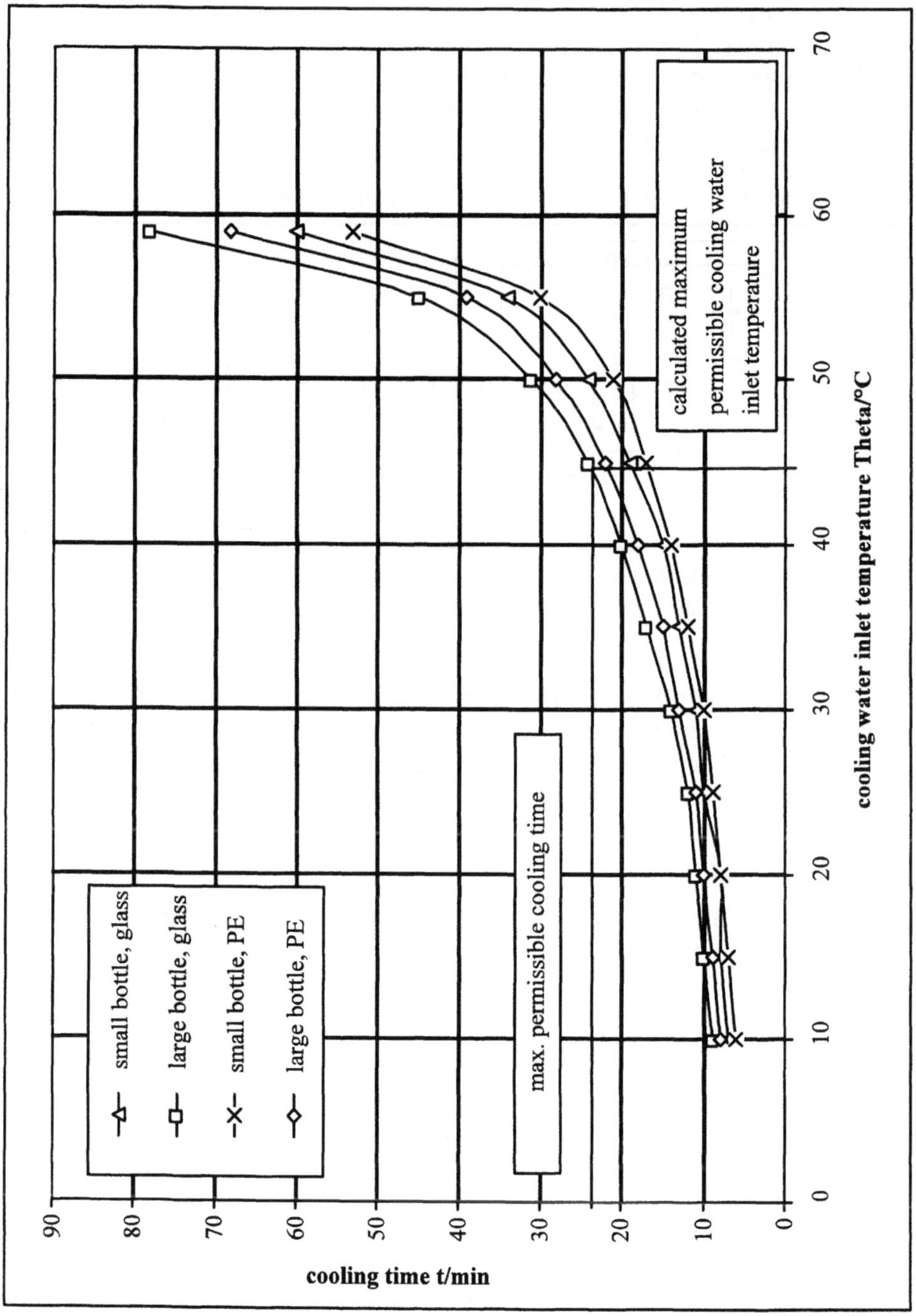

External Closed Loop

The data analyses of the process have shown that the cooling water can be essentially warmer than the fresh water from the local water supply (by about 12°C). The solution, which suggests itself here, is to conduct a part of the waste water stream via an external cooling loop back into the process to cool down the product in the autoclave.

Fig. 5: Block diagram of water flows with "external cooling loop"

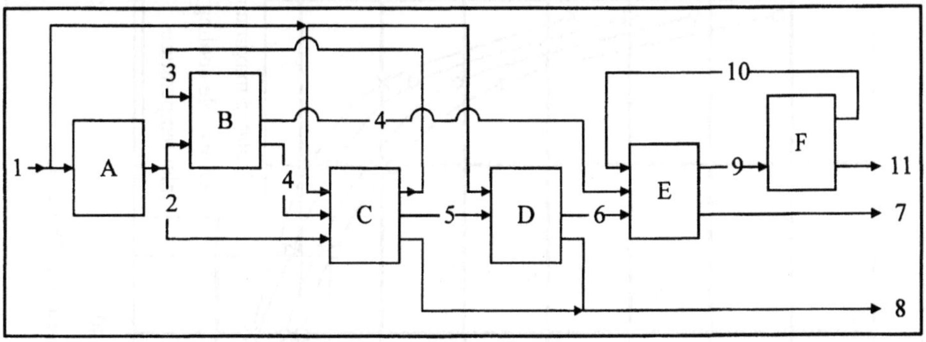

9 To cooling pond (40° C)
10 From cooling pond (30° C)
11 Overflow

F Cooling pond

The block diagram in Figure 5 shows the external cooling loop. The waste water (9) is cooled down by a pond (F) and is led back into the process for cooling the product in the autoclave E. Such a pond can even have an additional advantage: in case of fire, it could be used as a water resource by the local fire brigade.

Basics for the calculation of the pond:

Climate data:		Waste water specification:	
Temperature:	T = 10°C	Inlet cooling pond:	14,000 m³/Jahr
Wind velocity:	v = 4 m/s.		T = 40-60°C
Rel. humidity:	f = 40%	Outlet cooling pond:	14,000 m³/Jahr
			T= 30°C

Results of the calculation:

Required cooling capacity:	Required size of the cooling pond:
P_{cool} = 100 kW	A_{pond} = 300 m²

Internal Closed Loops

The following suggestions for internal closed loops have not been implemented yet, but are technically feasible.

- The distillate holding tank is cooled down by tap water (outlet temperature 20-30°C) which can still be used for the product cooling in the autoclave.

- After every charge, the batching tank is cleaned with distilled water. The waste water of this cleaning process can be used as boiler feed water for the steam boiler.
- Completely demineralised water from the Ion Exchanger (A) can be used first for cooling and afterwards as feed for the distiller.
- In addition, the industrial heat can be used as community heating in the surrounding area.

Perspective

Future projects of ZERIA will focus on realizing Zero Emission process in traditional industries, such as textile, food or metalworking industries. Secondly, we will specify and combine various technologies for process water streams to approach Zero Emissions and prove that Zero Emission processes are technically feasible and economically justifiable.

Sustainable Community Development in the Baltic Sea Region

Heather Worosz, Paulo Bento Maffei de Souza, Yury Kazhura
Blekinge Institute of Technology,
SE – 37179 Karlskrona, Sweden

Introduction

This research examines the issues of Sustainable Community Development (SCD) in the context of the Baltic Sea Region (BSR). The BSR was the first multi-country region in the world to develop a shared commitment towards sustainability. However, to date, there has not been a shared approach to sustainable development. There is also a great deal of diversity in the region in terms of culture and social and economic conditions. This diversity is perhaps the greatest between the Western Baltic Sea Region (WBSR) countries and the Eastern Baltic Sea Region (EBSR) countries. The countries in the east lag behind the countries in the west on many indicators [1]. The EBSR countries that have recently joined the European Union (EU) are now focused on rapid development in an attempt to narrow these differences. A focus on economic growth may result in less attention to sustainability issues.

Among the BSR countries, Sweden has a long tradition of SCD. Swedish pioneers in this area are currently piloting an SCD process in Robertsfors, Sweden that is said to be transferable anywhere in the world [2]. A specific aim is to transfer this model to communities in the EBSR.

The purpose of this research is to study the characteristics of the Robertsfors Model, how strategic and engaging it is, how transferable it is to the EBSR and how it could be improved.

To be strategic in working towards sustainability requires back-casting from a principled definition of success based on scientifically peer-reviewed principles of sustainability [3]. These principles must be general enough, non-overlapping, necessary to achieve sustainability and concrete enough to guide decision-making [4]. The sustainability principles used in this research were developed by 'The Natural Step', a non-profit sustainability organization in Sweden and are described as follows:

"in a sustainable society, nature is not a subject to systematically increasing:

(1) … concentrations of substances extracted from the Earth's crust;

(2) … concentrations of substances produced by society;

(3) … degradation by physical means and, in that society

(4) … people are not subject to conditions that systematically undermine their capacity to meet their needs." [5]

To be engaging, according to the definition used in this study, means to ensure broad-based citizen participation in the process.

The Robertsfors Model is said to be strategic towards sustainability and engaging. Verifying this was, therefore, one of the main objectives of this study.

This study also looked at the considerations for transferring the model to communities in the EBSR. To do so required an understanding of the BSR context and considerations for knowledge transferability. In order to assess the possibilities for improvement of the Robertsfors Model, related theories, concepts and best practices were examined. This included researching SCD best practices in communities in Canada, India and the BSR.

Research Questions

In order to guide this study, the following research questions were posed:

- What are the key concepts and tools used in the Robertsfors Sustainable Community Development Model?
- What elements in the Robertsfors Sustainable Community Development Model indicate that it is (1) strategic and (2) engaging?
- What might influence the success of transferability of the Robertsfors Sustainable Community Development Model to communities in the Eastern Baltic Sea Region?
- What additional concepts and tools can help to improve the Robertsfors Sustainable Community Development Model?

Methods

The methods used to answer these questions included both secondary and primary research. The secondary research encompassed a literature review about sustainable development (SD) theories and related concepts that were used by some communities or that might be helpful in addressing sustainability issues if applied. This helped to set the context for the primary research. The primary research consisted of focused interviews [6] with representatives of the Robertsfors community to establish the base case. Interviews were also conducted with community members within the EBSR to understand considerations for transferability and improvement of the model. This primary research was undertaken in Latvia, Estonia and Belarus. The community of Whistler in Canada, Storstrom County in Denmark and a SCD process used in southern India were also studied to learn from these best practices. The primary research was carried out in person whenever possible. Otherwise, an online questionnaire was sent to selected community members.

Results

The results are presented in two main streams: from the secondary research and from the primary research.

Findings from the secondary research are meant to establish the context within which the research is carried out and to bring a better understanding of the theories involved in SD. The following concepts were studied more in-depth: Triple Bottom Line [7]; The Natural Step Framework [8]; Robertsfors Sustainable Community Development Model [9]; Human Scale Development [10]; Appreciative Inquiry [11]; Knowledge Transferability [12]; Leverage Points of a Complex System [13]; and Community Based Social Marketing [14].

Findings from the primary research were meant to bring understanding of the practices of SD from the base case – Robertsfors – and from different communities. The findings for the primary research were structured using the Five-Level Model of The Natural Step Framework. The Five-Level Model is composed of the following levels: (1) system; (2) success; (3) strategy; (4) actions; (5) tools.

At the *systems* level, the primary goal is to understand the basic principles of the system within which planning is undertaken. In the case of sustainable development, the system to be understood is the biosphere with all its parts, processes, interrelationships, values, beliefs and functions. At the *success* level, a desired future is defined according to a principled definition of success. For sustainable development, the four sustainability principles as described above should bind this definition of a desired future to avoid planning against the laws of nature. At the *strategy* level, a plan to arrive at the desired future is defined. This is done by backcasting from the principled definition of success, rather than by forecasting from current trends. This results in a plan where every measure moves towards success and avoids dead ends. At the *actions* level the real movement towards the desired future begins. This is where the strategic plan comes to life by defining and prioritizing measures. At this level, it is recommended to choose actions that will lead to quick, visible and measurable wins. As well, it is important to define longer-term actions that will lead to sustained results overtime. Lastly, the *tools* level is where all-levels find the necessary support. There are three broad categories of tools. Strategic tools are used to ensure that actions are moving in the right direction. Systems tools are used to measure the direct impact on the system at a point in time. Capacity tools are used to build capacity for working towards sustainability, such as sustainability training and education etc.

Discussion

The discussion was a means to analyse and relate the primary and secondary research data to answer the research questions.

The discussion provided an understanding of the Robertsfors Model. While the Robertsfors project document emphasized a Triple-Bottom Line approach to SD, in practice, The Natural Step Framework was applied. The model also included many techniques to engage community members in the process. In order to achieve sustainability, no matter the chosen method, a SCD model includes both codified (more easily transferable through a communication mechanism) and tacit (not as tangible and more difficult to transfer) knowledge. Transferring the model, therefore, requires different approaches for each part. Members of the community should gain ownership of the SD process for a solution to have real meaning at the local level. How the model is applied and the results achieved may vary greatly depending on the local context.

In discussing how the Robertsfors Model can be improved, concepts from Human Scale Development, Appreciative Inquiry, The Leverage Points of a Complex System and Community Based Social Marketing play an important role. These theories may provide the Robertsfors Model with the trans-disciplinary foundation that is necessary to deal with the complexity of a social system. It is recommended to study these concepts further to determine how the existing model might be supported or improved. Ideas were also garnered from best practices in the communities studied in the BSR, Whistler, Canada and Southern India.

Conclusions, Propositions and Questions for Further Research

In order to move towards sustainability, a community is dependent on a shared vision for sustainability bound by valid sustainability principles. The vision is supported by appropriate strategies, actions, tools and indicators and the ability to engage citizens in working towards this vision. The range of knowledge required in a SCD project goes from the most explicit to the most tacit kind. It is important to understand this not only when planning to address sustainability in different cultures and settings, e.g., between the western democracies of the Baltic Sea region and the newly accepted EU members with extensive communist pasts, but also when dealing with the huge complexity of a single community.

What has been observed during this research is that community sustainable development is not at all a simple issue. Even when experienced practioners competently use a solid and strategic sustainability framework to guide the process, such as The Natural Step Framework, communities still change very slowly. Among the most difficult issues is achieving broad-based participation, which rarely happens fully. People's availability for the commons does not usually arrive as a result of engagement processes.

In general, this study indicates that some elements of the 'Robertsfors Model' can be considered as strategic and engaging. However, the broader research indicates that it does not seem that the model can be directly transferred to another community. The large percentage of tacit knowledge embedded in the process of applying the model and the differences in local context will impact on the success of the model elsewhere. What is more transferable is the objective, codified elements of the model, specifically, The Natural Step Framework. With respect to applying the model and community engagement, it would be best to create a tool-box of codified guidelines, processes, check-lists, engagement tools, suggested measures etc., that a community could draw from based on an understanding of the local context. It is also helpful to participate in an SCC practitioners' network to share ideas and to exchange support.

In reflecting on the results and conclusions of this study, many propositions emerged, such as:

- Proposition 1: In the process of engaging a community towards sustainability, start where people are;
- Proposition 2: The main barriers should be identified at an early stage;
- Proposition 3: Create a team (network) of early adopters (fire souls). Work with this network to create a "network of networks";
- Proposition 4: Ensure early and ongoing support from local, regional and national government and politicians;
- Proposition 5: It is important to ensure that local leadership emerges;
- Proposition 6. Broad-based community education in sustainability concepts is a very important part of the process;
- Proposition 7. A community sustainability plan based on The Natural Step Framework, with shared ownership across all municipal departments, is key for a science-based, systematic approach to SCD;
- Proposition 8. Rather than having one SCD model, use of the SSD framework might be best supported by a toolbox of principles, guidelines, codified processes and potential measures that can be used and adapted to the local context and needs.
- Proposition 9. When inviting citizens to create a shared community vision, it is important to find the balance between inviting too early, when awareness about sustainability and engagement are still not strong and too late, when the momentum may have been lost.
- Proposition 10. When defining strategies and conceiving and selecting measures, try to consider the best balance of leverage points.
- Proposition 11. It would be useful to have indicators for 'Completeness, Consistency and Decidability' to assess, respectively, a community's ability to: self-organize; avoid self-destruction; and learn from experience at all levels.
- Proposition 12. Efforts towards sustainability should be informed by the real understanding that results from deeper involvement in the community.

This study also led to questions for further research:

- Question 1. What are the enablers and barriers to community engagement?
- Question 2. In attempting to engage a community in sustainable development, what types of knowledge are needed and what are the best means of transferring this know-how and understanding?
- Question 3. How can an understanding of systems-level leverage points help to inform better development and evaluation of measures towards sustainability?

These questions for further research are meant to guide future studies towards a better understanding of Sustainable Community Development. Deeper approaches to this subject may lead to a more comprehensive way of working with communities making efforts to achieve sustainability.

References

1. Five Years of Regional Progress Towards Sustainable Development. A Baltic 21 Report to the Prime Ministers of the Baltic Sea States. In: Baltic 21 Series no. 1/2004. The text of the Report can be found on the Baltic 21 website at: www.baltic.21.org.
2. *James, S. & Lahti, T.* (2004), The Natural Step for Communities: How Cities and Towns Can Change to Sustainable Practices. Gabriola Island, BC: New Society Publishers.
3. *Holmberg, J. & Robèrt, K.-H.* (2000), Backcasting from non-overlapping sustainability principles – a framework for strategic planning. International Journal of Sustainable Development and World Ecology; 7(4): 291-308.
4. *Robèrt, K.H., Schmidt-Bleek, B., Aloisi de Larderel, J., Basile, G., Jansen, L., Kuehr, R., Price, Thomas P., Suzuki, M., Hawken, P. & Wackernagel, M.* (2002), Strategic sustainable development – selection, design and synergies of applied tools. Journal of Cleaner Production; 10(3): 197-214.
5. *Karl-Henrik, R., Basil, G., Broman, G., Byggeth, S., Cook, D., Haraldsson, H., Johansson, L., MacDonald, J., Ny, H., Oldmark, J. & Waldron, D.* (2004), Strategic Leadership towards Sustainability. Karlskrona: Psilanders grafiska. p. xxiv.
6. *Merton, R.K, Fiske, M. & Kendal, P.L.* (1990), The focused interview: a manual of problems and procedures. New York: The Free Press. A Division of Macmillan, Inc.
7. *Elkington, J.* (2001), The Chrysalis Economy: How Citizen CEOs and Corporations Can Fuse Values and Value Creation, Capstone Publishing.
8. *Robèrt, K.H., Schmidt-Bleek, B., Aloisi de Larderel, J., Basile, G., Jansen, L., Kuehr, R., Price, Thomas P., Suzuki, M., Hawken, P. & Wackernagel, M.* (2002), Strategic sustainable development – selection, design and synergies of applied tools. Journal of Cleaner Production; 10(3): 197-214.

9. *James, S. & Lahti, T.* (2004), The Natural Step for Communities: How Cities and Towns Can Change to Sustainable Practices. Gabriola Island, BC: New Society Publishers. p. 233.
10. *Max-Neef, M.* (1991), Human Scale Development: Conception, Application and Further Reflections. New York. The Apex Press.
11. *Bradbury, H. & Waage, S.* (2004), Organizational Development for Sustainability: Integrating Appreciative Inquiry and The Natural Step. Submitted to Michael Manning MManning@NMSU.EDU, ODC.
12. *Lam, A.* (1999), Tacit Knowledge, Organizational Learning and Societal Institutions: An Integrated Framework. University of Kent, Canterbury, Kent, UK. The text of this publication is also available on the website of the Conference on National Innovation System, Industrial Dynamics and Innovation Policy of the Danish Research Unit for Industrial Dynamics: www.druid.dk/conferences/summer1999/conf-papers/lam.pdf.
13. *Meadows, D.* (1999), Leverage Points: Places to Intervene in a System. The Sustainability Institute. Hartland, VT, USA.
14. *McKenzie-Mohr, D. & Smith, W.* (1999), Fostering Sustainable Behaviour. New Society Publishers.
15. *Karl-Henrik, R., Basil, G., Broman, G., Byggeth, S., Cook, D., Haraldsson, H., Johansson, L., MacDonald, J., Ny, H., Oldmark, J. & Waldron, D.* (2004), Strategic Leadership towards Sustainability. Karlskrona: Psilanders grafiska.

Integration of Cultural Heritage as a Resource for Sustainable Development in Municipality – A Case of Durbe Manor in Tukums

Agrita Ozola

Tukuma muzejs,
Harmonijas iela 7, LV – 310 Tukums, Latvia

Abstract

The paper deals with conservation of cultural heritage and emphasises the actuality of integration of monuments in the development strategies in municipality. The author has been responsible for the renovation project of Durbe manor situated in Tukums town. The case study describes the process of planning and conservation and integration of the monument of national significance in the cultural activities of municipality.

Introduction

Planning the development of municipality it is very important to understand what values and resources there are to ensure the sustainability of community. Mostly natural, financial and social resources are mentioned and cultural resources are forgotten. Economists and planners have paid little attention to the role of culture in the economic development of society. It is common to think that culture is something existing by itself and it is meant only for leisure therefore does not have any importance in the development processes. During the last few years the traditional paradigm is changing and there are several cases proving that the role of cultural heritage is growing if the monuments are conserved and well kept, if they are integrated in the life of the community.

Every village and town has some historic building or sites, works of monumental sculpture, art collections or even a museum. They could be monuments of national regional, or local significance. The definition of cultural heritage given in UNESCO Covention for Protection of the Worlds Cultural and Natural Heritage (1972) states that monuments, groups of buildings and sites should becoming a part of heritage if it has outstanding value from the point of view of history, art or science that is worth preserving for future generations [15]. Very often the buildings or sites in the single village or small town do not have any universal or national or even regional value but it has some special meaning for a place. It could also be that nobody takes care of the monument or the methods used to preserve it are not adequate for the specific character of it. Very often owners of monuments do not understand the value of the monument that they have and how to use it.

According to M. Trimarche [13] the value of heritage tends to increase if the historical building or monument is accessible for people and also if it is used for cultural activities. He is stressing the importance of integration of monuments and cultural activities, which derives much higher educational or cultural benefit, makes people become familiar with history, traditions, signs and symbols in their locality and helps to identify themselves with the place they are living.

According to the culture economist D. Trosby [14] the culture is central to the development process, providing both the context within which economical progress occurs and the very object of development when viewed from perspective of individual needs. D. Trosby stresses that opening up the model of sustainable development for a full range of economic, cultural and ecological concerns, provides the foundation for a new development paradigm, centred on human beings and recognising the totality of their material and non-material needs.

As M. Trimarshi states our heritage refer to a set of values that can be defined as cultural identity [13]. In many cases buildings become part of a heritage once they have lost their original function in some cases because they start to play a wider role as representative units of some culture.

The case of Durbe manor in Tukums town shows that with appropriate conservation methods used and integration the monument of national significance in the cultural actions of the town gives a new life for it and changes its meaning in the local community. The value of the place is growing due to its new function given in the process of renovation.

Materials and Methods

Case study research describes a process of planning and renovation of Durbe manor, which is situated in Tukums town and its integration in the life of municipality as well in the regional tourist routes.

The manor was a part of bigger estate called Šlokenbeka and has been known about since 1644 when a document about selling a piece of arable land was signed between two landlords: A living house dates back to 1671 but later it was rebuilt several times. Because of renovation of the building in 1820-1823 today the living house of the manor is known as a very good example of classicism architecture in Latvia [12]. Durbe manor is one of several typical manors in western Latvia.

The process of renovation was started in the beginning of 1990's when the manor was given to Tukums museum. During more than 350 years of its existence the centre of the manor served several different functions. It was built as a living house for a noble family but during the Agrarian reform of the Republic of Latvia in the 1920's it was nationalised and turned into a resort and later a hospital. All buildings were more or less rebuilt. The living house was used for the needs of the hospital and therefore halls were transformed into small chambers for patients, toilets and laboratories were erected in the writing room, a bedroom was transformed for physical trainings etc. The old barn was used for stores of food and a carpenters workshop. The house was very old and it was rebuilt during the 1970's. Even a small refrigerator was built in. A servant's house was used for offices and also for some flats and an old shed for washing premises. Only the old out-house was not in use and it was destroyed during the 1960's [7]. A so-called English Style Park was not cultivated; one of the stone bridges came down, and the rotunda was almost without the roof.

In 1991 when the hospital was closed and the museum took over the manor most of its romance and authenticity was lost and it was almost impossible to imagine how it looked in its best years. Luckily the exterior of the Durbe manor was preserved untouched. During the 1920's there was an idea to organise a school for farmers and a sketch design of a new house in the style of renaissance was made [11]. During the 1970's a plan for development of the hospital was worked out but was not realised.

The Durbe manor was not the only monument in Latvia in such a bad condition. During the soviet rule attention were not paid to the evidences of Baltic German culture in Latvia. An inventarisation of old castles, palaces and manors as well as churches in Latvia was started in the middle of 1960's but only a little was done to preserve them as many of the manors were used for schools, libraries and clubs, hospitals and offices for local municipalities and collective farms and re-planning of living houses to fulfil their needs. During centralisation of collective farms and optimisation of small schools and clubs in the 1970's and 1980's many manors were left and slowly collapsed.

To understand the history of the manor and the reasons of its change over the time author did analysis of literature and archive documents as well as old graphic works and photography's. Because the manor is a monument of architecture the visual evaluation of the exteriors and interiors of the living house was ordered and used to elaborate the program of renovation of the manor. The old outhouse was destroyed therefore the archaeological excavations were organised under the leadership of the author and archaeologist M. Ruša. During excavations the old fundaments were funded and the site for reconstruction of the outhouse was defined. The author was responsible for the whole project of renovation of the manor and development of the museum in it.

Planning of the Development in the Monument

The planning for the renovation of Durbe manor was started in 1994. The first three years were devoted to research using different methods: research of history of the manor, artistic and architectural research of the living house, archaeological excavation in the place of the old outhouse, research of the cultural environment of estates at the end of 19th century.

After all the necessary information was gathered architect I. Caunite worked out the sketch design and technical documentation. She devised a project for the living house in 1995 and a project of reconstruction of the outhouse in 2004. Many consultations and "brain storms" were used in the planning process [5]. During ten years much has changed in the philosophy of the museum and also in architecture and conservation strategies as well in the society. Starting the process the author, other museum workers and even architects involved tried to do their best but there was not enough experiences, consultations and expertise and therefore several mistakes were made which were corrected in the process of renovation.

In the 1990's the economical situation in Latvia was very hard and the regional authority that owned the manor was not able to fulfil all its functions and the museum was not the priority. Discussions about combining the museum and hotel for tourists in one living house of the manor took place but luckily soon everybody understood that it was not possible to combine such different things.

When the planning of the living house was started the program of the museum was quite simple. It was planned to have a premises for the museum including exhibition rooms, storerooms for collections, working rooms for museum workers and a museum restaurant. The idea was to have historical paint in the halls for permanent exhibitions on the first floor but so called "white interiors" on the second floor. During the renovation process so many details and fragments of historical paint were found and such interesting information about the building was collected that the decision was changed. In 1999 the author devised a new long-term program for the museum [10]. It defined that the living house of Durbe manor will be used for historical interiors and theme related exhibition premises. The idea to take storerooms out of the living house and reconstruct the old outhouse for museum stores and regional restoration centre was supported at both the professional and the political level. It was agreed also that after reconstruction of the outhouse two other buildings of the manor would be reconstructed and used for conference premises.

Restoration of Buildings and its Interiors

The main task of the project was to create a new product for tourists and at the same time to keep the authenticity of Durbe manor. The renovation was started with the restoration of the living house. A main principle taken into account was a principle of wariness. The approach of safeguarding the architectural, aesthetic and historic value of the building, their interiors and spirit was realised as well as it was possible [8]. The historical functions of the house were considered when the museum's permanent exhibitions were planned.

There are two different approaches used in the process of renovation of buildings: the conservation of old details, planning and paints as well as reconstruction of interiors where it was necessary to use the method of analogy. The photographs received from a family of previous landlords of Durbe were used in the reconstruction process. The furniture and other objects necessary for historical interiors were collected from very different places and restored in the museum workshops.

The next step is to reconstruct an old outhouse. It is planned for the museum's storage and regional restoration centre. The building will include a yard of manor, which should be closed but which was opened with destruction of the old outhouse during the sixties. The exterior of the building will be reconstructed as it was but the interior will be functional and appropriate for its new functions-storing museum objects and doing conservation work. The new function of the building will give new challenges for the museum and also local community. It is planned to have 11 new working places for museum professionals.

Integration of the Monument into Local Community

Since the planning of renovation was started Durbe manor has opened to visitors. At the beginning of the nineties people were just interested to see the buildings which were used for the hospital before. The museum decided to organise different events and exhibitions and the manor was not like a museum at all. Afterwards when reparation works started visitors were interested to see how the building was changing, they compared restored and unrestored rooms, constructions and details. Many local people come to he museum time after time to compare what has been done and they are interested to hear commentaries of museum workers. Some times they like to consult about the problems they have in their houses or old furniture.

The number of visitors in 2004 exceeded 14 000 [8]. The quantitative analysis of statistics proves that the number of visitors has increased year by year and it has doubled since 1997. The qualitative analysis and interviews shows that there are three main trends in the motivation of people visiting Durbe manor. 35% of visitors are coming on organised tours and they have chosen this trip because of the good product – A living manor house where historical interiors are reconstructed and historical costumes are used. 32% of visitors are coming as individuals and they are interested in the quality of restoration works, interiors, furniture etc. 33% of visitors are participants of different activities taking place in Durbe [8]. There are traditional festivities like the Art Festival in the third Sunday of May, the Day of Poems in the middle of September, and the Golden Sunday just before Christmas. The manor is popular because of very good concerts of Classical music and art exhibitions.

Model of Management

The project of renovation of Durbe manor is supported from the Council of Tukums District, which is a regional authority. The built centre of the manor is a property of regional authority but the park, which is more than 24 ha big, is the property of the municipality of Tukums town. The buildings and park in fact are a joint entirety and it is developed by the museum like one single complex. Parcelling out was done in the year 2000 because of economical reasons: the planned renovation of the monument was too expensive for one municipality. According to the laws and regulations of the Republic of Latvia municipalities can only finance their property. After long discussions at two levels of municipalities the decision was taken and the regional government took responsibility for the renovation of five old buildings but the local municipality took care of the park and its structures – like a stone bridge, rotunda, ponds and open-air stage.

The Durbe manor is a branch of Tukums museum, which is the regional museum and specialises in cultural environment [6]. Museum administration is responsible for management of the whole development of the monument and its sustainability. There is a team of four persons working for the project. Under the leadership of the author there are three professionals working: one responsible for technical, one for scientific and one for finances. Once a month there is a meeting of the board where architect, builder, restaurateur and artist participate. If necessary consultants like constructors, art historians, archaeologists, gardeners etc. are invited. To ensure successful work the museum provides a platform for cooperation between two municipalities and their officials as well as for professionals working in the field of restoration and planning of historical buildings.

The renovation process of the monument and its integration into the flow of cultural tourism is complicated therefore close cooperation between politicians and professionals is vital. The experience of Tukums museum proves that involvement of professionals with different competencies makes the process of decision making longer but ensures more advised results in the long term.

Long-term Conservation Strategies

The idea of having a museum in Durbe manor is not new. The most outstanding Latvian poet and philosopher Rainis (Janis Plieksans) became the owner of the manor shortly after its nationalisation. In 1923 during his first visit to Durbe he was confused by its spaciousness and lordliness. The manor was not appropriate for his requirements to have a place for creative work and he decided to donate his property to the Union of the Teachers (latv.- Skolotāju savienība). His desire was to have a museum of him and the resort for teachers and children. His dream was to also create a Natural Park in the surroundings of manor. The museum of Rainis was founded in 1929 shortly before the poet's death and existed until 1937 when the exhibits were moved to Riga [7].

Rainis was also one of the most famous politicians of his time. He was open to new ideas and spoke about the long-term perspective of education including what we now call life-long education and about the necessity to preserve natural and historical heritage. He was the first person to point out the urgency of documentation of present day events and processes. His idea about the museum in Durbe was not forgotten. The local society has been quite active in reminding people about the actuality of development of the museum and that the manor was used for a specialised hospital of national significance.

Only in 1991 the Council of Ministers of Soviet Republic of Latvia made the decision to close the hospital and give the manor to the regional authority whose will was to develop the existing regional museum. The decision was made only because the optimisation of the system of Health Protection was started in Latvia and the existing premises for the hospital in Durbe were so bad that it was more efficient to look for a more appropriate site.

The development of the museum is determined by national laws and regulations. The Law on Municipalities adopted in 1993 determines the responsibility of regional municipalities for heritage [3]. That was a platform for the decision of the Council of Tukums District to start a renovation process of Durbe manor and the development of the cultural environment museum. In 1995 the Council adopted the development program for Tukums museum [9] but there was not any program or any priorities set for development of the whole region. Because of uncertainty with administrative reform in Latvia the process of development of the manor was stopped in 2000 and it took four years to elaborate and adopt the program for regional development of Tukums District according to which the conservation of cultural heritage and development of infrastructure for cultural tourism is one of the priorities. The administrative reform is still not in force but the long-term development program has been worked out and adopted by Riga planning region of which Tukuma District is a part.

The conservation of Durbe manor, development of the museum and regional restoration centre is also planned in the National program "Culture" [4]. The author is a member of a national group of experts working on the new National Development plan 2008-2013 and also a member of a working group elaborating the new Strategy for Culture 2008-2013. The experience gained in the process of renovation of Durbe manor helps us to understand the importance of a holistic approach in the planning process and the significance of preservation of natural and cultural heritage in the sustainability of society.

Conclusions

Municipality is responsible for sustainable development of the territory and it also means that it is responsible for the preservation and use of cultural heritage and transmission of it to future generations. The management of the cultural environment in municipalitys concerned with monuments of national, regional or local significance and traditions, know-how, oral culture etc., what used to be called intangible heritage and people's creative activities. To make this system more effective it is suggested that we integrate experiences of management of cultural actions with methods and practises, used in preservation of heritage.

The problem is rather the need to choose among a wide range of techniques and conceptions of preservation since the decision about what and how to preserve is subjective and tends to change through time. Very often old buildings had lost their original function and only some times has the function and role has been given back. In these cases if the present use of the building coincides with its past function like at Durbe manor the visitors can better understand the historical and cultural significance of heritage more deeply.

The main principle used in the conservation policy is wariness. The experience gained in the process of renovation of Durbe manor proves that the consultations with professionals working in the field of monument protection, town planning and culture will never be too much and it is worth spending the time in discussions before making the decision because sometimes it is almost impossible to repair mistakes committed through negligence. One of the methods very often used in the management of cultural actions is teamwork. Integrating the principle of wariness and using method of team work in the renovation projects of cultural heritage and at the same time ensuring cooperation with all actors involved in the process will ensure the results are as good as possible.

Another aspect is connected with long-term planning. Starting the renovation process the first task is to research the object, its history and architecture. It is worth using the time and money to have professional research using different methods like study of the literature and documents, archaeological excavations of the land and house as well as artistic and architectural investigation After careful study it is possible to define the aim and the function of the monument and work on a development program. Only after that can one start to work on detailed designs and all other necessary documentation and then start its realisation.

In the situation where so many monuments in Latvia are in catastrophic technical conditions the long-term strategy should be worked out at a national and regional level and it is necessary to develop know-how of preservation and use of monuments.

The case of Durbe manor is one of several successful examples not only in terms of quality of the renovation of the house and interiors but also in terms of good choice of function for it. Museums are a public institution and they enable relationships between people, places and the past. [2] Taking a decision about what aspects of heritage are conserved, exhibited and interpreted to local people and visitors of the place.

References

1. Durbes pils. Arhitektoniski mākslinieciskās izpētes materiāli. AIG & Kroks, Riga 1994, 6.-9.lpp.
2. *Davis, P.* (2004), Ecomuseums and the Democratisation of Japanese Museology.//IJHS, vol. 10, no. 1 March 2004 pp. 93-110.
3. LR Likums "Par pašvaldībām" (1993).
4. National Programm "Culture".Ministry of Culture of Republic of Latvia, Riga 2001.
5. Materials of Working Discussions on Renovation of Durbe manor, 1997-2004.
6. *Ozola, A.* (2000); Durbes muižas centrs kā kultūrvides objekts.//Rakstu krājums.Durbes muižas atdzimšana mūsdienu kultūrvidē.Tukuma muzejs, Tukums, 21.-43.lpp.
7. *Ozolina, I.* (1996), Rainis un Durbe.//Rakstu krājums. Tukuma rajona kultūrvēsture. Tukuma muzejs, Tukums, 53.-54.lpp.
8. Statistics of Tukums museum, 1997-2004.
9. The Developmet Plan for Tukums museum (1999).
10. The Development Program for Tukums museum (1995).
11. The State Archive of History of Latvia. Fund 2145 Desciption 2 Things 198, 199.
12. The State Archive of History of Latvia. Fund 6999 Desciption 29 Thing 36.
13. *Trimarchi, M.* (2004), Regulation, Integration and Sustainability in the Cultural sector.//International Journal of Heritage Studies, vol. 10, no. 5, December 2004, pp. 401-415.
14. *Trosby, D.* (2001), Economics and Culture. Cambridge University Press, Great Britain, p. 160.
15. UNESCO Covention for Protection of the Worlds Cultural and Natural Heritage (1972).

B. Science for Sustainable Development

Baltic Sea Algae as a Resource for Renewable Plant Nutrition and Bio-Fuel

Fredrik Fogelberg

University College of Hedmark,
Faculty of Agriculture and Natural Sciences,
Blaestad, N – 2322, Ridabu, Norway

Abstract

A recent Swedish governmental study (Anon. 2005) has shown that despite efforts to reduce nitrogen and phosphorous leakage into the Baltic Sea the eutrophication is severe and an increasing problem. Additional national and international efforts to reduce N and P-leakage must be initiated shortly in order to stop the ongoing "slow death" of the Baltic Sea.

A wide range of research reports, newspaper debate articles and governmental investigations have been presented in Sweden in the last five years concerning the health of the Baltic Sea and suggestions to solve the pollution problem. One interesting solution presented in 2002 (Elofsson 2002) is to invest more in sewage treatment plants in low-cost countries such as Poland and to focus on reduction of either nitrogen or phosphorous.

Another possibility to stop or at least slow down, the present eutrophication is however seldom studied; to mechanically remove large amounts of N and P from the Baltic Sea by large-scale harvest of algae from the shores along the Baltic coast, thus removing nitrogen, phosphorous and partly heavy-metals from the coastal area.

Algae have traditionally been used as fertilisers in agricultural production in many coastal regions in Sweden and other countries. New ideas for the use of algae in production of paper and cellulose powder, or as a resource for energy production have been presented recently in Swedish investigations (Melin 2001).

Using harvested algae as fertilisers in agriculture is unfortunately not as easy as it may seem, although small-scale farming and private gardens still uses algae to some extent. One major concern is the content of toxic substances such as heavy-metals (Engkvist et al. 2001).

The concentration of N, P, K and important heavy-metals in algae found on the beaches of south-east Sweden has until now not been sufficiently investigated. In a pioneer study we have found that the algae mats drifting ashore contain large amount of plants nutrients (Table 1) which may be used as a resource for renewable plant nutrition (fertiliser) for agriculture or as a basis for bio-fuel.

The main issue is to develop strategies to concentrate and/or remove heavy-metals and other toxic or unwanted substances. Combustion can partly solve this problem (Obernberger et al. 1997) and advanced anaerobic treatment using biogenic sulfides can be used to precipitate heavy metals as sulfides (anon. 2003).

Tab. 6: Content of nutrient and heavy metals in algae (mainly *Polysiphonia sp.*) from the coast of Öland (south-east Sweden) and amount available nutrients per km shore (metric tonnes, 50% extraction, 11% dry matter) based on total available algae of 5000 -8000 tonnes per km shore

Plant nutrient and heavy metals	**Concentration (mean value of 4 sites) [mg kg^{-1} DM]**	**Theoretically extractable amount [kg km^{-1} shore]**
Nitrogen	37,500	10,300-16,500
Phosphorous	2,200	600-970
Potassium	5,000	1,380-2,200
Magnesium	5,900	1,600-2,600
Cadmium	4.4	1.2-1.9
Lead	5.2	1.4-2.3
Nickel	11.3	3.1-5.0
Zinc	148	40.7-65.1
Copper	17	4.7-7.5

Using the algae would result in multiple beneficial effects; first we will obtain a considerable resource of organic material for production of bio-fuels, compost, organics or fertilisers. Secondly, we will achieve a large-scale reduction of phosphorous and nitrogen from the Baltic Sea which today are a considerable threat to the environment. Thirdly we will achieve a cleansing of some heavy metals such as cadmium, lead and copper from the environment.

There will also be additional beneficial effects of a large-scale algae harvest; the tourism industry is heavily dependent on clean beaches and attractive environments for camping and hotel activities. The local fish industry is also facing problems with algae in fishing nets and reduced fish populations due to increased algae growth.

However, there are several technical problems which must be studied before an actual harvesting method can be presented. Moreover, we cannot yet predict the outcome of algae harvest on marine life in general and the effect on eutrophication in particular. However, a joint Swedish-Latvian project possibly including Estonia and Lithuania, would be of great interest to find solutions to turn an environmental problem into an environmental resource.

References

Anonymous (2003), Anaerobic Biotechnologies. www.uasb.org/ discover/ anaerobic_biotechnologies.htm.

Elofsson, K. (2002), Economics of marine pollution. Dissertation. Swedish University of Agricultural Sciences, Uppsala, Sweden.

Engkvist, R., Malm, T., Svensson, A., Asplund, L., Isaeus, M., Kautsky, L., Greger, M. & Landberg, T. (2001), Makroalgblomningar längs Ölands kuster, effekter på det lokala näringslivert och det marina ekosystemet. Report 2001: 2. Kalmar University, Kalmar, Sweden.

Melin, Y. (2001), Alternativ anvændning av marina fintrådiga alger. Report 2001: 41, EU Life Algae LIFE96ENV/S/380. Lænsstyrelsen Væstra Gøtaland, Sweden.

Obernberger, I.; Biedermann, F.; Widmann, W. & Riedl, R. (1997), Concentrations of inorganic elements in biomass fuel and recovery in the different ash fractions. Biomass and Bioenergy 12, 221-224.

European Rice Husks Studies with Focus on Utilization Opportunities

Janis Gravitis[1], Janis Grabis[2], Oskars Bikovens[1], Andris Veveris[1]

[1] Latvian State Institute of Wood Chemistry, Dzerbenes St. 27, Riga LV – 1006, Latvia

[2] Institute of Inorganic Chemistry of Riga Technical University, Miera St. 34, Salaspils, LV – 2169, Latvia

Summary

This paper presents European rice growers rice husks utilization in silicon-based materials. Husks have been studied as a natural nano-model of SiO_2 contained in the natural organic system. Other minor metallic impurities were analyzed. Preliminary experiments were performed using the low temperature plasma treatment technology. The possibility of obtaining superfine silicon carbide powder was demonstrated. Some first characteristics of synthesized powders were done. The ideas can be transformed into high tech utilization of Latvia biomass based materials impregnated with the nano-SiO_2. Rice husks were also treated with steam explosion autohydrolysis and analyzed to open new ways for applications. Steam explosion treatment could be combined with low temperature plasma technology.

Introduction

On 16 December, 2002, the United Nations General Assembly declared 2004 the International Year of Rice to focus the world's attention on the role that rice can play in providing food security and poverty alleviation [1]. "*Rice is life* for major populations of the world and is deeply embedded in the cultural heritage of societies. It is the staple food for more than half of the world population. Rice is not a major food crop in Europe. However, the rice-based production systems in Europe have a number of opportunities for sustainable development." Rice is grown in 10 European countries. The changes in the harvested area were pronounced between 1992 and 1997, with a significant increase in Spain and Greece. The rice-harvested area in other parts of Europe declined rapidly from about 370,000 ha in 1992 to around 250,000 ha in 1996, mainly caused by significant reductions in the Russian Federation and Romania. Since 1997 it remained stable at around 240,000 ha. Rice husks are one of the waste streams in rice production system. So, the Zero Emissions Concept would be applied to rice husks utilization using high nano-tech.

Problems of Rice Husks Silica Applications

Resource depletion, The Kyoto Protocol demands environmental contamination target technologies for all biomass waste stream recycling and recovery. Renewable Energy Sources currently contribute to only around 6% of the European Union's energy needs. The objective of the European Commission's White Paper is to raise the share of Renewable Energy Sources to 12% of the Union's gross domestic energy consumption by 2010. The production of one of the major food crops, rice, generates a lot of waste – rice husks. Rice husks (RHs) organic constituents make recycling difficult mainly because of the high content of silica (15-20 wt%) and accordingly generates high ash content in incineration. At the same time RHs is an advanced composite material containing organic and inorganic components. Organic components can be used as a carbon course in case of thermal decomposition and silica can serve as a silicon source for preparation of the silicon-based, for instance, silicon carbide SiC. In RHs silica is in a hydrated amorphous form. However, under special conditions, silica can be converted into highly reactive ultra fine silicon carbide particles. These particles can be used for production of advanced materials. Advanced ceramics of silicon carbide and silicon nitride represent a growing market for silica and silicon-based chemicals. The annual production of industrial-grade silicon carbide some years ago was in the order of 800,000 t but powder use for silicon nitride was about 300,000 t [2]. Dense reaction infiltrated silicon/silicon carbide ceramics (SiSiC) were extracted from wood based composites [3]. Therefore silicon-based material studies derived from rice RHs can generate new ideas for advanced materials preparations from wood materials. It is well known that wood is the main national organic resource of Latvia. Different silicon-based materials from rice husks and their applications are reviewed in [4], however we concentrated our emphasis only on steam explosion and low temperature plasma studies because plasma techniques allow us to intensify the process and had been successfully applied to producing nano-size powders of silicon carbide by reducing silica of evaporation of silicon [5, 6].

Experimental

Starting materials. Rice husks were kindly supplied by the Novamont Company (Italy).

Chemical methods. Air dried rice husks and steam exploded samples were milled finely using a vibratory ball mill (Retsch, MM200, Germany) for 5 min with a vibration of 30 sec^{-1} for microanalysis. The ground sample was extracted successively with ether in the Soxlet apparatus for 8 h followed by boiling in 80% ethanol for 1 h (3 times). Extractives were determined gravimetric. Aliquots were evaporated and dried in oven at 100±3°C.

The neutral sugar composition was determined by the alditol acetate procedure [5]. Samples (around 10 mg) were treated with 72% sulfuric acid (0.125 mL). After 1 hr at room temperature the acid was diluted with water (3.45 ml) to give 4% sulphuric acid and heated for 1 h at 121°C. After cooling, solutions were made neutral by adding a 15 M ammonia solution (0.32 ml). Myo-Inositol was added as an internal standard (0.05 mL of a 20 mg/mL solution). Aliquots (0.1 ml) were reduced with a solution of sodium borhydride in anhydrous dimethyl sulphoxide (20 mg/ml). Monosaccharides were reduced for 90 min at 40°C by adding 1 mL of the sodium borohydride. After reduction, the excess of sodium borohydride was decomposed by the addition of 18 M acetic acid (0.1 ml). Acetylation was carried out. 1-Methylimidazole (0.2 ml), followed by acetic anhydride (2 ml), was added to the reduced monosaccharides and the components were mixed. After 10 min at room temperature, water (5 ml) was added to decompose the excess of acetic anhydride. When cooled, dichloromethane (1 ml) was added and the mixture was agitated on a vortex mixer. After phases had separated, the lower one was removed with Pasteur pipette. Alditol acetates were separated on a glass-capillary column TC17 (GL Science, 30m x Ø 0.25mm) fitted to a gas chromatograph (Shimadzu 18A) equipped with flame-ionization detector. Mixtures of rhamnose, arabinose, xylose, mannose, galactose and glucose at five concentrations were used as standard. Neutral sugar composition of samples was recalculated to polysaccharides.

The lignin content of the extract free sample was determined by the Klason method, and involved treating the samples (1 g) successively with 72% sulphuric acid (15 ml) for 2 h at room temperature and boiling 3% (v/v) aqueous sulphuric acid for 4 h. Klason lignin content was corrected to ash. Acid-soluble lignin was determined in clear filtrate from second stage acid hydrolyzate. The absorbence at 205 nm was measured using 3% sulphuric acid as a reference solution. The specific absorption coefficient value $\varepsilon = 110$ g-1*L*cm^{-1} at 205 nm was used [6]. The lignin content was calculated as a sum of Klason lignin and acid-soluble lignin.

The lignin content was also determined by acetyl bromide procedure [7]. The samples (4-6 mg) were placed in a glass reaction bottle with solution of 25% (w/w) acetyl bromide in acetic acid (2.5 mL) and 70% $HClO_4$ (0.1 mL). The reaction bottles were placed in an oven at 70°C for 30 min. The bottles were shaken at 10-min intervals to promote dissolution of the samples. After digestion, the mixture was cooled and transferred to 50 ml volumetric flask containing 2 M sodium hydroxide (10 ml) and acetic acid (12 ml). The bottle was rinsed, and the solution was made up to 50 ml, with acetic acid. The UV absorption spectrum of the resulting solution was measured against a blank solution that was run in conjunction with the sample. The lignin content of the samples was determined by measuring the absorbance at 280 nm and using the specific absorption coefficient $\varepsilon = 20\ g^{-1}*L*cm^{-1}$.

Ash content was determined by combustion before extraction. 2 g of oven dry sample was weighted accurately (± 0.0001 g) in a melting pot. Carefully turn the matter to coal, avoiding an open flame. The melting pot with coaled sample was heated through 700±25 °C for 3 hours. After that the melting pot is cooled until ~100 °C and put in an exicator with silica gel. The melting pot with ash is weighed at room temperature.

SiO_2, Fe_2O_3, Al_2O_3, CaO, MgO, Na_2O and K_2O were determined at the Institute of Silicate Materials with standardized methods. Minor elements had been determined using atomic spectroscopy.

Infrared spectra of rice husks, steam explosion pulp and extracted lignin were recorded with PerkinElmer FTIR spectrometer SpectrumOne (Perkin Elmer) in KBr pellets at resolution 4 cm^{-1} and 32 times scan. 2 mg sample was mixed with 200 mg KBr.

The chemical and phase composition of plasma treated products where determined by the conventional chemical and X-ray analysis. The specific surface area of the produced powders was determined by the BETA argon adsorption-desorption method.

Steam Explosion: Steam explosion (SE) treatment was performed in a home-built reaction chamber. The equipment included an electrically heated steam generator, a 0.5 l reactor, a pulp collector and a condenser. The reactor was heated to the desired temperature by an electrical heater. The reactor was filled through the top trap and discharged through the ball valve on the bottom. The reactor allows great changes in the process parameters (steam pressure, temperature and duration of saturated steam treatment). Treatment time: 0.5-3 min, treatment temperature: 250°C (40 atm). Steam explosion pulp was extracted successively with water and 90% dioxan/water to remove soluble polysaccharides and lignin-like products. Ash, lignin and polysaccharides were determined before extraction and in extract-free pulp. Both the processes, i.e. hydrolysis and defibration, could be realized only with inner "tools" of the treated system, without any additional reagents (except for water steam). Therefore, under SE, the lignocellulosic material can provide "self-sufficient" conditions for chemical and physical transformations.

Low Temperature Plasma Treatment and Derived Materials Research

The low temperature plasma treatment was based on decomposition and evaporation of raw materials in inductively coupled nitrogen plasma. The elaborated experimental apparatus consists of radio-frequency (5.28 MHz) oscillator with maximum power of 100 kW, quartz discharge tube with induction coil, raw powder and gas supply systems, water cooled stainless steel reactor and heat exchanger, and cloth filter for collecting powders (Figure 1).

Fig. 1: Schematic view of plasma apparatus

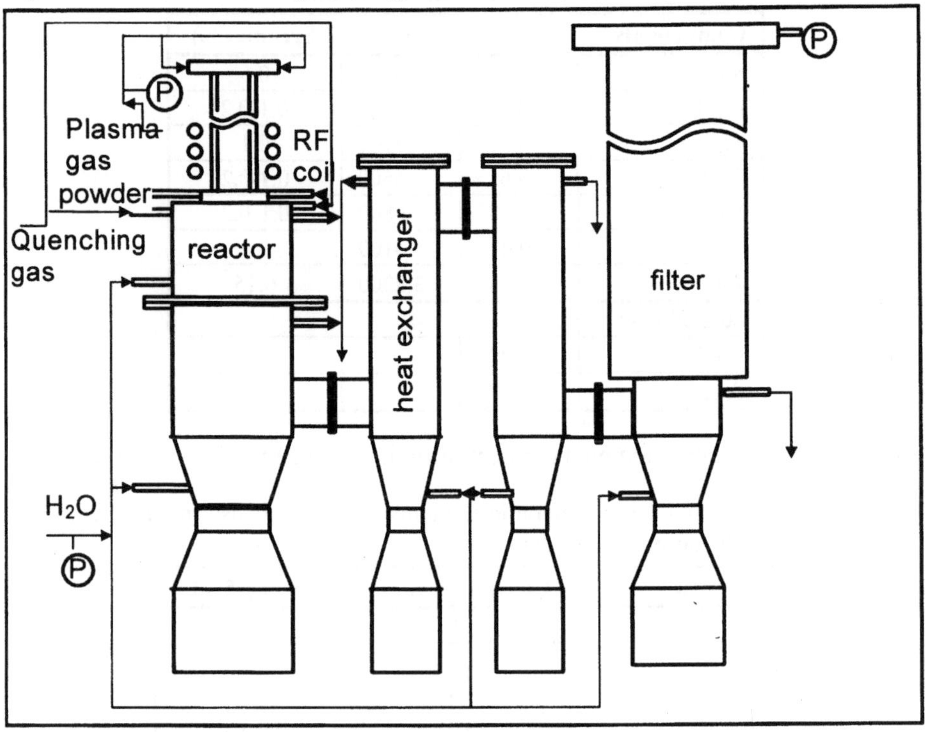

Optimal parameters of the radio-frequency oscillator and parameters of the plasma are determined by calorimetric methods. The grounded particles of rice husks were injected into the plasma flow through eight tubes by carrier gas. For improving particle heating and their introduction into the high-temperature region of the plasma in some experiments the rice husks were premixed with 10 wt.% of silica or silicon with particle size in the range of 20-40 µm. In this case an additionally corresponding amount of hydrocarbon was introduced in the vapour region of raw materials. The growth of product particles was controlled by introducing cold nitrogen as a quenching gas in the reactor.

Results and Discussion

RHs ash analysis shows (Table 1) high homogeneity of inorganic components. SiO_2 dominates, however content of other oxides is also high, around 10%. It is interesting that the next important oxide is K_2O.

Tab. 1: The basic inorganic components of rice husks ash

Components	**wt (%)**	**mg/kg**	**Error ± (%)**
SiO_2	90.50	90,5000	0.5
Al_2O_3	0.59	5900	0.1-0.2
Fe_2O_3	0.51	5100	0.1
CaO	0.65	6500	0.1-0.2
MgO	0.48	4800	0.1-0.2
Na_2O	0.41	4100	0.1
K_2O	3.83	3,8300	0.15
Loss of mass			
1,000°C	1.70		
Total	98.67		

Some minor inorganic elements concentration is demonstrated in Table 2.

Tab. 2: Concentration of minor metallic elements in rice husks ash (mg/kg); Sdev is standart deviation, t is Student factor

Element	**Content (mg/kg)**	**Sdev t (95%)**
Cd	0.347	
Cr	<0.7	
Cu	2.08	
Zn	15.1	± 1.0
Pb	<2.3	
Ni	<1.3	
Co	<1.3	
Hg	<0.2	

Table 3 reflects SE treatment influence on organic part of RHs. There is a significant change in solubility of polysaccharides and lignin after SE. In comparison with other SE treated biomass [10-13], particularly wood, there is sharp destruction of hemicelluloses with 5-carbon sugars (xylose content decrease from 14.4% to 2.8%). At the same time after extraction of lignin with dioxan/water content of polysaccharides is high – 52.8%. So, it would be a good substrate for acid or enzymatic hydrolysis. Extractives or whole exploded pulp could be used as self-binders for fiber- or particle- boards.

Tab. 3: Rice husks before and after steam explosion (230^{0}C/30 atm, 3 min)

	Rice husks	SE husks	SE, extracted husks*	
			water	water/dioxan
Extractives				
Ether soluble	0.4%			
Ethanol soluble	5.0%			
Extractives total	5.4%			
Polysaccharides				
Rha	0.1%	0.1%		
Ara	1.7%	0.2%		
Xyl	14.4%	2.8%		2.3%
Man	0.3%	0.2%		0.1%
Glc	33.4%	32.9%		49.2%
Gal	1.6%	0.8%		1.2%
Polysaccharides Total	**51.5%**	**37.0%**		**52.8%**
Lignin (AcBr)	25.5%	45.8%		20.0%
Klason Residual	23.5%	30.4%	33.8%	
Ash	15.5%	17.7%	18.6%	24.9%
Total	**98.0%**	**100.5%**		**97.7%**

Rha: rhamnose; Ara: arabinose; Xyl: xylose; Man: mannose; Glc: glucose; Gal: galactose (as anhydro sugars). Lignin (AcBr: lignin determined by acetyl bromide method

* – Extracted with water and dioxan (90%)

Unfortunately we have no data about changes of SiO_2 during SE. This is the target for future research. It would be possible to change Si atoms coordination with O atoms. That case gives the opportunity to get special sorbents.

FTIR spectra confirm changes during SE and extraction treatment (Figure 2). However the polymer system is complicate and resolution is not high in comparison with other biomass systems. Isolated lignin from dioxan/water extract as a cleaner sample shows all the typical lignin signals (Figure 3). It is interesting to note an intensive signal at 1726 cm^{-1}, which corresponds to conjugate carbonyls and is characteristic to grass lignins.

Fig. 2: FTIR spectra: A – natural rice husks; B – rice husks after SE; rice husks afer SE and dioxan/water extraction; D – lignin isolated from dioxan/water extract

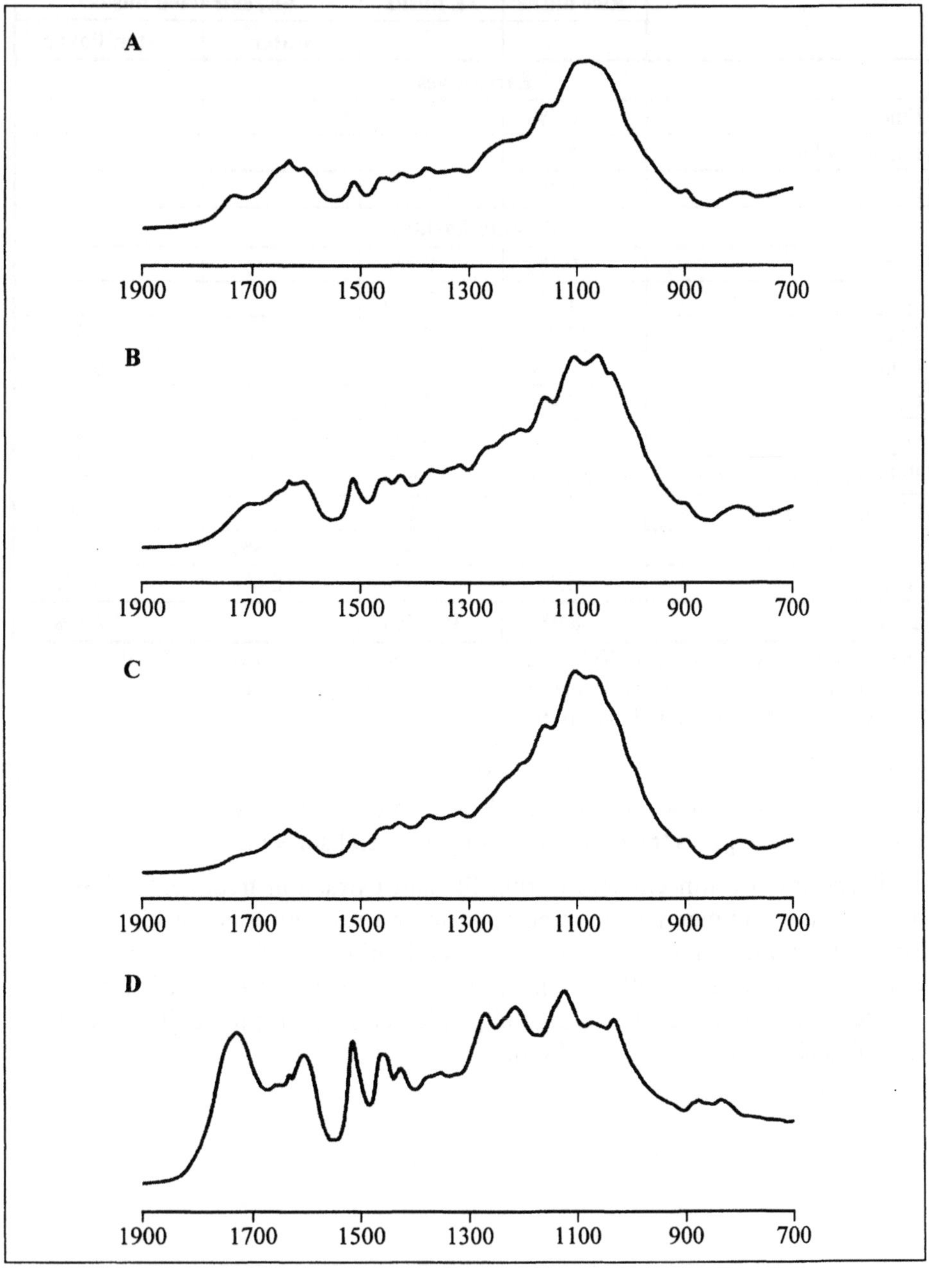

The characteristics of the produced by plasma treatment powders are shown in Table 4. Specific surface area of powders is in the range of 20-42 m^2/g depending on

the used raw powder composition. According XRD analysis all products contain only β-SiC. Intensity of β-SiC maxima in XRD patterns increases when silicon or silica is added to rice husks. Obviously the improvement of injection conditions of raw powder promotes reduction of oxygen compounds of silicon carbide.

Fig. 3: FTIR spectrum of steam exploded rice husks lignin

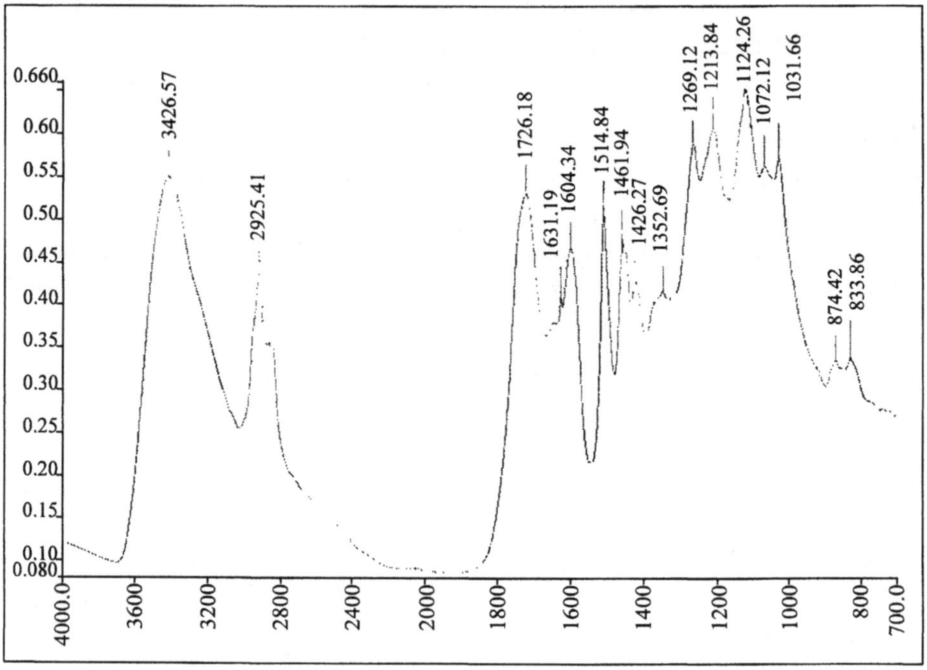

At the same time all samples contain some amount of nitrogen indicating on formation silicon nitride or oxynitride. Also a high ratio of carbon and total amount of silicon in the case of rice husks treatment evidences on presence of free carbon. The free carbon as well as possible silicon oxynitride or nitride is X-ray amorphous. The XRD patterns of additionally heat treated samples at 1350°C high temperature in argon show also weak maxima of silicon oxynitride beside intensive β-SiC maxima.

It seems that contact time of rice husks and silica particles with carbon in plasma flow is too short for full reduction of oxygen containing compounds.

Preliminary results show that plasma technique can bee applied for producing silicon carbide containing compounds from rice husks but it is necessary to increase reaction time decreasing velocity of plasma flow and injection powder at the angle against the plasma flow.

Tab. 4: Characteristics of produced products

Precursors	SSA $[m^2/g]$	N [wt.%]	C/Si	XRD
Rice husks	42	3.9	0.56	β-SiC
Rice husks+SiO2	21.8	3.1	0.37	β-SiC
Rice husks+Si	20.7	4.5	0.38	β-SiC

Conclusions

Steam explosion treatment of rise husks demonstrate difference from other biomass systems and open a way for new practical applications. Plasma technique can be applied for producing silicon carbide containing compounds from rice husks but it is necessary to increase reaction time decreasing velocity of plasma flow and injection powder at the angle against the plasma flow.

References

Blakeney, A.B., Harris, P.J., Henry, R.J. & Stone, B.A. (1983), A simple and rapid preparation of alditol acetates for monosaccharide analysis. Carbonhydrate Reserch, vol. 113, 291-299

Bolen, Wallace P. (1999), Silica. U.S. Geological Survey Minerals Yearbook – 1999. http://minerals.usgs.gov/minerals/pubs/commodity/silica/780499.pdf.

Grabis, J., Jankovica, Dz., Berzins, M., Chera, L. & Zalite, I. (2004), Coating of plasma-processed nanosized powders. J. Europ. Ceram. Soc., vol. 24, 179-184.

Gravitis, J. (1999), Biorefinery and Lignocellulosics Economy Towards Zero Emissions. In: Targeting Zero Emissions for the Utilisation of Renewable Resources (Biorefinery, Chemical Risk Reduction, Lignocellulosic Economy), eds. K. Iijama, J. Gravitis & A. Sakoda, Tokyo, Japan, Published by UNU/IAS, ANESC/UT and IIS/UT, 1999, pp. 2-11.

Gravitis, J., Della, Senta T. & Williams, E.D. (1997), The conversion of biomass into fuels, fibers and value added chemical products from the perspective of the zero emissions concept. Proc. Symp.Biomass Conversion, Sapporo, Japan, October 26, 1997, pp. 1-18.

Hofenauer, A., Treusch, O. & Troger, F. et al. (2005), Dence Reaction Infiltrated Silicon/Silicon Ceramics Derived from Wood based Composites. Advanced Engineering Materials, vol. 5, no. 11, 794-799.

Iiyama, K. & Wallis, F.A. (1990), Determination of lignin in herbaceous plants by an improved acetyl bromide procedure. J. Sci. Agric., vol. 51, 145-161.

Kallavus, U. & Gravitis, J. (1995), A Comparative Investigation of the Ultrastructure of Steam Exploded Wood with Light, Scanning and Transmission Electron Microscopy – Holzforschung, vol. 49, no. 2, pp. 182-188.

Kong, P.C. & Pfender, E. (1987), Formation of ultrafine β-silicon carbide powders in an argon thermal plasma jet. Langmuir, vol. 3, no. 2, 260-265.

Lin, S.Y. & Dence, C.W. (1992); Methods in lignin chemistry. Springer-Verlag.

Sun Luyi, Gong Kecheng (2001), Silicon-Based Materials from Rice Husks and Their Applications. Ind. Eng. Chem. Res. 2001, vol. 40, no. 25, 5861-5877.

Teeaar, R., Gravitis, J., Erins, P., Pugulis, J., Kulkevics, A., Gromovs, V.S. & Lippmaa, E. (1986), Changes in the Phase Structure and Conformation State of Wood Components During Steam Explosion Autohydrolysis – Dokl. Akad. Nauk SSSR, vol. 288, N 4, pp. 932-935 (in Russian).

www. Fao.org/docrep/meeting/008/J2133e.htm.

Removal of Hazardous Substances in Electronics – Processes and Techniques for SMEs[1]

Andrejs Lūsis

Institute of Solid State Physics, University of Latvia,
Kengaraga st. 8, Riga, LV – 1063, Latvia

Abstract

The "sustainable development" science and technology field topics have been under discussion all over the world for a long time. The ecological and quality of life issues are important elements in European public R&D support policy. EU R&D support policy is developed within the framework of the scientific field "Sustainable development, global change and ecosystem", where there has to be identified technologies that promote sustainable developments, and the ways in which their development should be supported. Identification of such technologies should be done for any country or region in the context of STEEP sectors (Society, Technology, Economy, Environment, Politics), i.e. technology versus society, economy, environment and politics. One of such technologies is lead free soldering technology. Of course clean technologies or zero emission issues are very important for any production process and product of any sector of industry. One such product group is : Electrical and Electronic Equipment (EEE), which is produced in large amounts and has a big impact on the environment (waste, water/air pollution) and natural resources. The European Commission to reduce that impact has issued Directive 20002/95/EC "Restriction of the Use of Certain Hazardous Substances in EEE (RoHS Directive, effective from 1st July 2006)", i.e. EEE on the market must not contain lead, mercury, cadmium, hexavalent chromium, polybrominated biphenyls and/or polybrominated diphenyl ether. The implementation of RoHS Directive causes demand for new materials and technologies as well as having a strong influence on economics (competitiveness of the economy).

The Collective Research Projects of FP6 are instruments to develop new technologies (like lead free soldering technology) for the sustainable development of certain sector of industry.

Below the author of the paper presented at conference will try to show as an example of the Collective Research Projects how science can help SME's to overcome technological and material problems caused by implementation of RoHS directive by short description of such project. The title of this projects is "Re-

---

[1] Collective Research Project of FP6 "GreenRoSE" # 500225. Duration: 36 month start from 1 july 2004, coordinator Knut Aune (Abelia, NO).

moval of Hazardous Substances in Electronics: Processes and Techniques for SMEs" and its acronym is "GreenRoSE".

Why GreenRoSE?

Environmentally sound electronic production is the new environmental and business challenge as reflected in European legislation

- RoHS (Restriction of the Use of Certain Hazardous Substances in Electrical and Electronic Equipment),
- WEEE (Directive on Waste Electrical and Electronic Equipment),
- EuE (Directive on establishing a framework for Eco-design of End Use Equipment)
- worldwide trend towards "Green Electronics".

GreenRoSE-partners

According to their roles the partners of Collective Research Projects are from 3 groups

- Indusrtrial associations/grouping (IAG)
- RTD institutions
- Smal and medium enterprises (SME's).

The GreenRoSE project partners are 29 partners from 8 Europe countries: DE, IT, LV, NO, PL, SE, SI, UK.

Tab. 1: Industrial associations/groupings (IAG)

Partic. no.	Participant name	Participant short name	Country
IAG-01	Abelia	Abelia	NO
IAG-02	Gospodarska zbornica Slovenije (Chamber of Commerce and Industry of Slovenia)	CCI	SI
IAG-03	Swedish Association of Electronics Industries	EIF	SE
IAG-04	Fachverband Elektronik Design e. V.	FED	DE
IAG-06	Polish Chamber of Commerce for Electronics and Telecommunication	KIGEIT	PL
IAG-07	Latvian Electrical Engineering and Electronics Industry Association	LEtERA	LV
IAG-08	Unione degli Industriali della Provincia di Padova	Unindustria	IT

Tab. 2: RTD institutions

Partic. no.	Participant name	Participant short name	Country
RTD-01	HIPOT-RR raziskave in razvoj tehnologij in sistemov, d.o.o.	HRD	SI
RTD-02	Jožef Stefan Institute	IJS	SI
RTD-03	Institute of Solid State Physics of University of Latvia	ISSP	LV
RTD-04	Instytut Tele- i Radiotechniczny (Tele and Radio Research Institute)	ITR	PL
RTD-05	ITRI Ltd	ITRI	UK
RTD-06	IVF Industrial Research and Development Corporation	IVF	SE
RTD-07	Politechnika Warszawska (Warsaw University of Technology)	PW	PL
RTD-08	SINTEF – Stiftelsen for industriell og teknisk forskning ved Norges Tekniske Høgskole	Sintef	NO
RTD-09	Technische Universität Berlin	TUB	DE
RTD-10	Let's train	Let's train	NO

Tab. 3: SME's

Partic. no.	Participant name	Participant short name	Country
SME-01	Crawford Hansford and Kimber Ltd	CHK	UK
SME-02	Cynel Unipress Sp. z o.o.	Cynel	PL
SME-03	ELDOS Sp. z o.o.	ELDOS	PL
SME-04	Hanzas Elektronika, SIA	HE	LV
SME-05	HIPOT-HYB proizvodnja hibridnih vezij, d.o.o.	HYB	SI
SME-06	HIPOT-P&EMS proizvodnja elektromehanskih sistemov, d.o.o.	PEMS	SI
SME-07	Radomska Wytwornia Telekomunikacyjna S.A.	RWT	PL
SME-08	SEMICON Sp. z o.o.	SEMICON	PL
SME-10	NewCad AS	NewCad	NO
SME-11	Prosol Electronics Ltd	Prosol	UK

Main Objective with GreenRoSE

To provide European SMEs of the electronics sector with the knowledge and tools to produce hazardous substances-free (in particular lead-free soldered and halogen free) electronic equipment with defined quality and reliability.

Important Topics

Many process-specific details must be examined. For example, the impact of changing materials and production processes on quality, reliability, supplies and product stewardship must be assessed. Analysis of product and process characteristics and their interaction with the environment is required. In-depth study of the environmental impacts of substitute materials and processes is needed because less toxic materials does not necessarily lead to reduced overall life cycle impact.

The project aims to:

- Maximise awareness of European electronics SME's to removal of hazardous substances from electronics via each consortium IAG, through regional workshops and distribution of at least 10 articles and a set of training materials in regional languages.
- Research and advance the available knowledge base on environmental, technical and business issues associated with removal of hazardous substances from electronics, producing and distributing reports, papers and presentations.
- Develop real-world customised solutions for SME's based on case studies carried out at all Core Group SME's, ensuring that all will comply with the RoHS Directive.
- Engage the SME community directly by creating a set of training materials for SME's in regional languages, holding on-site training courses at Core Group SME's and off-site training courses in each region represented in the consortium, involving at least 300 European SME's
- Provide practical technical support for SME's by creating a recognised and accessible facility for research and technology development, pilot line trials and product testing of electronics free of hazardous materials.

Final objective

The project aims for lasting proactive dissemination to help small and medium sized enterprises meet future challenges in the field of environmentally benign electronics in an expanding European and global market, including continuous improvement.

This will be achieved through:

- Holding at least one international conference on removal of hazardous substances in electronics eg. connected with the 'Productronica' international trade fair.
- Generation of at least one new proposal for cross IAG actions.
- Assimilation of the project outputs in creation of a self-funded GreenRoSE 'Virtual Academy' GVA continuing beyond the project timescale.

Fig. 1: Work plan structure (relationships between WP)

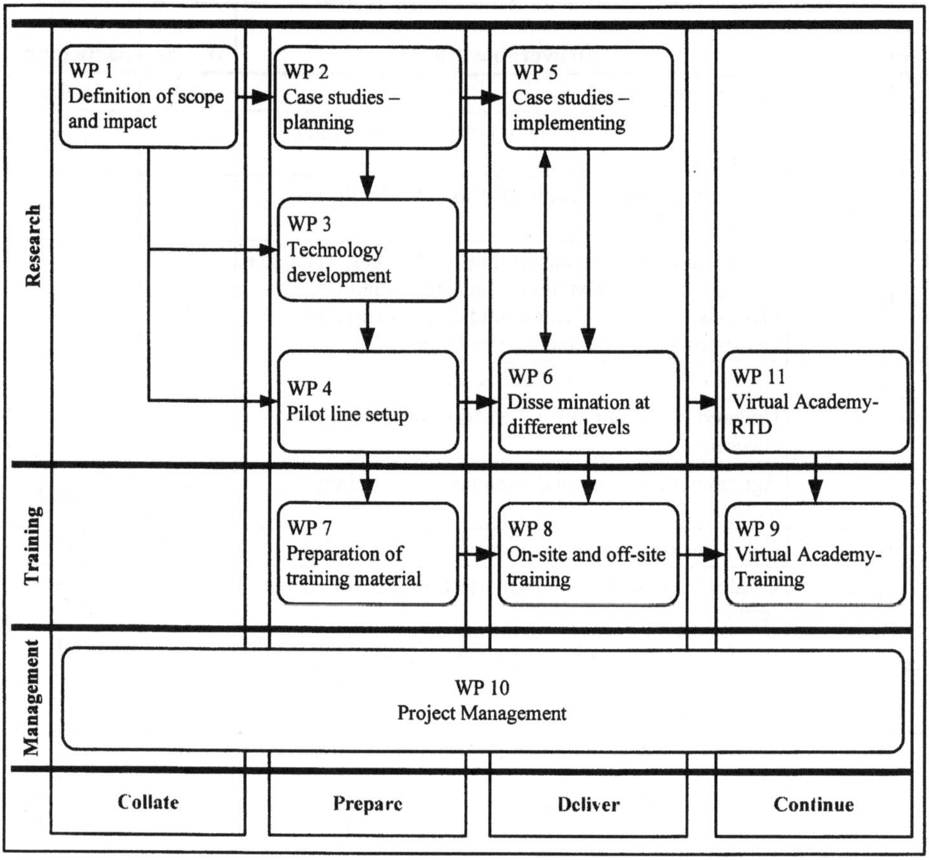

Short Description of GreenRoSE Project Activities

Work Packges

- WP1. Definition of scope and impact
- WP2. Case studies – planning
- WP3. Technology development
- WP4. Pilot line
- WP5. Case studies – implementing
- WP6. Dissemination at different levels
- WP7. Preparation of training materials
- WP8. On-site and off-site training
- WP9. GreenRoSE Virtual Academy GVA
- WP10. Consortium and Project Management
- WP11. GreenRoSE Virtual Academy GVA-RTD

Tab. 4: Deliverable List

Deliverable No.	Deliverable title	WP No.	Lead participant
D10.2	Project Presentation	10	Abelia
D10.1	Implementation of web-based management-system including training	10	Abelia
D1.1	Technical Report and checklist for SME's: Priority hazards in electronics	1	ITRI
D1.2	Technical Report on substitution substances: Technical data, quality and reliability issues, including test methods, availability (long-/mid-/short term) and environmental performance summary	1	ITRI
D1.3	Report of required additional research	1	ITRI
D2.1	Updated version of WP 1.1 SME implementation checklist	2	TUB
D7.1	Agreement on topics and contents of the training material	7	SINTEF
D3.1	Catalogue of available services for quality testing	3	PW
D3.2	Catalog of available reliability testing procedures and services	3	PW
D2.2	Implementation roadmaps for each Core Group SME	2	TUB
D2.3	Process Inventory for Core Group SME's with indication of "Hot Spots"	2	TUB
D6.4	Web site for distribution of case studies and reports	6	TUB
D4.1	Pilot line set-up report	4	TUB
D7.2	GreenRose Toolbox I – a set of training courses	7	SINTEF
D3.3	Test boards including advanced technology	3	IVF
D6.5	GreenRose multimedia based motivation and training	6	Lets Train
D5.1	Inventories for regarded enterprises – changed process	5	TUB
D6.2	One regional workshop about removal of hazardous substances from electronics in each region represented in the consortium	6	PW
D11.1	GreenRose Toolbox II – a set of manuals, checklists and data sheets	11	SINTEF
D11.2	Benchmarking guide	11	TUB
D3.4	Technical requirements and guidelines for quality and reliability	3	PW
D5.2	Certificate of the quantified environmental benefit of technology change for the investigated SME	5	TUB
D5.3	Handbook for environmentally sound SME technology implementation	5	TUB
D5.4	Case Study reports	5	ITRI

Deliverable No.	Deliverable title	WP No.	Lead participant
D8.1	On-site training at SMSs of the Core Group	8	ITR
D8.2	Evaluation report of the training activities	8	SINTEF
D3.5	Design guidelines for substitute advanced technology	3	IVF
D3.6	Handbook: Rating of environmental performance of SMEs	3	TUB
D11.3	Roadmap: "Green electronics beyond 2006"	3	TUB
D6.1	Nine articles, papers and presentations	6	
D6.3	Conference at the end of the project	6	Abelia
D10.3	Monthly Progress and Quality assurance Reports to co-ordinators	10	TUB
D10.4	Reporting to EU-commission every 6 month	10	Abelia
D10.5	Final project report	10	Abelia

Fig. 2: Dissemination Channels of GreenRoSE Project Results

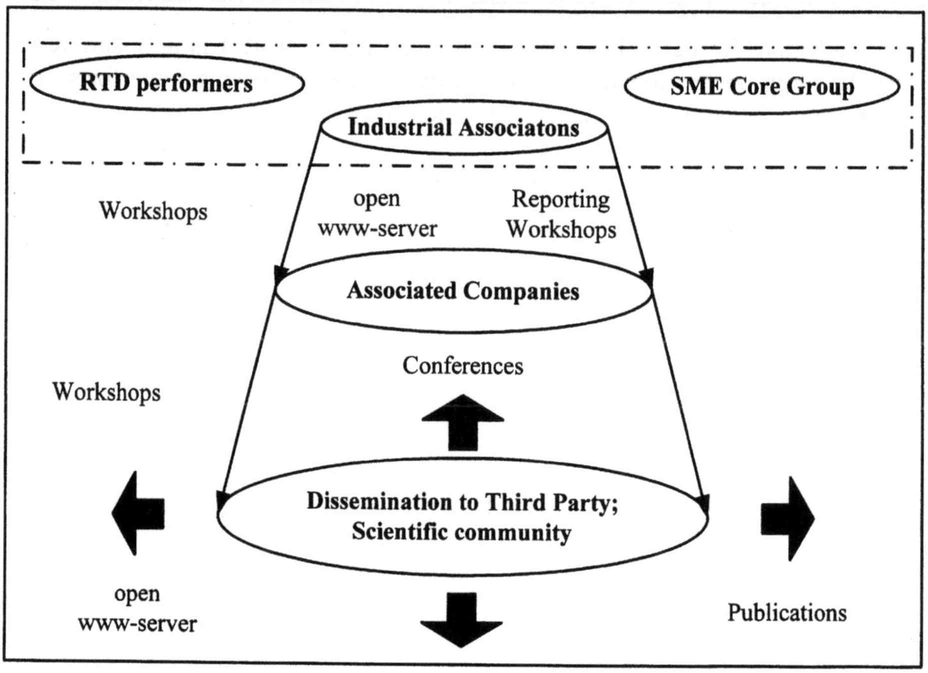

Tab. 5: Milestones List

Milestone	After Project Month
M1. Proposal of required additional research finished	6
M2.1 Technology implementation roadmaps produced for each Core Group SME	12
M2.2 Estimation of Core Group SME's current environmental performance	12
M3. Catalogs and guidelines for new technology available.	33
M4. Pilot line ready to training.	12
M5 All Core Group SME's ready to comply with RoHS Directive	30
M6. GreenRoSE conference	36
M7. Didactical Multimedia Training Material available for SMEs	18
M8. Evaluation report of training activities	30
M9. Roadmap: "Green electronics beyond 2006" finished	36
M10.1 Kick-off meeting	0
M10.2 Mid-term meeting	18
M10.3 Final meeting	36

Such Collective Research Projects as the GreenRoSE project is an example of integrative approaches towards sustainability, which is showing the efforts of a consortium of 29 partners to make electronics greener.

Collaboration Research on a Modified Biopolymer as a Soil Stabiliser and Amendment

G. Shulga[1], O. Nestore[1], V. Shakels[1], B. Neiberte[1], A. Verovkins[1], T. Betkers[2], O. Belous[3], A. Žukauskaite[3], D. Ambrazaitene[3], M. Lukošius[3]

[1] Latvian State Institute of Wood Chemistry,
27 Dzerbenes Str., Riga, LV – 1006, Latvia

[2] Riga Technical University,
1 Kalku Str., Riga, LV – 1658, Latvia

[3] Klaipeda University,
84 H. Manto, Klaipeda, LT – 5808, Lithuania

Introduction

In this work, the results of collaboration interdisciplinary laboratory research on the new lignin-based soil stabilisers conducted in the framework of the international "Eureka" project E!2622 by the Latvian State Institute of Wood Chemistry and Klaipeda University from Lithuania are discussed.

For the coastal zone of the Baltic Sea in the regions of Latvia and Lithuania, soil erosion is an especially urgent problem mainly in connection with the peculiarities of sandy soils. The diminishing soil erosion in Baltic countries is historically based on planting trees and erecting hedges around the crops. Unfortunately, these methods for soil erosion control are labour-intensive and not very effective. One of the attractive solutions to this problem may be a method for biological reinforcement of the sandy soil surface by hydroseeding, which combines simultaneous application of plant seeds, mineral fertilisers, mulch and a soil stabiliser, adhering the seeds to the sandy surface, forming reinforced surface layers and preventing water runoff and soil loss.

Lignin is the world's second most abundant natural substance performing the function of a polymeric glue in the wood matrix. In comparison with other bio-based polymeric products developed for soil amelioration, lignin-based soil improving agents offer advantages over them, since they can degrade and form humus substances in soil. It is also important that industrial water-soluble lignins are wide-scale and low-cost by-products.

Environmentally friendly non-expensive lignin-based soil stabilisers (LSC) with a pronounced adhesive and gluing ability have been developed by the Latvian State Institute of Wood Chemistry. They have been synthesised by modification of industrial lignin with various complementary environmentally friendly water-soluble polymers or oligomers, containing important nutrient elements for soil.

From the chemical point of view, LSC represent lignin-polymer complexes, in which the macromolecules of both components are linked together with the help of physico-chemical bonds. Earlier, we have shown that the developed lignin-based interpolymer complexes may be used as effective amendments for improving hydrophysical properties both for sandy and clay soils [1, 2].

The goal of the work is to show the first results of the study of the influence of the developed LSC and mulch on the properties of sandy soil as well as on seed germination and soil microbiological activity under laboratory conditions.

Materials and Methods

The interpolymer complex formed by sulphite lignin and water-soluble synthetic polymer has been applied as LSC. The composition of the lignin-based polycomplex (Z = polymer/lignin) was varied from 0.02 to 0.2.

The soil under study was a sandy soil from the Baltic coastal region of Latvia. It had the humus content 0.08 wt% and the pH value 5.9. The treatment of sandy soil for the formation of stabilized surface layers was carried out by spraying of 5-20 wt% water solutions of LSC on the sandy soil surface.

The sawdust and bark of different wood species such as aspen, birch, spruce and pine represented a heavy-tonnage waste of Latvia's woodworking enterprises. Fractions with the size from 1 to 2 mm prevailed in the mechanical composition of sawdust and ground bark. To study the ability of the soil conditioner to bind mulch, composite materials were made by mixing the mulch with LSC at different ratios and their subsequent compaction in the form of cylinders at the pressure 0.1 MPa.

The formation and drying of the stabilized soil layers and composite materials were performed at room temperature. The penetration resistance of the treated sandy layers was measured using a manual laboratory penetrometer. The ability of the composites to withstand the crushing force was characterized by the value of compressive strength, according to ASTM D695. Water resistance for the reinforced soil and the composites was evaluated according to the alteration of the penetration resistance of the layers after repeated wetting-drying cycles and as a ratio of the compressive strength of the composite material before and after its soaking for 1 hour in cold water.

Seeds germination was determined according to International seed analyzing regulations, ISTA, 1996. Results were obtained for seeds of *Gypsophila paniculata, Thymus serpyllum and Festuca rubra* plants.

Results and Discussion

The thickness and penetration resistance of the sandy layers treated with the developed LSC depend significantly on its water solution concentrations, the application rate and its composition (Figure 1).

Fig. 1: Dependence of the penetration resistance of sandy soil surface treated with Dustex (1), and the lignin-based soil conditioner with Z = 0.1(2) and Z = 0.2 (3) on the applied solution concentration; the application rate – 1.5 l/m^2

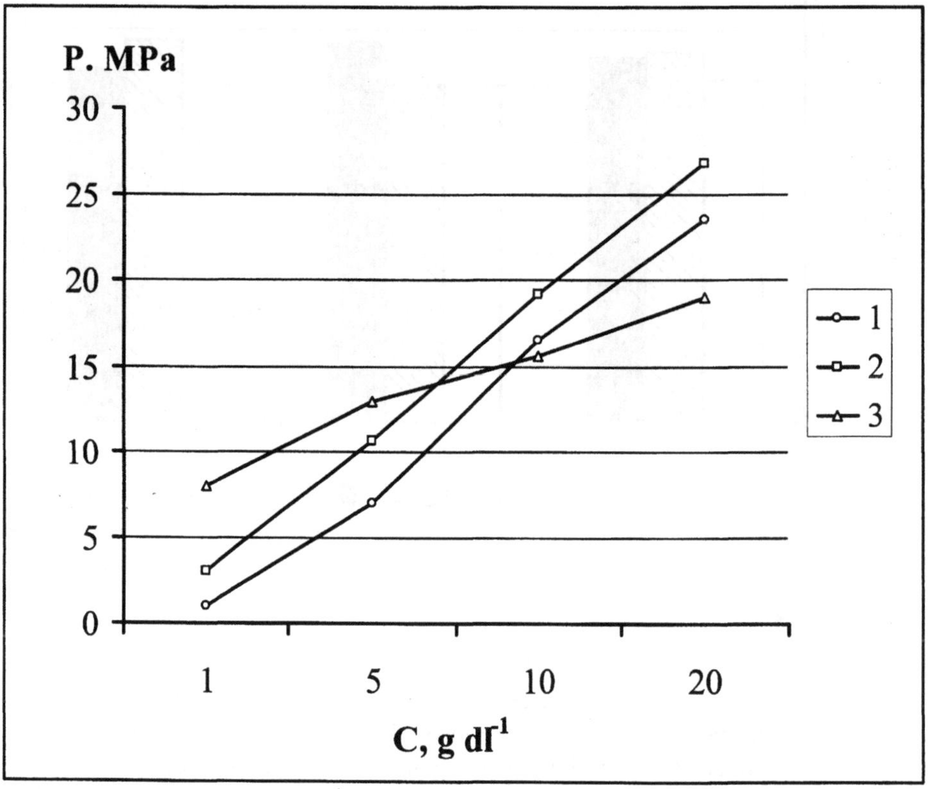

The studies carried out have shown that LSC is a universal adhesive. The compression strength and water stability coefficient of the formed composites depend both on the conditioner's concentration, its application rate and composition, and the wood species from which the mulch was obtained. At the same conditions, the composites filled with deciduous wood bark are characterized by better physico-mechanical properties than the composites from coniferous wood bark.

According to the obtained results, the aqueous solutions of the developed LSC with moderate concentrations have a beneficial action on the germination and development of perennial grass seeds, typical for the coastal dune strip, while the initial non-modified lignin inhibits seeds germination at the concentration of the aqueous solution ≥ 10 g dl^{-1} (Figure 2). In details, the results about the effect of the developed soil conditioners on seeds and plant has been reported in [3].

Fig. 2: Seed germination of *Gipsophila paniculate versus* the concentration of the applied solutions for LSC with Z = 0.5 (1), Z = 0.1 (2) and Z = 0.2 (3)

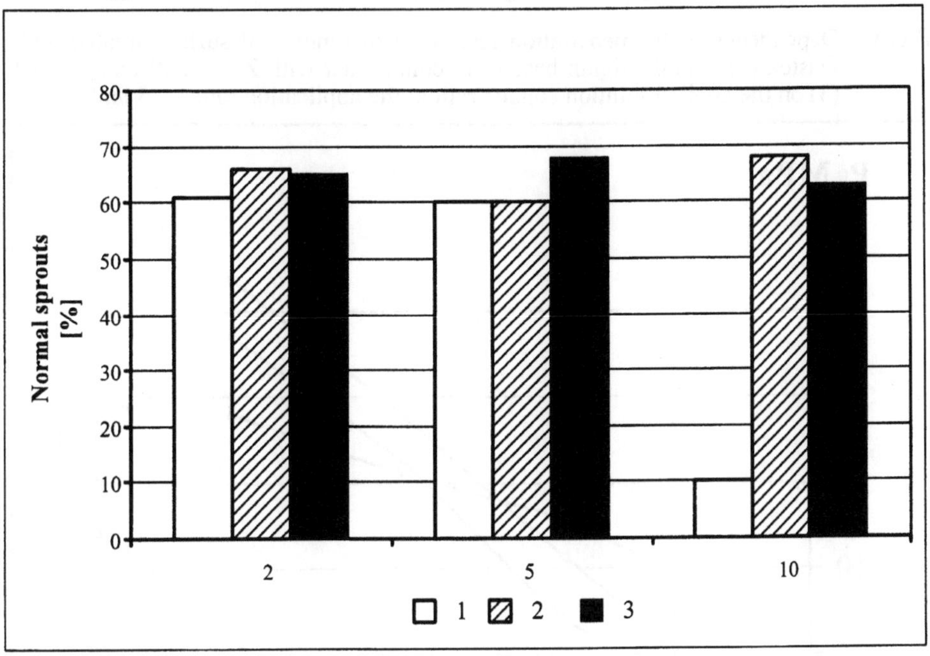

Conclusions

It has been shown that the newly developed soil conditioner has an adhesive affinity both for mineral soil particles and the organic surface of mulch. Its efficiency is determined by the composition of the interpolymer complex and depends on its concentration in the aqueous solution and the application rate. The new soil conditioner has a beneficial action on the development and germination of perennial seeds, typical for the coastal dune strip of the Baltic Sea.

Acknowledgement

The authors are grateful to the Latvian and Lithuanian Councils of Science for financial support of the Eureka project E!2622 "Prevention and Control of Forest Soil Erosion on the Coastal Strip of the Baltic Sea by Using the Lignin-Based Polymers".

Reference

Shulga, G., Nestore, O., Betkers, T., Neiberte, B., Verovkins, A., Belous, O., Žukauskaite, A. & Ambrazaitene, D. (2004), The new soil conditioner based on wood polymer for hydroseeding. 11th Int. Conf. on AgEng, Leuven, Belgium, CD Proceedings, 7 pp.

Shulga, G., Rekners, F. & Varslavans, J. (1998), Environmentally friendly lignin-based adhesives for reinforcement of drainage channel slopes. 8th Int. Conf. on AgEng, Oslo, Norway, pp. 989-990.

Shulga, G., Smilga, H. & Liepinsh, K. (2002), Effect of a lignin-based conditioner on some physical properties of clay soil at its deep loosening. 10th Int. Conf. on AgEng, Budapest, Hungary, pp. 273-274.

Emergy Life Cycle Assessment within a Zero Emission Framework

Sergio Ulgiati, Silvia Bargigli, Marco Raugei

Energy and Environment Research Unit, Department of Chemistry,
University of Siena
Via Banchi di Sotto 55, I – 53100 Siena, Italy

Abstract

Society uses environmental resources directly and indirectly from both renewable energy fluxes and storage of materials and energy that resulted from past biosphere production. The actions of society, its use of resources and the load this resource use places on the biosphere are of great concern. Clearly it is imperative that perspective be gained concerning the interplay of society and environment to help direct planning and policy for the third millennium. In this paper we deal with a systems view of society. The interaction and integration among a system's components, the internal exchange of resources and services, the identification of matter and energy flows to, from and within a system (LCA), the demand for environmental support (emergy), and finally the efficiency of resource use for maximum power output and decreased emissions, are discussed and their importance for more sustainable production patterns is highlighted. Mathematical expressions for total emergy demand, emergy savings from waste recycling, emergy cost of technology and information needed for integration, waste disposal and damage repair, are provided and a quantitative criterion for the evaluation of benefits from zero-emission patterns is finally discussed.

1. Introduction

We present in this paper a framework for an environmental assessment of production processes, aimed at decreasing their load on the environment by optimizing their energy and resource use efficiency, by exploring all possible patterns for the exchange of resources and information, by recycling still usable materials. This framework, indicated as Zero Emission Techniques and Strategies (ZETS 2004), can be regarded as an important evolution of traditional production patterns, based on the myth of unlimited availability of resources on both source and sink sides, towards a deeper understanding of natural cycles and a systems view of the relationship between human economies and environmental dynamics. We first summarize the concept of Life Cycle Assessment, as a widely accepted tool for describing and evaluating a process input and output flows as well as for understanding a processes relationship with the surrounding environment. Then Emergy Synthesis, a scientific measure of the environmental support provided by nature to ecosystems, production processes and human society as a whole, is shortly presented. This support is very often disregarded in conventional economic analyses, where environmental services are not accounted for due to their lack of market value. How Emergy Synthesis can be synergically coupled to well known methods for Life Cycle Assessment is discussed. The goal of this paper is to provide a theoretical basis to the construction of an Emergy Life Cycle Assessment tool, which can be used for environmental policy.

2. Life Cycle Assessment

Life Cycle Assessment is a technique for investigating the environmental aspects and potential impacts associated with a product, by:

- compiling an inventory of a system's relevant inputs and outputs;
- evaluating the potential environmental impacts associated to those inputs and outputs;
- interpreting the results of the inventory and impact phases in relation to the goals of the study.

LCAs conventionally account for each energy and matter flow to and from a chain of processes from raw material to the final product(s) (SETAC 1993, Ayres 1995; ISO 14040/1997, ISO 14041/1998, ISO 14042/2000, ISO 14043/2000). LCA is generally performed based on average values, since the different steps of a process occur in different locations and may be characterized by different performance, which is often location- and technology-specific. Yet LCA provides interesting information about the resource and environmental cost of a given product or service (Cederberg and Matsson, 2000: LCA of milk production; Ayres et al. 2000: LCA of Copper; Ecobalance 2000: LCA of Nickel; Schleisner 2000: LCA of a wind farm; Stripple 2001: LCA of a Road; IAI 2003: LCA of Aluminium; etc.), even if the individual case may differ from the average one due to the existing uncertainty about data collected and processes used. Output flows are also assigned to a specific LCA impact category, in order to better investigate how each flow can potentially affect the surrounding environment (Seppala et al. 1998; Herrchen & Klein 2000; Scheuer et al. 2003).

In general, the output of an LCA is a set of indicators related to specific impact categories (contribution to energy resource depletion, global warming potential, rain acidification potential, etc.), which can be used to suggest an appropriate use for each product, for resource use planning as well as for process optimization by means of step-by-step energy or exergy analysis procedures (Szargut et al. 1988). LCAs in general only account for matter and energy flows occurring under human control. Flows outside market dynamics (such as environmental services, cooling water, dilution of pollutants by the wind) and flows which are not associated to significant matter and energy carriers (such as labor, culture and information) are not generally included. When sustainability comes into play as a major concern, these flows become highly relevant and can no longer be disregarded. Emergy accounting (Odum 1996; Brown & Ulgiati 2004) helps expand the focus of LCA in order to properly account for the contribution of non-market flows and services to the a system/process dynamics. In fact, by means of emergy accounting, all resources are referred to the scale of the biosphere, their usefulness and quality is quantified on the same value basis (i.e. the scale of biosphere cycles beyond economic scales of value such as willingness to pay and cost-benefit evaluation) and then compared with the product(s) generated. In so doing, the option (or set of options) characterized by a lower demand for environmental support can be identified and choices among alternatives facilitated.

3. Emergy Synthesis. Concepts and Definitions

H.T. Odum introduced the concept of *emergy* (spelled with an *m*) in the nineteen-eighties, in order to properly account for the quality of matter, energy and information flows within systems, including their degradation due to second law losses (Odum 1988). Emergy accounts for the environmental services supporting a process as well as for their convergence through a chain of energy and matter transformations in both space and time (Odum 1996; Brown & Ulgiati 2004). By definition, emergy is the amount of *available energy* (exergy) of one type (usually solar) that is directly or indirectly required to provide a given flow or storage of energy or matter. The units of solar emergy are solar emjoules (abbreviated seJ) to distinguish them from actual energy joules (abbreviated J).

The total emergy driving a process becomes a measure of the self-organization activity of the surrounding environment, converging to make that process possible. When the emergy required to make something is expressed as a ratio to the available energy of the product, the resulting ratio is called (solar) *transformity* and is expressed in solar emergy joules per joule of output flow (seJ/J). The *transformity* is an expression of the quality of the output itself from a donor side point of view, for the higher the transformity the more emergy was required to make the product flow. Depending on the flow, transformities are sometimes expressed as seJ/g, seJ/$, seJ/hr, seJ/bit, which justifies the more general name of *emergy intensities*, which is also used.

The emergy method is used in the rest of this paper as a quantitative measure of the total environmental support to the flows of energy, matter and information involved in a system's dynamics. Since our focus is on human-dominated systems, emergy investigations complement results from Life Cycle Assessment (LCA), identifying those patterns which are characterized by lower or higher demands for environmental support and, at the same time, offer a better balance between renewable and non-renewable input resources.

More than an evaluation procedure aimed at just assigning numerical values to processes, flows and products, the emergy method is a conceptual framework, a window through which systems are investigated from a donor-side perspective (i.e. the perspective of the environmental work required to support a system's dynamics). Based on the recognition that "value" has different meanings depending on the scale and perspective of the evaluation, the emergy method assigns values according to what it takes directly and indirectly to drive a process and make the desired products. Emergy accounting (or Emergy Synthesis as it was recently renamed, Brown et al. 2001) has been applied to a large variety of environmental systems, processes, production sectors and economies, in order to highlight systems performance, sustainability, environmental integrity. A selection of the most recent results can be found in Brown et al. (2001, 2003, 2005).

4. Zero Emission Patterns

Ecosystems recycle every kind of waste. The concept itself of "waste" is no longer appropriate for ecosystems. The products from one component or compartment are always a useful resource for another component or compartment. Ecosystems self-organize in order to maximize the total product (e.g., biomass, stored exergy) by optimizing the resource throughput, according to Lotka-Odum's Maximum Empower Principle (Lotka 1922a, 1922b; Odum 1983). Self-organization for maximum empower ensures that all available resources are utilized to the maximum possible extent and no unused resources are left.

The detritus chain in ecosystems is a clear example of this statement. Human dominated ecosystems should be reorganized according to the same principle, for maximum resource use and decreased or zero emissions. In traditional linear production systems, resources are processed and passed on to the next step and some unused wastes are released to the environment. As a consequence, the energy and material cost of the product is higher and the efficiency is lower, as well as a higher load is imposed on the environment in the form of emissions. It is very unlikely for such systems to develop maximum-power-like behavior and therefore be successful in medium and long-term competition, when resources become scarcer.

Instead, in an integrated, "zero-emission" strategy, processes are reorganized and clustered in such a way that unused resources become the raw input to new production patterns. If resources become scarce, this behavior translates into a selective advantage as predicted by the Maximum Empower Principle. While in the conventional way of production the main resources are matter, energy and labor, "zero-emission" patterns mainly rely on knowledge, i.e. on better information about needs of and surpluses from each component as well as about technological tools for resource processing and exchange. The Zero Emission concept

> "... represents a shift from the traditional industrial model in which wastes are considered the norm to integrated systems in which everything has its use. It advocates an industrial transformation whereby businesses emulate the sustainable cycles found in nature and where society minimizes the load it im-

poses on the natural resource base and learns to do more with what the earth produces" (ZEF 1999).

A significant experience in this regard is the so-called Industrial Symbiosis in the Danish town of Kalundborg, (Evans 1995; Ehrenfeld & Gertler 1997; www.symbiosis.dk/), where careful planning around an oil refinery/oil power plant system and the local waste management agency allows huge savings of surface and ground water (3 million m^3/yr), fuel oil (20 000 t/yr), and other resources, supports smaller production processes which use co-product and waste flows, and contributes to a significant decrease in airborne and liquid emissions. It is important to underline that the Kalundborg Eco-Industrial Park was not initially designed as such, but gradually evolved over a number of decades when the participants discovered that the establishment of energy and waste exchanges resulted in economic benefits for all the involved parties. Further information about the development of industrial symbiosis experiences and eco-industrial parks can be found in Gertler (1996), Heeres et al. (2004), Desrochers (2004), Mirata and Emtairah (2005).

4.1 Information, Know-how and Technology Assessment

Since zero emission strategies are significantly dependent on information and know-how for the optimum use of resources, the cost of generating, testing, disseminating and storing information is of paramount importance for sustainability. For example, the information carried by DNA in living systems is undoubtedly generated and supported by direct and indirect solar emergy flows. The information carried by books, software, money, expertise is also generated and supported by direct and indirect emergy flows, but this is never accounted for in conventional Life Cycle Assessment procedures. The problem is that the information content of a specific input, design, or tool is very difficult if not impossible to quantify as such. The large effort put in for information accounting since Shannon firstly introduced a quantitative expression of this concept has not led to any unanimous consensus on information measures, especially when complex systems (ecosystems, societies, culture) are involved (Shannon 1948; Brillouin 1962; Tribus & McIrvine 1971). Instead, the amount of resources supporting the generation of information, i.e. how much it takes to support labor, generate innovation, test and spread new solutions and designs, although still difficult, are easier to quantify in emergy terms (Odum 1996, Chapter 12; Ulgiati & Brown 2005).

If emergy accounting procedures were used to expand the focus of LCA, such an integrated approach could become a valuable tool for developing, monitoring and improving sustainable production patterns in our economies, decreased emissions and a new role for knowledge and information.

4.2 Benefits from Zero Emission Patterns

Self-organization of technology and economies for the optimum use of resources makes them more similar to natural systems and increases the probability of success in a limited-resource world. The results of zero emission strategies is manyfold:

a) Fewer resources are required to drive the global multi-product process than would be needed if each sub-process were driven individually (just consider co-generation of electricity and hot water).
b) Fewer unused resources are released, which are potentially able to drive undesired environmental transformations; as a consequence, less load on the environment is generated and/or fewer resources need to be invested for the safe disposal of wastes.
c) Synergic effects (i.e., increase of benefits) become possible, due to exchange and collaboration links among components;
d) The total output is maximized, since additional products are generated by usefully degrading still usable resources, instead of releasing them unused.

As a last all-encompassing remark, a zero-emission designed system is much closer to a natural ecosystem than any conventional human-dominated chain of processes. Ecosystems along with their trophic levels, structure, complexity and metabolism have often been compared to human-dominated systems in order to find similarities in behaviour and trends (socio-economic metabolism). (Odum 1994; Adriaanse et al. 1997; Giampietro et al. 1998; Kleijn et al. 1999; Jørgensen & Müller 2000; Costanza et al. 2000; Haberl 2000; Kay 2000, Jørgensen et al. 2005).

The zero-emission concept is part of this way of looking at natural systems as a reference. In ecosystems, new organisms and species tend to develop where substrates are more abundant and where competitors and predators (occupying higher levels in the food chain) are scarcer. This results in the occupation of all the available ecological niches and in the development of a higher level of complexity (as the ecosystem gets more mature and reaches the climax stage) and, finally, in the increase of the total free energy (exergy) dissipation. This natural consumption behaviour, which is typical of all living dissipative structures, is also confirmed by the Maximum Power Principle. For instance, from this point of view even the seemingly unattractive detritus chain is a most important and necessary part of the main system and contributes to maximizing the power. Moreover, it offers abundant substrates for those organisms that are able to take advantage of them.

In the same way, a mature zero-emission system should develop a network of processes which are interlinked closely enough so as to work like a natural system and cover all the possible industrial and biological transformations of matter, including waste or by-product reuse/recycling/regeneration. The building of such an interconnected industrial cluster would also help less efficient but strategically important processes to become economically sustainable. Moreover, in the market there could still be vacant niches which could be successfully occupied by those side-processes which can effectively take advantage of specific substrates operating in an environment with less competition.

Fig. 1: Energy System Diagram of three Independent Production Processes Supported by Outside Resource Inflows.

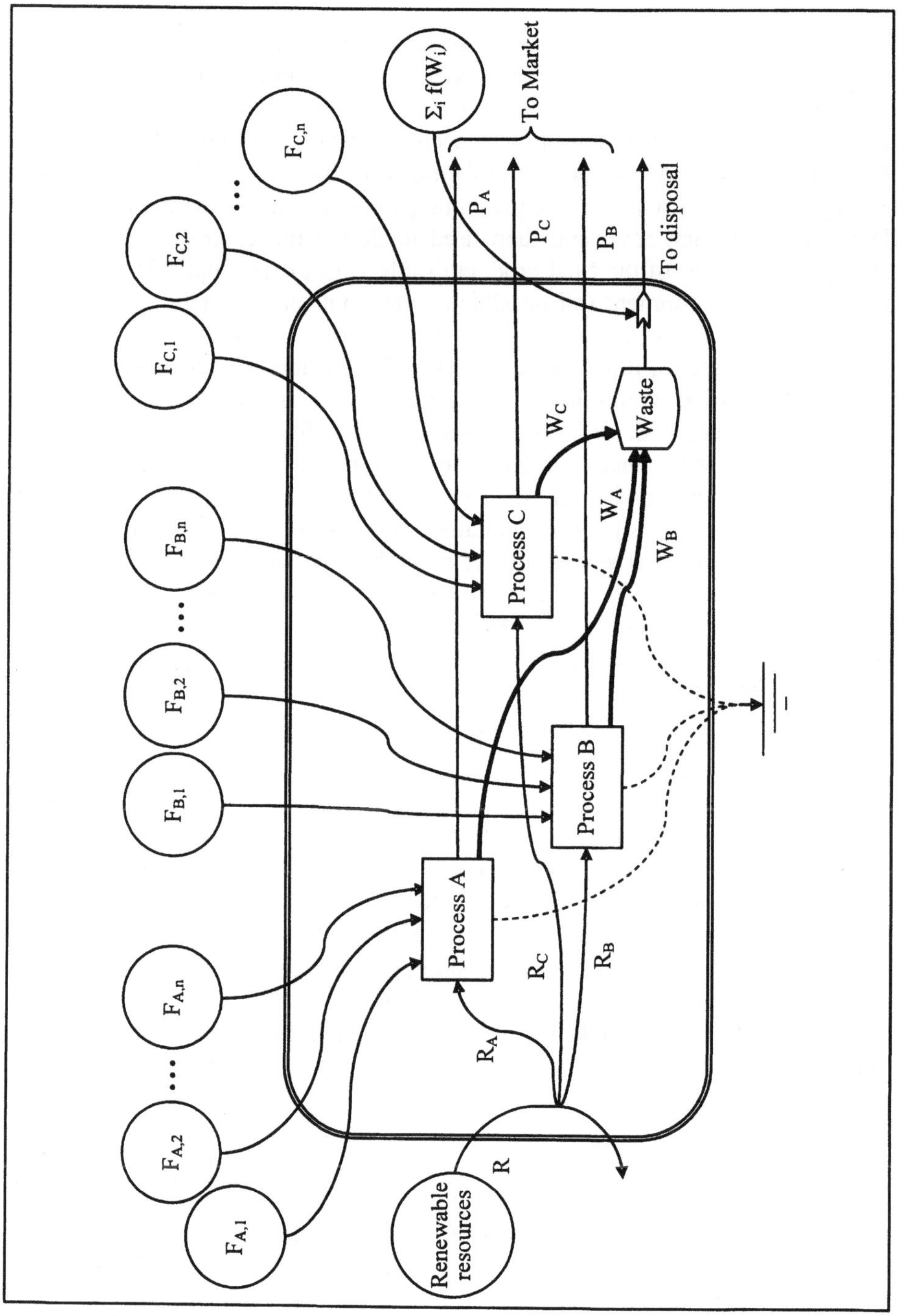

5. Modelling Matter and Emergy Flows within a Zero Emission Framework

Fig. 1 shows a local system consisting of three independent production processes A, B, and C, which are assumed not to exchange any resource flow. After performing an inventory of all input and output flows, the emergy supporting a given process can be calculated by multiplying each renewable and nonrenewable input flow of energy and matter by a suitable transformity value (Odum 1996; Brown & Ulgiati 2004). In so doing, the total emergy driving each process can be calculated (Table 1, first row) and then used to derive the emergy intensity (unit emergy cost, seJ/g) of the final product (Table 1, last row). Input flows are supplied by the environment (all possible sources outside of the process) to each process. Each process in turn delivers a final product and a given amount of waste material (solid, liquid or airborne) to the outside environment (including society), quantified within the life cycle impact assessment phase. Ulgiati et al. (1995) and Ulgiati et al. (2005) suggested that an additional emergy input for the safe disposal of waste materials and/or for the repair of the damage generated by them should be assigned to the final product, thus increasing its emergy intensity. We provide in this paper quantitative expressions for the assessment of the emergy cost of disposal and repair, which is then included in the expressions of Table 1, as a function $f(W_i)$ of the amount of waste material released. The function $f(W)$ consists of at least two kinds of contribution: the investment needed for abatement or disposal, $f_d(W)$, and the investment needed for damage repair, $f_r(W)$. The final expression is:

$$f(W) = f_d(W) + f_r(W) = \Sigma_j c_j * m_{wj} + \Sigma_k d_k * N_k$$

where c_j is the emergy investment needed (seJ/g) for safe disposal of a unit mass m_{wj} (grams) of the jth waste flow and d_k is the emergy investment needed (seJ/unit) for repair of one unit of the kth damaged storage, N_k. A large number of possible damaged items can be accounted for as N_k. If we deal with facades of buildings damaged by traffic emissions, N_k is m^2 of façade; if we deal with forests damaged by acid rain, N_k could be hectares of forest ecosystems to be restored; if focus is on land degraded by mining activities, N_k is hectares of land to be reclaimed; finally, if human health damaged by pollution is of concern, a rough estimate for N_k could be the amount of additional health services (e.g. beds in the hospital, increased number of physicians) needed to face the increased pollution.

In principle, the direct emergy loss due to the generation of irreversible damages that cannot be repaired should also be accounted for and charged to the final product, but its quantification is more difficult. It should be measured as the decrease of the related resource storages (decrease of topsoil layer due to erosion, decrease of standing biomass of forests due to pollution, decrease of biodiversity due to simplification of the ecosystem, decrease of fresh water reservoirs due to excessive withdrawal, etc.) all evaluated in emergy terms as a function of the intensity of the economic activity. Although some interesting results have been presented in this regard, further research activity is still needed.

Tab. 1: Emergy flows from and to the processes of a local economy and the surrounding environment, without emergy exchanges among processes

To⇨ / ⇩From	Environ-ment	Process A	Process B	Process C
Environment (seJ)	0	$R_A + \Sigma_{j=1,\dots,n} F_{A,j} + f(W_A)$	$R_B + \Sigma_{j=1,\dots,n} F_{B,j} + f(W_B)$	$R_C + \Sigma_{j=1,\dots,n} F_{C,j} + f(W_C)$
Process A (g)	$P_A + W_A$	0	0	0
Process B (g)	$P_B + W_B$	0	0	0
Process C (g)	$P_C + W_C$	0	0	0
Emergy Intensity of product (seJ/g)		$[R_A + \Sigma_{j=1,\dots,n} F_{A,j} + f(W_A)]/P_A$	$[R_B + \Sigma_{j=1,\dots,n} F_{B,j} + f(W_B)]/P_B$	$[R_B + \Sigma_{j=1,\dots,n} F_{B,j} + f(W_B)]/P_C$

P_i= mass of product(s) from the ith process; W_i= mass of waste(s) from the ith process; R_i = renewable emergy to the ith process; $F_{i,j}$ = non-renewable emergy of the jth input to the ith process; $f(W_i)$ = emergy invested for disposal of waste from the ith process or for damage repair. (i= A, B, C; j= 1, ...,n).

Fig. 2 shows a significantly different situation, in which processes A, B, and C exchange some of their still usable waste resources, in addition to delivering their main products. The effect of this clusterization of the system is multifold. First of all, a lower amount of resources is required from the environment, which is indicated in the Figure by the absence of flows $F_{i,n}$ replaced by the waste material exchanged between two processes. Second, a smaller emergy investment is needed for abatement, safe disposal, or damage repair, due to the smaller amount of waste released, Wi,r. Finally, additional emergy investments are needed, in order to make each process able to usefully receive waste resources from a nearby process. This last input, g(Wi,u), includes transport, infrastructures and information. In conclusion, the total emergy demand of a cluster system is:

$$Em_{(cluster)} = \overbrace{\Sigma_i R_i}^{\text{Renewable sources}} + \underbrace{\Sigma_i g(W_{i,u})}_{\text{Additional information, know-how, infrastructure}} + \overbrace{\Sigma_i\Sigma_{j=1,\dots,n-1} F_{i,j}}^{\text{Decreased emergy investment needed to drive the process}} + \underbrace{\Sigma_i\Sigma_k d_{kr}\times N_{kr}}_{\text{Emergy investment needed to fix damages from uncontrolled}} + \overbrace{\Sigma_i\Sigma_j c_{rj}\times m_{wr,j}}^{\text{Safe disposal of residual waste}} + \underbrace{?}_{\text{Emergy loss due to irreversible damages}}$$

Fig. 2: Energy Systems Diagram of three Production Processes Supported by Outside Resource Inflows and Clustered for Internal Resource Exchange.

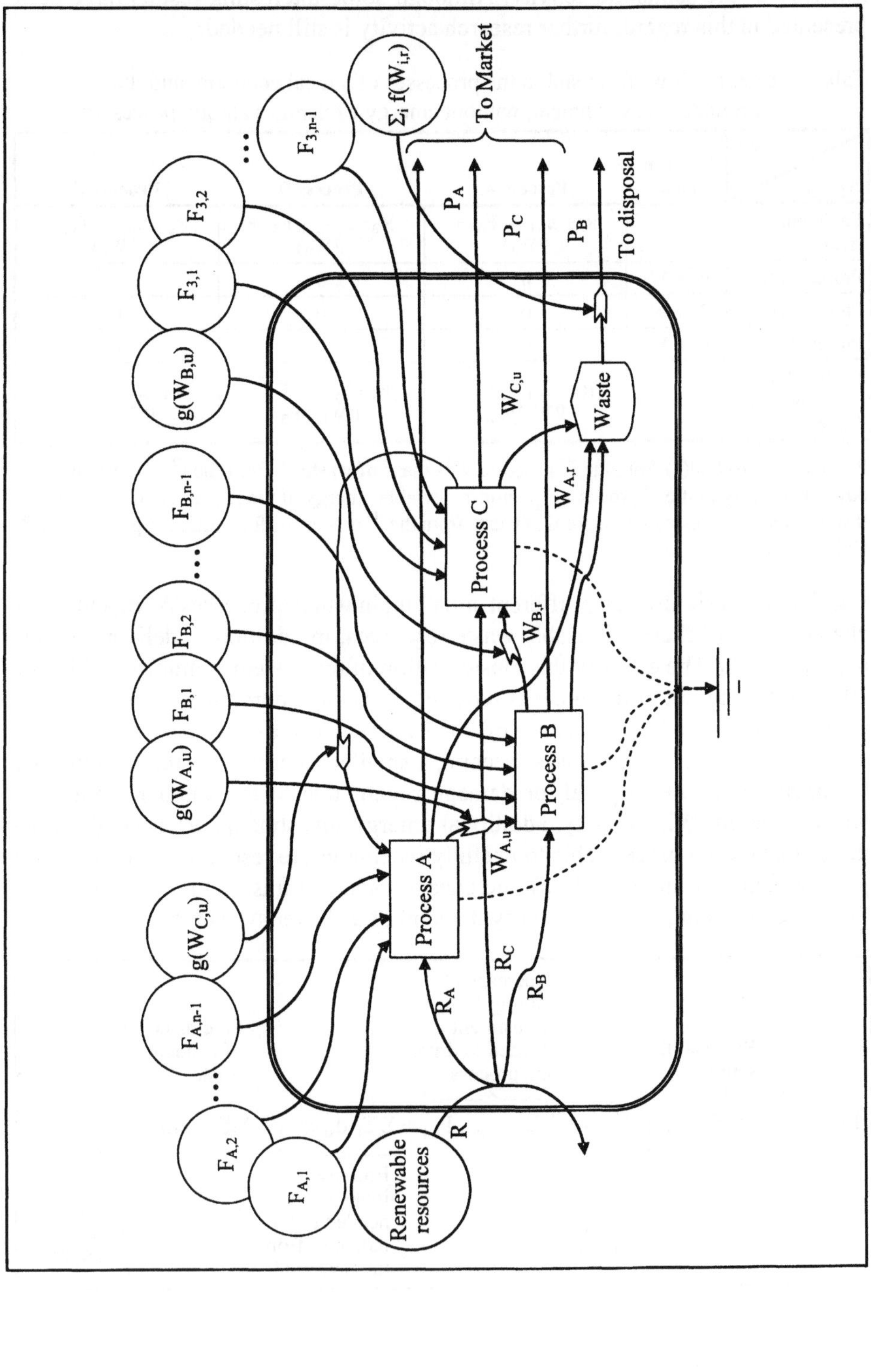

The new situation is summarized in Table 2.

Tab. 2: Emergy flows among the processes of a local economy and the surrounding environment, with emergy exchanges among processes.

To⇨ / ⇩From	Environment	Process A	Process B	Process C
Environment (seJ)	0	$R_A + \Sigma_{j=1,\dots,n-1} F_{A,j} + g(W_{Cu}) + f(W_{Ar})$	$R_B + \Sigma_{j=1,\dots,n-1} F_{B,j} + g(W_{Au}) + f(W_{Br})$	$R_C + \Sigma_{j=1,\dots,n-1} F_{C,j} + g(W_{Bu}) + f(W_{Cr})$
Process A (g)	$P_A + W_{Ar}$	0	W_{Au}	0
Process B (g)	$P_B + W_{Br}$	0	0	W_{Bu}
Process C (g)	$P_C + W_{Cr}$	W_{Cu}	0	0
Emergy Intensity of product (seJ/g)		$[R_A + \Sigma_{j=1,\dots,n-1} F_{A,j} + g(W_{Cu}) + f(W_{Ar})]/P_A$	$[R_B + \Sigma_{j=1,\dots,n-1} F_{B,j} + g(W_{Au}) + f(W_{Br})]/P_B$	$[R_C + \Sigma_{j=1,\dots,n-1} F_{C,j} + g(W_{Bu}) + f(W_{Cr})]/P_C$

P_i= mass of product(s) from the ith process; W_{ir}= residual mass of unused waste(s) from the ith process; W_{iu}= mass of waste(s) usefully transferred from the ith process; R_i = renewable emergy to the ith process; $F_{i,j}$ = non-renewable emergy of the jth input to the ith process; $g(W_{iu})$ = emergy invested to transfer still usable waste materials from the ith process to any other process; $f(W_{ir})$ = emergy invested for disposal of residual waste W_{ir} from the ith process or for repair of the related damage.. (i= A, B, C; j= 1, ...,n-1).

Table 3 shows the total emergy supporting the three processes (A+B+C) with and without a resource exchange among processes. The difference between the situation without exchange and the situation with resource exchange is calculated as:

$$\Delta = \Sigma_i F_{i,n} - \Sigma_i g(W_{iu}) + \Sigma_i f(W_i) - \Sigma_i f(W_{ir}) \qquad (i= A, B, C).$$

Since $W_{ir} < W_i$, then is $\Sigma_i f(W_i) - \Sigma_i f(W_{ir}) > 0$. If also were $\Sigma_i F_{i,n} - \Sigma_i g(W_{iu}) > 0$ or, at least, not too negative, so that can be the whole expression $\Delta > 0$, the creation of links and exchange flows among components translates into a global advantage to the local system in terms of less emergy invested. The expression Δ includes the savings of emergy input needed for process support (due to the use of resources supplied by another process) and the decreased emergy demand for waste management, as well as the cost of knowledge, information and technology that make the cluster possible. i.e. the function g(Wi). It should be underlined that decreasing the unused materials thanks to the cluster organization also abates the potential pollution at an emergy cost lower than without cluster, in so designing a trend towards zero emissions through process optimisation.

Tab. 3: Environmental support to the local system of production processes A, B, and C, with and without cluster-type organization (from Tab. 1 and 2)

System	Emergy (seJ)
$(A+B+C)_{no\ cluster}$	$\Sigma_i R_i + \Sigma_i\Sigma_{j=1,\dots,n} F_{i,j} + \Sigma_i f(W_i)$
$(A+B+C)_{cluster}$	$\Sigma_i R_i + \Sigma_i\Sigma_{j=1,\dots,n-1} F_{i,j} + \Sigma_i g(W_{iu}) + \Sigma_i f(W_{ir})$
Difference (no cluster-cluster)	$\Delta = \Sigma_i F_{i,n} - \Sigma_i g(W_{iu}) + \Sigma_i f(W_i) - \Sigma_i f(W_{ir})$

W_{ir}= residual mass of unused waste(s) from the ith process; W_{iu}= mass of waste(s) usefully transferred from the ith process; R_i = renewable emergy to the ith process; $F_{i,j}$ = non-renewable emergy of the jth input to the ith process; $g(W_{iu})$ = emergy invested to transfer still usable waste materials from the ith process to any other process; $f(W_i)$ = emergy invested for disposal of total waste from the ith process or for repair of the related damage; $f(W_{ir})$ = emergy invested for disposal of residual waste W_{ir} from the ith process or for repair of the related damage. (i= A, B, C).

Conclusion

The employment of the Emergy Accounting method within the framework of Life Cycle Assessment, leading to what might be termed 'Emergy Life Cycle Assessment', is suggested in order to provide information about input and output material flows as well as about the environmental support to the system, and thus to facilitate choices and policy-making towards Zero Emission Strategies and Techniques. We provide in the paper mathematical expressions describing the performance of a set of processes with and without interaction and resource flow exchange. The description is made possible by the use of emergy, in order to be able to express each kind of resources by means of the same unit, and therefore add or subtract flows of different nature. The comparison between the two situations with and without zero emission strategies provides a criterion for the assessment of the possible benefits on a quantitative basis.

References

Adriaanse, A., Bringezu, S., Hammond, A., Moriguchi, Y., Rodenburg, E., Rogich, D. & Schültz, H. (1997), Resource Flows, The Material Basis of Industrial Economies. World Resources Institute, Washington, DC.

Ayres, R.U. (1995), Life cycle analysis: a critique. Resources, Conservation and Recycling 14, 199-223.

Ayres, R.U., Ayres, L.W. & Råde, I. (2000), The Life Cycle of Copper, its Co-Products and By-Products, pp. 168.

Brillouin, L. (1962). Science and Information Theory Academic Press, New York.

Brown, M.T., Brandt-Williams, S., Tilley, D. & Ulgiati, S. (eds.) (2001), Emergy Synthesis. Theory and Applications of the Emergy Methodology. Book of

Proceedings of the First International Emergy Research Conference, Gainesville, FL, 2-4 September, 1999. The Center for Environmental Policy, University of Florida, Gainesville, FL, pp. 328.

Brown, M.T., Campbell, D., Comar, V., Huang, S.L., Rydberg, T., Tilley, D.R. & Ulgiati, S. (eds.) (2005), Emergy Synthesis. Theory and Applications of the Emergy Methodology – 3. Book of Proceedings of the Third International Emergy Research Conference, Gainesville, FL, 29-31 January, 2004. The Center for Environmental Policy, University of Florida, Gainesville, FL.

Brown, M.T., Odum, H.T., Tilley, D. & Ulgiati, S. (eds.) (2003), Emergy Synthesis. Theory and Applications of the Emergy Methodology – 2. Book of Proceedings of the Second International Emergy Research Conference, Gainesville, FL, 20-22 September, 2001. The Center for Environmental Policy, University of Florida, Gainesville, FL.

Brown, M.T. & Ulgiati, S. (2004), Emergy Analysis and Environmental Accounting. In: Encyclopedia of Energy, C. Cleveland Editor, Academic Press, Elsevier, Oxford, UK, pp. 329-354.

Cederberg, C. & Mattsson, B. (2000), Life cycle assessment of milk production — a comparison of conventional and organic farming. Journal of Cleaner Production, 8: 49-60.

Costanza, R., Cleveland, C. & Perrings, C. (2000), Ecosystem and Economic Theories in Ecological Economics. In: S.E. Jørgensen and F. Müller (eds.), Handbook of Ecosystem Theories and Management, Lewis Publishers, Boca Raton, Florida, pp. 547-560.

Desrochers, P. (2004), Industrial symbiosis: the case for market coordination. Journal of Cleaner Production 12: 1099-1110.

Ecobalance (2000), Life Cycle Assessment of Nickel Products. Final Report to Nickel Industry LCA Group. Prepared by: Ecobalance, Inc., 7101 Wisconsin Avenue, Suite 700, Bethesda, MD 20814, pp. 109.

Ehrenfeld, J. & Gertler, N. (1997), Industrial ecology in practice: the evolution of interdependence at Kalundborg. Journal of Industrial Ecology, 1(1): 67-79.

Evans, L. (1995), Lessons from Kalundborg. Business and the Environment, 6(1): 51.

Gertler, N. (1996), Industrial Ecosystems: Developing Sustainable Industrial Structures. Master Thesis. Massachusetts Institute of Technology.

Giampietro, M., Mayumi, K. & Pastore, G. (1998), A dynamic Model of Socio-Economic Systems Described as Adaptive Dissipative Holarchies. In: Advances in Energy Studies. Energy Flows in Ecology and Economy. S. Ulgiati, M.T. Brown, M. Giampietro, R.A. Herendeen & K. Mayumi (eds.), MUSIS Publisher, Roma, Italy, pp. 167-190.

Haberl, H. (2001), Societal Energy metabolism and Sustainable Development. In: Advances in Energy Studies. Exploring Supplies, Constraints, and Strategies. S. Ulgiati, M.T. Brown, M. Giampietro, R.A. Herendeen & K. Mayumi (eds.). SGE Publisher Padova, Italy, 2001, pp. 199-209. ISBN 88-86281-61-7.

Heeres, R.R., Vermeulen, W.J.V. & de Walle, F.B. (2004), Eco-industrial park initiatives in the USA and the Netherlands: First lessons. Journal of Cleaner Production 12: 985-995.

Herrchen, M. & Klein, W. (2000), Use of the life-cycle assessment for an environmental evaluation process. Pure Appl. Chem., vol. 72, no. 7, pp. 1247-1252. © 2000 IUPAC.

IAI (2003), Life Cycle Assessment of Aluminium: Inventory Data for the WorldWide Primary Aluminium Industry. Prepared by IAI – International Aluminium Institute, LCA Advisory Committee, March, 2003, pp. 52.

ISO (1997), Environmental management – Life cycle assessment – Principles and framework (ISO 14040: 1997). International Standard Organization, June 1997, Brussels. pp.16.

– (1998), Environmental management – Life cycle assessment – Goal and scope definition and inventory analysis (ISO 14041: 1998). International Standard Organization, October 1998, Brussels. pp. 27.

– (2000a), Environmental management – Life cycle assessment – Life Cycle Impact Assessment (ISO 14042: 2000). International Standard Organization, October 1998, Brussels. pp. 24.

– (2000b), Environmental management – Life cycle assessment – Life Cycle Interpretation (ISO 14043: 2000). International Standard Organization, October 1998, Brussels. pp. 24.

Jørgensen, S.E. & Müller, F. (eds.) (2000), Handbook of Ecosystem Theories and Management, Lewis Publishers, Boca Raton, Florida, pp. 584.

Jorgensen, S.E., Costanza, R. & Xu, F.L. (eds.) (2005), Handbook of Ecological Indicators for Assessment of Ecosystem Health, pp. 439. CRC Press, Boca Raton, Florida.

Kay, J.J. (2000), Ecosystems, Science, and Sustainability. In: Advances in Energy Studies. Exploring Supplies, Constraints, and Strategies. S. Ulgiati, M.T. Brown, M. Giampietro, R.A. Herendeen & K. Mayumi (eds.). SGE Publisher Padova, Italy, 2001, pp.319-328. ISBN 88-86281-61-7.

Kleijn, R., Bringezu, S., Fischer-Kowalski, M. & Palm, V. (1999), Ecologizing Societal Metabolism. Designing Scenarios for Sustainable Materials Management. Proceedings of a ConAccount Workshop, CML Report 148, Leiden, pp. 57-61.

Lotka, A.J. (1922a), Contribution to the energetics of evolution. Proceedings of the National Academy of Sciences, U.S., 8: 147-150.

– (1922b), Natural selection as a Physical Principle. Proceedings of the National Academy of Sciences, 8: 151-155.

Mirata, M. & Emtairah, T. (2005), Industrial symbiosis networks and the contribution to environmental innovation: The case of the Landskrona industrial symbiosis programme. Journal of Cleaner Production 13: 993-1002.

Odum, H.T. (1983), Maximum power and efficiency: a rebuttal. Ecological Modelling, 20: 71-82.

– (1988), Self-organization, transformity, and information. Science 242 (Nov. 25, 1988): 1132-1139.

– (1994), Ecological and General Systems. An Introduction to Systems Ecology. Revised edition of Systems Ecology, 1983. University Press of Colorado, Niwot, Colorado, pp. 644.

– (1996), Environmental Accounting. Emergy and Environmental Decision Making. John Wiley & Sons, N.Y.

Scheuer, C., Keoleian, G.A. & Reppe, P. (2003), Life cycle energy and environmental performance of a new university building: modeling challenges and design implications. Energy and Buildings, 35: 1049-1064.

Schleisner, L. (2000), Life cycle assessment of a wind farm and related externalities. Renewable Energy, 20: 279-288.

Seppälä, J., Melanen, M., Jouttijärvi, T., Kauppi, L. & Leikola, N. (1998), Forest industry and the environment: a life cycle assessment study from Finland. Resources, Conservation and Recycling, 23: 87-105.

SETAC, Society of Environmental Chemistry and Toxicology (1993), Life Cycle Assessment: A Code of Practice.

Shannon, C.E. (1948), A mathematical theory of communication. Bell System Technical Journal, vol. 27: 379-423.

Stripple, H. (2001), Life Cycle Assessment of Road. A Pilot Study for Inventory Analysis. Report to the Swedish National Road Administration, prepared by IVL – Swedish Environmental Research Institute, Gothenburg, Sweden, Second Revised Edition, 182 pp.

Szargut, J., Morris, D.R. & Steward, F.R. (1988), Exergy analysis of thermal, chemical and metallurgical processes, Hemisphere Publishing Corporation, London.

Tribus, M. & McIrvine, E.C. (1971), Energy and information. Scientific American, 225 (3), 179-184.

Ulgiati, S. & Brown, M.T. (2005), Emergy and Ecosystem Complexity. Int. Journal of Ecological Complexity, submitted.

Ulgiati, S., Bargigli, S. & Raugei, M. (2005), Dotting the I's and Crossing the T's of Emergy Synthesis: Material Flows, Information and Memory Aspects, and Performance Indicators. In: Brown, M.T., Campbell, D., Comar, V., Huang, S.L., Rydberg, T., Tilley, D.R. & Ulgiati, S. (eds.) (2004), Emergy Synthesis. Theory and Applications of the Emergy Methodology – 3. The Center for Environmental Policy, University of Florida, Gainesville, FL, 2004 (in press).

Ulgiati, S., Brown, M.T., Bastianoni, S. & Marchettini, N. (1995), Emergy-based indices and ratios to evaluate the sustainable use of resources. Ecological Engineering, 5: 519-531.

ZEF (1999), Zero Emission Forum, United Nations University. www.unu.edu/zef/.

ZETS (2004), 'Less Bad is Not Good Enough': Call for Papers for a Special Issue of the Journal of Cleaner Production. Journal of Cleaner Production 12: 439-441.

The Research of Biological Phosphorus and Nitrogen Removal from the Waste Water using new Wastewater Treatment Technology

G. Vabolienė, A.B. Matuzevičius

Vilniaus Gedimino Technikos Universitetas,
Saulėtekio al. 11, LT – 10223 Vilnius-40.

Abstract

Nitrogen and phosphorus removal is necessary to avoid eutrophication of water bodies. "BioBalance" technology as the newest way for nitrogen and phosphorus removal has been applied at Utena Waste Water Treatment Plant in Lithuania. "BioBalance Symbio" technology for nitrogen removal is based on the active sludge technology with simultaneous denitrification. The anaerobic zone is equipped before the nitrification/denitrification tank for the removal of biological phosphorus. "BioBalance" measuring equipment, a NADH fluorescensor is based on the process mentioned above. The technology has been evaluated and compared with technology before reconstruction during this scientific work. Influence of wastewater in the effluent has been evaluated on Vyžuona rivulet before and after reconstruction of Utena Wastewater Treatment Plant. The investigation for the BOD_7/P-total after mechanical treatment and the total phosphorus concentration in the effluent shows that the effect of removal of phosphorus was high when the BOD_7/total phosphorus after mechanical treatment was higher than 15.

1. Introduction

Nitrogen and phosphorus removal from wastewater is one of most important processes to avoid eutrophication of water bodies [1, 2, 3]. Some wastewater treatment plants were reconstructed, using traditional and new technologies for nitrogen and phosphorus removal following EU regulation at Lithuanian towns during the last decade. "BioBalance" technology as the newest way for nitrogen and phosphorus removal has been applied at Utena Waste Water Treatment Plant in Lithuania.

Technology is based on new equipment control systems which can directly control the biological process. The purpose of the control is to supply minimum air quantity to assure the constant oxygen amount necessary for the biological processes.

„BioBalance Symbio" technology for nitrogen removal is based on the active sludge technology with simultaneous (synchronous) denitrification) [5]. There is no special separation between the nitrification and denitrification zones in the equipment (the aeration tank). At the same time the denitrification is carried out, during the nitrification in the equipment and there are zones without the dissolved oxygen (anoxic zones) [5, 6].

The simultaneous transformation of the complex nitrogen to the free form is realized directly in the sludge flocks at the constant aeration. The same process conditions, i.e. good mixing and low oxygen concentrations (0,1-0,5 mg/l) are in the whole volume of the equipment The conversion of the nitrates and nitrites composed during the nitrification is carried out inside the sludge flock. Oxygen is directly injected and adapted to the sludge flock needs. Therefore inside the sludge flock there are aerobic and anoxic zones. The external zone is aerobic and the internal one is anoxic.

A C/N ratio is necessary for the denitrification process being reduced because of the little oxygen injected. Significantly less of the carbon is oxidised by free oxygen. Correspondingly a larger part of the carbon is suitable for the nitrates and nitrites reduction [7].

There commonly exists co-ferments functioning as suppliers of hydrogen and electrons. These are nicotinamide adenine dinucleotide NAD^+ and its phosphoresced compound derivative $NADP^+$. Reduced NAD(P)H form, can be oxidised once more during the formation of ATP [1].

During the processes of metabolism co-ferments NADH and NADPH are produced in all the micro organisms. The quantity of NAD(P)H produced in the micro organisms depends on their activity. As mentioned above activity depends on the existing sludge loads, supplied in the form of nutritious matters. It means that, for example, NAD(P)H quantity will remain stable under the conditions of constant quantities of nutritious matters supplied to the active sludge system, the constant load and sludge activity. The sludge activity and at the same time NAD(P)H production increases gradually with the increasing of the sludge load because of a more intensive supply of nutritious matter. Conversely NAD(P)H production decreases during decrease of the nutritious matter supply. During research linear dependences were determined. "BioBalance" measuring equipment NADH fluorescensor is based on the process mentioned above. This is a new control technique which can directly control the biological process. NADH fluorecsensor is a fluoremeter, installed in a stainless steel case and emitting a 340 nm shaft through the quartz glass. The light from the instrument penetrates to the micro organisms and detects the NADH compound that illuminated emits 460 nm shafts, and a NADH sensor registers them. NADH is being constantly converted from the oxidator NAD into the reducing agent NADH in the active sludge. The fluorecsensor fixes only ADH. This gives information on how much energy the micro organisms possess. The information on the energy is defined as the biological active potential (BPA). Monitoring BPA it is possible to control biological processes [7].

The dissolving oxygen reduction to the directly necessary quantity assurance for the biocenosis positively influences the active sludge flocks structure and nitrificant quantity in the system. The nitrificant dependence on the oxygen usually is laid out on the external side of the active sludge flock that is always sufficiently supplied with oxygen. Higher oxygen concentration and consequently stronger oxidation influence lead towards the more intensive decomposition of the external membrane substance. Because of that the nitrificant can hardly make combinations in the active sludge flock. Besides, intensively dissolving oxygen, the sludge is influenced by decomposing power, consequently flocks can be decomposed mechanically or their surface can be broken. The nitrificant, suspended on the sludge flock surface, can be decomposed and washed out [8].

Using "BioBalance" technology, oxygen is probably decreased and the decomposing power and oxidation effect are decreased simultaneously. Therefore, the increased nitrificant quantity can settle in the system [5].

2. Object of the Research

The Wastewater treatment plant of Utena is situated at Palijoniškis County, near the rivulet Vyžuona.

The estimations mentioned below were carried out during establishment of treated wastewater outlet impact into the rivulet Vyžuona (after and before reconstruction). The estimations of wastewater mixture with the water body of the rivulet Vyžuona are for wastewater pollution concentration of the rivulet below the outlet:

$$K_{up}=(Q\times\gamma\times K_F\times K_{nuot.})/(Q\times\gamma+q) \quad (1)$$

Here: Q – flow of rivulet (95% probability during January), m^3/s; q – flow of wastewater in the rivulet, m^3/s; K_{nuot} –pollution of wastewater in the effluent, mg/l; K_{up}- pollution of rivulet below outlet, mg/l; K_f-background pollution of rivulet, mg/l; γ- mixing coefficient. The mixing coefficient depends on: hydraulic coefficient (α), ratio of flow rivulet and wastewater (Q/q) [13]. Following rivers size classification, Vyžuona is assigned by medium flow, basin area and river length [11, 12]. Characteristics of rivulet Vyžuona are: flow – 3.80 m^3/s, length – 26.0 km, area of basin – 415 km^2, then hydraulic coefficient is 0.5 [11, 12]. Wastewater mixture with water of Vyžuona rivulet is good, because the rivulet is small. The mixing coefficient depends on the ratio of flow rivulet and wastewater (Q/q), because the hydraulic coefficient is constant. There is a 95% probability during January of Vyžuona rivulet being 0.506 m^3/s [13]. Average flow of wastewater was estimated according to 1990-1997 data and accredited 0.209m^3/s. Then Q/q=0.506/0.208=2.4; when Q/q=2÷5, mixing coefficient is 0.95 [13].

In 1997 the average concentration of BOD_5 in the rivulet Vyžuona was 5.3 mg/l, usual concentration of nitrogen – 2.2 mg/l, usual concentration of phosphates –

0.1 mg/l. So, rivulet Vyžuona belonged to cleanness class number III in 1997 (Table 1).

Tab. 1: Category of rivers according cleanness [11, 12]

Cleanness class	**1**	**2**	**3**
Describing characteristics	Very clean	Clean	Slightly polluted
BOD_5 mgO_2/l	<2.3	2.3-4.0	4.1-5.8
Total-N, mg/l	<0.3	0.3-.0.7	0.8-2.5
Phosphates, mg/l	<0.03	0.03-0.05	0.06-0.12
Cleanness class	4	5	6
Describing characteristics	Medium polluted	Extremely polluted	Highly polluted
BOD_5 mgO2/l	5.9-9.2	9.3-20.7	>20.7
Total- N, mg/l	2.5-7.5	7.5-15.0	>15.0
Phosphates, mg/l	0.13-0.29	0.30-0.50	>0.50

The average pollution of rivulet after reconstruction during 2002-2004 decreased (Figures 1, 2, 3): usual concentration of BOD_5 was 3.18 mg/l, usual concentration of nitrogen – 2 mg/l, usual concentration of phosphates – 0.05 mg/l. It seems that the rivulet can be classed as cleanness class number II, according to BOD_5 and phosphates, the positive impact of reconstruction to rivulet Vyžuona is significant.

Fig. 1: Influence of BOD5 in the effluent on Vyžuona rivulet

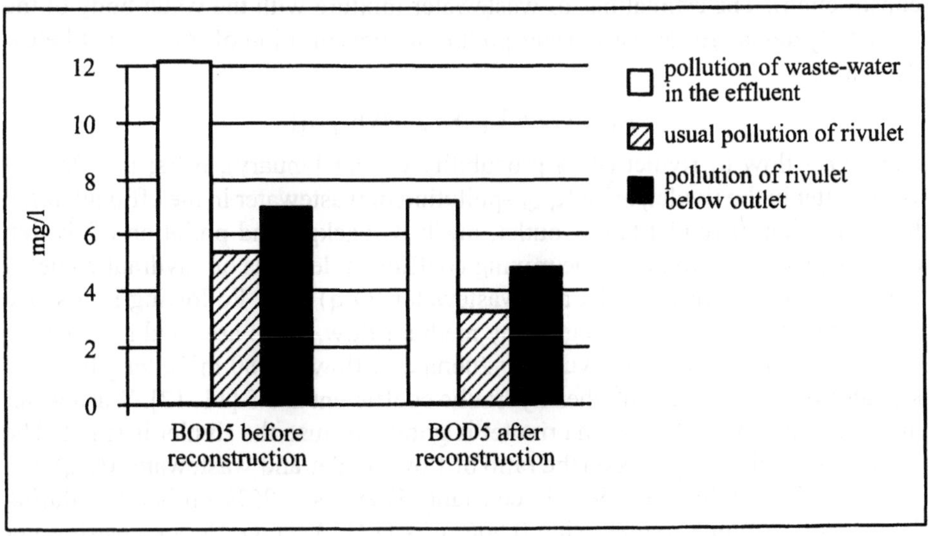

Fig. 2: Influence of nitrogen in the effluent on Vyžuona

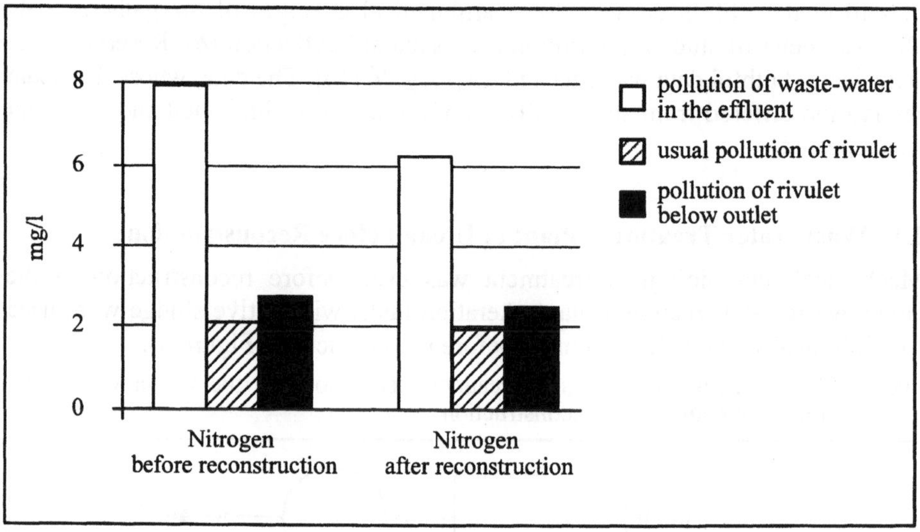

Fig. 3: Influence of phosphates in the effluent on Vyžuona rivulet

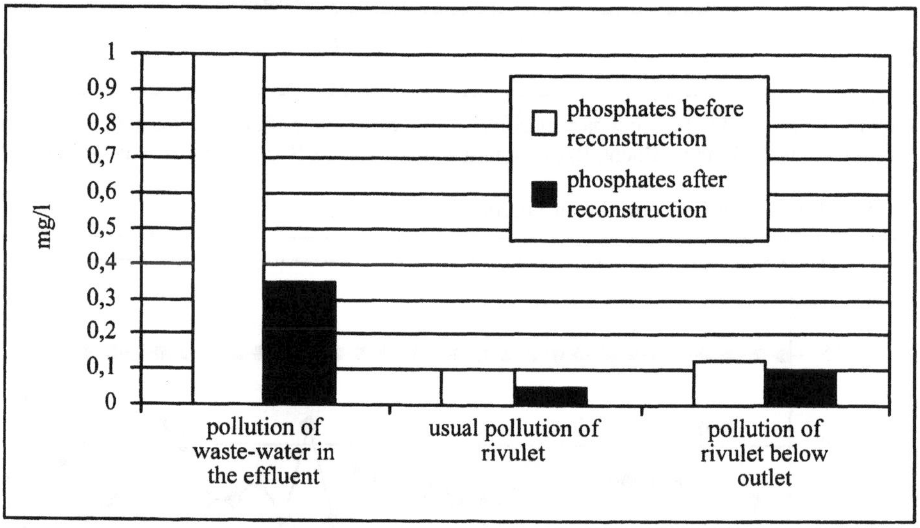

Investigations were carried out at Utena Waste Water Treatment Plant during the three – year period 2002-2004. The information about nitrogen and phosphorus removal before reconstruction has also been gathered.

Once a week pollutions of wastewater: biochemical oxygen demand (BOD_7), suspended solids (SS), total phosphorus (total-P) and total nitrogen (total-N) concentration after the mechanical treatment and in the effluent had been measured. The experimental data were statistically estimated. The experimental data is statistically faithful if the likelihood is from 92.5% to 97.5%. For the gathered data, checking of statistical faithfullness was done: arithmetical averages of the gathered data (X), parameter of student repartition (t_{95}), standard deflection (δ). Reservation of statistically faithful data was checked: $[X_i - X] < 2\delta$ [14]. The data, whose dissonant reservation differed from statistically faithful ones, was eliminated and the count was repeated.

2.1 Wastewater Treatment Plant of Utena before Reconstruction

Mechanical and biological treatment was used before reconstruction in the Utena wastewater treatment plant. Aeration tanks with active sludge were used for biological wastewater treatment before reconstruction (Figure 4).

Fig. 4: Basic schema of biological treatment technology in the Utena wastewater treatment plant before reconstruction

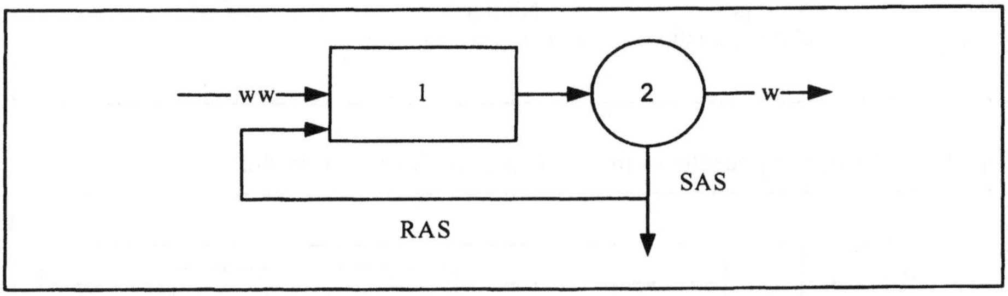

here: 1 – aeration tank, 2 – secondary clarifiers, ww – wastewater after mechanical treatment, w – wastewater in the effluent, RAS – return activate sludge, SAS – surplus activate sludge

Fig. 5: BOD_7 in the effluent before reconstruction

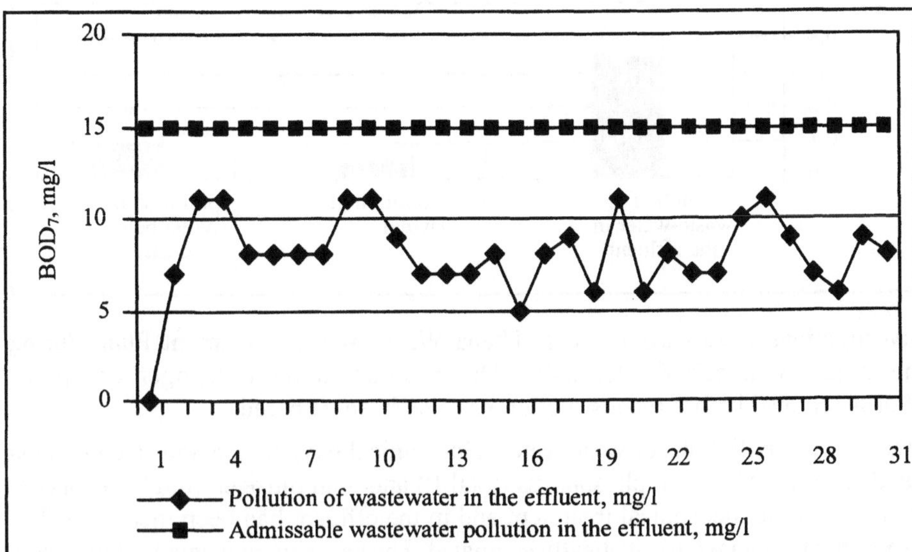

Fig. 6: Suspended solids in the effluent before reconstruction

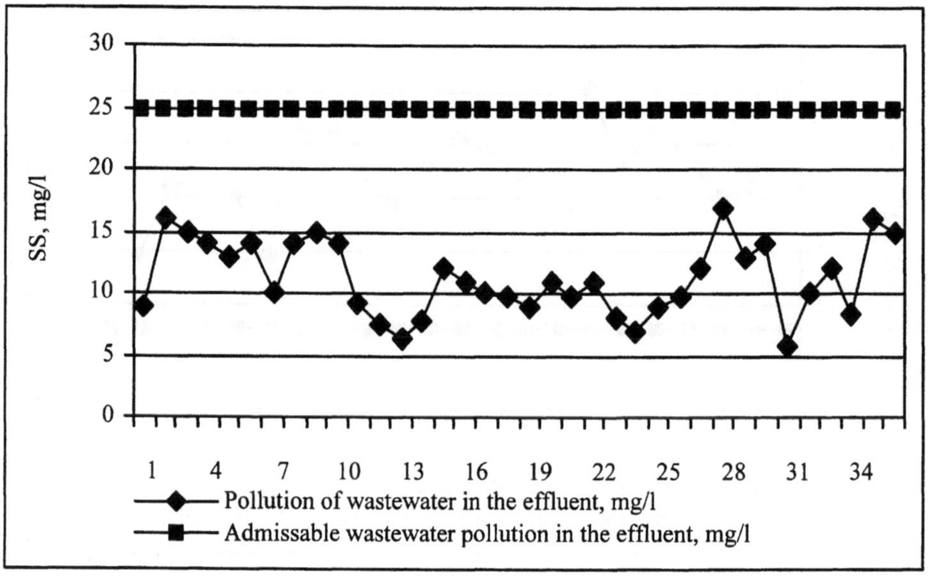

Fig. 7: Total nitrogen in the effluent before reconstruction

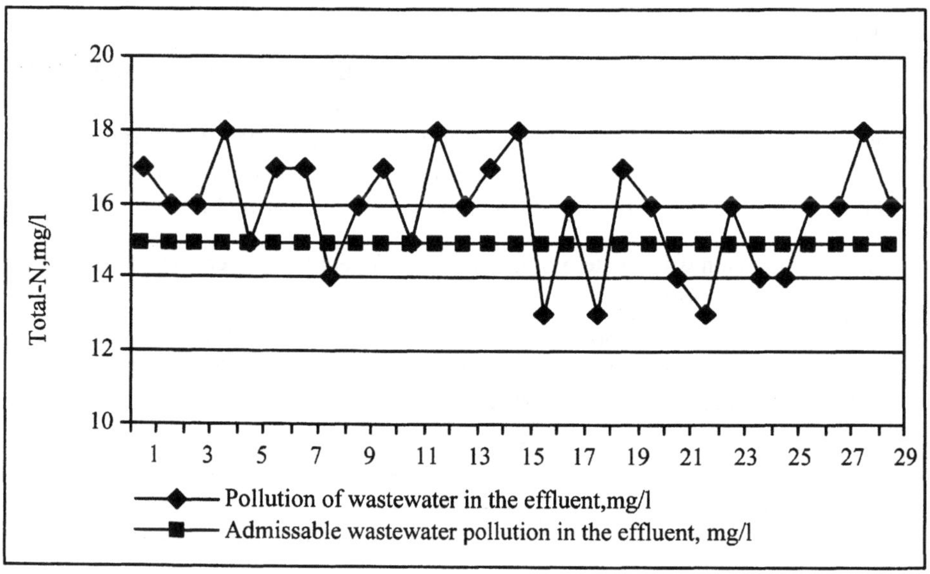

Fig. 8: Total phosphorus in the effluent before reconstruction

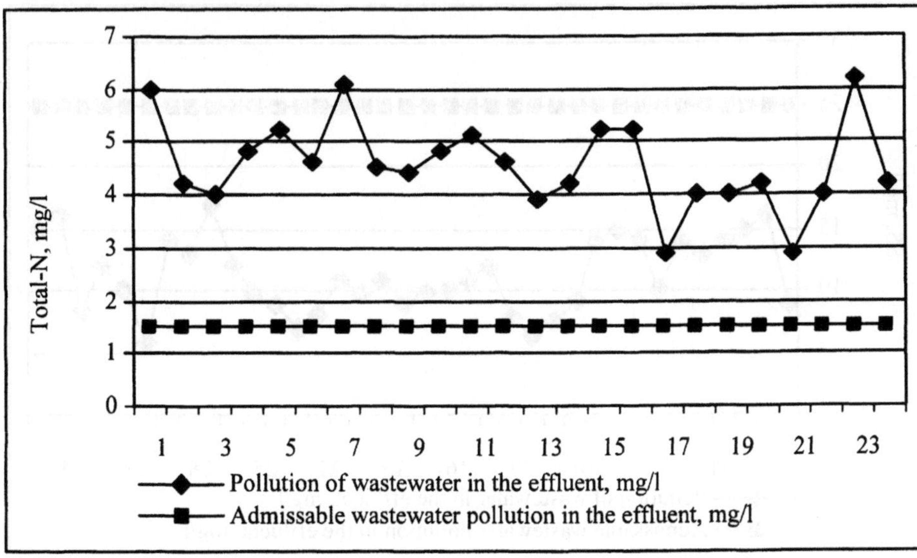

Using this technology for biological wastewater treatment, total nitrogen and total phosphorus concentrations were higher than normal, because nitrogen and phosphorus were used only by microorganisms of active sludge in biochemical processes.

The experimental data shows that (Figures 5, 6), BOD_7 and suspended solids in the effluent were not higher than the allowable norm. During evaluation of wastewater outlet impact on the Vyžuona rivulet the need to increase BOD_7 removal effectiveness was established. In the meantime, concentration of total-P and total-N in the effluent was mostly higher than the allowable norm (Figures 7, 8).

Tab. 2: Comparison wastewater pollution in the effluent before reconstruction with allowable norm

Pollution	**Concentration of pollution in the effluent before reconstruction, mg/l**	**Admissible concentration of pollution in the effluent, mg/l**
BOD_5	5÷11	15
SS	5.6÷16	25
Total –P	2.9÷6.2	1.5
Total- N	13 ÷18	15

2.2 Wastewater Treatment Plant of Utena after Reconstruction

A new wastewater treatment plant started operation in 1999 at Utena after reconstruction (Figure 9).

After reconstruction the capacity of Utena wastewater treatment plant was on average -10000 m^3/d, wastewater pollution treatment: biochemical oxygen demand (BOD_7) – 9 mg/l, suspended solids – 12 mg/l, total phosphorus – 1.2 mg/l, total nitrogen -7 mg/l.

Two of three aeration tanks are used in parallel (Figure 10). The intake part in each aeration tank is used as anaerobic zone for biological phosphorus removal. The remaining part is used as an aeration tank. Accumulating polyphosphates microorganisms are growing in the anaerobic zone. They can accumulate a large quantity of phosphates in their own cells.

Fig. 9: Wastewater Treatment Plant of Utena

Fig. 10: Basic scheme of the aeration tank

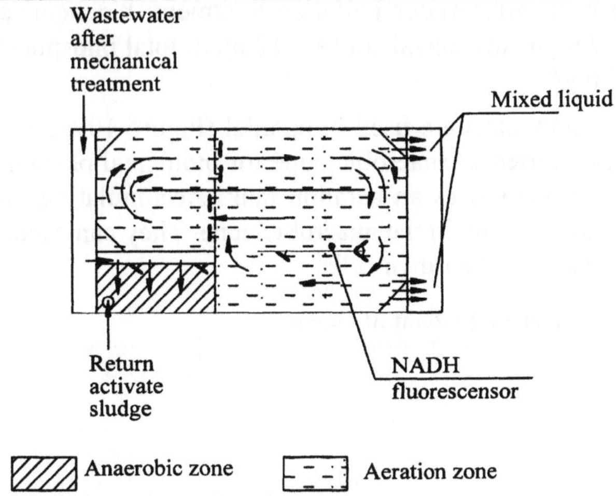

NADH fluorescensor controls denitrification and nitrification processes according to the sludge activity in a single aeration tank (Figure 10). Ammonia is oxidised to nitrate by nitrifying bacteria during the nitrification phase.

The nitrate is reduced to molecular nitrogen (N_2) by denitrifying bacteria during the denitrification phase. These processes are carried out periodically. Organic matter is oxidised bacterially during both the nitrification and denitrification phases, with oxygen and nitrate, respectively, as the oxidising agents and phosphorus is absorbable with special microorganisms. A NADH fluorescensor and oxygen meter is mounted on each aeration tank.

A low oxygen concentration (0.1-0.5 mg/l) is needed for the denitrification to proceed and the higher oxygen concentration (0.5-1mg/l) has been maintained for the nitrification processes to proceed periodically. If the organic load decreases, the oxygen concentration should be decreased to avoid a "sloop" in the process that would cause an increase of ammonium nitrogen concentration in the aeration tank.

Conditions to carry out the nitrification were approximately 2-3 hours, after that 2-2.5 hours were spent carrying out the denitrification process.

Tab. 3: Comparison wastewater pollution in the effluent after reconstruction with allowable norm

Pollution	Concentration of pollution in the effluent on the 2002 – 2004 [mg/l]	Admissible concentration of pollution in the effluent [mg/l]
BOD_7	5.5÷7.5	17.25
SS	4.8÷12	25
Total-P	0.3÷1.5	2
Total-N	5.6÷10	15

Fig. 11: BOD_7 in the effluent after reconstruction

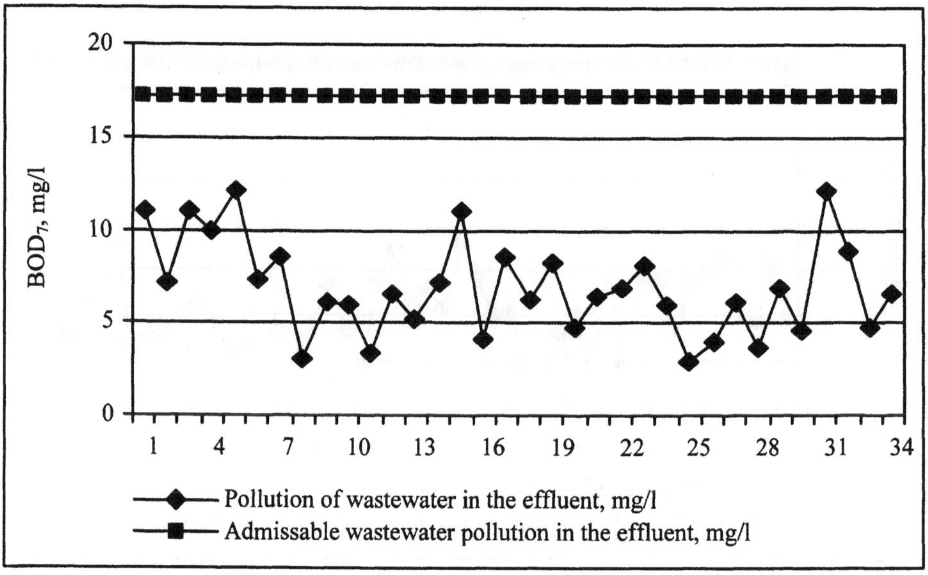

Fig. 12: Suspended solids in the effluent after reconstruction

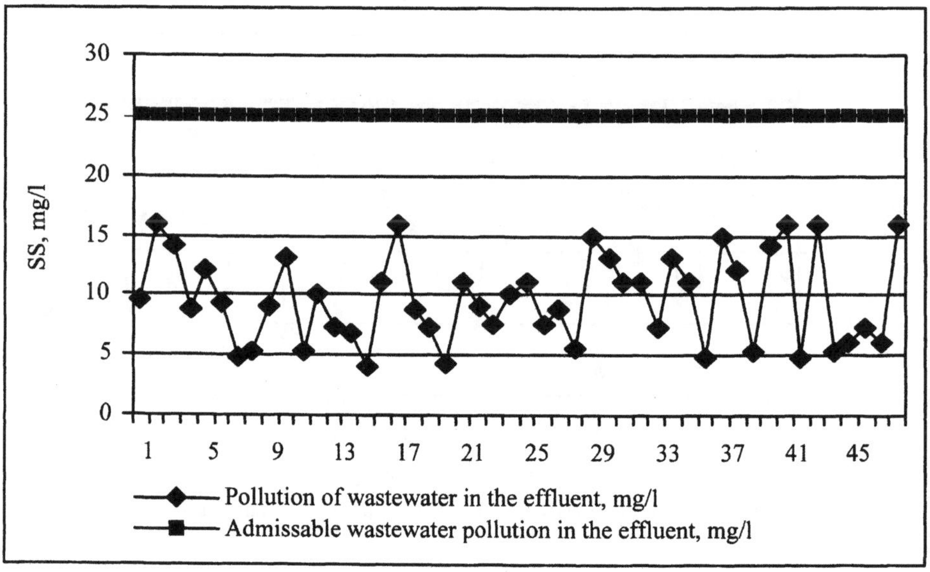

Fig. 13: Total nitrogen in the effluent after reconstruction

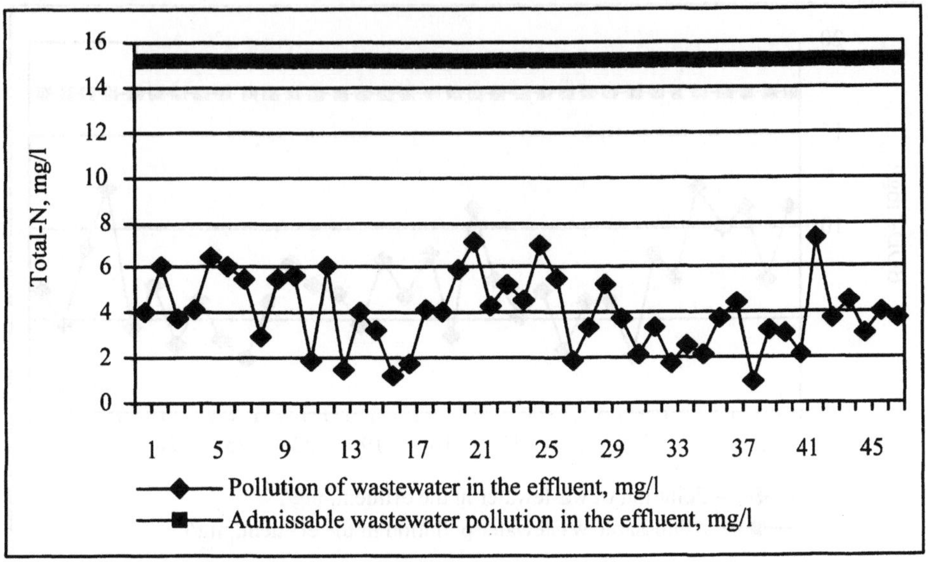

Fig. 14: Total phosphorus in the effluent after reconstruction

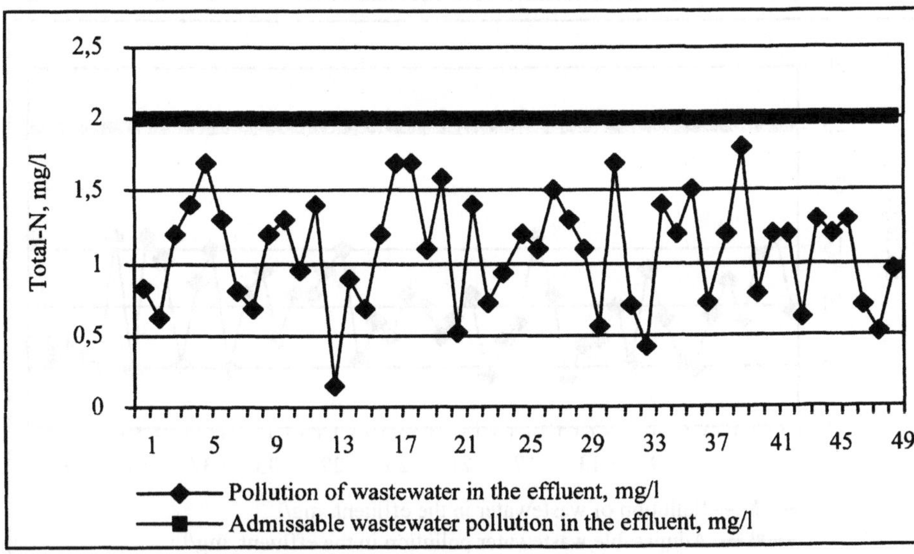

The experimental data shows that (Figures 11, 12, 13, 14) wastewater treatment increased after reconstruction, when "Biobalance" technology was used. Concentration of total-N and total-P in the effluent decreased significantly after reconstruction. Concentration of BOD_7 and suspended solids in the effluent decreased also.

84.6% of total nitrogen and up to 97% of total phosphorus are removed after reconstruction.

The possibility of removing nitrogen and phosphorus biologically was also estimated. The ratio of BOD_5/Total phosphorus shows the possibility of biological phosphorus removal. The higher this ratio is, the more effectively biological phosphorus removal processes are carried out. Information sources show that biological phosphorus removal can be carried out effectively when the ratio of BOD_5/Total phosphorus is higher than 20 [4, 9, 10]. The ratio of BOD_7/Total phosphorus in wastewater after mechanical treatment went from 9 to 47 in 2002-2004 after reconstruction.

Tab. 4: The ratio of BOD_7/Total phosphorus and Total phosphorus in the effluent during the year 2004

BOD_7/Total –P ratio	Average of BOD_7/ Total-P mg/l	Total -P in the effluent mg/l	Average of Total-P in the effluent mg/l
9÷15	11	1.4÷2.2	1.88
15÷20	18	0.23÷1.2	0.75
20÷47	26	0.24÷0.99	0.64

When the ratio of BOD_7/Total-P in wastewater after mechanical treatment went from 9 to 15, the effect of biological phosphorus removal was reduced. Concentration of total phosphorus in effluent went from 1.4 to 2.2 mg/l. When the proportion of BOD_7/ Total phosphorus went from 15 to 20, phosphorus was removed effectively. Concentration of total phosphorus in effluent went from 0.23 to 1.2 mg/l. When the proportion of BOD_7/Total-P was over 20, phosphorus was removed effectively too. Concentration of total phosphorus in effluent went from 0.24 to 0.9 mg/l.

4. Conclusions

1. The evaluation of Utena town wastewater treatment plant before reconstruction showed that total nitrogen and total phosphorus in the effluent was higher than allowable norm.
2. The reconstruction was necessary, because of the negative impact of wastewater onVyžuona rivulet which needed to decrease.
3. The experimental data shows that total nitrogen and total phosphorus decreased in the effluent after reconstruction and was not higher than the allowable norm
4. Pollution of Vyžuona rivulet has significantly decreased after reconstruction.

5. After reconstruction biological phosphorus removal was carried out more effectively, when the ratio of BOD_7/Total phosphorus is higher than 15, using "BioBalance" technology.

References

1. *Beržinskienė, J.* (1999), Water microbiology. Textbook Vilnius: Technika, 144 p.
2. *Henze, M., Harremoes, P., Jansen, J. la Cour & Arvin, E.* (1995), Wasterwater Treatment. Biological and Chemical Processes. Springer-Verlag, 383 p.
3. *Droste, Ronald L.* (1997), Theory and practice of water and wastewater treatment, 954 p.
4. *Matuzevičius, A. & Paulauskienė, Z.* (1998), Experimental researches of phosphorus and nitrogen consumption and nitrogen removal from the wastewater in the biological treatment plant. 3rd International Conference. Cities Engineering and Environment, VGTU, "Technika", 137 p.
5. BioBalance (1996). SymBio technology, 51 p.
6. Verlag fur Architektur und technische Wissenschaften (1997), Lehr- und Handbuch der Abwassertechnik. Vereinigung, e.V., Berlin, 144 p.
7. BioBalance (1999), On-line contact with biological processes. Wastewater purification, 14 p.
8. *You, S. J., Hsu, C.L., Chuang, S.H. & Ouyang, C.F.* (2003), Nitrification efficiency and nitrifying bacteria abundance in combined AS-RBC and A2O systems. Water Research 37: 2281-2290 p.
9. *Bilitewski, Bernd, Hardtle, Georg & Marek, Klaus* (1994), Waste Management. Springer-Verlag, Berlin, Heidelberg, 621 p.
10. *Casey, T.J.* (1997), Unit Treatment Processes in Wastewater Engineering, Wiley, New York, 482 p.
11. Lithuanian Ministry of Enviroment Protection (1997), Lithuanian Rivers water quality chronicle 1996, 64 p.
12. Lithuanian Ministry of Enviroment Protection (2002), Lithuanian Rivers water quality chronicle 2001, 69 p.
13. VTU, Laboratory of water research (1993), Report on wastewater research at Utena town.Vilnius, 55 p.
14. *Martinėnas, B.* (2004), Statistical analysis of experiment's data. Vilnius: Technika, 101 p.

Modelling the Behavior of Heavy Metals Releasing from Sewage Sludge to Soils

Péter Sipos

Institute for Geochemical Research, Hungarian Academy of Sciences, Budaörsi út 45, H – 1112 Budapest, Hungary

Introduction

Sewage sludge is widely used for soil-amelioration primarily due to its high nutrient and organic matter content. Additionally, deposition of sewage sludge on cultivated lands results in improvement of physical conditions of soils, such as cultivability, hydraulical conductivity, or available water content. However, in case of its long-term application the mobilization of toxic metals results in potential environmental hazard (Logan & Chaney 1983).

According to Juste and Mench (1992) more than 40% of total metal (such as Cd, Mn, Cu, Zn, Pb etc.) amount in sewage sludge amended soil can migrate below 20 cm even in a few years. The enhanced mobility of metals in soil may cause increasing bioavailability, which may have a harmful effect on crops depending on the plant species and chemical conditions. Metals may form chelates or complexes with compounds having active functional group in sewage sludge resulting in an increase in their solubility (Bender et al. 1970). In addition, significant amounts of sewage sludge enhance the acidity of soils causing further mobility. Despite liming used for buffering the increasing acidity, some metals (such as Cd) mostly remain in solution (Pepper et al. 1983).

In Hungary, 75% of sewage sludge is used in agriculture after certain treatments. However, crops for human ingestion are not produced on those lands (Vermes 1989). Both Hungarian and European directives determine the maximum toxic metal content of sewage sludge used in agriculture. According to the 86/278/EEC EU Council Directive (1986), these values for toxic metals are as follows: Cd 20-40 mg/kg, Cr 1000 mg/kg (proposed), Cu 1000-1750 mg/kg, Ni 300-400 mg/kg, Pb 750-1200 mg/kg, and Zn 2500-4000 mg/kg. Among these metals, Cu, Ni, Pb and Zn are characterized by both high concentration and high mobility in sewage sludge, while Cd and Cr by relative low concentration and low mobility, respectively (McBride et al. 1999).

Our purpose was to study the immobilization and re-mobilization of some heavy metals (Ni, Cu, Zn and Pb) released from sewage sludge in soils. Soil characteristics that play an important role in heavy metal retention (such as organic matter and carbonate content, as well as clay mineralogy) were compared in respect of immobilizing capacities. The laboratory experiments were performed on natural soil samples from forest soil profiles, which cover more than a third of Hungary's area.

Materials and Methods

Four natural soil samples from two brown forest soil profiles (Haplic Luvisols) were used in the laboratory experiments. Samples were chosen according to their specific characteristics, which may be significant at heavy metal retention. The sample from the A_0 horizon of a forest soil profile (P9A) is characterized by relative high organic matter content (6.74% total organic carbon). Two samples are from the B_t horizons of forest soils, one of them is characterized by relatively low clay mineral content (15%) and a clay mineralogy dominated by vermiculite (P9B), while the other one by high clay mineral content (45%) and a clay mineralogy dominated by montmorillonite (P13B). The sample from the C_k horizon of a forest soil profile (P9C) is characterized by high carbonate content (25% calcite). The detailed description of the two soil profiles can be found in the paper by Sipos (2004).

Batch adsorption experiments were carried out in the concentration interval of potentially mobile metal fractions in sewage sludge. According to Li et al. (1995) the following proportions of total metal content can be considered as mobile in a sewage sludge amended soil: 10% for Cu, 10% for Ni, 20% for Pb and 30% for Zn. The calculation of metal concentrations in solutions was performed using these proportions and the maximum metal values determined in the EU Council Directive. The highest initial metal concentrations in solution used in adsorption experiments were as follows: 1200 mg/l for Zn, 250 mg/l for Pb, 200 mg/l for Cu and 50 mg/l for Ni. Maximum metal adsorption was calculated from the Langmuir adsorption isotherms. The pH change during the adsorption process was followed up.

Sequential extraction after Li et al. (1995) was performed on samples treated by the solutions that contained the highest initial heavy metal concentrations. Four fractions were separated in four steps as follows: (1) exchangeable with $MgCl_2$, (2) specifically adsorbed and bound to carbonates, extracted by NaOAc + HOAc, (3) bound to Fe and Mn oxides, extracted by $NH_2.OH.HCl$ + HOAc, and (4) bound to organic matter and sulfides, extracted by H_2O_2 + NHO_3 + NH_4OAc. Metal concentrations in solutions were analyzed by the ICP-AES method (Jarrel-Ash Atomscan 25). The changes in sample composition were studied by XRD analyses (Philips 1710).

Results and Discussion

The results of adsorption and desorption studies are presented in Table 1. The affinity of metals to the studied soil samples is generally according to the order of Cu > Zn > Pb >Ni, except in the case of the sample containing montmorillonite, when Zn shows higher affinity than Cu. However, this sequence is strongly influenced by the initial metal concentrations. To avoid this effect, the adsorbed metal amounts can be expressed as the percent of total available metal concentrations („relative adsorption"), and in this case the following sequence can be observed: Pb > Cu > Zn > Ni. This sequence is similar to that shown by other studies (e.g. Welp & Brümmer 1999; Gomes et al. 2001, etc.). The sequence is due to the differences among the adsorption characteristics of the studied metals on the one hand, and to the pH conditions during the adsorption process on the other. According to Berti and Jacobs (1996), the adsorption of Ni and Zn in soils depends on electrostatic forces, while the Cu and Pb adsorption depends on covalent interaction with mineral surfaces, but the adsorption of the latter two metals is always preferred compared to the previous ones. During the metal adsorption process at high metal concentrations the pH decreases resulting in mobilization of Ni and Zn, while Cu and Pb still adsorb (McKenzie 1980; Brümmer et al. 1988; Naidu et al. 1997). The above sequence changes in the case of samples containing vermiculite because of a higher affinity of Cu to soil vermiculite than Pb, as was shown also by Das and Bandyopadhyay (2001).

Tab. 1: Adsorbed and desorbed heavy metal amounts (mg/kg) in the studied soil samples.

Sample	Metal	Adsorbed			Desorbed by							
					$MgCl_2$		NaOAc		$NH_2.OH.HCl$		H_2O_2	
		mmol/kg	mg/kg	%*	mg/kg	%**	mg/kg	%**	mg/kg	%**	mg/kg	%**
P9A	Zn	64	4189	14	3173	74	452	11	655	15	n. d.	
	Pb	43	8832	58	1547	18	4500	51	2085	24	653	7
	Cu	82	5236	51	834	15	1674	31	2200	41	674	13
	Ni	2	145	4	n. d.		n. d.		n. d.		n. d.	
P9B	Zn	58	3766	13	2137	59	560	15	952	26	15	<1
	Pb	25	5154	31	2742	52	1703	32	738	14	99	2
	Cu	60	3822	37	427	11	907	24	2288	62	100	3
	Ni	2	128	4	n. d.		n. d.		n. d.		n. d.	
P13B	Zn	124	8120	16	6858	85	574	7	547	7	100	1
	Pb	32	5520	32	5005	77	1144	17	318	65	90	1
	Cu	38	2401	19	1113	48	288	12	833	36	99	4
	Ni	2	124	4	n. d.		n. d.		n. d.		n. d.	
P9C	Zn	113	7405	23	849	12	5158	70	1354	18	14	<1
	Pb	63	12965	92	43	<1	10324	82	2073	16	191	2
	Cu	187	11888	91	126	1	8500	71	3161	27	113	1
	Ni	6	350	13	188	55	108	32	45	13	n. d.	

* in the percentage of total available metal

** in the percentage of maximally adsorbed metal

The sample containing carbonate (P9C) immobilizes the most significant amount of metals (Figure 1). In this sample it is not the adsorption, but the precipitation that is the most important retardation process, which is also supported by the determination of Pb and Cu carbonates by the XRD studies. In the other three samples, Zn and Ni show similar extents of adsorption (their "relative adsorption" is around 14% and 4%, respectively), while Cu and Pb exhibit large variations. These differences are due to the above mentioned impeded adsorption of Zn and Ni by Cu and Pb, and they suggest the strong effect of the competitive situation on the adsorption of Zn and Ni, as many other studies reveal (e.g. Zhu & Alva 1993; Fontes et al. 2000, etc.). Among the carbonate free samples the most significant metal amounts are adsorbed in the organic matter rich sample, which immobilizes 58% and 51% of total available Pb and Cu, respectively (Figure 1). A somewhat lower amount was adsorbed by the sample containing vermiculite (37% of total Cu and 31% of total Pb). Contrarily, in the sample characterized by montmorillonite dominated clay mineralogy the "relative adsorption" of Cu (19%) is lower than that of Pb (32%).

Fig. 1: The "relative metal adsorption" by the studied soil samples. P9A – rich in organic matter, P9B – contains vermiculite, P13B – contains montmorillonite, P9C contains carbonate

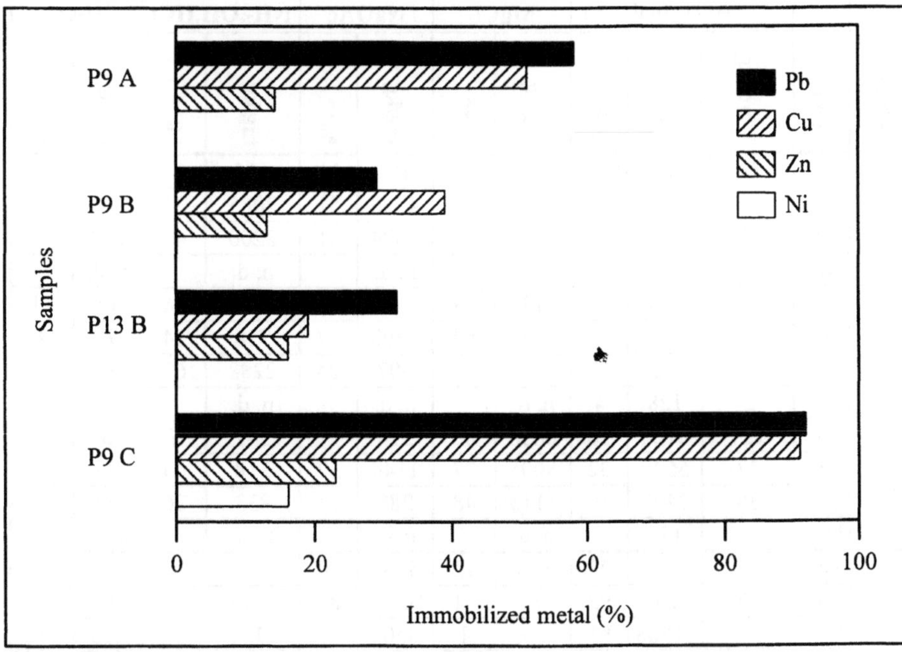

The adsorption process is strongly affected by the pH (Brümmer et al. 1988; Gomes et al. 2001). The "relative adsorption" of the studied metals decreases with decreasing pH, and there is a narrow pH range for each metal, where the adsorption starts to fall back. This pH range differs from metal to metal, and as follows in our case: 6.5-7.5 for Ni, 6-7 for Zn, 5.5-6.5 for Pb and 5-6 for Cu (Figure 2).

The exchangeable metal amounts show high variety among the studied samples (Figure 3). The role of ion exchange in the metal adsorption is the most important in the case of Zn. 74, 59 and 85% of adsorbed Zn is adsorbed by ion exchange into samples characterized by high organic matter content, as well as vermiculite and montmorillonite dominated samples, respectively. According to Han and Banin (1997) the ion exchange is the most important process at immediate adsorption of zinc into soils. Pb adsorption in this form is significant in the clay mineral rich samples: 52 and 77% of adsorbed lead is exchangeable in vermiculite and montmorillonite containing samples, respectively, while Cu is adsorbed by ion exchange mostly in the case of montmorillonite containing samples (48% of adsorbed Cu). Many studies (Shumann 1977; Sidle & Kardos 1977; Harter 1979) reported the adsorption of Cu, Pb or Zn by ion exchange in soils, but the most significant adsorption process can depend on the chemical properties of soils (Han & Banin 1999). This process is significant mostly in the samples of high clay mineral content, and primarily in the case of copper and zinc in samples characterized by montmorillonite, as well as in the case of lead in samples containing vermiculite. The ion exchange of zinc can also be considerable in the case of samples rich in organic matter, because the competition has a strong effect on the adsorption of this metal.

Fig. 2: The change of the "relative metal adsorption" in the function of pH for copper.

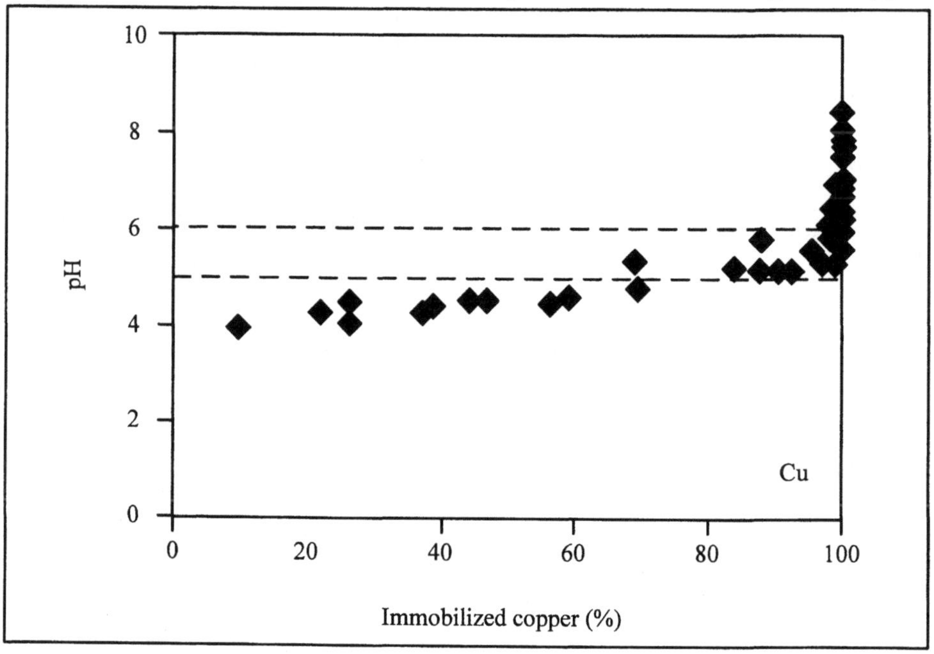

The specific adsorption is important mostly for Pb and Cu, and primarily in samples rich in organic matter and containing vermiculite (Figure 3). In the sample rich in organic matter 51% of adsorbed Pb and 32% of adsorbed Cu is specifically adsorbed, while in the sample containing vermiculite this ratio is 32% for Pb and 24% for Cu. The specific adsorption plays an important role in the case of the sample having high organic matter content. Metals characterized by high affinity to soils (Pb, Cu) adsorb mainly at these sites. According to the general experience, metals tend to adsorb by specific adsorption and into organic matter. However, there are "preferred" elements (Cu, Pb) occupying the better adsorption sites from before the "non-preferred" elements (Zn, Ni) (Han & Banin 1999).

Fig. 3: Extractability of adsorbed metals from the soil samples characterized by different composition.

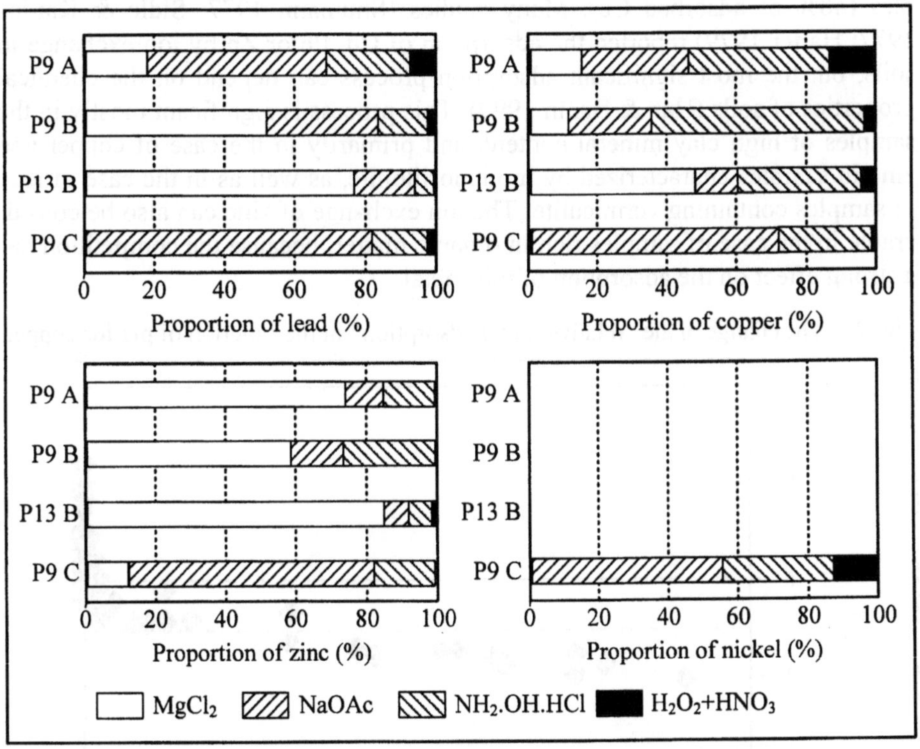

The co-precipitation of metals with Fe-Mn oxides is significant primarily for Cu (36-62% of total Cu is in this fraction), but in some cases Pb (in organic matter rich sample 24% of adsorbed Pb) and Zn (in vermiculite containing sample 26% of adsorbed Zn) can be also found in this fraction. This suggests the high affinity of Cu to bind to stable phases in soils even within a short time. According to Han et al. (2001) redistribution of added heavy metals under a wetting-drying cycle regime starts immediately, and significant changes can be found even after a few days.

The most significant metal amounts can be immobilized by the carbonate containing sample due to the carbonate dissolution in a slightly acidic medium. Ni amount above the detection limit was measured only in this case, probably because of competitive situation and impeded adsorption of this metal by the others.

Conclusions

The adsorption characteristics of heavy metals vary almost similarly independent of soil composition suggesting that the same soil properties control their adsorption. The ion exchange, specific adsorption and precipitation are the most important immobilizing processes in the case of soils characterized by high clay mineral, organic matter and carbonate content, respectively.

Soils with high amounts of organic matter and clay minerals, as well as with significant amount of carbonate minerals may be suitable media for the immobilization of heavy metals. The immobilizing effect of a soil decreases with the increasing number of metals, and a very small pH change can result in loss of an important part of this capacity. The advantage of complex soil composition is that soil acts as a selective buffer for the change of natural conditions, but this buffering capacity is finite. The loss of the buffering capacity of one of the soil compounds does not result in the loss of effect of the other components.

The preliminary study of buffering capacity of soils is necessary to use the sewage sludge in the agriculture, and it may form the base of the sustainable land management.

References

Bender, M.E., Watson, W.R. & Jordan, R.A. (1970), On the significance of metal complexing agents in secondary sewage effluents. Environmental Science and Technology 4: 520-521.

Berti, W.R. & Jacobs, L.W. (1996), Chemistry and phytotoxicity of soil trace elements from repeated sewage sludge applications. Journal of Environmental Quality 3: 250-253.

Brümmer, G.W., Gerth, J. & Tiller, K.G. (1988), Reaction kinetics of the adsorption and desorption of nickel, zinc and cadmium by goethite. I. Adsorption and diffusion of metals. Soil Science 39: 37-52.

Council Directive 86/278/EEC of 12. June 1986, on the protection of the environment, and in particular of the soil, when sewage sludge is used in agriculture. Official Journal L 181, 04/07/1986, pp. 6-12.

Das, N.V. & Bandyopadhyay, K. (2001), Selective sequence of adsorption of heavy metals by vermiculite. Asian Environment 13: 13-20.

Fontes, M.P., Matos, A.T. de, Costa, L.M. da & Neves, J.C.L. (2000), Competitive adsorption of zinc, cadmium, copper, and lead in three highly weathered Brazilian soils. Communications on Soil Sciences and Plant Analysis 31: 2923-2958.

Gomes, P.C., Fontes, M.P., Silva, A.G. da, Mendoca, E.S. de & Netto, A.R. (2001), Selectivity sequence and competitive adsorption of heavy metals by Brazilian soils. Soil Science Society of America Journal 65: 1115-1121.

Han, F.X. & Banin, A. (1997), Long-term transformation and redistribution of potentially toxic heavy metals in arid-zone soils. I. Under saturated conditions. Water, Air, and Soil Pollution 95: 399-423.

– (1999), Long-term transformation and redistribution of potentially toxic heavy metals in arid-zone soils. II. Incubation at the field capacity moisture content. Water, Air, and Soil Pollution 114: 221-250.

Han, F.X., Banin, A. & Triplett, G.B. (2001), Redistribution of heavy metals in arid-zone soils under a wetting-drying cycle soil moisture regime. Soil Science 166: 18-28.

Harter, R.D. (1979), Adsorption of copper and lead by A_p and B_2 horizons of several northeastern United States soils. Soil Science Society of America Journal 47: 47-51.

Juste, C. & Mench, M. (1992), Long-term application of sewage sludge and its effect on metal uptake by crops. In: Adriano, D.C. (ed.): Biogeochemistry of trace metals, Lewis Publishers, Boca Raton, pp. 159-193.

Li, X., Coles, B.J., Ramsey, M.H. & Thornton, I. (1995), Sequential extraction of soils for multielement analysis by ICP-AES. Chemical Geology 124: 109-123.

Logan, T.J. & Chaney, R.L. (1983), Metals. In: Page, Al.L. (ed.): Utilization of municipal wastewater and sludge on land. University of California, Riverside, pp. 235-326.

McBride, M., Richards, B.K., Steenhuis, T. & Spiers, G. (1999), Long-term leaching of trace elements in a heavily sludge-amended silty clay loam soil. Soil Science 164: 613-623.

McKenzie, R.M. (1980), The adsorption of lead and other heavy metals on oxides of manganese and iron. Australian Journal of Soil Research 18: 61-73.

Naidu, R., Kookana, R.S., Summer, M.E., Harter, R.H. & Tiller, K.G. (1997), Cadmium sorption and transport in variable charge soils: A review. Journal of Environmental Quality 26: 602-617.

Pepper, I.L., Bezdicek, D.F., Baker, A.S. & Sims, J.M. (1983), Silage corn uptake of sludge-applied zinc and cadmium as effected by soil pH. Journal of Environmental Quality 12: 270-275.

Shuman, L.M. (1977). Adsorption of Zn by Fe and Al hydrous oxides as influenced by aging and pH. Soil Science Society of America Journal 41: 703-706.

Sidle, R.C. & Kardos, L.T. (1977), Adsorption of copper, zinc and cadmium by a forest soil. Journal of Environmental Quality 6: 313-317.

Sipos, P. (2004), Geologic and pedogenic effects on heavy metal distributions in forest soils from the Cserhát Mts. and the Karancs area, NE Hungary. Acta Geologica Hungarica 47: 411-429.

Vermes, L. (1989), A szennyvíziszap hasznosításának Kelet-európai tapasztalatai. Meloriáció, öntözés és talajvédelem 2: 48-66.

Welp, G. & Brümmer, G.W. (1999), Adsorption and solubility of ten metals in soil samples of different composition. Journal of Plant Nutrition and Soil Science 162: 155-161.

Zhu, B. & Alva, A.K. (2003), Differential adsorption of trace metals by soils as influenced by exchange cations and ionic strenght. Soil Science 155: 61-66.

Feasibility Study of Wood-fired Boiler House Construction for Heat and Hot Water Supply of Slavutich City

Konstantin Drozd, Georgiy Geletukha

Scientific Engineering Centre Biomass (Kiev, Ukraine),
2a Zheliabova str, 03057 Kyiv, Ukraine

Introduction

This feasibility study (FS) was performed by SEC Biomass (Kiev, Ukraine) and Sumitomo Corporation (Tokyo, Japan). The performing of FS began at the beginning of February 2005 and finished at the end of May. Sumitomo Corporation and SEC Biomass had meetings with all interested parties, with Slavutich City Mayor and with the representatives of forestry companies that are going to be wood suppliers for the project operation. Consultants have estimated the economical viability of the project implementation. All information obtained from companies, which are interested in project realization have been taken into account in project viability estimations.

Current Situation in District Heating System of Slavutivh City

Heat supply in Slavutich City is provided by the Central Boiler-house (for industrial consumers as well as for the population) and also by small scale industrial boilers installed directly at industrial sites that produce heat for the needs of their own industrial enterprises. Some years ago the Utility Enterprise that owned the existing Central Boiler-house of Slavutich City (see Figure 1) was the monopolist of the heat supply market. But now the decentralization of the heat supply sector is in progress, so a number of private companies have already installed the boilers, and produce heat for their own needs at the moment (Private transport company ATASS, Medical-sanitary area, etc).

The existing boiler-house was constructed in 1987 has three hot water boilers KVGM-50-150 and two hot water boilers DEV-10-14, switched from a steam generating regime to hot water regime. In winter time two KVGM- 50-150 are in operation, and in summer time one or two DEV-10-14 are in operation, depending on heat demands.

Fig. 1: Central boiler-house of Slavutich City and pine forests surrounding Slavutich City

In 2004 the total heat consumption was 156,615 Gcal and 2,295 Gcal was consumed by the boiler-house for its own needs, official losses were 13%. 112,072 Gcal were provided to the population and 44,543 to state enterprises and to commercial industrial consumers (private companies). The pattern of heat supply was as follows: 75% of produced heat was delivered to the population (heat supply tariff is 74,39 UAH/Gcal), 8% of the heat was consumed by industrial consumers (heat supply tariff is 114-116 UAH/Gcal), 15% – to state enterprises (Heat supply tariff is 82,6 UAH/Gcal). The level of payments for heat and hot water supply services by the population was 92%, by commercial consumers it was 100%.

Project Idea

It is planned to construct a wood-fired boiler-house with a capacity that is able to cover summer heat demands (see Figure 2). It is expected that the new wood-fired boiler-house and CHP units (2 MW already installed) will provide nearly 80-90% of the summer heat demand. The other 10-20% of summer heat demand will be covered by the peak natural gas boiler (one of the existing DEV boilers at Slavutich Central Boiler-House. The new wood-boiler will be operated in summer time as well as during the heating season. In such case the capacity of existing gas boilers will be reduced by the amount equal to the new wood-fired boiler capacity. The produced heat energy would be sold to consumers by the existing heat pipe-lines. Project implementation also allows us to reduce CO_2 emissions by switching from natural gas to biomass fuel, to increase heat supply safety, to decrease probability of forest fires, to create new jobs.

Project Participants and their Roles

Slavutich Utility (full name is Utility Enterprise of Housing and Communal Services of the town of Slavutich) was established in 1998. Its activities include water supply, waste water treatment, road service, power distribution, heat generation and supply, wastes utilization. This enterprise is an owner of the heat network, power grid and sewerage of Slavutich. Utility Enterprise will not be able to implement the project using its own funds because of lack of finances. Probably as Utility Enterprise does not have the money to take part in the project implementation, it will take part in project realization by guarantying the site for the project or granting on lease (or maybe free) its heat network to supply heat to consumers, or by providing the trucks for biomass transportation from the forests.

Ukrainian company "LiV ELEKTRA Slavutich" is a private company that is interested in project implementation in the case of its viability. LiV ELEKTRA Slavutich is the new player in the energy supply sector of Slavutich City. There is a program of decentralization of utility services in Slavutich City and it is expected that very soon the tender will be announced for leasing the heat network and water network. It is a good opportunity for LiV ELEKTRA Slavutich to implement a biomass project and to lease the existing heat network and to provide heat to consumers. LiV ELEKTRA Slavutich has intentions to invest some money into the project implementation but it does not want to be the only investor.

SEC Biomass was registered in January 1998 and it is the first and most likely the only private Ukrainian company working in the sector of energy production from biomass (combustion, gasification, pyrolysis, anaerobic digestion, landfill gas extraction and utilization, liquid fuel production). The staff of SEC Biomass are 16 persons (including 5 Ph.D.) now. The main activities are consultancy, R&D, feasibility studies, identification of sites, and participation in implementation projects in the field of bio-energy technologies. Since the Kyoto protocol has been in force and Ukraine may take part in emissions trading, SEC Biomass has started to provide customers with consultancy services focusing on joint implementation projects. At the moment SEC Biomass is one of the experienced companies in this field in Ukraine. A number of JI projects have already been identified, prepared and submitted to several International JI/CDM programs (Austrian, Dutch, German, Japanese, Belgium) by SEC Biomass. In this project SEC Biomass has taken part as a consultant and has prepared a feasibility study together with Sumitomo Corporation for Japanese JI/CDM program.

Sumitomo Corporation. By the order of Japanese government Sumitomo Corporation performs the feasibility studies of biomass energy generation installation implementation in Ukraine. All the projects must be directed for the reduction of CO_2 emissions. Sumitomo Corporation is going to purchase emission reduced units and to invest some share of investments depending on the profitability of the project.

Wood Fuel Supply for the Project

As Slavutich City is surrounded by the forests of rather a large area, there are good possibilities to supply biomass for the new boiler-house operation. There are a good number of forestry facilities and some sawmills close to the Slavutich City. We have estimated the possibilities of wood supply from 4 forestry companies. The results of investigations are summarized and presented in Table 1. As the Forestry companies and sawmills have no experience of permanent delivering of large quantities of wood (the consumers comes to get the wood with its own transport, and the rest of the wood residues are combusted directly in the forests), they have not provided us with data about transportation prices. But transportation cost has been calculated based on the company ATASS's offer. In such a case transportation cost for 20 km distance is 6,2 UAH/m^3, and transportation cost for 50 km distance is about 16 UAH/m^3. The price of firewood is from 25 UAH/m^3 (Dobriansk Leshoz) to 41-44 UAH/m^3 (other forestry companies). The price of residues is from 0.5 UAH/m^3 (Dobriansk Leshoz) to 20 UAH/m^3 (Chernigivleshos Company

Tab. 1: Potential of forestry companies to supply wood fuel for new biler-house

Company	Area [hectares]	Fuel firewood [m^3]	Technological firewood [m^3]	Wood residues [m^3]
Dobriansk Leshoz	41,092	1,600	1,800	6,490
Company Chernigivleshoz	35,449	22,000		4,120
Chernigiv Military Forestry	27,596	6,000		700
Chernigivraiagroleshoz	114,652	31,400	1,800	11,310

Technology to be Applied

As a basic scenario we have estimated the installation of a 5 MW reciprocating-grate hot water boiler manufactured by Polytechnik Company (Austria). Also we have analyzed the variants of 3 and 4 MW Polytechnik boilers. We have chosen the 5 MW project as the basic scenario due to existing heat supply patterns in summer time and due to its better economical indexes despite LiV Elektra not being very interested in the installation of more than a 3 MW boiler. We based our estimations on universal commercial offers we received from Polytechnik directly for this feasibility study (www.polytechnik.com)

Expected Project Operation Performance

For a 5 MW wood-fired boiler it is necessary to provide about 16,940 t (or 24200 m^3) of wood fuel. Calculations have been done assuming that density of wood is 700 kg/m^3, moisture content is about 40% and low heat value is 10.5 MJ/kg, annual boiler-house load rate is 8400 hours, boiler efficiency is 85%. The new boiler-house will generate about 36,000 Gcal of heat per year and taking into account 13% losses that take place in Slavutich DH system, about 32,300 Gcal of heat will be supplied to consumers. Implementation of the new biomass boiler-house will reduce the consumption of natural gas by 4,630 thousands m^3 yearly. That will allow a reduction correspondingly in CO_2 emissions by 9,220 t/a.

Project Profitability Calculations

Investments costs where calculated based on commercial offers obtained during FS and experience of SEC Biomass. A 5 MW Polytechnik hot water fire tube boiler with moving step grate furnace costs 525,000 €, pumps (main and reserve) and chippers (also main and reserve) total cost is 15,000 €, boiler-house building and fuel storage building construction costs are 40,000 €, transportation cost is 10,000 €, assembly and start up – 42,000 €, Ukrainian design and certification costs – 20,000 €. Taking into account contingencies as 30,000€, the total project investments are 685,000 €. This figure was calculated without consideration of VAT and custom fees that generally must be paid if imported equipment is going to be installed. But in Slavutich a Special Economical Area was in force while the FS was being performed. Thus if the project is implemented through the Free Economical Area it is possible to avoid VAT and custom fees.

Operational costs consist of wood fuel costs, chipper, pump, and boiler reciprocating grate operational costs (power must be purchased, tariff is 41€/MWh), wage payments, maintenance costs. Total operational costs are 244,000 €/a including the biggest share – wood fuel costs – 175,000 € (price for wood fuel is 7.25 €/m^3).

Annual revenue was calculated based on the following prices and tariffs: tariff for supplied heat is 11.6 €/Gcal (determined based on consumers pattern – 75% – population, 15% – state enterprises, 10% – commercial enterprises), Emission Reduction Unit price is 6.0 €/t CO_2 eq. (payments for ERUs are expected during commitment period 2008-2112). Annual revenue is expected to be 430,000 € with carbon payments and 375,000 € without carbon payments.

Fig. 2: IRR of the project for different capacities depending on heat supply tariff[1]

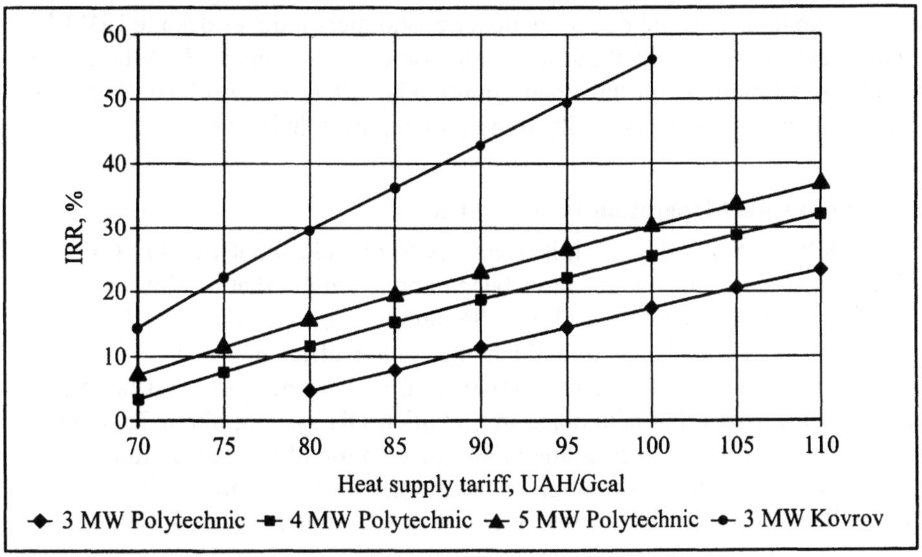

Economical indexes were calculated taking into account a discount rate of 10% and assuming that 50% of investments are loan investments (loan conditions: 5 years, interest rate is 10%). Also the sensitivity analyzes was carried out. The main reason for that was the possibility of increasing natural gas prices. At the moment the Ukraine government is in the process of negotiations with the Russian government concerning further natural gas supplying. Russia is interested in a three times natural gas price increase (from 60 up to 180 \$/1000 m^3). As natural gas

1 The option of Kovrov boiler (Russia) was considered as alternative one to Polytechnik boilers by request of LiV Elektra company.

prices increasing will lead to corresponding heat supply tariff we have evaluated project viability depending on heat supply tariff. The results are presented at Figure 2, 3.

Fig. 3: Payback period of the project for different capacities depending on heat supply tariff

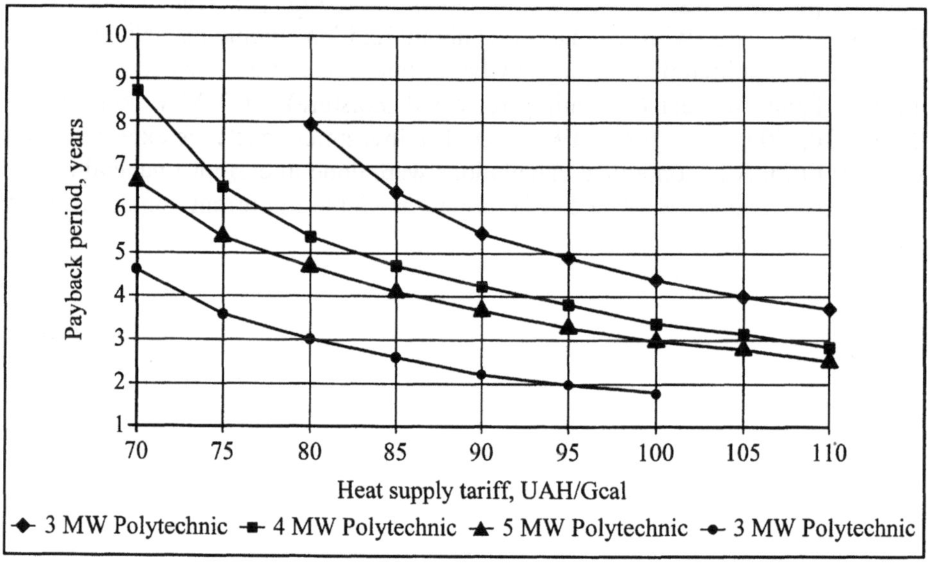

Conclusions

It is expected that the new boiler-house will have a capacity of not more than 5 MW (3-5 MW) due to the summer heat demand pattern. The new boiler-house with the peak gas boiler (one of the existing DEV Boilers of 6 Gcal/h capacity) and CHP already installed will cover all heat demands in summer time and substitute part of the existing boiler-house load in winter. We estimated different variants of capacity to be installed (3,4,5 MW Polytechnik boilers, 3 MW Russia Kovrov Boiler). So as independent consultants we recommend a Polytechnik boiler to be installed in this project despite the fact that Kovrov boilers are cheaper making the project economically quite attractive. But Polytechnik boilers are fully approved in operation and have modern control and automatic systems. An additional strong point of Polytechnik boilers is that they have the following advantages: completely automatized ash removal and ash handling system and tube cleaning system. We have chosen as the basic scenario the implementation of 5 MW Polytechnik boiler as this variant is most suitable for current summer heat demand and has most attractive economical indexes. The 5 MW project will allow them to produce 36,086 Gcal per year and deliver to consumers 32,307 Gcal per year of heat (taking to the account 13,0% losses). The project implementation also will reduce the CO_2 emissions by 9,220 tons per year. In the case of ERU price 6,0 Euro/t, it would be possible to realize an additional profit at the rate of about 276,000 Euro during 2008-2012 that covers about 40% of total investments. Four forestry companies in Slavutich region were questioned about the possibility of supplying wood and wood residues. The total biomass potential was estimated to be about 42,000 m^3 per year of wood residues and firewood (with average price about 7.25€/m^3 including delivery cost). It should be admitted that for the new boiler house operation it is necessary from 14,000-18,000 m^3/year (depending on wood moisture) for 3 MW boiler up to 24,000-30,000 m^3/year for 5 MW boiler. The investment costs are expected to be 685,000 Euro. All economical calculations were done in current financial conditions, tariffs, and prices. The 5 MW project is viable in current conditions with CO_2 investments.

Ambient Air Quality Mapping for Integrated Urban Traffic Flows Management

L. Kliučininkas, D. Martuzevičius, J. Mockuvienė

Kaunas University of Technology,
Department for Environmental Engineering
Radvilenu 19, LT – 50254 Kaunas, Lithuania

Abstract

The study addresses issues related to assessment of environmental impacts from urban transport systems in order to provide decision makers with decision support tools for sustainable traffic flows managent.

The investigation gives an example of seasonal air quality maps formation. The formation of maps integrates air quality monitoring data, intensity counts of traffic flows and estimations of meteorological conditions. The AIRVIRO v. 2.20 software is applied to make modeling for chosen weather conditions. To address traffic flows management and minimize air pollution in the Kaunas city, ambient air quality modelling/mapping based on the concept of scenarios has been applied. User-defined scenarios are defined in terms of an appropriate transportation infrastructure.

1. Introduction

Rapidly growing urban traffic flows have been identified as major contributors to the ambient air quality in a number of medium and large size cities. Many cities in Central Eastern European (CEE) countries have faced consequences of a substantial increase of private cars and crisis of public transport through the last decade. In most cases, the used cars from Western Europe were imported into the CEE countries.

The central part of Kaunas city, where administrative, commercial, and cultural activities are performed, is heavily affected by the traffic flows. The traffic jams became an endemic feature of the city centre, due to the insufficient street capacity. The majority of the 3-5 storage buildings situated in the city centre make up unfavorable conditions for distribution of the pollutants.

Integration of data on traffic flows and air quality by modeling/mapping systems is considered an efficient instrument in transport systems planning. The major objective of such systems is to provide analytical results to the urban planners and decision makers in order to meet current needs of city development. Mapping and scenario results can be compared with air quality limits. Impact assessment of traffic air pollution abatement measures can also be carried out (Jensen et al.

2001). The probabilistic mapping technique has been applied to a review of air quality in central London in 1996/97 and an assessment of future air quality against air quality objectives in 2005 (Colvile et al. 2002). The British authors report on development of a program designed to provide detailed estimates of vehicle emissions on urban road networks, and so act as a precursor to urban air quality modelling (Namdeo et al. 2002). The programme includes emission and air quality mapping, evaluation of associated transport policies and scenarios, and preparation of inputs to other (e.g., epidemiological) models.

2. Measurements and Basic Data Properties

Six sampling sites (see Figure 3) were selected to carry out measurements relevant to modeling the impacts of traffic flows on ambient air quality. The sites were situated in a radius of 2.5 km to examine the spatial pollution variations within the neighborhoods. The measurement results from five sampling sites (No. 1-3,5,6) were obtained using a mobile laboratory, the sampling site No.4 represented a permanent monitoring station. To conform with EU siting criteria, sampler inlets were placed at least 2.3 m above ground level. The sampling sites were primarily selected to obtain representative measurements of the study area; other considerations included the security of equipment and the availability of an uninterrupted power supply.

Fig. 1: The location of the sampling sites

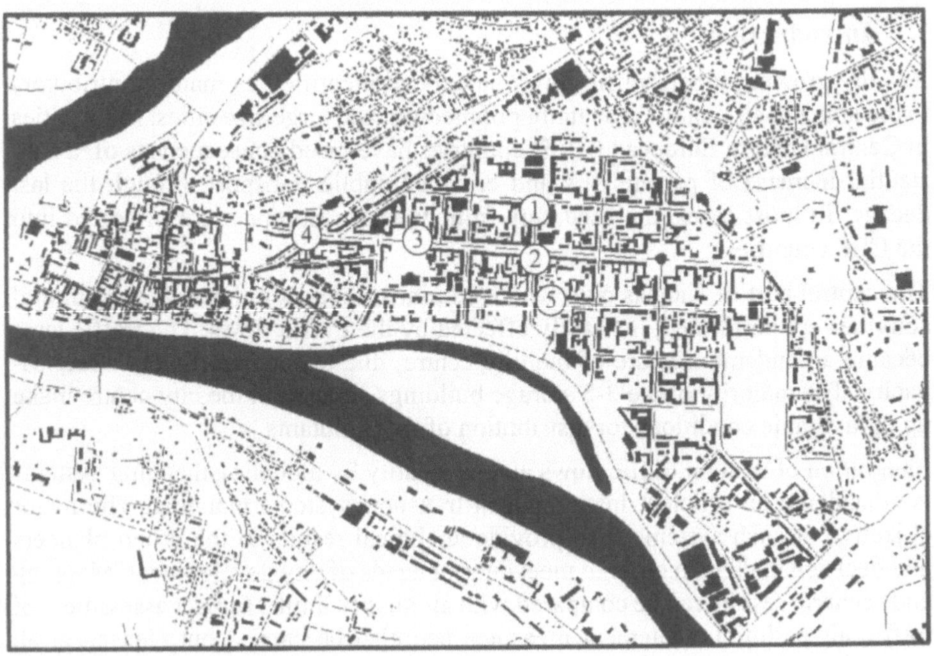

The measurement installations from Environnement S.A. (Paris-Poisy, France) composed chemiluminescent $NO/NO_X/NO_2$ analyzer – model AC31M and IR gas filter correlation CO analyzer – model CO11M. The measurement error for AC31M analyser makes up 0.17 ppb, for CO11M – 0.05 ppm respectively. *In situ* measurements for selected sites were conducted during the working days from 9.00 AM to 5.00 PM. Resultant *in situ* data is obtained from the hourly averages of the continuous measurements. The 24-hour continuous sampling of data from the stationary monitoring station was averaged on a 1/2 hour basis. Table 1 presents descriptive features of the measurement sites in respect to degree of impact.

Tab. 1: Descriptive features of measurement sites in respect to degree of impact

No	Site	Type of impact
1	University, Main building	35 meters away from K. Donelaicio street. Moderate impact from the street
2	University, Faculty of Management	160 meters from Kestucio street. No direct impact from the street
3	Municipality	165 meters from K. Donelaicio street. No direct impact from the street
4a	Monitoring station *Centras*	12 meters from the street. Strong impact from the street.
4b	Bank	4 meters from St. Gertruda street. Very strong impact from the street
5	Police	4 meters from Kestucio street. Very strong impact from the street
6	Fire brigade	4 meters from King Mindaugas street. Very strong impact from the street

Estimation of urban background concentrations of NO_X and CO was performed using data series from *Silainiai* stationary monitoring station. The station was situated in the highly populated district of Kaunas city and was not directly influenced by emissions from transport and industries. Urban background concentration of NO_X was attributed – 15 $\mu g/m^3$, respectively urban background concentration for CO was estimated 0.05 mg/m^3.

3. Formation of Seasonal Ambient Air Quality Maps

3.1 Model

The Airviro software developed by the Swedish Meteorological and Hydrological Institute (Airviro 2.20) integrates different functions required in Air Quality Management. In the central part of Kaunas city, the dominant part of the air pollution is induced by the street traffic emissions. Emissions from the streets were assumed evenly distributed point emissions along the lines and were a good approximation to the real situation (Russel et al. 1983). In order to understand how air quality in the central part of Kaunas will be affected by changes in meteorological conditions, authors have applied the following Airviro functional blocks – Emission Database, Dispersion Modeling and Indico data analysis and presentation package.

The Traffic Emissions Database comprises data on traffic speed, traffic intensity and variations as well as information on the types and proportions of vehicles operating on that street. Consequently, the variations as well as proportions of different vehicles on a street could be defined by the structural scheme presented within Figure 2.

Fig. 2: Structural scheme of traffic emission database

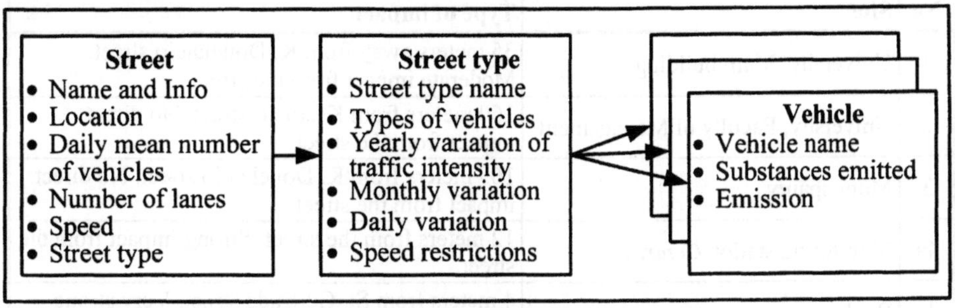

For dispersion modeling authors have applied the Gauss model, based on a Lagrangean Gaussian formulation. For linear emission sources (streets) introduced individually, the horizontal dispersion coefficient σ_y is set to

$$\frac{dxy}{2},$$

where

$$dxy = \sqrt{width^2 + length^2} \tag{1}$$

However, in built up areas where the Gauss model calculations are evaluated at roof level, σ_y is set to 0.667* building height.

The vertical dispersion coefficient σ_z is defined with an empirical formula and in open areas is set to 3.57-0.53 U, where U – horizontal wind velocity (Airviro 1998).

3.2 Meteorological Conditions

When forming the seasonal pollution dispersion maps, the meteorological conditions were considered driving variables of the model. The complex meteorological observations were performed in the vicinity of Kaunas.

The wind field calculations are based on the concept first described by Danard, where small scale winds can be seen as a local adaptation of large scale winds (free winds) due to local fluxes of heat and momentum from the sea or earth surface. The two most important parameters for the scaling of the atmospheric stability and turbulence in the surface boundary layer are the vertical flux H and the friction velocity u. From these, a characteristic length scale – the Monin-Obukhov length is calculated:

$$L = \frac{\bar{T}}{g} * \frac{u^3 \rho c_p}{\kappa H} \tag{2}$$

where $\bar{T}$ is the average surface air temperature, g is gravity, κ is the von Karman constant and ρc_p is the specific heat capacity. In the Airviro dispersion model, the Monin-Obukhov length is extensively used as a discriminator of different meteorological regimes (Panofsky et al. 1984).

The boundary layer stability is estimated in the form of calculated 50 and 98 percentiles of Monin-Obukhov length. The calculated Monin-Obukhov length (L) was used to classify the stability conditions in the boundary layer (see Figure 3). Corresponding computations for chosen weather and emission scenarios were carried out.

Fig. 3: Essential meteorological parameters used to develop seasonal air quality maps

Season	Percentile	Monin-Obukhov length [m]	Wind direction [degree]	Mixing height [m]
Cold	50	2222	240	1666
	98	7	217	24
Warm	50	1786	264	840
	98	7	260	24

Emission Factors for Vehicles

To calculate emissions produced by the exhaust systems of road vehicles, the authors have applied EMEP/CORINAIR emission factors. Emission factors were worked out by the members of the CORINAIR Working Group (Eggleston et al. 1993), taking into account the results of comprehensive studies carried out in France, Germany, Greece, Italy, the Netherlands and the United Kingdom. In addition, some data measured in Austria, Sweden and Switzerland was incorporated.

Street Traffic Flows

The street traffic flows analysis was performed by the Transport and Road Research Institute. The study has involved measurements of traffic intensity, composition of flows and vehicle speed (TKTI 2001).

Analysis of the data sets revealed that the average traffic intensity (ATI) for the central part of Kaunas city based on a 24-hour time series, with the exceptional case of summer vacation period, is rather stable throughout the year. Analysis of street traffic flows in respect to the driving speed has showed that the characteristic speed ranges between 40-60 km/ h. The stationary counter-classifier Marksman 660 (Golden River Traffic, United Kingdom) was used to identify 24-hour yearly flow averages in the different vehicle classification groups.

4. Mapping Ambient Air Quality Scenarios

Scenarios were applied to the central part of Kaunas city and an estimated presence of new M. K. Ciurlionis bridge: reference scenario (not developed traffic infrastructure around the bridge), scenario with re-distributed traffic flows (developed traffic infrastructure around the bridge). The predictive mapping was based on perspective data of traffic flows elaborated by municipal enterprise "Kauno planas".

The initial NO_x and CO concentration mapping (scenario 0) was based on recent traffic flows data. The regressive data analysis was used to verify the model. The linear regression equation $y = 0.9589x + 4.2318$ for NO_x, and respectively equation $y = 0.118x + 0.2034$ for CO, were used to correct the model output data. Figures 4 and 5 illustrate distribution of NO_x concentrations.

Fig. 4: Distribution of NO_x concentration: present situation

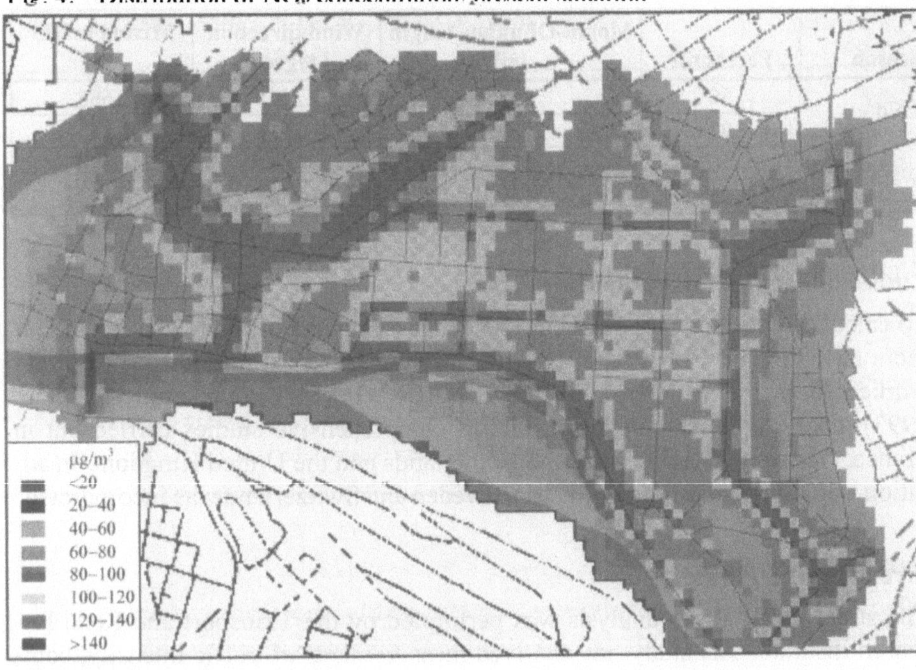

The comparative studies of NO_x maps for the warm (from April 1 to September 30) period revealed that a major part of the investigated area falls into the concentration interval 20-60 µg/m³. The NO_x concentrations in the close vicinity of the main streets comprise 100-120 µg/m³. Due to the considerable distances from the main streets, the areas with high concentration of pedestrians are not effected by the enlarged concentrations of NO_x.

Fig. 5: Distribution of NO_x concentration: scenario with predicted traffic flows

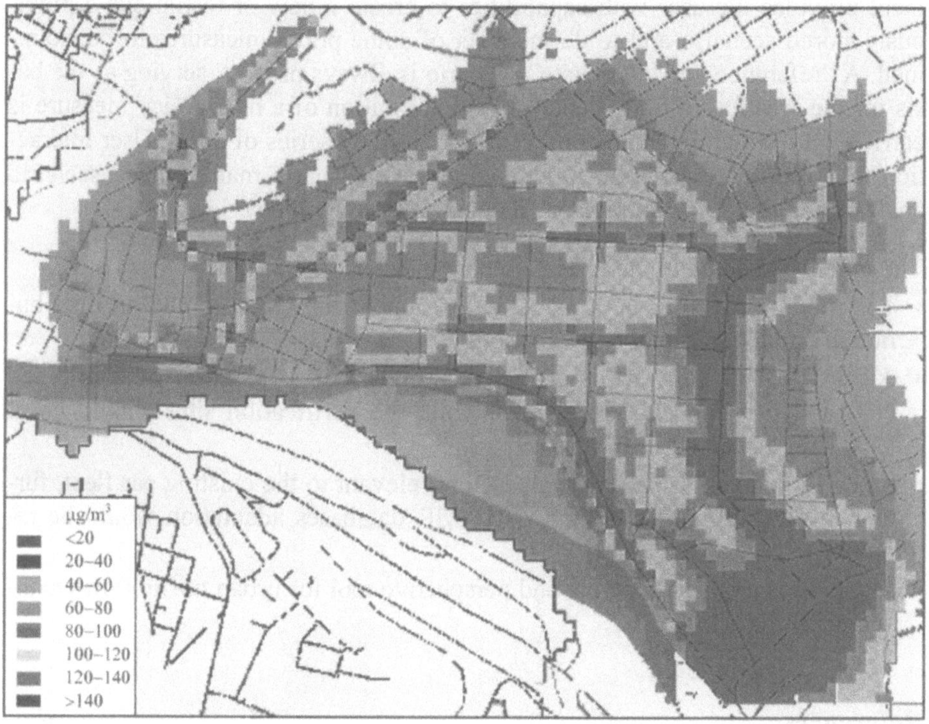

The trends of NO_x concentration levels for re-distributed traffic flows have showed that in the vicinity of the main streets the concentrations would decrease to 80 – 100 µg/m3, in the area around M.K. Ciurlionis bridge concentrations would rise from 70 to 100 µg/m3. At the same time the high concentration spot would cover a bigger area. This will also have an effect on NO_xµg/m3 concentration increase at the streets directly connected to the bridge.

As the verification of the model showed, the hourly street concentrations could considerably differ from those seasonal concentrations modeled by the authors, especially on the street crossings. Therefore, the above considerations should be estimated when interpreting the ambient air pollution maps.

Discussion and Conclusions

Scenario comparison, with the emphasis on the changes in emissions and exposure are often of far more value in a planning context than absolute values. The magnitude and location of change can convey information to planners and decision-makers, highlighting the consequences of trends and of planned actions, and focussing attention both on where interventions might have the greatest effect and on where flexibilities may exist (Affum et al. 2003). Scenario management provides the user with capabilities to create a new or to retrieve a previously stored scenario and to define a set of some policy measures to be examined. A "reference" or "zero-state" scenario is always present, serving as the basis for the creation of new scenarios. The definition of a new policy measure is carried out through interactive procedures based on series of maps. User interaction falls into three different functional groups: scenario management, scenario analysis and scenario comparison (Kliucininkas et al. 2005).

The others had come to the following conclusions:

- The model proved to be a valuable instrument for spatial pollution distribution in ambient air quality evaluation;
- Airviro V.2.20 software is sufficiently precise in evaluation of distributed NO_x and CO concentrations, meanwhile the verification sites should be at least 25m distance from the street;
- Since Lithuania has no emission factors relevant to the existing car fleet, further investigations of EMEP/CORINAIR databases adaptation would be required;.
- Mapping is a quick, accurate and perspective tool for urban traffic flows management.

References

Affum, J.K., Brown, A.L. & Chan, Y.C. (2003), Integrating air pollution modeling with scenario testing in road transport planning: the TRAEMS approach. The Science of the Total Environment, vol. 312, issue 1-3, p.1-14.

Colvile, R.N., Woodfield, N.K., Carruthers, D.J., Fisher, B.E.A., Rickard, A., Neville, S. & Hughes A. (2002), Uncertainty in dispersion modeling and urban air quality mapping. Environmental Science & Policy. 5, 207-220.

Jensen, S.S., Berkowicz, R., Hansen H.S. & Hertel, O. (2001), A Danish decision-support GIS tool for management of urban air quality and human exposures. Transport Research Part D. 6, 229-241.

Kliucininkas, L., Balkeviciene, S. & Mockuviene, J. (2005), Modelling of traffic policy measures for ambient air quality management. Proceedings of the International Conference Environment, Technology, Resources. Rezekne, 187-207.

Namdeo, A., Mitchell, G. & Dixon, R. (2002), TEMMS: an integrated package for modeling and mapping urban traffic emissions and air quality. Environmental Modelling & Software. 17, 179-190.

Panofsky, H.A. & Dutton, J.A. (1984), Atmospheric Turbulence. John Wiley & Sons. Eggleston, S., Gaudioso, D., Girisen, N., Jourmard, R., Rijkeboer, R.C., Samaras, Z. & Zierock; K.-H. (1993), CORINAIR Working Group on Emissions Factors for Calculating 1990 Emissions from Road Traffic, vol. 1: Methodology and Emission Factors. Final Report, Document of the European Commission.

Russel, A., McRae, G.J. & Cass, G.C. (1983), Mathematical modeling of the formation and transport of ammonium nitrate aerosol. Atmospheric Environment. 17, 949-964. Airviro User Documentation. Swedish, 1998. Hydrological Meteorological Institute, Norrkoping, Sweden.

Transporto priemoniu eismo intensyvumo skaiciavimas automatiniais eismo intensyvumo skaitikliais-klasifikatoriais (Trafic intensity measurements using automatic counters-classifiers). Report, 2001. Transport and Road Research Institute, Lithuania.

Development of Ecologically Friendly Technology for Gasification of MSW

Maxim Slenkin, Nikolay Zhovmir

Institute of Engineering Thermophysics of NASU,
POB 78, 01010 Kiev, Ukraine

Ukraine as well as other countries all over the world has a problem with municipal solid waste utilization. One of the most popular technologies is combustion of MSW. However, MSW combustion cannot provide permissible level of harmful matters emission without very difficult and expensive gas cleaning systems. This especially applies to polychlorinated dibenzo-(p)-dioxins and furans – global superecotoxicants, which have strong mutagenic, immunosuppressive, carcinogenic, teratogenic and embryotoxic activity. Dioxins are always generated with the presence of oxygen, chlorine and organic compounds at high temperatures (above 300°C). In the waste chlorine is mainly present in such compounds as kitchen salt and polyvinylchloride.

Existing technologies of MSW combustion do not decrease dioxin emissions with 100% effectiveness. The most powerful method of dioxin formation suppression is based on deactivating the potential of its formation precursors, such as: chlorine, oxygen, and catalysts. One of the possible technological approaches is gasification of chlorine-containing fuels or wastes: absence of oxygen effectively blocks dioxin formation routes. Product gas cleaning from ashes and halogens eliminates dioxin reformation after combustion. In the Institute of Technical Thermophysics the experimental fluidized bed installation for chlorine containing waste gasification is under development.

Dioxins and Furans

Dioxins – global superecotoxicants, which have strong mutagenic, immunosuppressive, carcinogenic, teratogenic and embryotoxic activity.

- They are crystal substances with high melting temperatures (200-400°C).
- They are easily dissolved in organic solvents, fats, not distilled water.
- They have strong adhesion properties.
- The period of degradation in soil is 10-20 years, in water up to 2 years.
- They have high thermal resistance. Thermal degradation under temperature less than 1250°C is reversible.

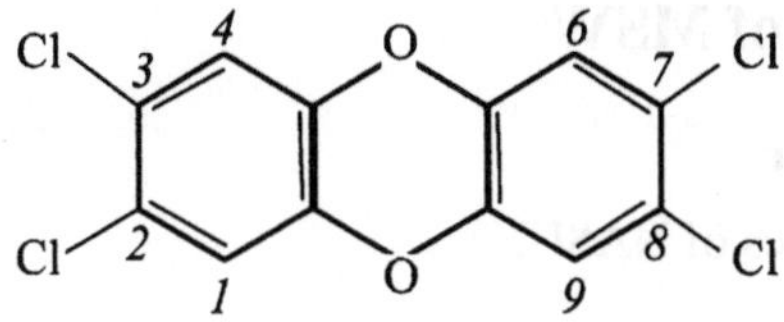

Dioxin
2,3,7,8 tetrachloro dibenzo-p-dioxin

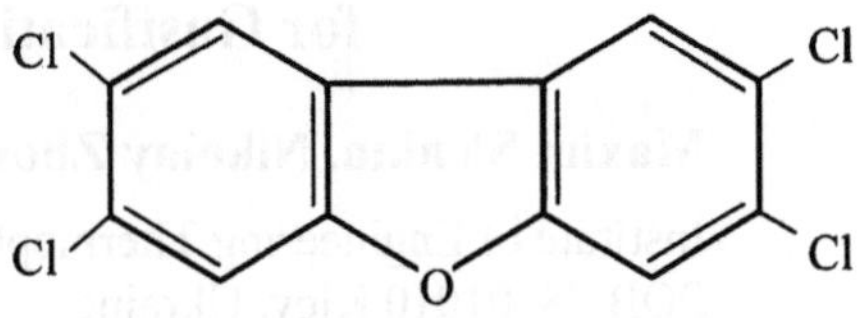

Furan
2,3,7,8 tetrachloro dibenzo-p-furan

The Causes of Decreasing of Dioxins and Furans Emission by Gasification Process

In the case of combustion dioxins they are always generated in presence of oxygen, chlorine and organic compounds at high temperatures (above 300°C).

Decon reaction is one of the main stages in the process of dioxins generation:

$$2HCl+1/2O_2 \Leftrightarrow Cl_2 + H_2O$$

The reaction happens in the presence of Cu, Pb, Sn catalysts and Cl_2 formation.

$$Cl_2+O_2+C_xH_y + \text{time and } 250\text{-}350°C = \text{DIOXINS/FURANS}$$

The main idea of dioxin formation suppression is based on deactivating the potential of its formation precursors, such as: Cl_2, O_2 or catalysts.

Gasification of chlorine-containing fuels or wastes in a reduction atmosphere leads to formation of HCl. Product gas cleaning from ashes and HCl eliminates dioxin reformation after combustion.

Gasification Unit with two Fluid Bed Reactors

The basic idea of the gasifier concept is to divide the fluidized bed into two zones, a gasification zone and a combustion zone. Between these two zones a circulation loop of bed material is created but the gases should remain separated. The circulating bed material acts as heat carrier from the combustion to the gasification zone. The principle is shown graphically in Figure 1.

Fig. 1: The principle of the gasifier concept.

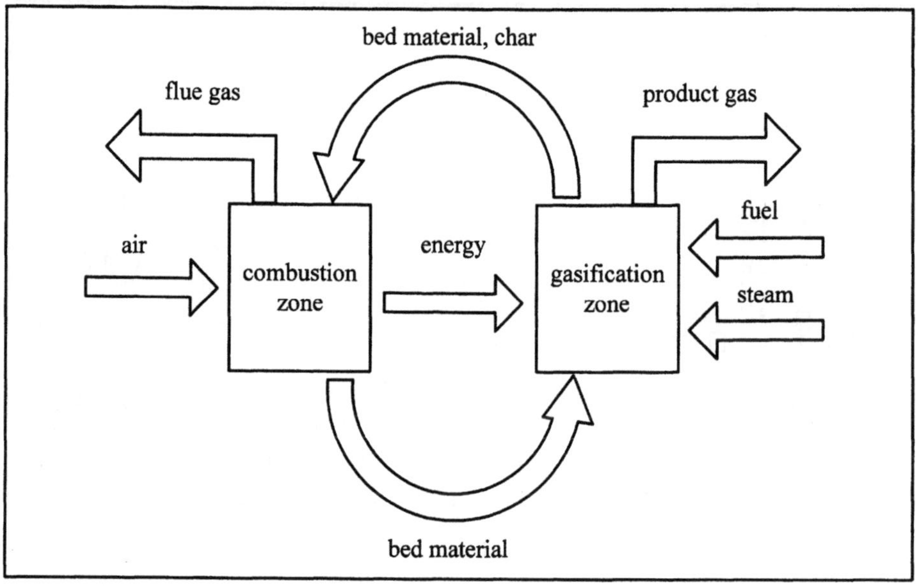

General Description of the Unit

The MSW gasification unit which is under development in the project consists of two fluid bed reactors (Figure 2). The first reactor is a gasifier itself, the second reactor is a combustion chamber for charcoal. To obtain producer gas of middle calorific value water steam is applied as a blowing. The fluid bed is organized by supplying water steam to the gasifier (inert material is sand) and air to the combustion chamber. The installation is equipped with all necessary devices to measure rate, temperature and pressure.

Fig. 2: General flowchart of two fluid bed gasification unit

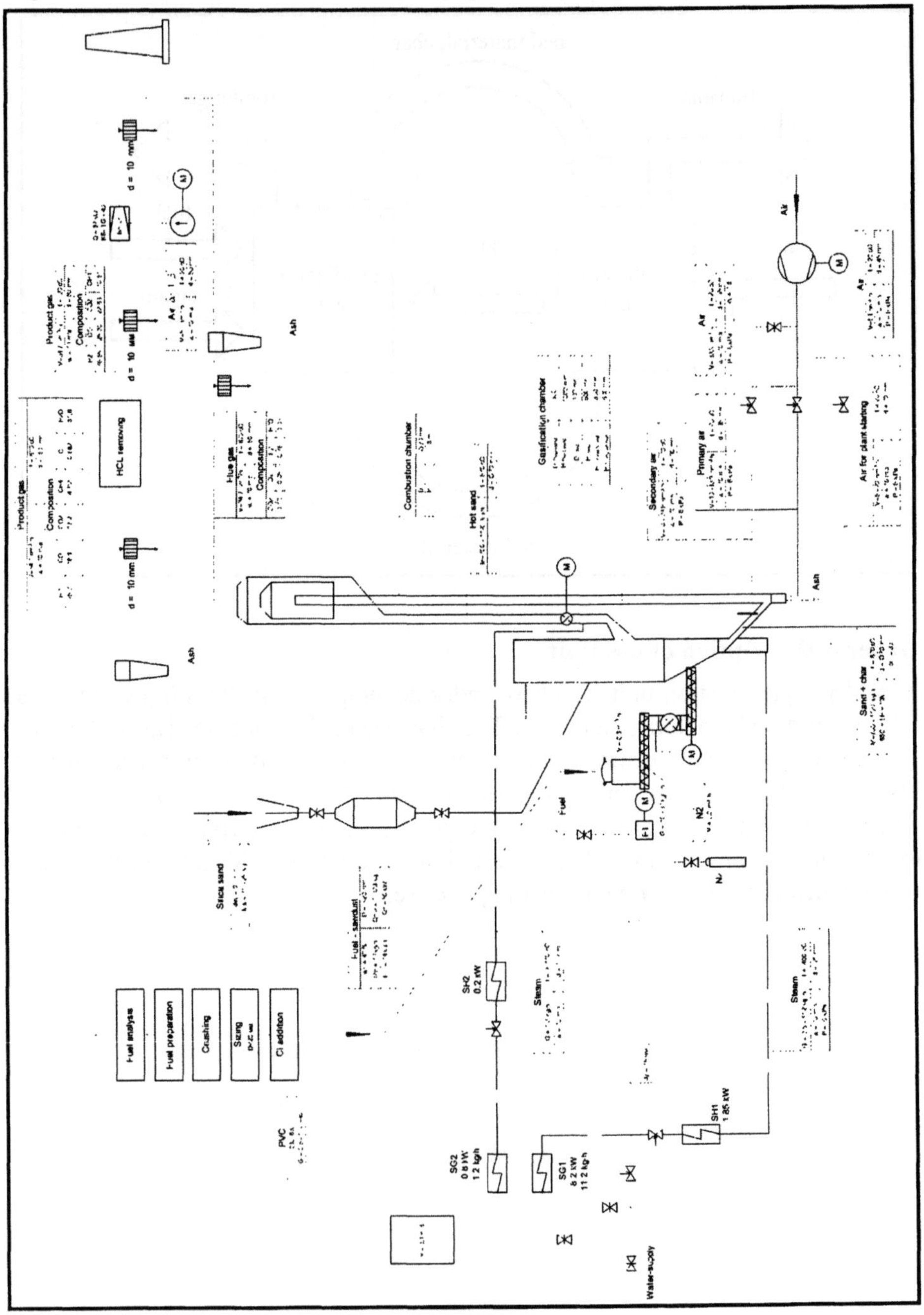

Feature of the Experimental Unit

Heat capacity – 50 kW;	Gasifier chamber:
Fuel – Model fuel, separated MSW;	diameter – 300 mm
Bed material – silica sand	height – 1500 mm
Bed weight – 40 kg	Combustion chamber:
Mean size of solid material – 0.6 mm	diameter – 70 mm
Fuel input- 19 kg/hour	height – 3800 mm

Information about Project

Authors: Institute of Engineering Thermophysics, SEC "Biomass".

Patent: under preparation.

Present status: detail design is developed.

Support of Science and Technology Centre in Ukraine (Project Agreement N 3036, $ 118,830).

October 2004-September 2006

Collaborators: Pacific Northwest National Laboratory (USA); Biomass Technology Group BV (the Netherlands).

Field of application: utilization of MSW in urban and rural areas.

Further steps: manufacture of production prototype; scaling-up.

Chloride and Sodium Concentrations in Snow Cover on Roadsides of Highway Panevėžys – Šiauliai in Lithuania

Agnė Kazlauskienė, Pranas Baltrėnas

Department of Environment Protection,
Vilnius Gediminas Technical University,
Saulėtekio al. 11, LT – 10223 Vilnius-40, Lithuania

Summary

This paper presents data on chloride and sodium concentrations in the snow cover on the roadside of the highway Panevėžys – Šiauliai.

1. Introduction

The roads of the Republic of Lithuania are maintained in winter conditions (almost 5 month of the year). It is caused by geographical position and climate. Snow sweeping, spreading of powdery and ice–dissolvent substances and their mixtures on carriageway are the main jobs to ensure road safety. Chemical substances often used for pavement maintenance are salts (chlorides). Usage of chlorides is a direct and indirect pollution of the environment and has a negative influence on ecosystems (soil, groundwater, surface water, vegetation and fauna) [1, 2].

2. Methods

The aim of the research is to find out the concentrations of chloride and sodium in the snow cover on the roadside of the highway Panevėžys – Šiauliai in Northern Lithuania. To analyze the distance that Cl^- and Na^+ spread away form the road and the distance at which the maximum amount of chloride and sodium accumulates.

Snow samples were taken, cross-sectioning them. A cross-section consists of 7 points: Point 1 is located immediately at the edge of the carriageway, Point 2 is located 1 meter away from the carriageway, Point 3 is located 2 meters away and Point 4 is located 4 meters away and so on. Thus, this enables the determination of the alteration in the distribution of chloride and sodium concentrations transverse to the road.

Chloride (Cl^-) concentration in the snow melt water is found out by applying the titrimetric method [3]. Flame photometry (analyzer ПАЖ-2) was used for sodium (Na^+) concentration detection. The molar concentration of hydrogen ions was measured by pH-meter WTW pH 538.

3. Research Results

The highway Panevėžys – Šiauliai was chosen for the determination of chloride and sodium concentrations in the snow cover. This highway is a national road of Lithuania. The length of the highway is 78,94 km, its marked as A9. The highway is also marked as E272, which means that the highway is a part of the network of European roads. The E272 highway connects Vilnius – *Panevėžys – Šiauliai* – Palanga – Klaipėda. This highway has a protective zone of 70 meters. The highway A9 crosses a preserved area of Lithuania – Daugyvenė landscape reserve (area 3870 ha, established in 1992). The highway Panevėžys – Šiauliai in winter is maintained according to the 2nd maintenance level. Snow sampling was carried out on January 27, 2004 at 4 spots.

Fig. 1: Distribution of chloride concentration in transverse profile in Berčiūnai (Panevėžys)

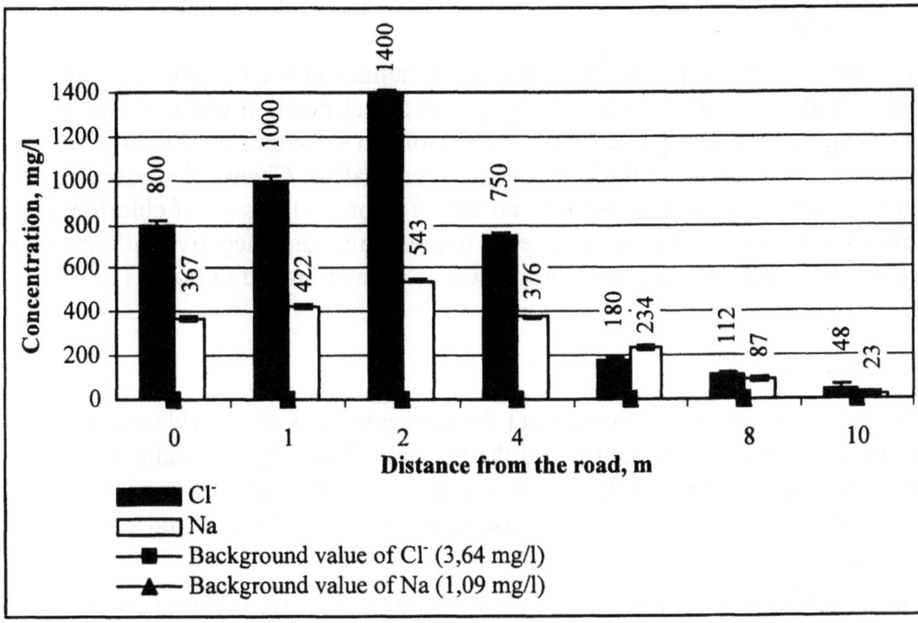

Receding from the road along the cross-section chloride and sodium concentrations are spread very unevenly. The highest chloride and sodium concentrations are found 2 meters away from the road and reach 1400 mg/l and 543 mg/l. Such uneven distribution of chlorides is determined by the flows of the passing vehicles, which cause windflows and precipitates the salt away from the carriageway. The acidity of the snow melt water used for the determination of Cl^- and Na^+ concentrations, was pH = 7,12-7,54. The snow covering at the spot of sampling was 15 cm.

Fig. 2: Distribution of chloride concentration in transverse profile in Šeduva

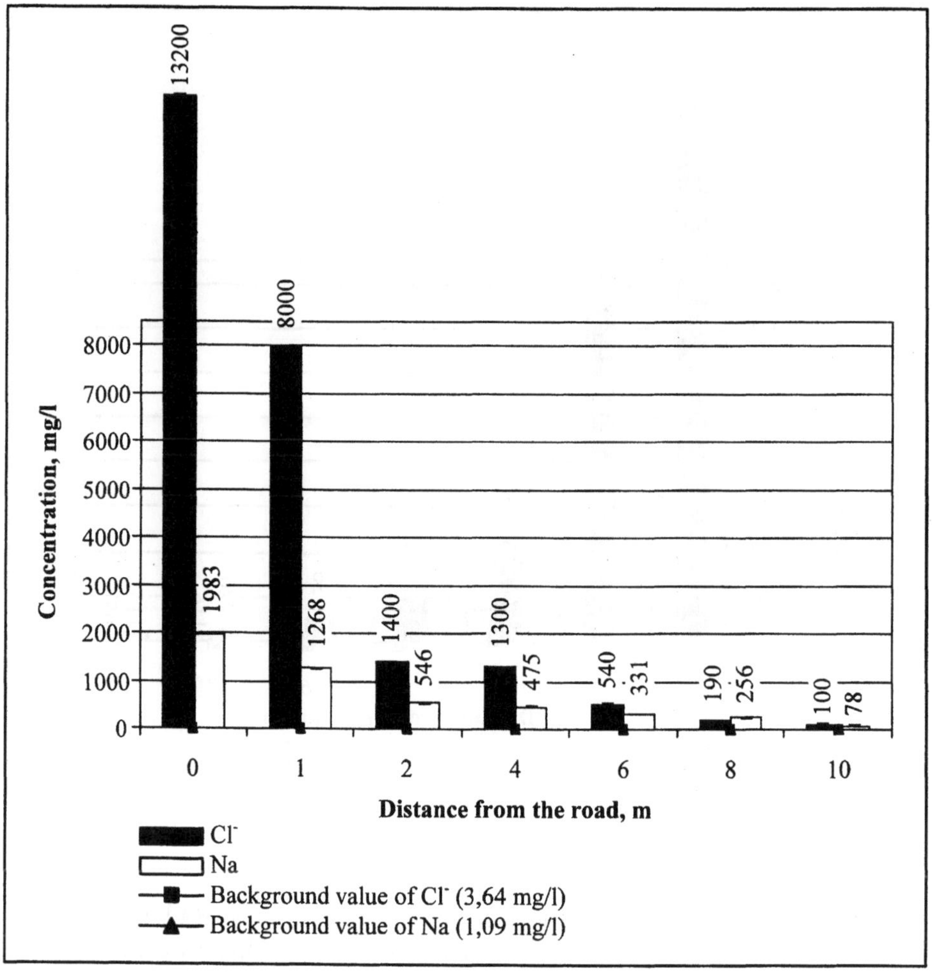

Fig. 2 shows that chloride and sodium concentrations receding from the road along the cross-section distribute depending on the distance, i.e. it tends to go down. At the first spot near the road the concentrations of chloride and sodium is highest and reaches 13200 mg/l and 1983 mg/l. Such concentrations are the highest of all the concentrations in snow cover found. pH = 7,45-7,49. The snow covering at the spot of sampling was approximately 14 cm.

Fig. 3: Distribution of chloride concentration in transverse profile in Radviliškis

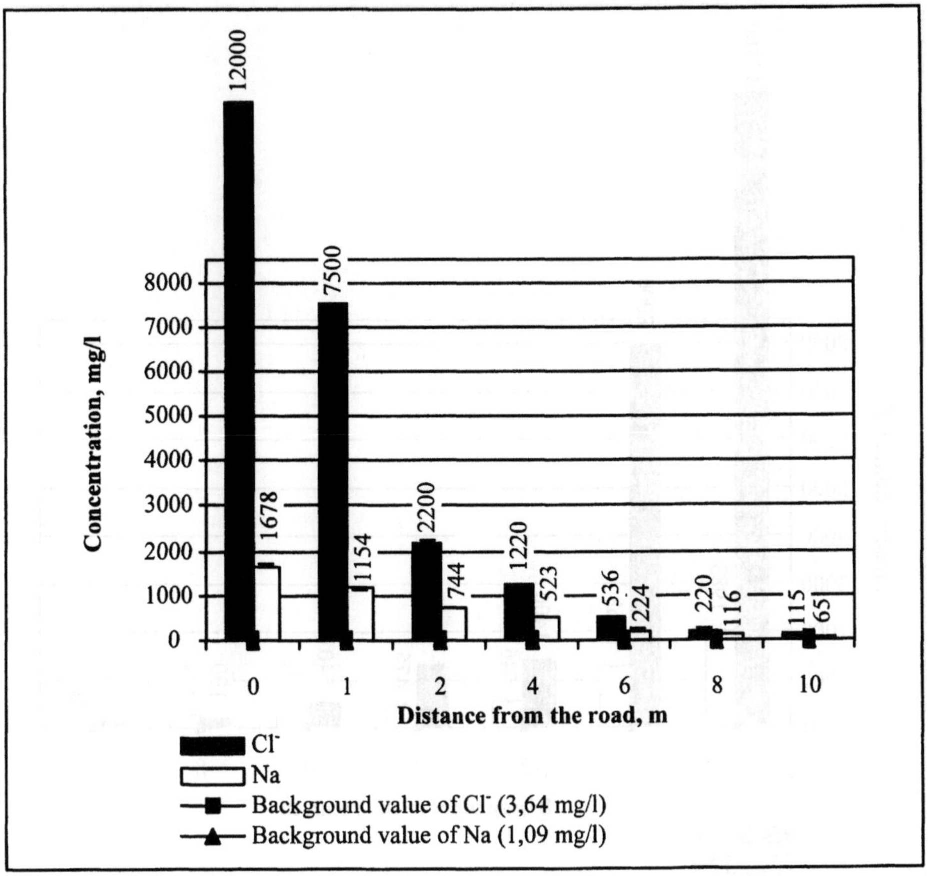

Chloride and sodium concentrations in transverse profile in Radviliškis distribute analogous concentrations in snow cover in Šeduva. Highest concentrations are 0-1 m from road. pH = 7,42-7,48. The snow covering at the spot of sampling was approximately 15 cm.

Fig. 4: Distribution of chloride concentration in transverse profile in Šiauliai

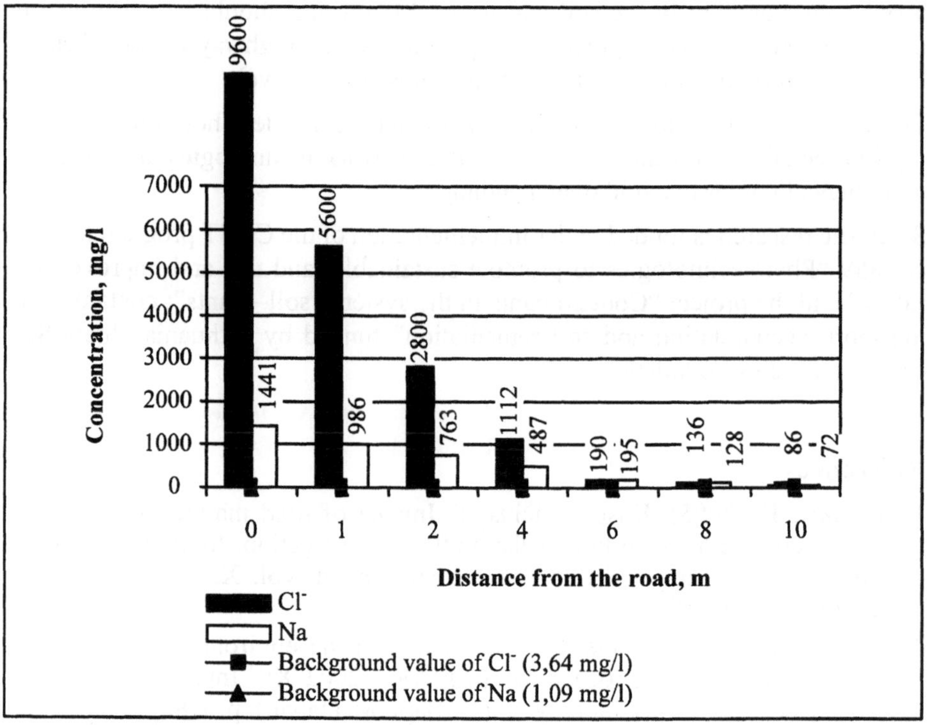

Cl^- and Na^+ concentrations concentrate in the trench. Cl^- and Na^+ are spread according to the geographical peculiarities of the location. pH = 7,11-7,52. The snow covering at the spot of sampling was approximately 17 cm.

4. Conclusions and Recommendations

Chloride and sodium concentrations dependence on the main variables (traffic level, road category, road maintenance level, geographical and climate conditions) is shown in the transverse profile of chloride and sodium concentration distribution. The analysis results show that maximal chloride and sodium stocks are accumulated near the road edge and at a distance of 1 m from roads. The highest concentration of chloride (Cl^-) was 9600-13200 mg/l, of sodium (Na^+) was 1441-1983 mg/l. The amount of chloride in snow melt water in Lithuania is 3,64 mg/l, of sodium – 1,09 mg/l [4].

The indicator of hydrogen ions (pH) of the snowmelt water that was used for the determination of sodium chloride concentration fluctuated from 7,11 to 7,54 making a slightly alkaline medium. The background value pH = 6,75 [4] (close to the neutral one) was on the average exceeded by 0,7.

Defining chloride concentrations in the snow melt water is the first step in observing their amount in the soil and water. Monitoring should be performed in the preserved areas having their own specifics, as the highway crosses them or borders them (in this instance Daugyvenė landscape reserve).

To reduce a negative impact on the environment, chlorides should be used taking into consideration the norms of their use, work technologies and other requirements in order to avoid overspending.

Scientific research is funded under implementation of the COST program activity no. 859 "Phytotechnologies to promote sustainable land use and improve food safety" and the project "Contaminants in the system "soil–plants": contaminants transport, accumulation and soil remediation" funded by Lithuanian State Science and Studies Foundation.

References

1. *Oškinis, V.* (2005), Kasperovičius, T. Impact of road maintenance salts on water ecosystems according to diatom flora investigation. Journal of Environmental Engineering and Landscape Management, vol. XIII, no. 1, Vilnius: Technika, p. 51-55.
2. *Blomqvist, G.* (2002), De-icing salt and roadside environment – Strategies for impact analyses. Proceedings of PIARC 2002 XIth International Winter Road Congress, 28-31 January 2002 – Sapporo (Japan). p. 1-8.
3. *Storpirštytė, I., Kazlauskienė, A. & Ščupakas, D.* (2004), Investigations of chloride concentration in snow cover on roadsides of Lithuanian roads with intensive traffic. Journal of Environmental Engineering and Landscape Management, vol. XII, Suppl 2, Vilnius: Technika, p. 60-66 (in Lithuanian).
4. *Kadūnas, V.* et al. (1999), Geochemical Atlas of Lithuania. Vilnius: LGT, 90 p.

Utilization of Pine Bark as Waste Wood for Soil Improvement

V. Biziks, B. Andersons, Z. Cakste, G. Zakis, B. Neiberte

Latvian State Institute of Wood Chemistry,
27 Dzerbenes St., LV – 1006, Riga, Latvia

Introduction

A great amount of waste including bark, which makes up 10% of the tree mass, is formed in the woodworking processes. In the past decade, the export of wood (sawn timber, chips) has increased. As a result, great amounts of wood bark are piled up, more than 1 million t according to experts' calculations, and these volumes are expected to grow.

If the wood composition in all tree species is similar, then the chemical composition of bark is different for each tree species. Bark, in comparison with wood, has a high content of extractives, especially water-soluble ones, and lignin. Bark extractives contain a whole range of organic compounds of different classes, namely, phenols, alcohols, flavonoids, etc. Bark also contains relatively high levels of phenolic acids – substances related to lignins, with lower molecular weight and methoxyl group (- OCH_3) content. As a result of a prolonged storage, the moisture content of bark increases, its chemical value is lost, and the environment is polluted with water-soluble degradation products. Bark can be used for mulching to inhibit weeds and to improve the soils thermal and moisture conditions. However, from all aspects, the most high-value product can be obtained, realizing bioconversion of bark. Stable humus is formed from bark, similar to the case of other materials with a high lignin content, which assists in renewing and maintaining the soil structure, improving its fertility.

Materials and Methods

Experiments were carried out in laboratory and outdoor conditions, testing different compost variants, namely, bark without additives (control), bark and bark-peat with nutrients (nitrogen, phosphorus, calcium (NPCa), and also both these combinations with the addition of a biocomposting agent or whey (a nitrogen and sugar source readily accessible for microorganisms). Moisture was kept at 50-60% both in laboratory and outdoor conditions. Outdoors, the stacks were protected against leaching with precipitation and drying, covering them with a film. A constant ambient temperature of 20-25°C was maintained in the laboratory. The compost temperature (for outdoor samples – in the vegetation period) was measured at regular intervals, samples were taken, their moisture and extract pH were determined, and the chemical composition was analysed. The compost readiness for seedling growing was checked at regular intervals, taking for comparison purposes the peat substrates available on sale.

Fig. 1: Schematic representation of the compost mass initial composition fractions and fraction analysis

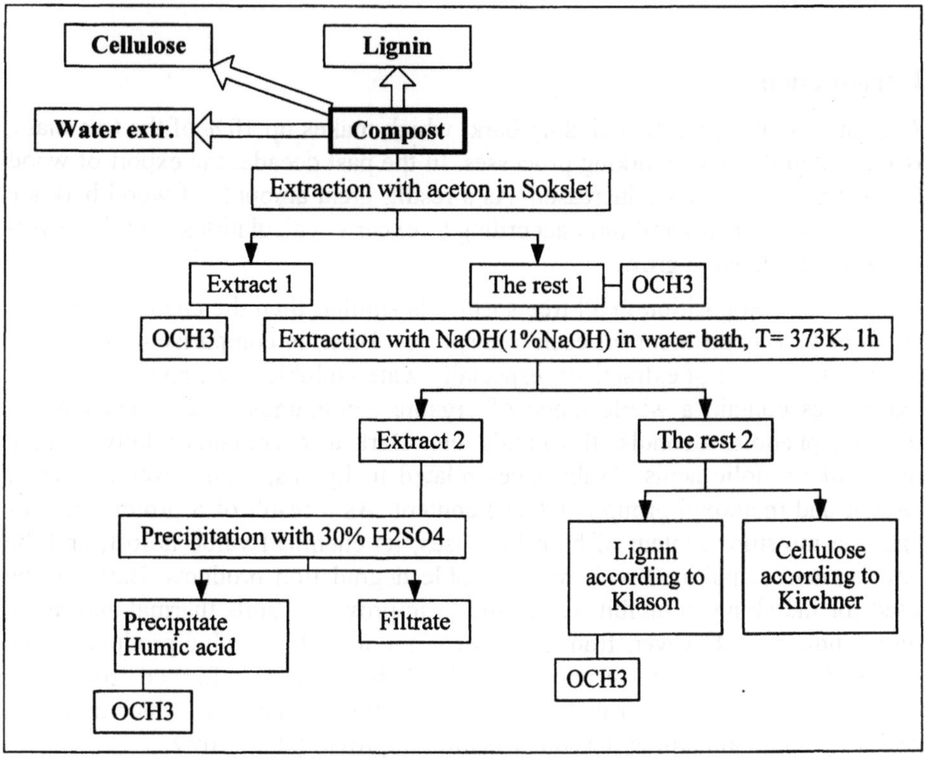

Results and Discussion

The obtained results testify that bark composts, after a 2-month bioconversion in outdoor conditions, contain toxic substances, which adversely affect the seedlings and inhibit their growth (probably, low-molecular phenolic compounds). At least 7 months are necessary for bark composting in outdoor conditions at mean temperatures about +10°C. When composting at a stable ambient temperature about 20°C (indoors), the compost for plants does not contain harmful compounds after 5 months. In outdoor conditions, the active composting processes for the investigated compositions cease after approximately a year.

If nutrients (NPCa) are not added to bark, the microbiological degradation activity is very low. The addition of peat to bark in the ratio 1:1 promotes the biological processes, ensuring a more even distribution of moisture in the compost. The biocomposting agent promotes the biodegradation of bark and maintains the optimum moisture of the substrate and a stable moderately high temperature (30-35°C on the average), ensuring an even activity of microorganisms. In the composting process, the amount of humic acids grows, the content of cellulose decreases, while the lignin content relatively increases. The highest increase in the content of humic substances is observed for substrates with a biocomposting agent, which testifies its positive role in the intensification of bioconversion.

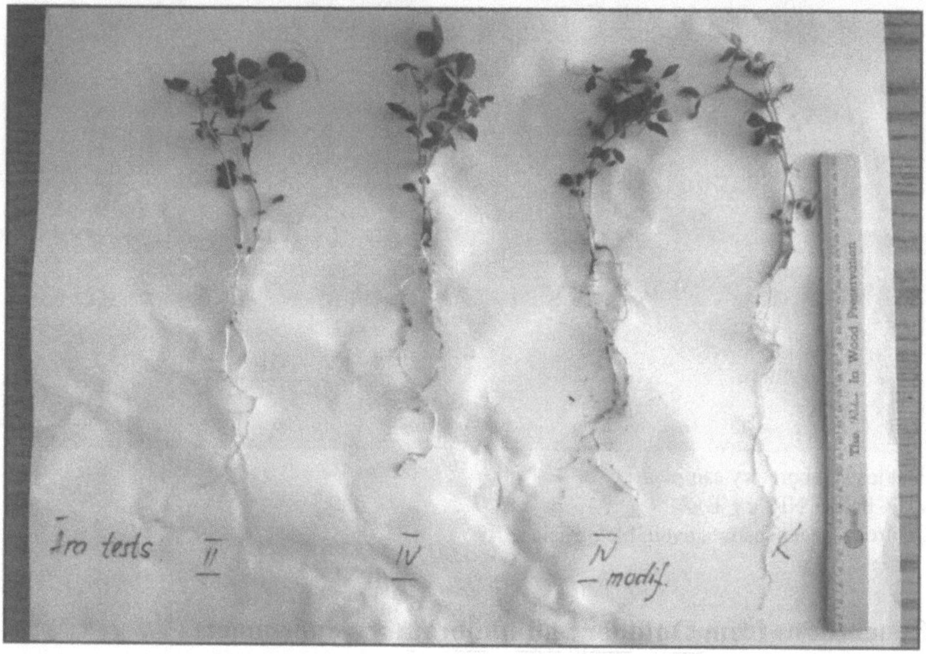

Root system, outdoor samples
II – bark + NPCa;
IV – bark + NPCa + bioagent;
IV – bark + peat + NPCa + bioagent;
K (Control) – peat substrate available on sale

Peat tests, laboratory samples
VII – bark + NPCa + E -2
Control – peat substrate available on sale

Conclusions from Outdoor and Laboratory Experiments

1. For bark and bark-peat composting, nutritives are necessary, first of all, nitrogen, also phosphorus and calcium compounds. Without additives, bark degrades slowly.
2. Peat additive for bark promotes biological processes, creates optimum composting conditions, obviously, owing to the more even distribution of moisture.
3. The biocomposting agent "Trichodermin" promotes bark biodegradation and provides the optimum moisture of substrate as well as a stable moderately high (30-35°C on the average) temperature, ensuring an even activity of microorganisms. Judging from the material's structure and the specific odour of humus, the substrate bark + peat + NPCa + biocomposting agent could be one of the most promising composting variants.
4. For bark composting in outdoor conditions at the mean temperature above +10°C, at least 7 months are necessary. In the case of composting at a stable ambient temperature about 20°C (indoors), the compost does not contain any compounds harmful for plants after 5 months.

5. The obtained results testify the high value of bark as a composted material and the prospects for its widespread use.

References

Aiken, G.R., McKnight, D.M., Wershaw, R.L. & MacCarthy, P. (eds.), Humic substances in soil, sediment, and water: GeoChemistry, Isolation, and Characterization, 1-9. Wiley, USA.

Tuomela, M., Vikman, M., Hatakka, A. & Itävaara, M. (2000), Biodegradation of lignin in a compost environment: a review. Appl. ELSEVIER. Bioresource technology 72, 169-183;

www.adeq.state.ar.us/solwaste/branch_technical/compost.htm.

Impact of Water-Soluble Substances of Birch Lignocellulose on the Mechanical and Sorption Properties of Granular Activated Carbon

J. Rizhikovs, M. Puke, J. Zandersons, N. Vedernikovs, G. Dobele, A. Tardenaka, B. Spince
Latvian State Institute of Wood Chemistry,
27 Dzerbenes St., LV – 1006, Riga, Latvia

Abstract

The carbonised granules of birch lignocellulose (LC) were activated at 850-870°C with superheated steam for 90 minutes in a fixed bed. The velocity of steam flow in the bulk of deciduous wood chips in the prehydrolysis reactor had a definite impact upon the yield of furfural, but did not influence the sorption properties of the activated carbon prepared from the LC specimens.

The granular activated carbon prepared from the leached LC demonstrated minimum sorption activity differences in comparison with untreated LC granular activated carbon specimens. The respective activated carbon characteristics obtained during the studies by using a sorptometer and dyes are as follows: Lengmuir surface area – 1095 m^2/g and 1205 m^2/g; total pore volume 522.5 mm^3/g and 671 mm^3/g; micropore volume 250.2 mm^3/g and 266.0 mm^3/g. The difference of iodine adsorption was negligible: 1054 and 1047 mg J_2/g, respectively. Unfortunately, the granular activated carbon prepared from the leached LC demonstrated low mechanical strength properties.

The low ash and sulphur contents of the granular activated carbon from leached LC has a very serious positive advantage, but in this case it makes practical sense if the activated carbon is used in a powder form.

Granular activated carbon from untreated lignocellulose retained high mechanical strength and had a high bulk density.

Introduction

All characteristic lignocellulose (LC) specimens made from birch veneer shorts during prehydrolysis and furfural production were studied as carbonisation starting materials. To elucidate the impact of water-soluble substances of LC on the potential of pore formation, two parallel sets of samples of LC were prepared (sulphuric acid modulus 0.02, temperature 167°C, process duration 60 minutes, four different steam velocities in the wood bulk), from which one set was leached with water.

Experimental

Obtaining of furfural. Veneer shorts containing 22.6% of pentosans, 36.8% of cellulose and 19.2% of lignin (calculated on dry mass) were used as the raw material. The furfural obtaining experiments were performed on a pilot plant with the reactor's volume 10 L and height 2 m. The initial material was mixed with sulphuric acid in a special mixer and treated in a steam flow at a temperature of 167°C for 60 min.

Lignocellulose. To elucidate the impact of water-soluble substances of lignocellulose on the potential of pores forming, two parallel samples of lignocellulose were prepared, from which one was leached with water. Wet lignocellulose containing 45 to 50% water (wet basis). It was dried at ambient temperature to 20 – 25% moisture content and at 50 to 60°C 7-8% moisture content. The chips measuring from 5 to 20 mm were milled to particles measuring less than 2 mm.

Granulation. Granulation of both samples of lignocellulose (moisture 7-8%, wet basis) was carried out in an extruder type apparatus.

Carbonisation. The specimens of granules were carbonised in an electrically heated reactor with stirrer at maximum reactor wall temperature 600°C for 60 min.

Activation. The granules of the carbonised lignocellulose were inserted in a stationary kiln mounted with a steam-superheating device. The kiln was placed in an oven, preheated to 850°C. The superheated steam, heated-up to 850°C was let into the kiln at a speed corresponding to the chosen oxidizing agent – charcoal ratio and the time of activation. The sorption capacity was determined by using conventional methods of dyes sorption (methylene blue, methylviolet, Congo red and iodine) and a sorptometer "KELVIN 1042".

Results and Discussions

The data demonstrated in Table 1 do not testify any connectedness between the velocity (or rather the consumption) of steam in the wood bulk in the prehydrolysis reactor and the water-soluble matter amount and composition (at least within the chosen parameters of steam feeding). In all cases, the yield of reducing substances is 11-13%, and that of the total dry matter yield 24.5-27.7% on the o.d. LC mass basis.

Tab. 1: Proximate analysis of a water-soluble matter of birch lignocellulose

Velocity of steam in wood bulk m/s	Dry matter, % on the o.d. LC mass	Sulphuric acid		Non-volatile acids (levulinic acid)		Reducing substances (sugars)		Unidentified matter (ash, phenols etc)	
		% on the o.d. LC mass	% on the o.d. extractives mass	% on the o.d. LC mass	% on the o.d. extractives mass	% on the o.d. LC mass	% on the o.d. extractives mass	% on the o.d. LC mass	% on the o.d. extractives mass
0.040	24.5	1.8	7.5	3.7	15.3	12.4	50.8	6.6	26.4
0.080	27.1	1.9	6.9	3.6	13.5	12.8	47.2	8.8	32.4
0.120	27.7	1.8	6.7	4.5	16.4	11.8	42.6	9.6	34.3
0.243	25.7	1.6	6.2	4.3	16.5	10.7	41.6	9.1	35.7

The velocity of steam in the wood bulk during prehydrolysis has a considerable impact on the yield of furfural and acetic acid because it means a different amount of the steam feed. At the same time, its influence upon the charcoal yield does not demonstrate any regularity.

Since approximately 25% of the LC mass is lost due to leaching, the difference between the actual mean charcoal yield obtained from leached LC and untreated LC is considerably greater than 5.2% in favour of untreated LC (Table 2). The fixed carbon yield is almost the same in both cases – from untreated LC – 34.2%, and from leached LC – 33.2% o.d. LC mass basis. Therefore, the greater activated carbon yield from leached LC charcoal calculated on the amount of the fixed carbon mass (49.8 and 40.7% respectively) does not compensate the LC mass loss caused by the removal of water-soluble components and the lower charcoal yield during the carbonisation due to the elimination of sulphuric acid being a carbonisation catalyst. Testing of the carbonised granules made from washed LC demonstrated the same high mechanical strength as in the case of unleached LC, which is containing water-soluble extractives acting as some kind of binders.

Tab. 2: Carbonisation and activation of granulated birch lignocellulose

Velocity of steam in wood bulk m/s	Carbonisation T_{max}=600°C						Activation (T=850-870°C)			
	Untreated			Leached			Untreated		Leached	
	Yield, % o.d. LC mass	Fixed carbon,%	Ash, %	Yield, % o.d. LC mass	Fixed carbon,%	Ash, %	Yield,% char mass	Burn-off% fixed carbon mass basis	Yield,% char mass	Burn-off% fixed carbon mass basis
0.040	47.1	69.3	2.3	42.8	79.8	0.74	37.9	45.3	44.0	44.9
0.080	51.4	71.0	2.3	42.9	75.6	0.84	41.7	41.3	55.2	29.4
0.120	45.7	74.1	2.4	42.5	78.2	0.62	39.5	46.7	53.3	29.5
0.243	48.1	71.0	2.8	43.6	75.6	0.84	43.8	38.3	46.7	39.6
Mean yield	48.1	71.3		42.9	77.3			42.9		35.8

Tab. 3: Impact of prehydrolysis conditions and lignocellulose pre-treatment on the sorptivity of activated carbon

Pretreatment of LC	Velocity of steam in wood bulk, m/s	Burn-off, % o.d. fixed C mass	Yield of activated carbon,% o.d. LC mass	Adsorption capacity of activated carbon, mg/g			
				Iodine (<1.48nm)*	Methylene blue (1.52nm)*	Methylviolet (2.16nm)*	Congo red (2.26nm)*
non	0.040	45.3	20.1	1105	88	36	22
	0.080	41.3	19.7	1003	72	38	13
	0.120	46.7	19.0	1114	80	32	19
	0.243	38.3	20.0	993	62	27	13
Leached with water	0.040	44.9	15.6	1083	99	32	10
	0.080	29.4	23.5	1005	33	8	2
	0.120	29.5	23.2	1022	27	5	1
	0.243	38.2	20.1	1080	92	23	12

* critical pore diameter in molecular sieving adsorption [5]

The carbonised granules were activated at 850-870°C with superheated steam (mass ratio 1: 3, respectively) during 90 minutes in a fixed bed. The speed of steam in the prehydrolysis reactor did not influence the sorption properties of the activated carbon. The sorption activity of the granulated activated carbon prepared from the LC specimen obtained at the maximum steam speed did not differ from the mean granulated activated carbon prepared at a lower speed of steam in the prehydrolysis apparatus (respective dyes' sorption activities in mg/g are demonstrated in Table 3, and the characteristics of the micropores, and the total porosity of activated carbon in Table 4).

Tab. 4: Porosity characteristics of activated carbon prepared from granulated untreated and leached lignocellulose (m = 0.02; 167°C; 60 min; w = 0.120 m/s)

Characteristics	Units of measure	Untreated LC	Leached LC
BET surface area	sq.m/g	925.03	841.52
Langmuir surface area	sq.m/g	1205.77	1095.45
Total pore volume	cub.mm/g	671.31	522.54
Micropore volume	cub.mm/g	265.99	250.18

The granulated activated carbon prepared from the leached LC demonstrated minimum sorption activity differences during the studies by using dyes and a sorptometer (Tables 3 and 4, Figure 1) in comparison with untreated LC granulated specimens.

Fig. 1: Pores' characteristics of granulated activated carbon from lignocellulose

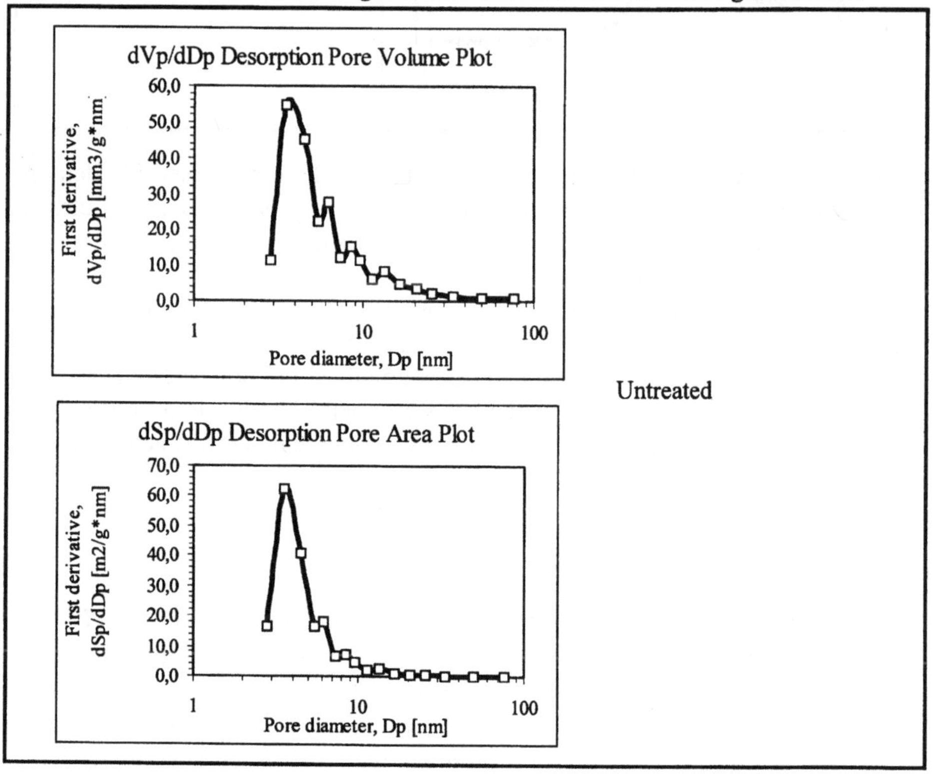

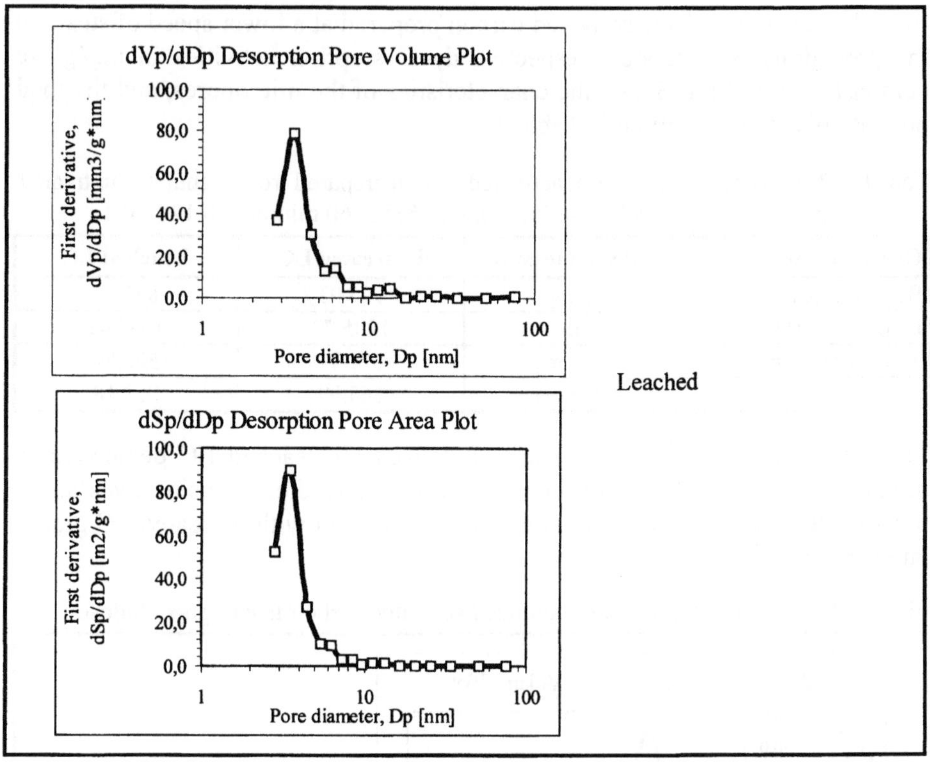

Unfortunately, the granular activated carbon prepared from leached LC demonstrated low mechanical strength properties. In contrast, carbonised granules of untreated LC retained the initial mechanical strength during the activation and the granular activated carbon had a high bulk density: 320-390 g/l. The charcoal of leached LC had low ash and sulphur contents: ash – 2.3-2.8% and 0.6-0.8%; sulphur – 0.2-0.3 and 0.01-0.04%, respectively. During the carbonisation 75-81% of the initial amount of sulphur in LC is volatised in both cases, but the rest is fixed in ash. The low ash content of the granular charcoal is a very serious positive characteristic but, in this case, it makes practical sense if activated carbon is used in a powder form.

Conclusions

1. The LC of the direct furfural production process contains approximately 25% of water-soluble matter: reducing substances, non-volatile acids (mainly levulinic), sulphuric acid, the soluble part of ash, etc. acting as binders of granules.
2. Leached LC demonstrates good enough granule forming properties yielding mechanically strong carbonisation products with a low ash and sulphur content, also without the presence of water-soluble extractives.

3. The steam activation of LC carbon granules from leached and untreated LC resulted in microporous activated carbon with similar porosity and sorption activity.
4. The activated carbon from leached LC specimens contained less ash and sulphur than those from raw LC but, unfortunately, its weak mechanical properties made it unacceptable for production of granulated sorbents.
5. The negative impact of the lack of water-soluble substances on the mechanical properties of granules was demonstrated during the activation process.

References

Kasaoka, S., Sakata, Y., Tanaka, E. & Naitoh, R. (1989), Design of molecular-sieve carbon. Studies on the adsorption of various dues in the liquid phase. International Chemical Engineering, vol. 29, no. 4, pp. 734-742.

Vedernikov, N., Kruma, I., Zandersons, A., Tardenaka, A., Spince, B. & Chirkova, J. (2002), Production of furfural and carbon materials from deciduous tree wood wastes. Proc. 7th European Workshop on Lignocellulosics and Pulp, Turku, Finland, pp. 51-54.

Vedernikov, N.A. (1991), Production of furfural. In: Hemicelluloses, Zinatne, Riga, pp. 209-222. (in Russian)

– (1996), Latvian patent 11032.

Zandersons, J., Vedernikov, N., Kruma, I., Puke, M. & Rizhikovs, J. (2004), Impact of birch wood prehydrolysis conditions upon the yield and properties of activated carbon from lignocellulose. Proc. 8th European Workshop on Lignocellulosics and Pulp, Riga, Latvia, pp. 543-546.

Variation of Chemical Composition of Essential Oils in *Thymus serpyllum L. subsp. serpyllum* growing wild in Lithuania

Kristina Ložienė[1], **Petras R. Venskutonis**[2]

[1] Institute of Botany,
Žaliųjų Ežerų g. 49, Vilnius, LT – 08406, Lithuania

[2] Department of Food Technology, Kaunas University of Technology,
Radvilėnų pl. 19, Kaunas, LT – 2021, Lithuania

Summary

Twenty-six individual plants of *Thymus serpyllum* subsp. *serpyllum* were collected from fourteen different habitats in Lithuania. The composition of their essential oils was analyzed by capillary gas chromatography (GC) and GC/MS.

In total, 68 compounds were identified. The study showed infraspecific variation in chemical composition of essential oils. The most variations were found in the amount of (E)-β-ocimene, 1,8-cineole, α-cadinol, caryophyllene oxide, β-caryophyllene, borneol, *cis-p*-menth-2-en-1-ol, germacrene B, germacrene D and (Z)-β-ocimene in separate samples. However this study proved that *T. serpyllum* subsp. *serpyllum* plants from the Northern climatic area do not accumulate volatile phenolic compounds thymol and carvacrol in remarkable amounts. The content of thymol in the oils of the analyzed plants did not exceed 2.3%.

Five main groups of the plants were observed using hierarchical cluster analysis based on the major essential oil constituents. Consequently five chemotypes of *T. serpyllum* have been defined, namely 1,8-cineole, germacrene B, (E)-β-ocimene, α-cadinol and *cis-p*-menth-2-en-1-ol.

Introduction

Previously published data showed differences in the oil composition of *Thymus serpyllum*. Four chemotypes of *T. serpyllum* with germacra-1(10),4-dien-6-ol, germacra-1(10),5-dien-6-ol, linalool and linalyl acetate as type-characterizing compounds were defined in Finland (Stahl-Biskup, 1990). *T. serpyllum* chemotype containing 60.3% of geraniol were reported in Siberia (Kalinkina et al. 1994). Mostly, the main phenolics of *Thymus*, thymol and carvacrol were not found or found in minor amounts in *T. serpyllum* growing in the temperate northern climatic zones. For instance, thymol and carvacrol were not detected in the oils of *T. serpyllum* from Sweden (Lundgren & Stenhagen, 1982), Finland (Stahl-Biskup, 1990) and Poland (Osiska et al. 2002); in the oil from *T. serpyllum* growing in Siberia these components were present in traces (Kalinkina et al. 1994).

The purpose of this study was to determine chemical composition of the oil of *T. serpyllum* subsp. *serpyllum* plants collected in various localities in Lithuania and to define the chemotypes within this subspecies growing wild in Lithuania.

Material and Methods

In total 26 *T. serpyllum* subsp. *serpyllum* individual plants (samples) were collected at the flowering stage in July from 14 different habitats in Lithuania (Figure 1). The plants were dried at room temperature. The oils were isolated from the selected individual plants by hydrodistillation in a Clevenger-type apparatus (European Pharmacopoeia) during two hours. GC and GC/MS carried out the analyses of the essential oils. The Joining Cluster Analysis was applied to the study of the similarity of individuals on the basis of constituent distribution. Euclidean distance between individuals was used as an index of their similarity and clustering was performed according to the weighed average linkage method. The statistical software package STATISTICA 7 was employed in the work.

Fig. 1: Collecting sites of *Thymus serpyllum* subsp. *serpyllum* samples

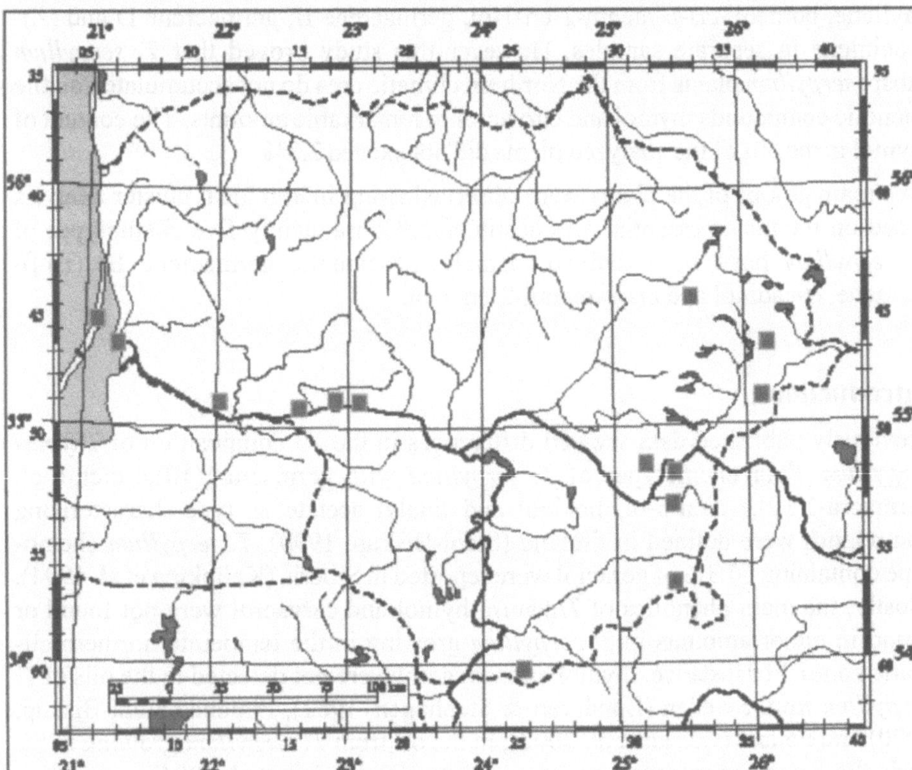

Results

In total, 68 compounds were identified by capillary GC and coupled GC/MS. The results show significant infraspecific variation in chemical composition of essential oils. The range of the amounts of the main essential oil components from different *T. serpyllum* samples is given in Figure 2. The most remarkable variations were found in the amount of (E)-β-ocimene (0.0-34.8%), 1,8-cineole (0.0-30.3%), α-cadinol (0.0-28.6%), caryophyllene oxide (0.0-27.2%), β-caryophyllene (0.0-27.2%), borneol (0.0-27.1%), *cis-p*-menth-2-en-1-ol (0.0-24.1%), germacrene B (0.0-25.4%), germacrene D (0.0-20.4%) and (Z)-β-ocimene (0.0-20.0%) (Figure 2). This study proved that *T. serpyllum* subsp. *serpyllum* plants growing wild in Lithuania does not accumulate volatile phenolic compounds thymol and carvacrol in remarkable amounts. The content of thymol in the oils of the analyzed samples did not exceed 2.3%.

Fig. 2: Variation of the amount (%) of the some main essential oil components in different *Thymus serpyllum* subsp. *serpyllum* plants

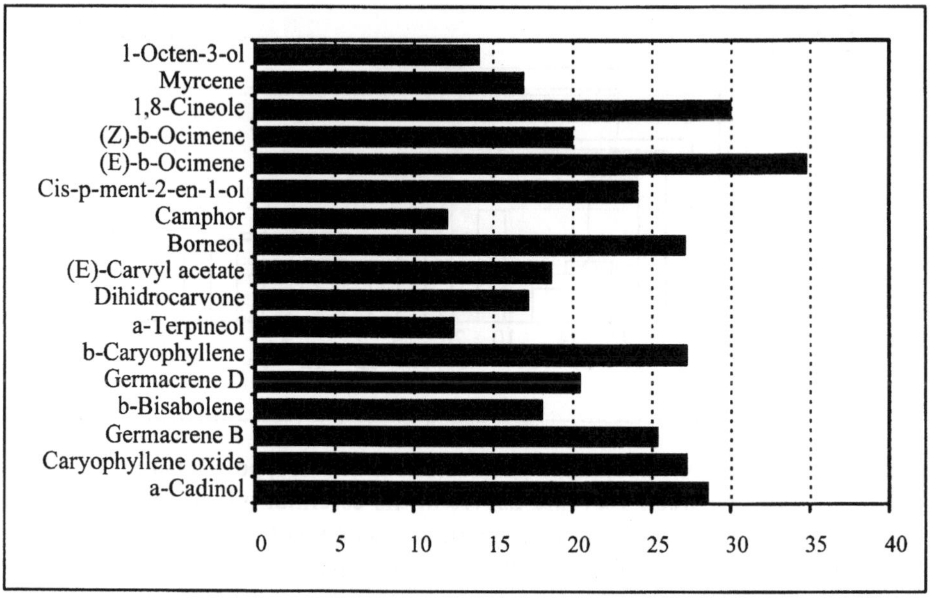

Five main groups of the plants can be observed using hierarchical cluster analysis based on the major oil constituents (Figure 3). Consequently five chemotypes of *T. serpyllum* can be defined within the investigated plant samples: 1,8-cineole, germacrene B, (E)-β-ocimene, α-cadinol and *cis-p*-menth-2-en-1-ol.

Six samples were attributed to 1,8-cineole chemotype (Figure 3). These samples contained from 13.9 to 30.3% of 1,8-cineole in the oil. This component was not detected in other investigated *T. serpyllum* subsp. *serpyllum* plants, or its content was very low (0.0-1.5%). Other important constituents in the oil of 1,8-cineole chemotype plants were myrcene (5.6–12.2%), β-caryophyllene (4.5–16.1%) and germacrene D (3.2–13.8%). It should be pointed out that one sample belonging to 1,8 cineole chemotype produced exceptionally high amount of *trans*-carvyl acetate (18.7%), which was absent in almost all other plant oils. 1,8-cineole was reported as quantitatively important component in *T. serpyllum* from Central Poland (11.64%) (Osiska et al. 2002), Finland (19.7%) (Stahl-Biskup 1990).

Figure 3 Two dimensional dendrogram of the essential oils of individual plants of *Thymus serpyllum* subsp. *serpyllum*: 1,8-cineole (samples 21, 17, 16, 23, 24, 11), germacrene B (samples 6, 3, 1), α-cadinol (sample 8), E-β-ocimene (sample 14), *cis*-p-menth-2-en-1-ol (sample 20)

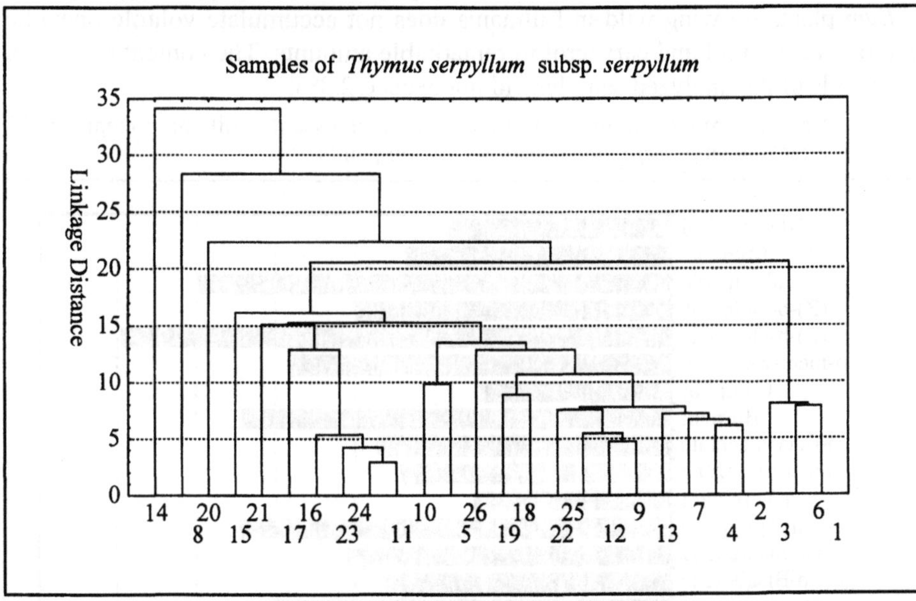

Three samples can be allocated to germacrene B chemotype. The amount of germacrene B was between 10.4 and 25.4% in the essential oils of this chemotype plants. 1-Octen-3-ol was also characteristic component constituting 6.2-8.7%. However, the content of some other quantitatively major components (α-terpineol, dihydrocarvone, borneol) in the plants of this chemotype were different.

(E)-β-Ocimene, α-cadinol and *cis-p*-menth-2-en-1-ol chemotypes were represented by one analyzed plant sample, however the content of prevailing components in these plants was very high, 34.8%, 28.6% and 24.1%, respectively.

It was rather difficult to ascribe to any definite chemotype other 13 investigated *T. serpyllum* subsp. *serpyllum* plants. The plants in this group contained large amounts of some chemical components ((Z)-β-ocimene, 1-octen-3-ol, germacrene D, dihydrocarvone. However, hierarchical cluster analysis did not separate these samples into individual groups.

References

Kalinkina, G.I., Tihonov, V.N., Guskova, L.A., Zarubina, L.A. & Taran D.D. (1994), Chemical composition and pharmacological properties of *Thymus serpyllum* L. s. l., essential oil grown in Central Siberian Botanical garden SO RAN. – Rastitel'nye Resursy 30: 66-70.

Lundgren, L. & Stenhagen, G. (1982), Leaf volatiles from Thymus vulgaris, T. serpyllum, T. praecox, T. pulegioides and T. x citriodorus (Labiatae). – Nord. J. Bot. 2: 445-452.

Osiska, E., Weglarz Z. & Angielczuk, M. (2002), Preliminary investigation on the variability of *Thymus serpyllum L.* growing wild in Central Poland. – Proceed. XVIth EUCARPIA Genetic Resources: 253-256.

Stahl-Biskup, E. (1990), Essential oil polymorphism in Finnish *Thymus* species. – Planta Med. 56: 464-468.

SCIAMACHY
–
Limb Measurements as a New Tool for Stratospheric Ozone Studies

J. Pukite[1,2], S. Kühl[1], W. Wilms-Grabe[1], C. von Friedeburg[1,3], T. Deutschmann[1], S. Sanghavi[1], J. Hollwedel[1], S. Beirle[1], C. Frankenberg[1], M.F. Khokhar[1], M. Grzegorski[1], T. Marbach[1], B. Kirchhof[1], S. Kraus[1], U. Platt[1], T. Wagner[1]

[1] Institute of Environmental Physics, University of Heidelberg, Im Neuenheimer Feld 229, D – 69120 Heidelberg, Germany

[2] Institute of Atomic Physics and Spectroscopy Faculty of Physics and Mathematics, University of Latvia 19 Rainis Blvd., LV – 1586 Riga, Latvia

[3] Now at the Wiley-VCH/Editorial Office Physics, Bühringstr. 10, D – 13086 Berlin, Germany

Abstract

Ozone, being a strong absorber in the UV spectral range and therefore shielding the Earth surface from baneful influence that could be caused by transmitted UV-B and UV-C radiation, is highly impacted by chemical processes in the stratosphere.

At latitudes close to the polarregions where low stratospheric temperatures can occur halogen activation can take place that causes significant ozone depletion.

The SCIAMACHY (Scanning Imaging Absorption Spectrometer for Atmospheric Chartography) on ENVISAT measures solar UV-VIS-NIR radiation transmitted, backscattered and reflected by the atmosphere and the ground. SCIAMACHY scans the atmosphere alternatively in nadir (radial view) and limb (tangential view) modes. From the limb scans, which are provided with approximately 3 km elevation steps, stratospheric concentration profiles of several trace gases such as ozone, nitrogen dioxide, bromine oxide and chlorine dioxide can be retrieved.

First, slant column densities (SCDs) for the trace gases are derived from the SCIAMACHY limb measurements by means of the Differential optical absorption spectroscopy (DOAS) method. In the second step the trace gases SCDs are converted into vertical concentration profiles by means of radiative transfer modelling. The inversion is constraint by the optimal estimation method using seasonal and latitudinal dependent climatology of the respective trace gas profile.

The SCIAMACHY limb observations allow us for the first time to derive information on the stratospheric profiles of BrO, and OClO (besides that of O_3 and NO_2) on a global scale.

Introduction

Along with the discovery of the Antarctic ozone hole there exists high scientific interest in the investigation of chemical processes taking place in the stratosphere (see detailed review in Solomon, 1999). These processes are associated with heterogeneous reactions involving highly reactant halogen and nitrogen oxide radicals forming at sunrise on polar stratospheric clouds (PSCs) that have formed in polar night at low stratospheric temperatures.

To understand the responsible processes, various instrumentation based mainly on spectroscopic techniques have been utilised. As a high advantage space born remote sensing instruments for atmospheric spectroscopy have been designed.

The SCanning Imaging Absorption SpectroMeter for Atmospheric CHartographY (SCIAMACHY)

SCIAMACHY is a spectrometer onboard ENVISAT (ENVIronmental SATellite) flown in a sun synchronous near polar orbit at a mean altitude of 800 km and an equator crossing time of 10: 00 in descending node mode. The satellite is designed for monitoring of the evolution of environmental and climatic changes. SCIAMACHY measures sunlight backscattered from Earth's atmosphere over a large wave-length range (240 nm-2380 nm) with moderate spectral resolution (FWHM: 0,2 nm-1,5 nm). For more details see Bovensmann et al. 1999.

Measurements at the dayside of orbit are performed alternatively in sequences of nadir and limb scans.

The limb scans are performed in the direction of flight of the satellite in swatches of 960 km with approximately 3.3 km elevation steps (vertical field of view 2.6 km) projecting entrance slit of the spectrometer by azimuth and elevation mirrors.

In the limb measurement mode different air masses under different elevations are scanned enabling us to retrieve vertical profiles of several trace gases.

It should be mentioned that besides the trace gases profile retrievals from SCIAMACHY (see for instance Rozanov et al. 2005) other satellite instruments such as the Optical Spectrograph and Infrared Imager System (OSIRIS) on Odin satellite (Haley et al. 2004; Krecl et al. 2005), the Stratospheric Aerosol and Gas Experiment (SAGE III) on the Meteor 3 (Rault 2005) are successfully used to retrieve several trace gases from light scattered measurements in limb geometry. However SCIAMACHY provides for the first time an opportunity to derive a variety of trace gases among them ozone, nitrogen dioxide, and also the halogen oxides as bromine oxide and chlorine dioxide.

Retrieval Method

For the retrieval a two step approach (Figure 1) is used (see also Krecl et al. 2005; Haley et al. 2004).

First, the integrated concentration of the considered absorber along the lightpath (the slant column density (SCD)) is derived, and in the second step the concentration profiles are calculated applying radiative transfer properties of the atmosphere.

Fig. 1: Diagram of the retrieval algorithm for profile derivation

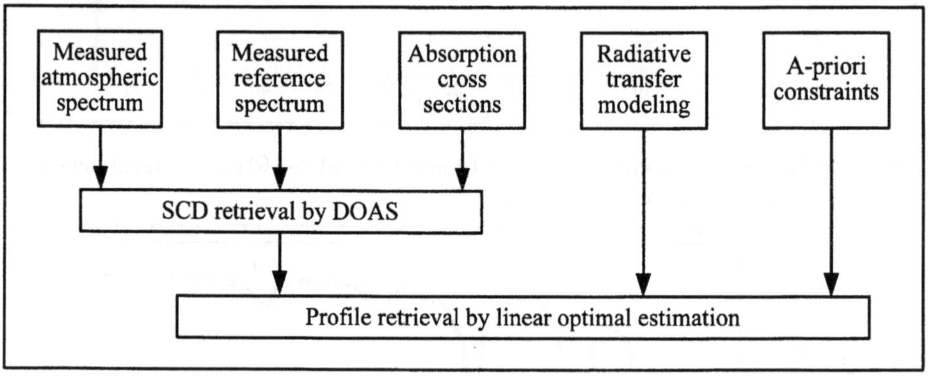

To derive the SCDs the measured and calibrated spectral information from SCIAMACHY is analysed using DOAS (Differential Optical Absorption Spectroscopy) method (Platt 1994). The main idea of the method is to perform a least squares fit for measured spectrum normalised by a reference spectrum containing almost no absorption signatures of considered trace gases. The fit is realized including differential absorption cross sections of various absorbers simultaneously, also taking into account atmospheric and instrumental spectral features.

As a result, the SCDs are obtained as a function of tangent height. SCD values derived at higher elevations are less precise because of less intensity of backscattered light measured (Figures 2 and 3).

The DOAS retrieval is done in the near UV spectral range for trace gases of interest (Table 1). In this region ozone absorption is assumed as not leading to considerable non-linear effects in radiative transfer.

Tab. 1: Wavelength ranges used for DOAS retrieval.

Trace gase	**Wavelength interval used for analysis**
Ozone and BrO	336-357 nm
NO_2 and OClO	363-391 nm

Fig. 2: Retrieved BrO SCDs as function of tangent height for example orbit

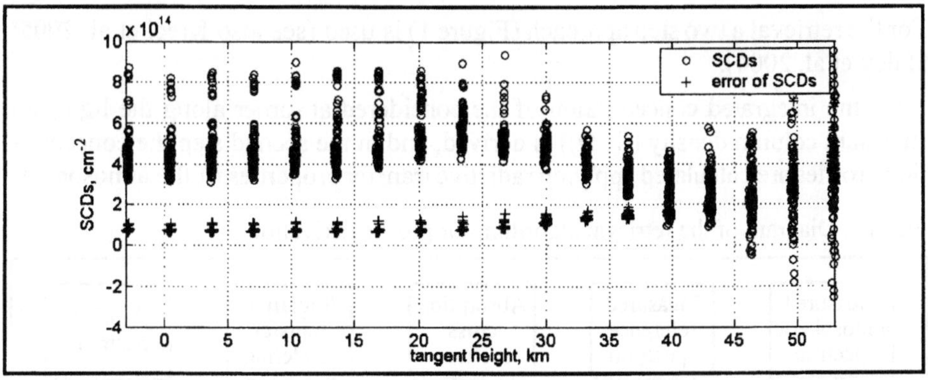

Fig. 3: Fitted ozone absorption spectra (①) and fit residual (②) at a tangent height of 10 km (top) and 50 km (bottom).

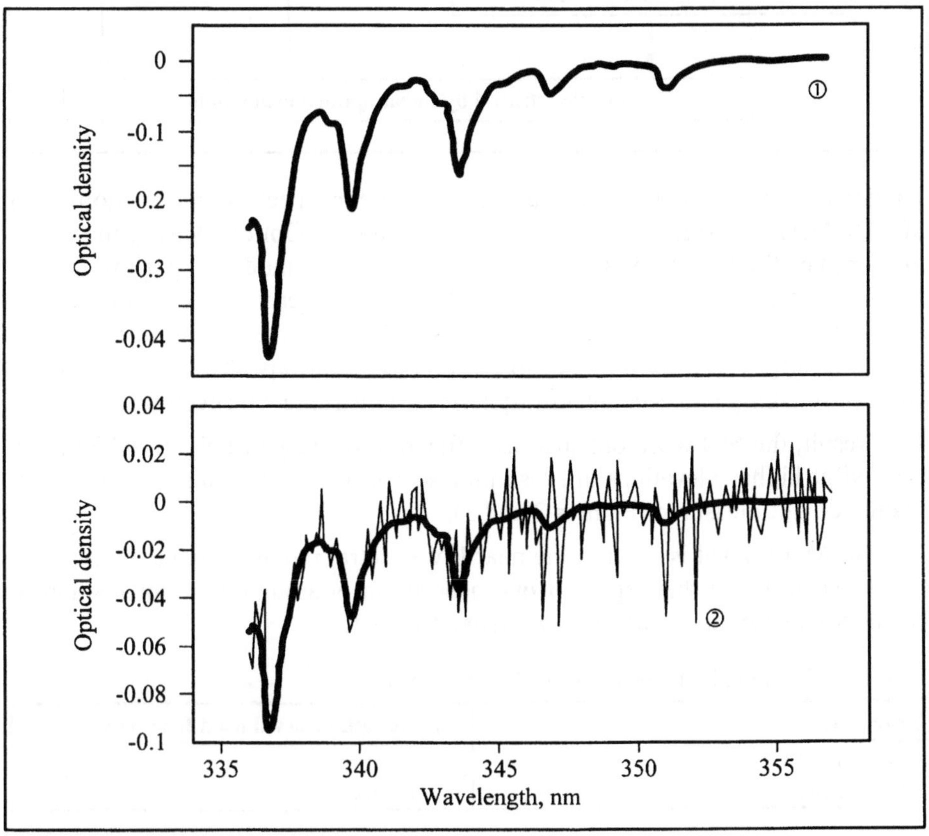

To perform the second step and to derive profile information, radiative transfer modelling (RTM) has to be performed and applied: box air mass factor (AMF)

of light paths through defined layers is calculated (Figure 4). This geometrical interpretation of AMF definition is correct for trace gases having low optical densities. For strong absorbers intensity change is no longer linear to change in concentration of the absorber (Marquard et al. 2000). In the selected wavelength interval relatively strong optical densities (up to 25%) for ozone can occur. However in this study the geometrical definition is used for the trace gases. Even for ozone the errors due to this effect are relatively small (~4%).

Fig. 4: The box AMFs for a given line of sight and atmospheric layer is defined as intensity units of measured light (i.e. after extinction in the atmosphere) that is crossing the given layer multiplied by path of this light through the layer and divided by total light measured and layer thickness.

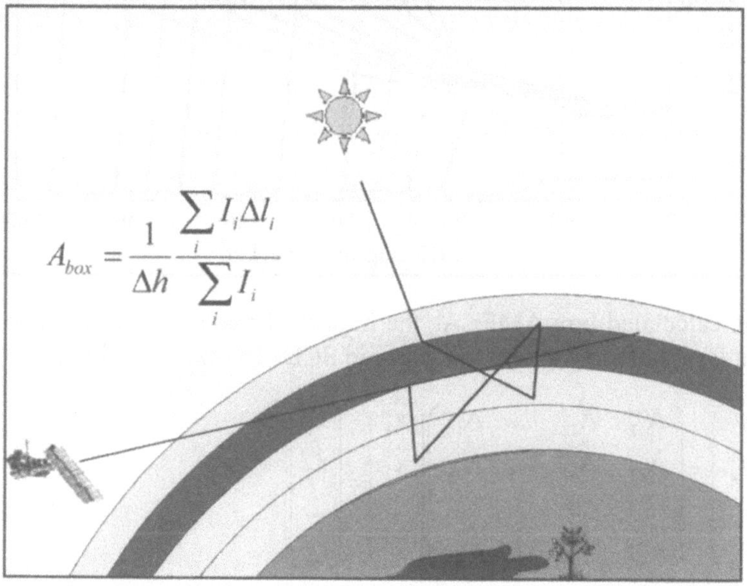

For the box AMF calculation the full spherical 3-dimensional Monte Carlo model "Tracy" (von Friedeburg 2003) is applied (Figure 5).

Fig. 5: Box AMFs modelled as function of line of sight (LOS) elevation for 3 km "boxes" located at different altitudes (example). The peak value usually is located at elevation equal to the altitude of the "box" simulated. Low sensitivity for tropospheric part of atmosphere can be seen.

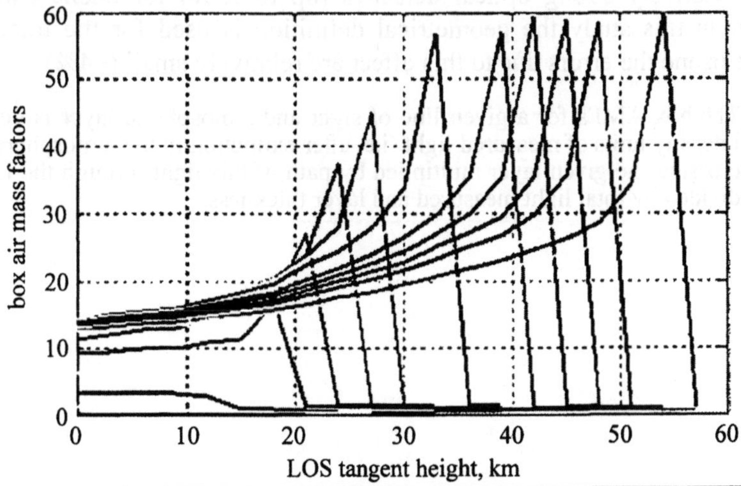

With the calculated box AMFs A_{ij} the impact of trace gas concentration (x_j) at different layers of atmosphere to measured SCDs (y_i) can be written:

$$\begin{pmatrix} y_1 \\ y_2 \\ \dots \\ y_m \end{pmatrix} = \begin{pmatrix} A_{11} & A_{12} & \dots & A_{1n} \\ A_{21} & A_{11} & \dots & A_{2n} \\ \dots & \dots & \dots & \dots \\ A_{m1} & \dots & \dots & A_{mn} \end{pmatrix} \begin{pmatrix} x_1 \\ x_2 \\ \dots \\ x_n \end{pmatrix} + \begin{pmatrix} \varepsilon_1 \\ \varepsilon_2 \\ \dots \\ \varepsilon_n \end{pmatrix} \tag{1}$$

To retrieve **x** the inversion should be made. Since the problem is not exact (SCDs have error **ε** and corresponding error covariance S_ε) and underdetermined, a-priori knowledge x_a and its covariance S_a are used to exclude unrealistic solutions.

Using Bayesian approach the maximum a-posteriori solution can be found. According to Rodgers (2000):

$$\hat{x} = (A^T S_\varepsilon^{-1} A + S_a^{-1})^{-1} (A^T S_\varepsilon^{-1} y + S_a^{-1} x_a) \tag{2}$$

and its covariance:

$$\hat{S}^{-1} = A^T S_\varepsilon^{-1} A + S_a^{-1} \tag{3}$$

Averaging kernels are characterising retrieval quality:

$$R = (A^T S_\varepsilon^{-1} A + S_a^{-1}) A^T S_\varepsilon^{-1} A \tag{4}$$

Results

Figure 6 shows latitudinal cross sections for sample orbits at austral winter – spring for ozone, BrO, NO_2 and OClO and corresponding averaging kernels.

Decrease of ozone and OClO and increase of NO_2 concentrations in the south polar region is visible for 30.9.2004 compared to 24.8.2004, in agreement with expectations from the stratospheric chemistry (notwithstanding some uncertainties in retrieval for polar and near polar regions that could appear because of the ozone absorption cross-section temperature dependence or imperfections in radiative transfer modelling). Regarding the retrieval uncertainties, one possible aspect already mentioned above is the optical density impact. Another possibility that could be mentioned is strong three-dimensional gradients of trace gas (for instance rapid change in ozone concentrations at the boundary of the polar vortex). In the limb geometry, the line of sight of the spectrometer has a long enough path through the atmosphere that the trace gas concentration at the place where line of sight actually is crossing some atmospheric layer can considerably differ from that at the tangent point where measurement have assumed to be made. Hence just one dimensional (altitude) discretisation in the RTM as in this study can lead to some uncertainties.

The high OClO concentrations correspond to large latitudes at the winter pole and low temperature conditions. Those are allowing formation of polar stratospheric clouds (PSCs) that catalyse ozone depletion. During spring the stratosphere warms up and the effect of OClO activation is reducing.

At low polar stratospheric temperatures, nitrogen (in form of HNO_3) is deposed in PSCs that explain low NO_2 concentrations at high south latitudes in August comparing to September.

The BrO distribution is determined by the dependency of troposphere-stratosphere transport of precursors and bromine chemistry on latitude leading to smaller BrO concentrations in the tropics (Pundt et al. 2002).

The Ozone hole formed can be identified in graphs shown for orbit from 30.9.2004.

From averaging kernel graphs can be seen the better possibility to retrieve ozone and nitrogen dioxide as they are major absorbers in the stratosphere. As it can be concluded from Figure 6, it is impossible to retrieve information about tropospheric part of atmosphere because of high Rayleigh scattering probability along the LOS and also increasing cloud probability.

Fig. 6: Left side: Latitudinal cross sections of retrieved trace gases – ozone, bromine oxide, nitrogen dioxide and chlorine dioxide for selected orbits from 24.8.2004 and 30.9.2004. (For altitudes from 10 to 15 km the retrieved profile consists mainly of the a-priori information (for ozone retrieval the climatology of Lamsal et al. 2004 is used).

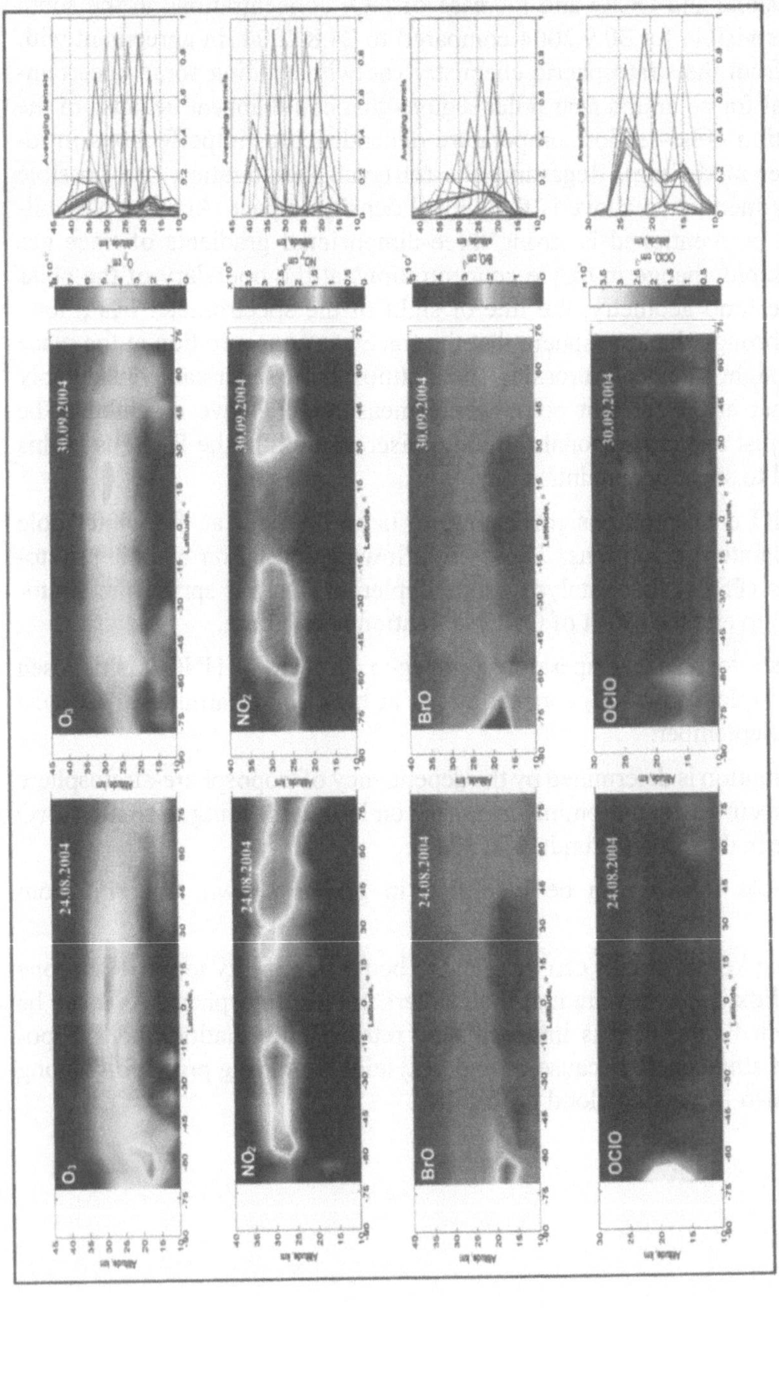

Right side: Characteristic averaging kernels. The retrieval is more corresponding to the real state of atmosphere at given altitude if the averaging kernels at this altitude are more close to unity.

Conclusions

SCIAMACHY provides for the first time backscattered light measurements in limb geometry from which atmospheric trace gas profiles of O_3, NO_2, BrO and OClO can be successfully retrieved. The obtained dataset is forming a unique time series of measurements for investigating atmospheric chemistry.

A two step retrieval for O_3, NO_2, BrO and OClO is performed for selected orbits of SCIAMACHY limb measurements enabling to study separately the impact of spectroscopy and radiative transfer.

It is possible to retrieve trace gas concentrations for altitudes 18-36 km (BrO), 15-39 km (ozone and NO_2), 15-24 km (OClO).

There are still uncertainties in retrieval for polar regions probably caused by ozone absorption cross-section dependence from temperature or radiative transfer modelling shortages.

References

Bovensmann, H., Borrows, J.P., Buchwitz, M., Frerick, J., Noel, S. & V.V. Rozanov (1999), SCIAMACHY: mission objectives and measurement modes. J. Atmos. Sc. 56(2), 127-150.

Haley, C.S., Brohede, S.M., Sioris, C.E., Grioen, E., Murtagh, D.P., McDade, I.C., Eriksson, P., Llewellyn, E.J., Bazureau, A. & Goutail, F. (2004), Retrieval of stratospheric O_3 and NO_2 profiles from Odin Optical Spectrograph and Infrared Imager System (OSIRIS) limb-scattered sunlight measurements, J. Geophys. Res., 109, D1630310.1029/2004JD004588.

Krecl, P., Haley, C.S., Stegman, J., Brohede, S.M. & Berthet, G. (2005), Retrieving the vertical distribution of stratospheric OClO from Odin/ OSIRIS limb-scattered sunlight measurements. Atmos. Chem. Phys. Discuss., 5, 2989-3046.

Lamsal, L.N., Weber, M., Tellmann, S. & Burrows, J.P. (2004), Ozone column classified climatology of ozone and temperature profiles based on ozonesonde and satellite data, J. Geophys. Res., 109, D20304, doi: 10.1029/2004JD004680.

Marquard L.C., Wagner, T. & Platt, U. (2000), Improved air mass factor concepts for scattered radiation differential optical absorption spectroscopy, J. Geophys. Res., 105, 1315-1327.

Platt, U. (1994), Differential Optical Absorption Spectroscopy (DOAS), in Air Monitoring by Spectroscopic Techniques, edited by M.W. Sigrist, Chemical Analysis Series vol. 127, John Wiley, New York.

Pundt, I., Pommereau, J.-P., Chipperfield, M.P., Van Roozendael, M. & F. Goutail (2002), Climatology of the stratospheric BrO vertical distribution by balloon-borne UV–visible spectrometry, J. Geophys. Res., 107(D24), 4806, doi: 10.1029/2002JD002230.

Rault, D.F. (2005), Ozone profile retrieval from Stratospheric Aerosol and Gas Experiment (SAGE III) limb scatter measurements, J. Geophys. Res., 110, D09309, doi: 10.1029/2004JD004970.

Rodgers, C.D. (2000), Inverse methods for atmospheric sounding: Theory and practice. World scientific publication, Singapore.

Rozanov, A., Bovensmann, H., Bracher, A., Hrechanyy, S., Rozanov, V., Sinnhuber, M., Stroh, F. & Burrows, J.P. (2005), NO2 and BrO vertical profile retrieval from SCIAMACHY limb measurements: Sensitivity studies, Adv. Space Res., accepted March 2005.

Solomon, S. (1999), Stratospheric ozone depletion: A review of concepts and history, Rev. Geophys. 37(3), 275-316.

von Friedeburg, C. (2003), Derivation of Trace Gas Information combining Differential Optical Absorption Spectroscopy with Radiative Transfer Modelling. Ph.D. Thesis, IUP Heidelberg, Heidelberg.

Behaviour of Pinewood (*Pinus Sylvestris*) in Various Mixtures

Pille Meier[1], T. Kaps[1], U. Kallavus[2]

[1] Department of Polymer Materials,
Tallinn University of Technology,
Ehitajate tee 5, EE – 19086 Tallinn, Estonia

[2] Centre for Materials Research,
Tallinn University of Technology,
Ehitajate tee 5, 1 EE – 9086 Tallinn, Estonia

Introduction

The swelling of wood in liquids is of fundamental importance both from a scientific viewpoint and in the context of commercial processes. Since swelling is a complex process, which is affected by both the type of wood and the solvent substrate, the influence of each of them on the swelling of wood is difficult to predict. Scientists have studied the phenomenon of wood swelling attempting to determine the factors influencing that process [1-6]. The aim of this study was to investigate the influence of some mixtures on the structure of pine sapwood.

Materials and Methods

Some sapwood samples of pine (*Pinus Sylvestris*) as typical representatives of softwood were selected for investigation. The samples were cut from green wood in the form 20 (radial) x 20 (tangential) x 10 (longitudinal) mm pieces and oven dried at 103°C to a constant weight. The longitudinal direction was selected as the shortest dimension to achieve complete penetration of the liquids in a reasonably short period of time. The density of pine varied from 0.4 to 0.55 g cm^{-3} and the number of annual rings per cm was 8 to 10 in the oven-dried condition.

Organic solvents used in this study were ethanol (EtOH), acetone (Act) and dimethylformamide (DMF). The aqueous solutions of organic solvents were prepared in different concentrations (molar fractions 0.1-0.8).

The samples of pinewood were prepared as described above for treatment with each solution – water, organic solvents and aqueous solutions of these solvents, at different concentrations. All experiments were performed at room temperature (20°C). All dimensional measurements were made with a Vernier caliper, accurate to ±0.05 mm. The oven-dried weights and radial, tangential and longitudinal dimensions were measured as quickly as possible (to avoid humidity uptake from the air) and samples were immediately immersed in the swelling agents and the containers sealed. The swelling of wood in the liquids was determined after 60 days. In order to study the multiple swelling the samples were oven-dried again at 103°C and placed in containers with new solutions for next 60 days.

Investigation of cell wall structure was carried out with a scanning electron microscope JEOL JSM 840A at 15 kV. Small sticks were cut out (3x3x7mm) from the swelled wood blocks and frozen at in liquid nitrogen. One end of the stick was cut flat by cryoultramicrotome using water as intermediate filling. The samples were air dried, and attached to the stubs with double-sided adhesive tape. The samples were then coated with gold by ion sputtering.

Results and Discussions

It is well known that some organic liquids swell wood less than water [5]. Some similarities exist between the swelling behaviour in ethanol and acetone (Figures 1 and 4). A strong effect of synergy became evident at MF_{Act} 0.4-0.5 and MF_{EtOH} 0.4-0.6.

Noticeable damping in swelling values occurred during re-swelling (Figures 1 and 4). Most significant damping appeared in ethanol and acetone aqueous solutions (MF 0.5). In spite of that, synergistic effect emerges also in case of re-swelling – wood swells in aqueous solutions more than in pure substances.

Fig. 1: Swelling of pine sapwood in aqueous solutions of ethanol

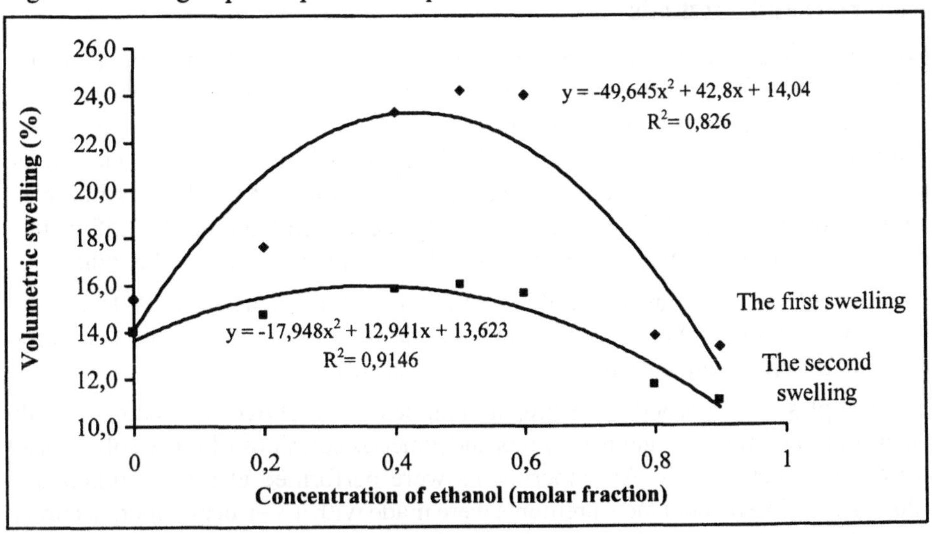

Figure 2a shows that the swelling of wood in ethanol (96%) caused the weakening of bonds between the CML and S1 and detachment of cells due to the forces applied during the sectioning of samples has occurred.

In Figure 2b it is evident that latewood cells have been so tremendously swelled in pure ethanol (96%) that the normal matrix of cells has been completely distorted. The single swelling in ethanol did not generate a similar effect.

The swelling of wood in ethanol-water mixtures has a dramatic effect on the wood structure. The wood cells (Figure 3a) were separated from each other and liberated CML was left hanging between the detached cells. The ethanol-water mixture dissolved parts of lignin from the CML and released individual cells at sectioning. The shape of the cells was distorted which indicates a loss in wood cell wall rigidity.

Fig. 2: The structure of pine sapwood swelled in the ethanol once (a) and two times (b)

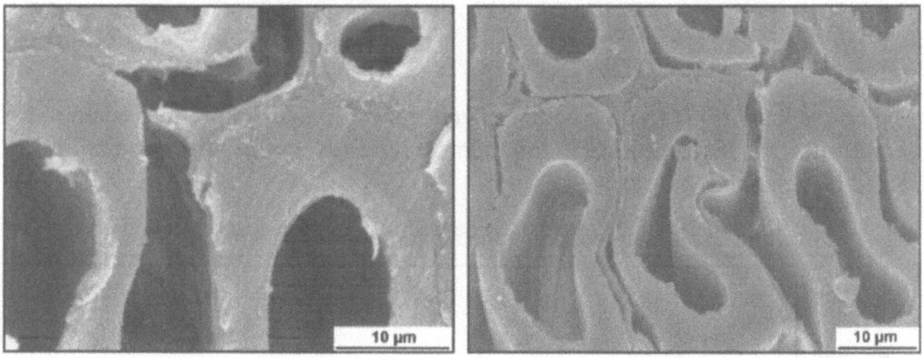

a b

In Figure 3b the disruption of wood architecture due to the second swelling in ethanol-water is very similar to Figure 3a. Only the rate of swelling is more expressive and the cell walls more distended.

Fig. 3: The structure of pine sapwood swelled in the ethanol aqueous solution (MF_{EtOH} 0.5) once (a) and two times (b)

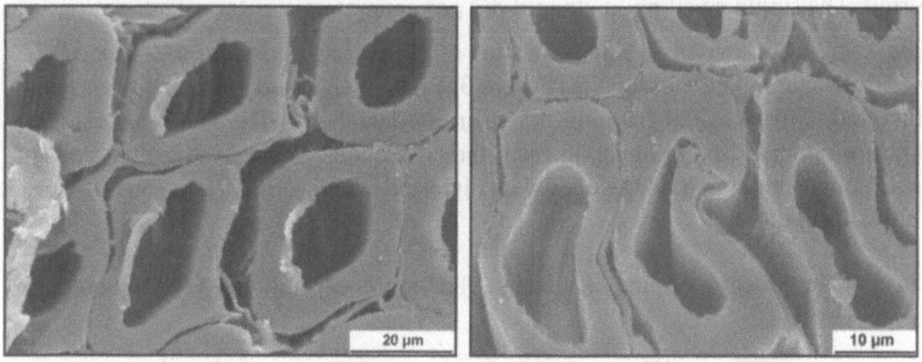

a b

Acetone is known [7] to be a better solvent for separated lignin than ethanol, especially for the high molecular fragments. Currently lignin was not directly affected by acetone.

Fig. 4: Swelling of pine sapwood in aqueous solutions of acetone

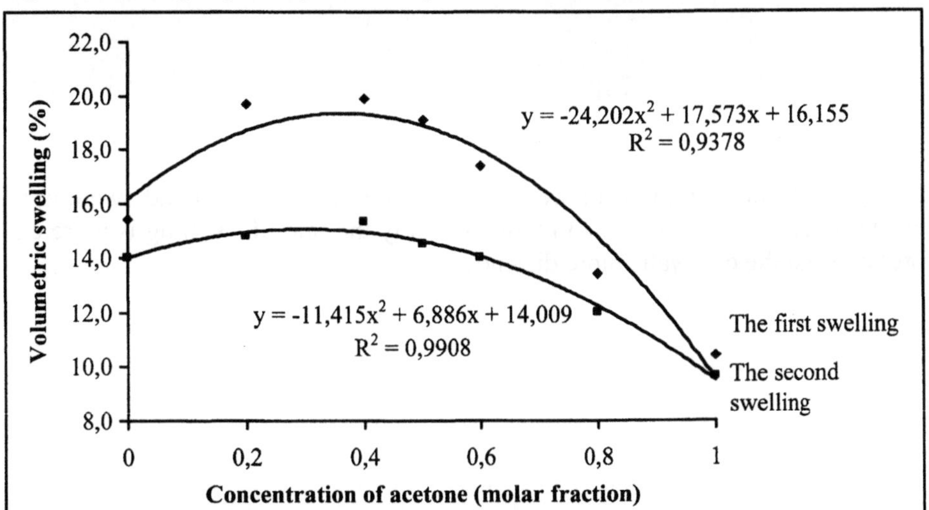

As it seen in the Figures 5 and 6, acetone and acetone-water mixture have a very mild effect on the cell wall structure and its bonding to the CML. Here the lignin is not directly accessible and the water does not fulfil the role of preliminary swelling.

Fig. 5: The structure of pine sapwood swelled in the acetone once (a) and two times (b)

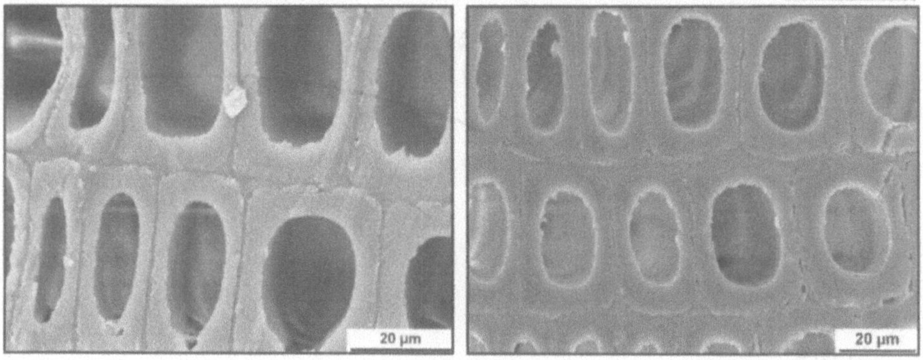

a b

Fig. 6: The structure of pine sapwood swelled in the acetone aqueous solution (MF_{Act} 0.5) once (a) and two times (b)

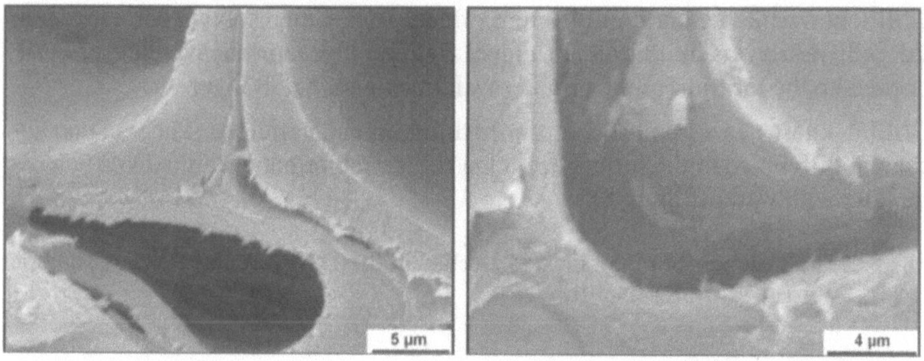

a b

Totally different swelling was detected on pinewood in DMF. There is no obvious synergy and after a rapid rise in swelling values after MF_{DMF} 0.2 the rate of increase becomes less evident.

Fig. 7: Swelling of pine sapwood in aqueous solutions of DMF

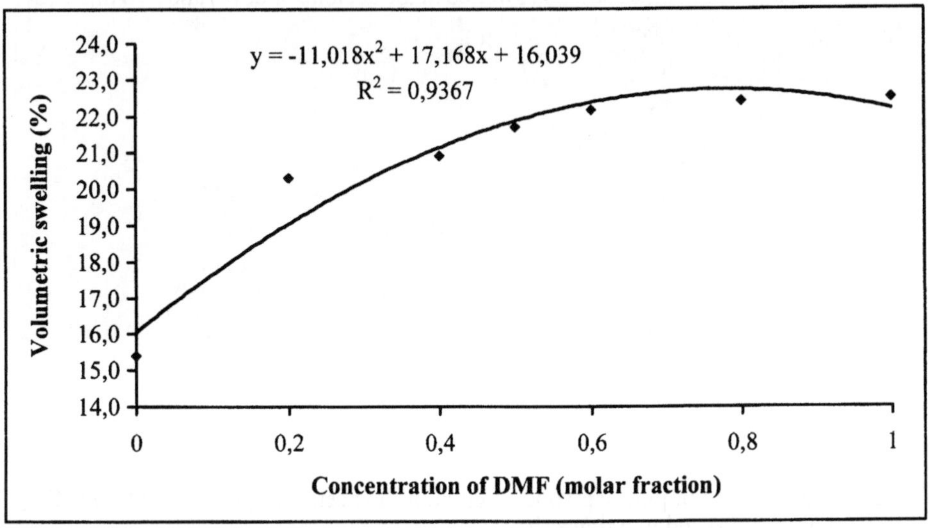

DMF shows interesting and different effects from other solvent influences on the cell wall layers and CML. In Figure 8 it can be seen that swelling in DMF opened up the location of the primary wall layer and the S3 layer.

DMF is known as a good solvent for both lignin and cellulose. There is no substantial difference in the influence on the cell structure between the DMF and its mixture with water (Figure 8).

Fig. 8: The structure of pine sapwood swelled in the DMF (a) and aqueous solution of DMF – MF_{DMF} 0.5 (b)

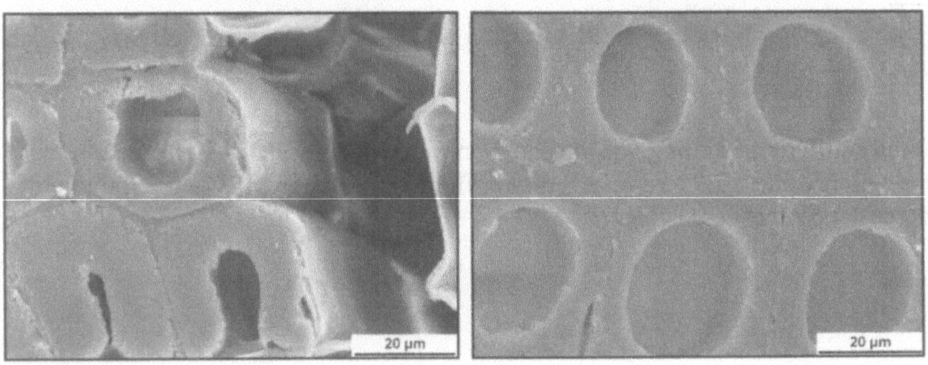

a b

Conclusions

1. Confirmation of the synergistic effects of binary solutions has been obtained. Anomalous properties of binary systems have a major influence on the swelling of the wood.
2. In wood a complex mechanism takes place where the water plays an important role in the swelling and solubility of the wood matrix in different solvents and water solvent mixtures.
3. From binary mixtures wood swelled most in ethanol aqueous solutions (molar fraction 0.4 – 0.6). Swelling values in acetone aqueous solutions were lower compared to ethanol aqueous solutions. Synergistic effect occurred also with swelling in acetone aqueous solutions.
4. In ethanol (96%) the latewood cells were so tremendously swelled that the normal matrix of cells had been completely distorted. The enlargement of the cross-section of latewood cells swelled in the ethanol showed the deliberation and disruption of the CML.
5. Acetone had a mild effect on the cell wall structure and it's bonding to the CML. The multiple swelling in acetone-water mixtures caused noticeable swelling of the S_3 layer of the cell wall.
6. The swelling in DMF opened up the location of the primary wall layer and S3 layer.

References

1. *Guyer, V.L., Hossfeld, R.L.* (1990), Interaction of Methanol-Water Binary Solutions with Wood. Holzforschung 44: 157-161.
2. *Ishimaru, Y. & Adachi, A.* (1988), Swelling anisotropy of wood in organic liquids I: External swelling and its anisotropy. Mokuzai gakkaishi 34: 200-206.
3. *Ishimaru, Y. & Maruta, T.* (1996), Wood swelling and its transverse anisotropy in organic liquids having two or more functional groups in a molecule. Mokuzai Gakkaishi 42: 234-242.
4. *Mantanis, G.I., Young, R.A. & Rowell, R.M.* (1994), Swelling of wood 1. Swelling in water. Wood Science and Technology 28: 119-134.
5. *Mantanis, G.I., Young, R.A. & Rowell, R.M.* (1994), Swelling of wood 2. Swelling in organic liquids. Holzforschung 48: 480-490.
6. *Morisato, K., Ishimaru, Y. & Urakami, H.* (1997), Adsorptsion of Liquids and Swelling of Wood. Part III. Adsorption of some organic liquids onto dried wood. Mokuzai Gakkaishi 43: 986-992
7. *Fengel, D., Wegener & G. Wood* (1989), Chemistry, Ultrastructure, Reactions. Walter de Gruyter, Berlin.

A Method Supporting Sustainable Landuse Management Applying Soil Erosion Models and Ecological Investigations

Adrienn Tóth

Geographical Research Institute, Hungarian Academy of Sciences,
Budaörsi út 45, H – 1112 Budapest, Hungary

Introduction

As the importance of nature and environmental protection became a central problem recently, the significance of natural resources came into the limelight as well. The most valuable natural resources of Hungary are the soil and, because of its turistical attraction, the landscape, the natural environment. The situation is the same in the case of the study area, the catchment of the Tetves stream: the main living source of its population is the agriculture, and, due to the closeness of the Lake Balaton, the tourism.

The soil regenerates in close relationship and interaction with the flora and fauna, if the circulation of the substances is undisturbed. But if it, for any reason they are destroyed they cannot be regenerated (Stefanovits 1992). Keeping this circulation undisturbed and protecting the soil from erosion has therefore also a great economical importance. In addition to its direct effects, soil erosion has also serious indirect, off-site effects, making this phenomenon even more dangerous. Accelerated eutrophication is such an effect, forming from the eroded soil and allowing organic matter to reach streams and lakes. In the case of the Balaton this process, accompanied by the considerable water level decrease, can lead to devaluation of the lake's attraction, threatening the living of many people.

Soil erosion caused by the cultivation and, as a consequence of it, the eutrophication endangers agriculture and tourism, in other words the living of the people living in the studied area. On account of this complex problem the following aims were set: systematization and mapping of the landscape types and habitats of the catchment, as well as working out and applying a method using erosion models, supporting the protection of the two main natural resources, the soil and the landscape.

Methods

Ecological Survey of the Study Area

To achieve the above aims an ecological investigation of the study area was performed. In addition their landscape types and habitats were identified. Landscape type mapping and the classification of the landscape types were carried out using existing landscape type arrangements, soil maps, digital elevation models, landuse maps and soil erosion maps. The data from the literature and maps were corrected on field observations. Also the habitat types, their proportion and patterns in the area was studied and mapped. (Pécsi et al. 1972; Marosi & Szilárd 1979; Kertész 2003).

Field Measurements

Wischmeier-Smith soil erosion plots were built and data from the plots were collected. Using the collected data, the USLE model was applied and the K (erodibility) factor was determined.

Modeling: Application of the MEDRUSH and the WEPP Models

Representative test sites, pairs of slopes, were selected, which demonstrate all the soil types, land use types and characteristic slope conditions of the catchment.

Soil samples from each soil type of the chosen slopes were taken and analysed in laboratory for the modeling. The necessary meteorological data series were collected and meteorological data series supposing climatic change for long term modeling were created. After collecting and systematizing the data needed for the models, the MEDRUSH model (Kirkby 1999) was calibrated with data from the test sites. Finally the MEDRUSH and WEPP models (Nearing et al. 1989) were run on representative slopes and the results were processed and evaluated.

Results

The following landscape types were determined on the Tetves catchment (Figure 1).

Fig. 1: Landscape types of the Tetves catchment

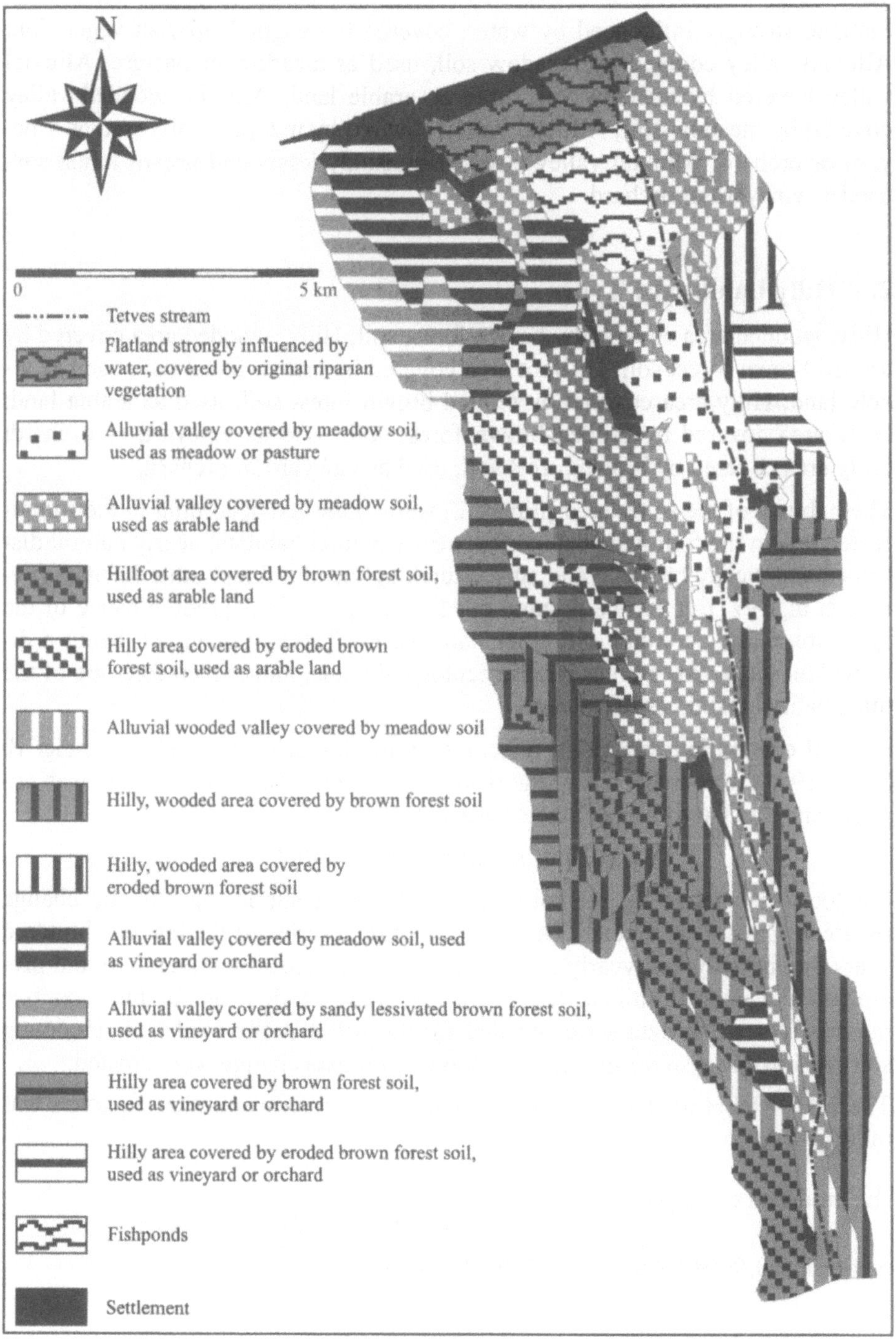

1. Flatland Landscape Types

Flatland strongly influenced by water, covered by original riparian vegetation; Alluvial valley covered by meadow soil, used as meadow or pasture; Alluvial valley covered by meadow soil, used as arable land; Alluvial wooded valley covered by meadow soil; Alluvial valley covered by meadow soil, used as vineyard or orchard; Alluvial valley covered by sandy lessivated brown forest soil, used as vineyard or orchard.

2. Hilly Landscape Types

Hilly, wooded area covered by brown forest soil; Hilly, wooded area covered by eroded brown forest soil; Hillfoot area covered by brown forest soil, used as arable land; Hilly area covered by eroded brown forest soil, used as arable land; Hilly area covered by eroded brown forest soil, used as vineyard or orchard; Hilly area covered by brown forest soil, used as vineyard or orchard.

The habitats of the study area (Figure 2) were classified according to their unaffectedness in the following categories: nearly natural habitats; nearly natural disturbed and weedy habitats; and agricultural, sylvicultural and other habitats (Fekete et al. 1997). It can be pointed out that in spite of the predominance of the agricultural, sylvicultural and other habitats, the Tetves catchment, being a diverse landscape, has a considerable ecological value, and is therefore one of the main natural resources of the area.

For soil erosion investigations Wischmeier-Smith erosion plots (Wischmeier & Smith 1978) were built. Using the data from the plots the USLE model was applied, and the K (erodibility) factor was determined:

$$0{,}034\ t \times ha^{2} \times h \times ha^{-1} \times MJ^{-1} \times mm^{-1}.$$

For long term modeling meteorological data series supposing climatic change (Kertész 1996) were created. On the basis of data series of the last one hundred years the monthly and yearly tendencies of the changes in temperature and precipitation were determined. Supposing the same tendencies for the next one hundred years, the data series needed for the MEDRUSH model for modeling soil erosion and also for the case of supposed climate change were created.

The MEDRUSH model was calibrated and its evapotranspiration equation was modified.

The modified equation: $$PET = \frac{R}{3{,}5 + B},$$

where $B = 0{,}0004714 \times T^2 - 0{,}0425 \times T + 0{,}0771$.

Fig. 2: Habitat types of the Tetves catchment

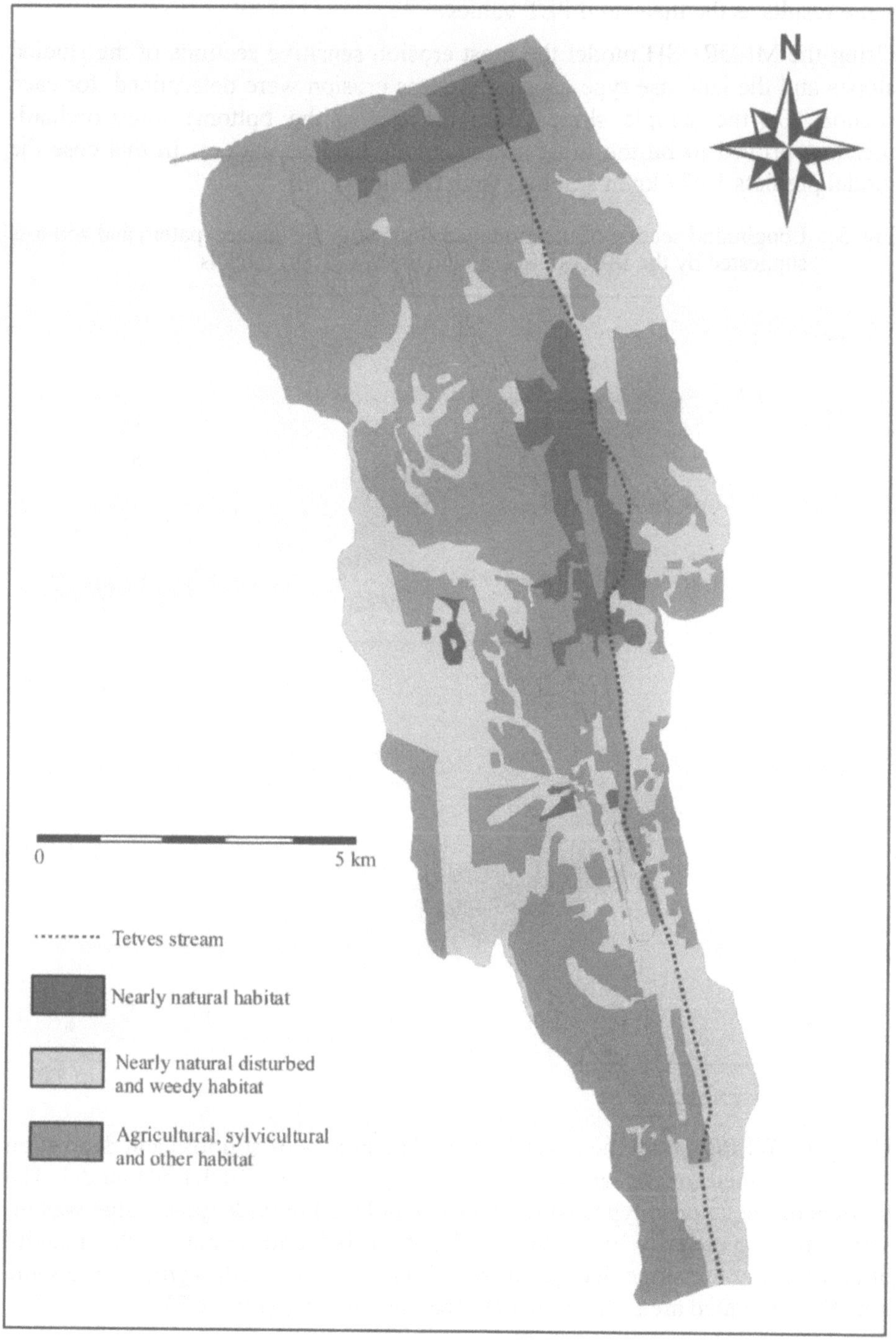

Using the modified equation, all the test size and time scales gave almost the same results as the measured PET values.

Using the MEDRUSH model the most erosion sensitive sections of the studied slopes and the land use type causing the less erosion were determined, for each section. On the sample slope (from the top to the bottom) forest-orchard-meadow proved to be the most advantageous landuse pattern. In this case the model predicts 1,477 kg/m soil loss from the slope.

Fig. 3: Longitudial section of the modelled slope with the landuse patern and soil loss suggested by the MEDRUSH (a) and the WEPP (b) models

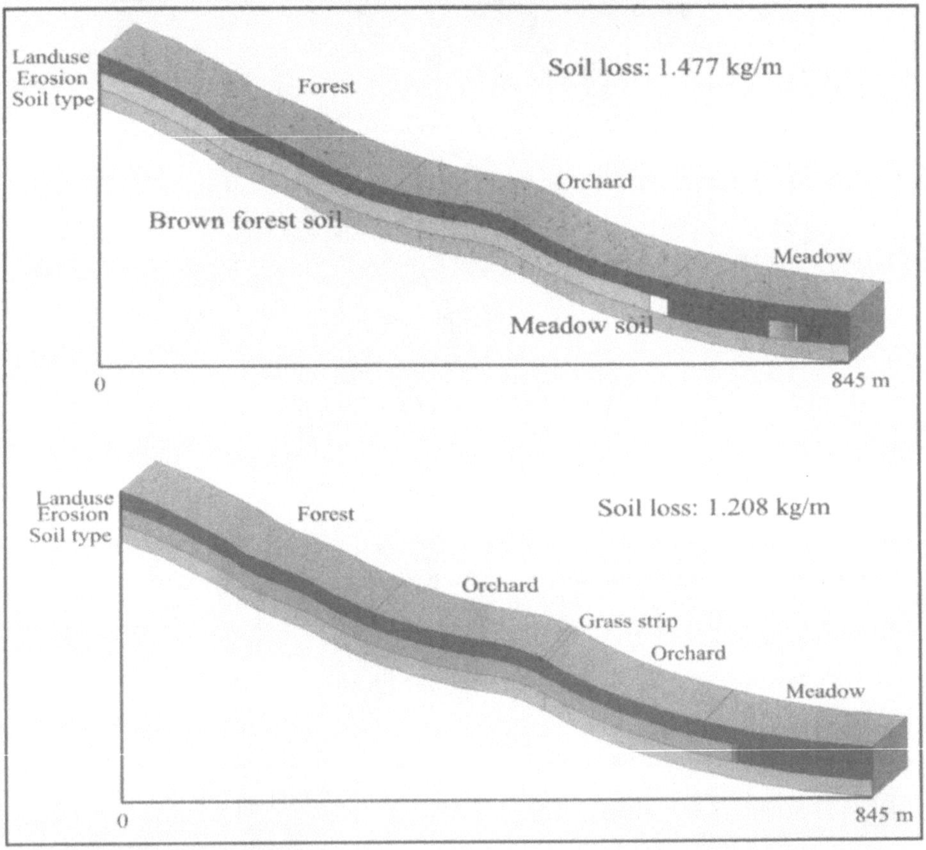

Using the WEPP model the exact spots and quantities of soil loss and deposition were determined for the landuse pattern set up with the MEDRUSH model. The borders of the landuse types were modified and a 20 m wide grass stripe was inserted into the orchard. In the case of this modified landuse pattern the quantity of soil leaving the slope decreased from 1,477 kg/m to 1,208 kg/m. In the same time the cultivated area (the orchard) expanded by 19% (Figure 3).

Thus a method supporting landuse planning was worked out, which helps to find the landuse pattern causing less soil loss and providing the most advantageous landuse type distribution. It was demonstrated that it is possible to increase the proportion of cultivated land and in the same time reduce the quantity of soil loss, with a relative trifling intervention. This offers the opportunity to withdraw from cultivation the fields mostly endangered by erosion, while keeping the given proportion of landuse types.

Conclusions

Overall landuse planning and regulation for whole catchments could help to solve the problem of harmful effect of not suitable landuse. Because of the fact, that agriculture is a main living source of the population, the given proportion of landuse types must be kept in mind. The most obvious solution: forestation or grassing of the erosion sensitive spots can not be realized, since the farmers have no choice but to choose the most remunerative cultivation system. The conditions for this should be created by governmental intervention, by supporting favourable landuse patterns. Another problem, which must also be taken into consideration during landuse planning, is that the present structure of estates with many long, narrow parcels does not always allows contour plowing.

The suggested regulation and the changes originating from it will not be free from difficulties anyway, so they should be planned for the long-term. Thus the weather, the climate and their possible change, which are essential factors of agriculture, cannot be neglected.

Finding a solution which takes erosion conditions, local features, and probable climatic changes into consideration is a complex task. On the basis of my work I propose to support landuse planning with two erosion models, in addition to field survey and deep knowledge of the area.

MEDRUSH model makes long-term study of erosion and water balance conditions, as well as the prediction of the effects of probable climatic changes possible. Modeling ten years under the present conditions and ten years under the changed ones on representative slopes of the catchment under different landuse types gives an overall view about the erosion conditions of the area. With the help of the obtained results the spots can be determined, where soil loss can be significantly reduced by landuse change, and thus intervention is very important. Fields can be found, where no changes are necessary, and non-cultivated areas (usually grasslands) can also be designated, which can be used as arables or orchards without causing damages, instead of the fields which must be withdrawn from cultivation.

The WEPP model supports the detailed, exact planning of the areas to be changed according to the results of the MEDRUSH model. It makes possible the study of the effects of single precipitation events as well as the very detailed setting and modification of the cultivation methods. Running the model for extreme storm events we can learn how much soil leaves the certain slope sections in the most extreme situations. The most suitable solution can be found changing the landuse type, the cultivation method, the cultivated plant, moving the borders of the landuse types, inserting protective grass- or forest stripes.

Supporting landuse planning with the above method, the complementary use of the two models, it become possible to find out the most erosion sensitive areas and on each slopes the best solution. By this means this method helps to preserve soil and landscape, the important natural resources.

References

Fekete, G., Molnár, Zs. & Horváth, F. (1997), Nemzeti Biodiverzitás-monitorozó Rendszer II. A magyarországi élőhelyek leírása, határozója és a Nemzeti Élőhely-osztályozási Rendszer. Magyar Természettudományi Múzeum, Budapest.

Kertész, Á. (1996), A fokozódó szárazság természetföldrajzi következményei Magyarországon. Környezet- és tájgazdálkodási Füzetek 1996/4. pp. 49-54.

– (2003), Tájökológia. Holnap Kiadó, Budapest, 166 p.

Kirkby, M.J. (1999), Application and further development of the MEDRUSH model. Book od MEDALUS III. Final Report. pp. 528-559.

Marosi, S. & Szilárd, J. (1979), Somogyi tájtípusok jellemzése és értékelése, Földrajzi Értesítő 28, pp. 51-85.

Pécsi, M., Somogyi, S. & Jakucs, P. (1972), Magyarország tájtípusai, Földrajzi Értesítő XXI. 1. 5-12. p.

Stefanovits, P. (1992), Talajtan. Mezőgazda Kiadó Budapest 380 p.

Wischmeier, W.H. & Smith, D.D. (1978), Predicting rainfall erosion losses. A guide to conservation planning. USDA Agriculture Handbook. No. 537. Washington.

Landform Map and Sediments of North Vidzeme

Dainis Ozols

North Vidzeme Biosphere reserve
NVBR Administrative center
Riga Street 10a, LV – 4033 Salacgriva, Latvia

Introduction and Aims

North Latvia and South Estonia have very special landform patterns – lowlands and elevations developed by surging Pleistocene glaciers on soft clastic Devonian bedrock.

The territory represents terrain of prevailing glacial erosion processes, situated proximally of main marginal and stationary lines (Straume 1982). Long-standing praxis is to interprete the bulk of sediments as glacigene and periglacially deposited (Zelchs & Dreimanis 1997; Kalm & Kadastik 2001).

During the last decades lot of publications stress the significance of subglacial meltwater flow, high hydrostatic pressure and plastic properties of subglacial drift (Shaw 1998; Cox 1998; Piotrowski, Geletneky & Vater 1999; Breemer, Clark & Haggerty 2002; Lian, Hicock & Dreimanis 2003, etc.).

Understanding of processes that created landforms and sediments is essential for prospecting of mineral resources (e.g. sand and gravel deposits) for evaluation of landscapes that contains values of geodiversity, biodiversity and cultural heritage, and for improving our knowledge of paleogeography.

Methods

A landform map of part of North Latvia is being originated. For examining detail topographical maps of scale 1:10 000 are used.

Because of the dominance of erosion processes around the territory special appointment is being paid to determination of negative landforms.

Glacial sediments were studied on outcrops along the coastline of the Baltic sea, Gulf of Riga and within the quarries. The Current study kept a watch on macro-scale features of the deposits.

Landforms

Strong evidence of action of subglacial currents is the pattern of medium scale erosion landforms (Figure 1).

Fig. 1: Character of landforms at the northwest part of lowland Burtnieku. Hue shows relative (not absolute!) elevation; positive landforms are lighter, negative – darker.

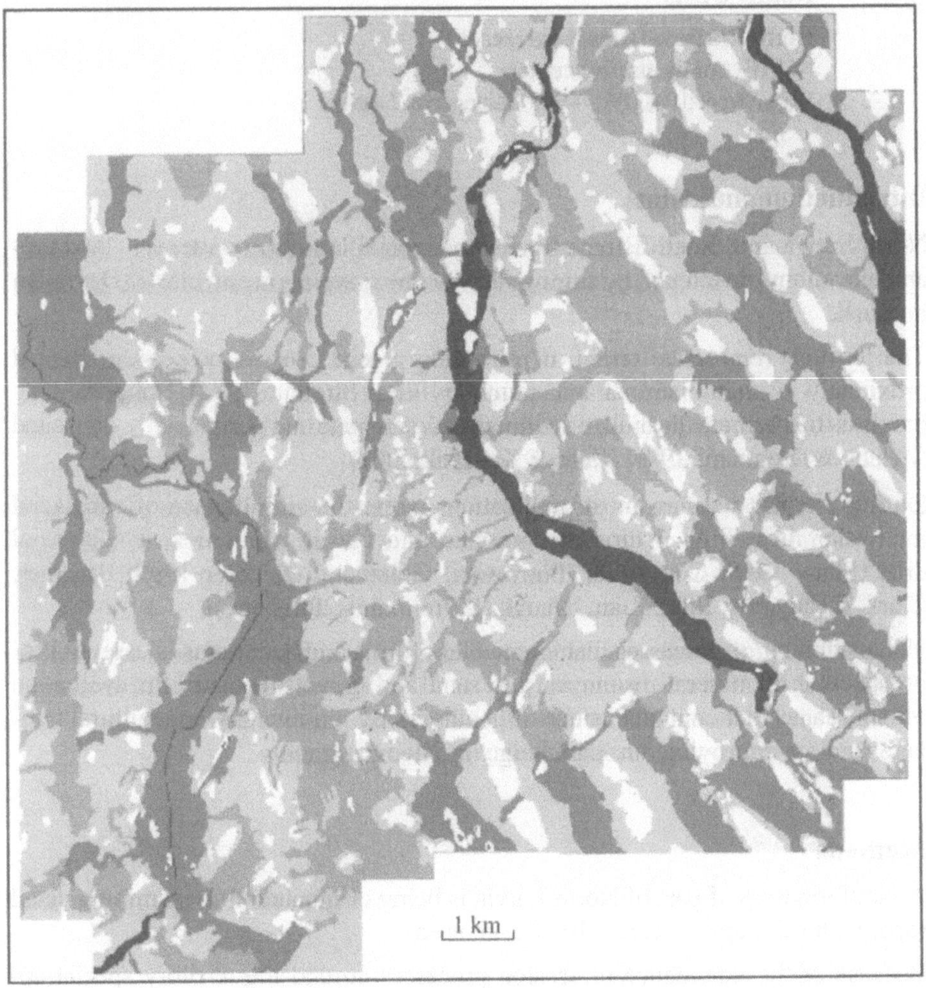

Maps obtained demonstrate the pattern of landforms that are bearing evidence of subglacial watercourses and were formed during two stages of last glacial surge: gentle forms of initial stage (valleylike depressions, drumlins)

Valleylike depressions are shallow gentle landforms interconnected within a common network. These forms are divided from each other by drumlins and similar forms. Sections of land surface across the set of valleylike depressions express a line that resemble sine curve. Similar *transverse valleylike depressions* are narrower, oriented in transverse direction and used to be situated on the slopes or upper parts of gentle positive forms, the more articulate forms of the final stage – subglacial valleys, eskers, dome like hills.

Subglacial valleys (channel valleys, "senlejas") – are the best shaped negative landforms of the territory. They are 5-20 m deep with steep slopes, with various morphology. Some subglacial valleys have tributaries and parallel channels (Svētupe), others have eskers and accumulative terraces within the valleys (Rūja, Lake Saruma). Some valleys have erosion's terraces covered by boulder pavements (Salaca, Rūja, Vitrupe).

Sediments

Mobilisation and Transportation

The beginning of the surge cycle is connected with the origin of the water saturated layer of unfrozen ground beneath the glacier. When the upward moving boundary of unfrozen ground reached the level where glacier buoyancy exceeded the gravity, the overlying glacier rose together with the *appended sediment layer*, frozen to the bottom of glacier. The glacier lost any rigid connection with the ground and started moving in the direction of lesser resistance. A sediment layer frozen to the bottom of the moving glacier seems to be the most prevalent way for long distances and prolonged time intervals of material transportation.

The most common way is transportation with *normal currents* where saturation of particles isn't high enough to change the properties of the flow. Particles were mobilised directly by current action and moved by tracing, saltation and as suspension.

Due to movement of the glacier some conduits are opening and some closing. That leads to the formation of sudden abundant subglacial currents that were able to carry great amount of clastic material. Dense water and clasts masse was supported by fluid turbulence and had great eroding possibilities. The mechanism of transportation allows us to relate such flows to *turbidite currents*.

Floating earth conditions became possible when mobilisation of material was triggered thanks to glacier movement. Moving over highly water saturated sand or silt sediment that was under very high pressure the glacier made it unstable. Material loosened grain to grain contact and friction strengths (became liquefied) because water pressure compensated the pressure of overlying material load (e.g., Geology 305).

Liquefaction requires small sizes of material particles (sand and silt with allowable small amount of pellitic and gravel/pebble material). Internal cohesion of clays and good permeability of coarse sediments doesn't allow them to be liquefied. Floating earth conditions derive from

1) differences of pressures within subglacial conduits

2) presence of liquefied sandy or silty material.

When the fracture or channel connected the area of high pressure with one where pressure is lower, water saturated masse began to flow. Conditions favourable for floating earth existed along the ice divides and zones of fracturing. There subglacial flow off from subglacial high pressure zones of moving glacier persisted. The mechanism described is similar to genesis of sand (sandstone, clastic) intrusions – "more frequently developed in tectonically active settings where applied tectonic stresses facilitate development of high fluid pressures within the sediments" (Shoulders & Cartwright 2004).

Different authors include different meanings of terms *slurry flow*, *debris flow* and *mud flow*. We are going to use the term *slurry flow* meaning the moving of water saturated masse of high density with an interstitial fluid of enhanced viscosity. The last indication differs the slurry from the liquefied sand or silt. Although we concede the possibility of using the term slurry flow (*sandy slurry flow*) concerning liquefied sand if it is moving by glacier traction (not by pressure gradient).

Slurry flows into the conditions of subaerial or subaqual environments are considered as inertia or mass flows that are moving due to gravity. Within the subglacial environment 2 kinds of driving forces exist – 1. gradients of pressure; 2. traction by moving glacier. Sometimes it may be difficult to distinguish which force prevails.

Slurry on average is viscous and is not able to move long distances due to differences of pressure. More usual is movement of slurry by glacier traction with subsequent penetration into cavities and fractures because of lower pressure.

Laminar flow of viscous substances can exist due to glacier movement only if saturation of clayey and clastic material is high enough. The situation which was favourable for the involvement of underlying sediments into the movement persisted during the final phases of the surge cycle, when pressure of subglacial water diminishes and the moving glacier was coming into the closer contact with the substratum. Mixing of underlying sediments generated uniform, clayey, viscous aggregate with high water content.

Mechanism of glacier movement on deforming layer is in close association with slurry flows and liquefied sand and silt. When the bottom of the moving glacier involves underlying substratum into the movement by glacier traction within the short distance, the sediments may keep remains of its previous structure and may be considered as a *deforming layer*.

Deposition, Deposits and Evidence

During the transportation of the sediment layer *frozen to the bottom* of glacier matter it experienced forces of flowing currents and glacier traction.

The later forces and melting caused *flaking off* of pieces and clasts of previously frozen material. Usually pieces of such subglacial drift are buried within coarse gravel and pebble deposits – formations of normal currents and turbidites.

Deposits of *normal currents* were observed in most places as parallel or cross laminated sands of various grain size, with or without admixture of pebble and gravel. Observation of drop direction of cross laminated layers is essential, because it denotes the direction of subglacial currents. It was established that during the active phase of glacial surge, currents flowed in the same direction as the glacier was moving. During the final phase currents within subglacial valleys flowed in the direction to the closest glacier margin, under the angle 30-70° compared to previouly.

Deposits of *turbidite* currents are less widespread. We regard coarse gravel, pebble and cobble mixtures with a dash of rounded pieces of local devonian clays and siltstones as sediments of turbidite origin. It has coarse parallel or cross lamination or is without lamination. Some have normal grading. Some layers have radical bimodal distribution of grain size with modes within intervals of sands and pebbles. In the quarries Veseta and near Krišjāņi on the western end of rampant Numierne we found the well rounded pebbles and cobbles of local dolomites transported only a few kilometres. That testifies to the very active abrasive interaction of particles within turbidite current.

The subglacial environment has some features similar to deep sea – high pressure, unconsolidated sediments, unstable slopes etc. Because of this we are granting a possibility of persisting fine grained turbidite deposits within the territory, but it must be difficult to distinguish it from weakly laminated sediments of normal currents, sediments of floating earth and from slurry flow sediments of sandy composition.

Our study testifies that sediments of *floating earth* (Figure 2) are widespread within the territory. They are mainly connected with local ice divides and flanks of uplands as well as side slopes of subglacial valleys and valleylike depressions. Characteristic features of floating earth deposits are massive composition, absence or weakly developed lamination, composition of fine fractions (sand and silt) with low content of gravel and coarser particles and usual reddish to ginger hue of colour (the latter due to its origin from local Devonian sandstone and siltstone). Deposits of floating earth make up thick layers and lenses and great parts of sections of landform complexes consisting of steep dome like hills and hilly elevations (near Augstroze, Ērģeme, Smiltene etc.). Also numerous drumlins include bodies of massive sand. Floating earth deposits associate with normal current, turbidite and slurry flow deposits, forming interlayers. Associated sediments of normal currents have structures showing evidence of high saturation of debris, e.g., ascending current ripples.

The stoppage of the moving glacier halted underlying *slurry flow*. Viscous matter before halting moved in a laminar manner rotating clasts and soft sediment pieces (Figure 3). High overburden causes high porewater pressure that considerably exceeds interrelation stresses of clasts and brings them to naught.

Our opinion is that a greater part of clayey varieties of glacial diamicton („till") of territory were formed as slurry flow deposits. Strong testimony of such a viewpoint is lack of thrust slices, massive or slightly bandy composition and abundant presence of rotated clasts both solid and loose. If the glacier continued some movement when the slurry lost high water content, cohesion strengths arose and solid clasts got orientation while loose ones became flattened and extended in the mode of interlayers or lenses.

Fig. 2: Floating earth deposits – fine silty sand with the textures caused by liquefaction. Slurry flow sediment (stony loam) – on the top. Quarry Romas near Valmiera

Within the regions of wide spreading sandy deposits (marine deposits or outwash plains) we observed *sandy slurry flow* deposits – massive beds of sand with inclusions of pebble or rounded pieces of laminated sand (Figure 3-1) and coarse matrix-supported gravel-pebble deposits (Figure 3-2).

Fig. 3: Sandy slurry flow deposits with rounded rotated clasts of loose sediments. Unit of the ruler =10cm. Cliff of Baltic sea near Jūrkalne.

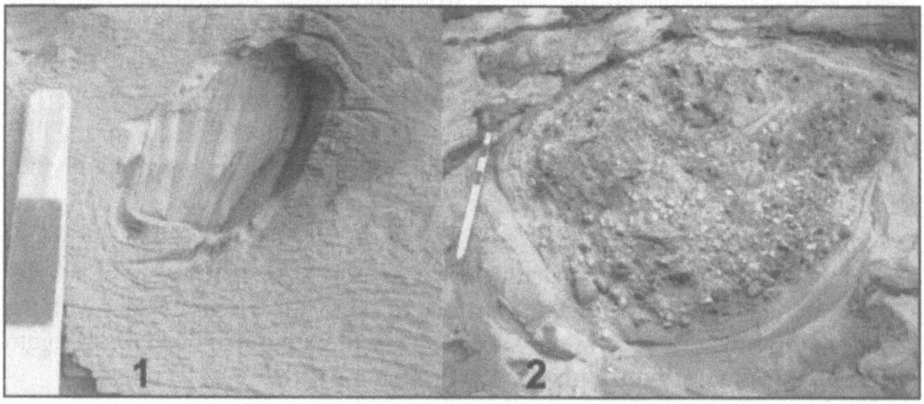

Slurry flow deposits has transition to *deforming layer* formations.

The later ones preserved some structures of initial sediment and got new ones due to movement by glacier traction within the slurry flow conditions. Deforming layer indications are characteristic for the upper part (2-3m) of a clastic Devonian sequence (loose sandstones and siltstones), seen almost on every Devonian outcrop.

Diagenesis. After sedimentation deposits experienced action of the final movement activities of the surging glacier it lead to an origin of deformations – thrust slices and folding. These changes we regard as diagenetic processes of glacial lithogenesis.

Only in someplaces did generation of *local till* by heavy modification of underlying sediments take place.

It must be noted that postsedimental treatment of glacial deposits is in close connection with landforms position. Sediments of uplands are much more reworked by direct action of glacier than sediments of lowlands.

Discussion. The theory of *glacial surges* is best at explaining the features described above.

It is hard to establish how to consider surge landforms – as glacigenic or glaciofluvial and erosion or accumulative because both sides of the processes are involved.

Each cycle of glacial surge consists of 4 phases.

1. Meltwater at the base of the glacier couldn't find free leakage. Due to the pressure of the overlying ice sheet the subglacial water was under very high pressure (10-100 atm) and made the overlying glacier *buoyant*. That led to disappearance of friction between glacier and underlying sediments. The glacial surge was *beginning*.

2. *Active stage*, when the pressure beneath the glacier was highest. The water saturated layer separated the glacier from underlying substratum. Meltwater erosion incised within numerous shallow valleys connected in fanlike network. Elongated hills arose between channels as erosion remains with some involvement of accumulation on the top of the forms (drumlins, eskers). The buoyant part of the glacial lobe and tongue, moved in the direction of the glacier margin. Clastic material had been carried in the direction of ice-divides (mainly – as floating earth) and ice margin.
3. *Ending*. A cinematic wave approached the ice margin and subglacial water got free outflow from the glacier. That led to a decrease in pressure of subglacial water and the base of the glacier tongue came into closer contact with the underlying substratum creating viscous slurry flow that had been pressed into the emptiness amongst the glacier and underlying sediments.
4. *Movement of the glacier gradually stopped*. The pressure of the water column was falling and the network of subglacial channels was reformed to a converging pattern – with the main meltwater streams that lay in deeper subglacial valleys oriented in the direction of the closest glacier margin. As a rule one deep valley lay within the axial part of each lobe. The final stoppage of the glacier marked the end of the surge cycle and the *transition to the next surge cycle*.

Surges were alternating with longer periods of glacier stagnation. Glacial surges must be considered as processes of *normal evolution of glacier* at the middle latitudes – not the events of catastrophic character.

Conclusions

North Latvia is located within the terrain that during the glaciations experienced prevailing erosion processes. Due to its negative landforms are determining the character of landscapes and geomorphologic mapping concentrates on negative landforms.

Glacial landforms and deposits are attributed to formation of glacial surges.

Glacial sediments in the majority of cases may be determined as subglacial deposits of normal currents, turbidites, floating earth, slurry flow and flaking off drift material.

During the surge cycle very high hydrostatic pressure prevailed beneath the glacier.

Direct glacier impact is interpreted as diagenetic modification of deposits. Four phases of glacial surge has been divided.

Proposed model of the evolution of surge processes is in good consonance with sections of deposits of lowlands and small uplands of North Latvia with sandy subglacial waterlain sediments at the base and clayey slurry flow deposits on the top of sections.

Acknowledgements

This study was undertaken as part of a research program of North Vidzeme Biosphere reserve (NVBR) and doctoral thesis of the author. I would like to thank my colleagues of NVBR for support, colleagues of Latvian University for constructive criticism and my family for assistance.

References

Breemer, C.W., Clark, P.U. &, Haggerty, R. (2002), Modeling the subglacial hydrology of the late Pleistocene LakeMichigan Lobe, Laurentide Ice Sheet Department of Geosciences, Oregon State University, Corvallis, Oregon 97331, USA www.geo.oregonstate.edu/people/faculty/clark_publications/breemeretal.-gsab-2002.pdf.

Cox, D. (1998), Drumlins and Subglacial Meltwater floods. www.sentex.net/~tcc/sgfcrit.html.

Geology 305 – Introduction to Engineering Geology.

Kalm, V. & Kadastik, E. (2001), Waterlain glacial diamicton along the Palivere ice-marginal zone on the West Estonian archipelago, Eastern Baltic sea.

Lian, O.B., Hicock, S.R. & Dreimanis, A. (2003), Laurentide and Cordilleran fast ice flow: some sedimentological evidence from Wisconsinan subglacial till and its substrate Boreas, vol. 32, no. 1, pp. 102-113(12) Abstract www.ingentaconnect.com/ content/ tandf/ sbor/ 2003/ 00000032/ 00000001/ art00005.

Ozols, D. (2004), Research into Pleistocene Deposits within the North Vidzeme Biosphere Reserve and South Estonia, in "Integrative approaches towards sustainability in the Baltic Sea Region", pp. 475-479, Frankfurt am Main, Peter Lang GmbH, Europäischer Verlag der Wissenschaften.

Piotrowski, J.A., Geletneky, J. & Vater, R. (1999), Soft-bedded subglacial meltwater channel, from the Welzow-Sued open-cast lignite mine, Lower Lusatia, eastern Germany. Boreas 28, 363-374.

Proc. Estonian Acad. Sci. Geol., 2001, 50, 2, 114-127.

Shaw, J. (1998), Sedimentary evidence favouring the formation of rogen landscapes by outburst floods Department of Earth and Atmospheric Sciences University of Alberta Edmonton, Alberta, Canada, T6G 2E3 www.sentex.net/~tcc/ rogen/.

Shoulders, S.J. & Cartwright, J. (2004), Constraining the depth and timing of large-scale conical sandstone intrusions Geology; August 2004; v. 32; no. 8; p. 661-664; DOI: 10.1130/ G20654.1.

Straume, J. (ed.) (1982), Geomorphologic map of the republics of Soviet Baltic, Scale 1: 500 000; map and explanatory text (in Russian), Vilnius.

www.earth.usc.edu/ geol305/ lab/ Lab11sp05.pdf.

www.kirj.ee/ esi-l-g/ g50-2-4.pdf.

Zelchs, V. & Dreimanis, A. (1997), Morphology, internal structure and genesis of the Burtnieks drumlin field, Northern Vidzeme, Latvia. – J. Sedimentary Geology, vol. 111, pp. 73-90.

Composition of Weed Community Depending on Soil Compaction in Barley Field

E. Reintam, J. Kuht, K. Trükmann, J. Puust

Estonian University of Life Sciences,
Department of Soil Science and Agrochemistry,
u64 Kreutzwaldi Str., EE – 51014 Tartu, Estonia

Abstract

The main objective was to investigate the effect of soil compaction on the composition of the plant community, on barley and weeds shoots dry weight and density in spring barley (*Hordeum vulgare* L.) fields. Data was collected from Estonian Agricultural University research field with different levels of soil compaction (4 levels) on sandy loam soil between 2001-2004. The soil was compacted by tractor MTZ-82 (with front loader, total weight 4.9 Mg) by multiple tyre-to-tyre passing 1, 3 and 6 times. The results of the experiment showed that soil compaction increased the number of barley and weeds shoots but decreased their dry weight. In the first few years the most widespread weed species were common lambsquarters (*Chenopodium album* L.), field pennycress (*Thlaspi arvense* L.) and common fumitory (*Fumaria officinalis* L.). After four years direct soil compaction and no use of fertilizers the number of great plantain (*Plantago major* L.) and coltsfoot (*Tussilago farfara* L.) plants on dense soil increased and the number of common lambsquarters decreased. From the community disappeared common fumitory in the earing phase of barley. The total number of weed species decreased from 19 species on uncompacted soil in the first year to 12 species on 6-times compacted soil in the fourth year.

Introduction

On the estimation of productivity of plants in agroecosystems the greatest attention is paid on cultural plants. Little attention is paid to weeds, whose production and competition ability depend also on changes in conditions of habit, including soil compaction. On the arable land, different communities of weed species appear not only due to the different type of soil, but also because of cultural plant diversity in agroecosystem and around that, durability of different cultures, intensity of management and isolation of agroecosystem from natural vegetation (Altieri 1999). Tillage systems affect weed emergence, management and seed production. Changing tillage practices consequently changes the composition, vertical distribution and density of weed seed banks in agricultural soils. (Ghosheh & Al-Hajaj 2004). Derksen et al. (1993) suggested that changes in weed communities were influenced more by environmental factors (location and year) than by tillage systems. Swanton et al. (1999) concluded, that long-term changes in weed flora would be driven by interaction of disturbance (tillage), environment (soil type and moisture), crop rotation, crop type and timing and type of weed management practice. In intensive management systems the composition of weed community is confirmed. In experiments in Norway there were no changes in weed community during 5 years even at the highest doses of herbicides (Fykse & Wærnhus 1999). Changing the tillage or management intensity and soil physical parameters (during compaction) will result in changes in weed flora. Without regular ploughing annual weeds disappear and perennial plants take over (van Elsen & Günther 1992). Similar changes in weed community and production of cultural plants are taking place during soil compaction (Reintam & Kuht 2002) due to higher soil bulk density and penetration resistance. In central Iowa a single wheel-tracking pass at crop sowing increased the cumulative number of seedlings of common lambsquarters (*Chenopodium album* L.) by 30% (Jurik & Zhang ShuYu 1999). Common lambsquarters is the most dangerous competitor for the nutrients to cultural plants because of its high mass and nutrient content. Common lambsquarters and wild mustard (*Sinapis arvensis* L.) are the most widespread weed species on mouldboard ploughed nutrient rich neutral soil (Zanin et al. 1997). However, common lambsquarters and wild mustard will leave the leading position to other weed species in case of low fertilizing and denser soil (Bischoff & Mahn 2000).

The aim of this work was to investigate the long lasting soil compaction effect on composition on weed community in barley (*Hordeum vulgare* L.) field without using herbicides and fertilizers.

Material and Methods

Data was collected from Estonian Agricultural University research field (58°23'N, 26°44'E) on sandy loam Stagnic Luvisol at Tartu County between 2001-2004. By tractor MTZ-82 (with front loader, total weight 4.9 Mg) soil compaction was done to get different levels of compaction before sowing time in spring 2001, 2002, 2003 and 2004. The inflation pressures in the wheels of the tractor were 150 kPa. For all that traffic applied uniformly to cover the entire experimental plots: 1 time, 3-times and 6-times. Four compaction treatments on the experimental field formed, one of them control (without special compaction). The size of every experimental plot was 3x9 m (27 m^2). Drilling of barley (crosswise to compaction variants) with 450 germinating seeds per m^2 was done. No fertilizers and herbicides were used for better investigation of weed flora. Every autumn the soil was ploughed at 0.21-0.22 m depth. The samples of soil and plants were taken in earing phase of the barley. Data regarding the content of the plant community was obtained from taking vegetation samples (plant shoots) from a plot of 0.25 m^2 (n=4). The types of component (barley and different weed species) were determined, counted, measured, weighed and dried. The dry weights of plants were calculated. Soil bulk density was measured with 50 cm^3 (h=5 cm, Ø=3.5 cm) cylinders in 10 cm layers up to 40 cm in four replications. From the same samples the soil moisture was measured. Penetration resistance was measured with a cone penetrometer (cone angle 30°, stick diameter 12 mm) in every 5 cm layer up to 45 cm in 6 replications. To compare the differences between values the least significant differences (LSD) at significance $p<0.05$ were founded. Because of weather conditions the growing period (from May to August) of barley was rainy and cold in 2001 and 2003. The precipitation totals were 373 mm and 450 mm, respectively. Average air temperature was 15.8°C in 2001 and 15°C in 2003 during the barley growing period. In 2002 it was warm and dry. During the vegetation period only 163 mm of rain fell and the average air temperature was 17.4°C. The rainiest year was 2004, when it was 475 mm precipitations during the growing period; average air temperature was 16.2°C.

Results

Soil compaction by 4.9 Mg tractor increased soil bulk density significantly ($p<0.05$) already after one single pass in 0-0.1 m depth in all the years of investigation (Figure 1). In 0.1-0.2 m depth the increase of soil bulk density was caused mostly by 3 and 6 passes in the first year, but also by one pass in the second and third years of soil compaction. Below the plough layer (0-0.2 m) compaction by 4.9 Mg tractor did not cause any significant differences in soil bulk density in the first year, where the soil moisture content was only 110 g kg^{-1} at the moment of compaction. The significant increase of bulk density with 0.15 Mg $m^{-3}$was observed in the second year by 6 passes in 0.2-0.3 m depth, where the soil moisture content was 200 g kg^{-1} at the moment of compaction. Third and fourth year continuous compaction (moisture content 190 g kg^{-1}) caused a significant increase in bulk density already after one pass by 0.1 Mg m^{-3} up to 0.3 m depth. No significant increase of bulk density was detected in 0.3-0.4 m depth in most experiment years.

In the first year of soil compaction the soil penetration resistance was significantly higher in 6-times passed soil until 0.35 m depth compared to the control and other compaction treatments (Figure 2). No differences were detected between control and one time and three times passed treatments.

Fig. 1: Effect of soil compaction on soil bulk density in earing phase of spring barley (*Hordeum vulgare* L.) in years 2001, 2002, 2003 and 2004; 0, 1, 3, 6 – number of passes; $LSD_{0.05}$ – least significant differences at significance $p<0.05$; ns – no significant differences between treatments

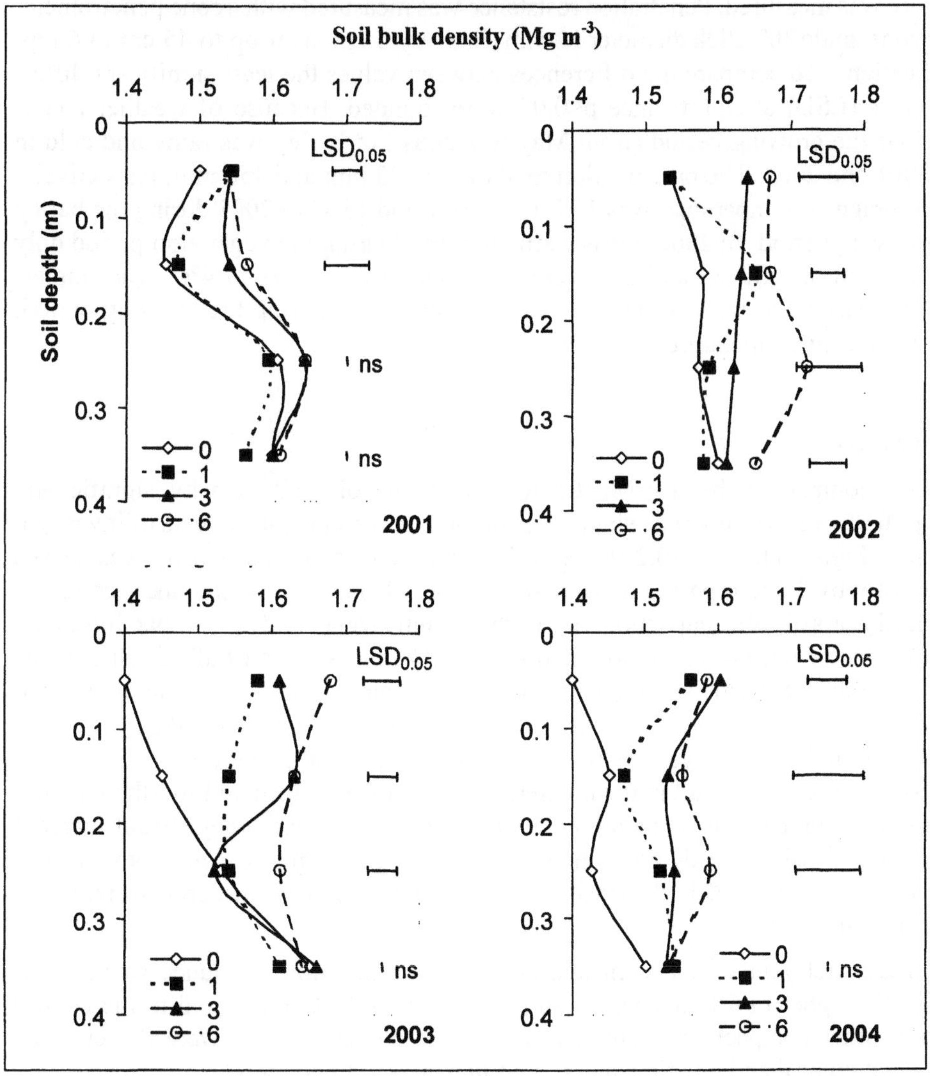

Fig. 2: Effect of soil compaction on soil penetration resistance in earing phase of spring barley (*Hordeum vulgare* L.) in years 2001, 2002, 2003 and 2004; 0, 1, 3, 6 – number of passes; $LSD_{0.05}$ – least significant differences at significance $p<0.05$; ns – no significant differences between treatments

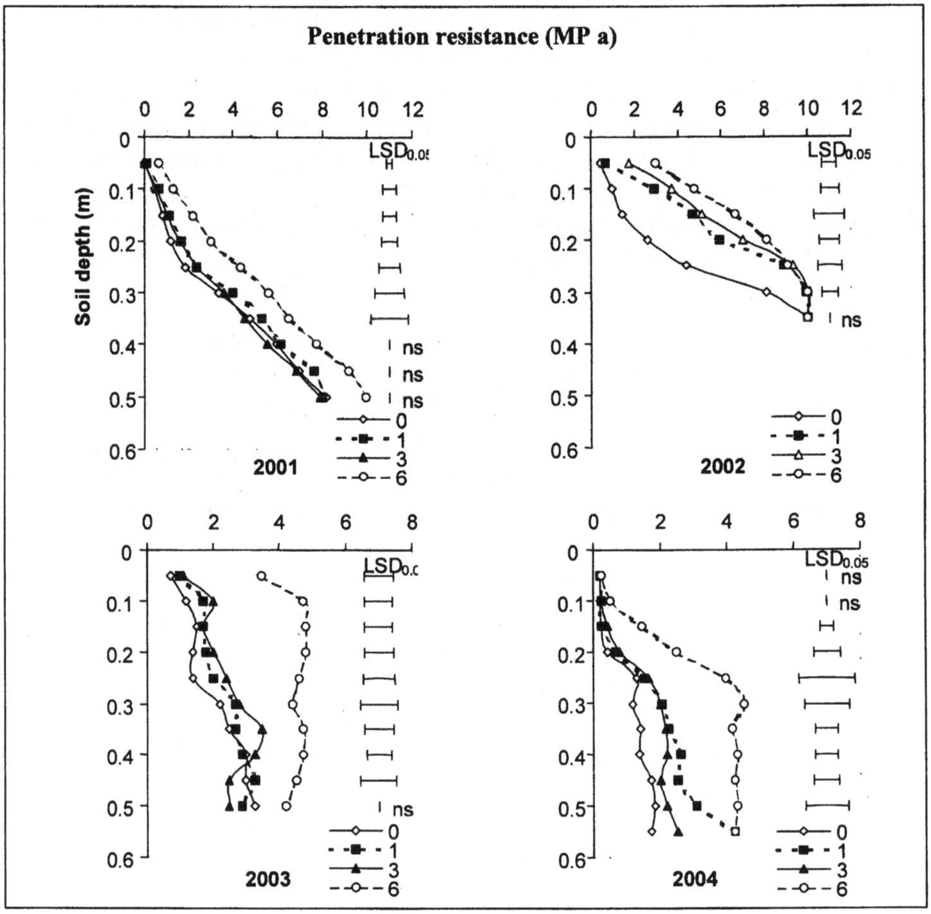

In the second year the soil was drier than in other years of the experiment and also the values of soil penetration resistances were higher. As the soil was very dry (moisture less than 100 g kg^{-1}) it was impossible to measure penetration resistance deeper than 0.35 m. After the third year direct soil compaction measuring of penetration resistance indicated compaction effect also below 0.3 m and development of hardpan in 3- and 6-times passed soil (Figure 2, 2003). 2004 was very rainy and the moisture content in the soil was high during the sampling period (210 g kg^{-1}). For this reason there were no significant differences of penetration resistance between compaction treatments in the upper part of the soil (Figure 2, 2004). Below the plough layer soil was drier and significant differences between uncompacted and 1-time, 3- and 6-times passed soil were detected.

Though the changes in soil bulk density (Figure 1, 2001) and penetration resistance (Figure 2, 2001) due to compaction were exiguous in the first year, there were still decreases in plant shoots dry weight and number of plants (Table 1, 2001). A significant decrease was observed in barley dry weight on the treatments of 6 passes and on barley plants density in the treatments of one and 6 passes, respectively. Compaction had a higher effect on annual weeds than perennial weeds. In the first year weeds formed 6.5% of the total plant shoot dry weight and 23.4% from plants density on uncompacted and 2.8% and 17.2% on 6-times passed soil, respectively. The most widespread weed species were common lambsquarters, field pennycress (*Thlaspi arvense* L.), common fumitory (*Fumaria officinalis* L.) and common chickweed (*Stellaria media* (L.) Vill; Table 1, 2001). Compaction decreased common lambsquarters dry weight, but did not affect it's density.

Second year compaction increased weeds share in plant shoot dry mass from 12% in the control up to 52% in 6-times compacted treatment and decreased significantly the share of barley (Table 1, 2002). Changes in perennial weed density and weight were not statistically significant. In 2002 one, 3 and 6 passes had a positive effect on annual weed shoots dry weight. It was caused mostly by the changes in weight and density of common lamb's quarters, whose share in weed community was more than 50% (Table 1, 2002). Compaction had no strongly positive or negative effect on pennycress and chickweed. If one pass decreased their weight and density, then 3 and 6 passes increased it again.

After three years of soil compaction without fertilizer use, the total shoot dry weight and also barley yield was only ¼ from the first year shoot dry mass and this even on uncompacted soil (Table 1, 2003). Changes in barley shoot dry weight were similar to earlier years, but if one and three passes decreased barley density, then six passes increased it again. Weeds formed between 9.7% and 14.9% from total shoot mass. There was a significant increase in perennial plants dry weight and density, mostly because of great plantain, whose share increased from 0.5% on uncompacted soil to 12.2% on 6-times compacted soil in weed community (Table 1, 2003). The dominating weed species was common chickweed, which formed from weed shoot mass 43.2% in control and 17.6%, 36.1% and 38.8% in one, 3- and 6-times compacted treatment, respectively. With compaction increased the share of field pennycress. Common fumitory disappeared on the 6-times passed soil in earing phase of barley.

In the fourth year due to a rainy year and lower penetration resistance (Figure 2, 2004), the impact of soil compaction to the barley dry weight and density was lower than in the previous 2 years (Table 1, 2004). The lowest weed dry weight and density were detected on one time passed soil, where their share from total shoot dry weight was 8.6% and from total plant density 37.9%. On 6-times passed soil weeds formed 27% from total shoot dry weight and 51.2% from total plant density. From total weed dry weight 29% to 67% formed perennial weeds. Also in the fourth year the annual weed mass increased with the increase of soil bulk density, but in the fourth year, because of higher plant density on compacted soil not due to higher plant shoot mass as in the second year.

Tab. 5: Soil compaction effects on some weed species dry weight and density in weed community in earing phase of spring barley (*Hordeum vulgare* L.) during experiment

	Dry weight (g m^{-2})				$LSD_{0.05}$[a]	Density (plants m^{-2})				$LSD_{0.05}$
Treatment	0[b]	1	3	6	(comp.)	0	1	3	6	(comp.)
2001										
Chenopodium album L.	8.5	9.8	7.2	5.2	2.1	68	88	71	68	ns[c]
Fumaria officinalis L.	5.8	0.8	2.7	1.4	4.7	25	12	20	6	14
Thlaspi arvense L.	2.0	1.4	0.3	0.8	ns	16	18	8	14	ns
Stellaria media (L.) Vill.	4.8	4.3	1.5	1.7	ns	43	63	35	24	ns
Plantago major L.	0	0.1	0	0	ns	0	3	0	0	ns
Annual weeds	54	31	21	15	18	245	236	187	146	86
Perennial weeds	1.4	1.2	1.2	1.0	ns	24	14	13	7	ns
Hordeum vulgare L.	797	632	758	551	190	883	778	859	738	104
2002										
Chenopodium album L.	21.8	30.6	32.7	36.0	8.4	91	95	68	36	25
Fumaria officinalis L.	1.0	1.0	1.1	0.1	ns	14	8	6	1	7
Thlaspi arvense L.	0.7	0.2	1.9	1.5	ns	11	4	7	8	ns
Stellaria media (L.) Vill.	1.0	0.3	5.8	1.7	ns	26	3	19	10	16
Plantago major L.	1.6	0	0.2	1.8	ns	4	0	1	2	ns
Annual weeds	33	39	58	60	22.6	213	148	182	80	73
Perennial weeds	8.3	16	2.2	5.5	ns	6	32	5	7	19
Hordeum vulgare L.	302	262	169	61	78	450	228	266	137	132
2003										
Chenopodium album L.	1.8	3.3	3.7	1.2	0.7	91	97	109	56	20
Fumaria officinalis L.	0.3	0	0.2	0	0.1	16	0	13	0	5
Thlaspi arvense L.	0.6	1.7	0.6	1.1	ns	22	38	9	19	ns
Stellaria media (L.) Vill.	9.2	3.2	10.7	3.8	5.8	109	50	63	41	45
Plantago major L.	0.1	2.4	0.2	1.2	0.1	3	3	6	13	7
Annual weeds	21	17	27	9	7.2	462	350	344	181	107
Perennial weeds	0.3	1.2	2.6	0.8	1.7	15	34	19	19	15
Hordeum vulgare L.	186	167	169	91	21	550	466	463	475	58

Treatment	Dry weight (g m^{-2})				$LSD_{0.05}$[a] (comp.)	Density (plants m^{-2})				$LSD_{0.05}$ (comp.)
	0[b]	1	3	6		0	1	3	6	
2004										
Chenopodium album L.	0.4	0.3	0.6	0.4	0.2	25	21	17	33	3
Fumaria officinalis L.	0	0.8	0.3	0	0.3	0	22	9	0	9
Thlaspi arvense L.	3.0	2.4	5.1	3.1	1.7	66	34	40	48	11
Stellaria media (L.) Vill.	1.0	1.4	0.8	3.7	1.4	50	25	30	150	86
Plantago major L.	0.02	0.3	0.6	25.2	0.4	1	2	3	6	1
Annual weeds	9	11	12	19	2.4	264	264	210	464	101
Perennial weeds	19	4.5	15	15	ns	88	38	74	22	22
Hordeum vulgare L.	151	164	132	92	18	573	496	429	462	97
$LSD_{0.05}$ (year)										
Chenopodium album L.	9.8	11.4	20.4	18.5		25	31	19	28	
Fumaria officinalis L.	4.3	ns	2.4	1.3		14	11	ns	2	
Thlaspi arvense L.	ns	ns	2.3	1.7		10	15	12	20	
Stellaria media (L.) Vill.	ns	3.2	6.9	ns		ns	38	ns	81	
Plantago major L.	ns	0.1	0.1	0.6		ns	ns	5	2	
Annual weeds	44.2	20.2	34.1	24.0		120	72	59	112	
Perennial weeds	8.6	16.5	3.7	6.5		24	19	18	7	
Hordeum vulgare L.	161	124	143	55		134	99	112	100	

a Least significant difference at $p<0.05$

b Number of passes

c No significant differences between treatments

After four years without fertilizer use the dominating annual weed species were field pennycress and common chickweed. Also increased was the density and mass of corn spurry (*Spergula arvensis* L.) and corn mayweed (*Matricaria inodora* L.) in the weed community, especially on 6-times compacted soil. From perennial weeds dominated sow thistle (*Sonchus arvensis* L.) on uncompacted and one and 3-times passed soil and coltsfoot (*Tussilago farfara* L.) on one time and 3-times passed soil. On 6-times passed soil the most widespread weed species was great plantain by 74.1% from the weed community. No one common fumitory plant was detected on uncompacted and 6-times compacted soil in the earing phase of barley after four years. Year had less impact on field pennycress shoot mass and common chickweed shoot mass and density. The most affected weed species were common lambsquarters and common fumitory. The total number of weed species decreased from 19 species on uncompacted soil in the first year to 12 species on 6-times passed soil in the fourth year.

Discussion

Though compaction by 4.9 Mg tractor with tyre inflation pressure 150 kPa in first and second year soil bulk density and penetration resistance increased, no hardpan was formed in the subsoil, mostly because of deep-freezing of the soil (up to 0.5 m) in those years, but also the moderate weight of tractor. Still, after the third year a continuous direct compaction hardpan below the plough layer and even deeper was formed even after one pass. The soils of the experiment area with their medium fine texture are moderately susceptible and moderately vulnerable to soil compaction by moist soil and not particularly vulnerable by dry soil according to the classes of vulnerability and susceptibility. Recommended maximum tyre inflation pressure to those soils is 120 to 160 kPa (van den Akker 2002). The highest values of soil bulk density in 2002 in all compaction treatments, compared to the other experiment years were caused from low moisture content (110 g kg^{-1}) in the soil and the lowest values in 2001 and 2004 from high moisture content (210 g kg^{-1}) in the soil. In experiments by Pickering and Veneman (1984) soil dry density increased to the soil moisture content 0.11-0.12 kg kg^{-1} and then started to decrease again.

Changes of dominant weed species in the plant community during four years were caused by changes in soil conditions: decrease of available nutrients in the soil, higher soil penetration resistance and soil bulk density. Low supplies of major nutrients such as N, P and K play an essential role in maintaining species richness in weed communities. Like cultural plants weeds also have different preferences on environmental parameters. In the experiment the main reason for the decrease of common lambsquarters dry weight was the nutrient availability but also high soil resistance to root growth. Common lambsquarters can grow well in a wide range of climates and soils, especially those with high organic matter content (Mitich 1988). It possesses a prolific rooting system, which allows it to resist adverse environmental conditions. Emergence rates of common lambsquarters increase with temperature and decrease with increasing soil penetration resistance and depth (Vleeshouwers 1997). Common fumitory favours well-drained soil. The plant prefers sandy and medium loamy acid, neutral and alkaline soils (Mitch 1997). For this reason individuals of common fumitory have only limited ability to compete with other weeds and suffer strongly from intraspecific competition. There were some plants of common fumitory in the tillering phase of barley, but they did not survive in competition with other weeds on compacted soil till the earing phase of barley. Field pennycress also prefers fertile soils but the plant also tolerates dense soils. If the intensity of tilling is reduced, the field pennycress share in the weed community increases (Stevenson et al. 1998). Great plantain and corn mayweed are common on edges of field and waysides, corn spurry on soils with low fertility (Trivedi & Tripathi 1982). Great plantain is characteristic of relatively fertile, disturbed habitats and its root system is restricted by compaction (Whitfield et al. 1996). The plants of great plantain were in weed flora already at the beginning of the experiment but were suppressed by other weed species. Compaction inhibited most other existing weed species and great plantain started to dominate in weed flora. Common chickweed is a cosmopolitan species, common in cereal and broad-leaved arable crops (Lutman et al. 2000). Walter et al. (2002) found that chickweed was positively cross-correlated with clay and negatively cross-correlated with pH and potassium content. In our experiment common chickweed tolerated moderately compacted soil more than severely compacted soil in most years of investigation. In Poland Borowiec et al. (1985) found that Ca favoured common lambsquarters and plots given N+ Ca were dominated by chickweed, and field pansy (*Viola arvensis* L.). In spring barley, mineral fertilizer eliminated corn spurry. Liming increased density of lambsquarters.

Conclusion

Soil compaction inhibits growth of cultural plants in the first place and decreases their ability to compete with weeds. By changing the field conditions the competitors for the cultural plants will also change. On compacted soils and without fertilizer use easier controlled annual weed species will be replaced with harder preventive perennial weed species.

Acknowledgements

The study was supported by Estonian Science Foundation grant no. 5418 and 4991.

References

Altieri, M.A. (1999), The ecological role of biodiversity in agroecosystems. Agriculture, Ecosystems and Environment 74, 19-31.

Bischoff, A. & Mahn, E.-G. (2000), The effect of nitrogen and diaspore availability on the regeneration of weed communities following extensification. Agriculture, Ecosystems, Environment 77, 237-246.

Borowiec, S., Kuszelewski, L., Kutyna, I., Leśnik, T. & Lukaszuk, T. (1985), The effect of long-term fertilizer use on soil properties and weed infestation of rye, spring barley, Sinapis alba and potato crops. Part II. The effect of different fertilizer treatments on weed infestation of individual crops. Zeszyty Naukowe Akademii Rolniczej w Szczecinie, 37, 71-86. (Polish, summary in English)

Derksen, D.A., Lafond, G.P., Thomas, A.G., Loeppky, H.A. & Swanton, C.J. (1993), Impact of agronomic practices on weed communities – tillage systems. Weed Science 41, 409-417.

Fykse, H. & Wærnhus, K. (1999), Weed development in cereals under different growth conditions and control intensities. Acta Agric. Scand., Sect. B, Soil and Plant Sci. 49, 134-142.

Ghosheh, H.Z. & Al-Hajaj, N.A. (2004), Impact of soil tillage and crop rotation on barley (Hordeum vulgare) and weeds in a semi-arid environment. J. Agron. Crop Sci. 190, 374-380.

Jurik, T.W. & Zhang, ShuYu (1999), Tractor wheel traffic effect on weed emergence in central Iowa. Weed Technol. 13, 741-746.

Lutman, P.J.W., Bowerman, P., Palmer, G.M. & Whytock, G.P. (2000), Prediction of competition between oilseed rape and Stellaria media. Weed Research 40, 255-269.

Mitch, L.W. (1988), Common lambsquarters. Weed Technol. 2, 550-552.

– (1997), Fumitory (Fumaria officinalis L.). Weed Technol. 11, 843-845.

Pickering, E.W. & Veneman, P.L.M. (1984), Strength characteristics of three indurated horizons in Massachusetts. Soil Sci. Soc. Am. J. 48, 133-137.

Reintam, E. & Kuht, J. (2002), Impact of soil compaction and sowing density of barley on uptake of nutrients. In: Turka, I., Kreismane, Dz., Selegovska, E. & Adamovich, A. (eds.), Scientific aspects of organic farming: Proceedings of the conference. Jelgava, Latvia 21-22 March, 2002. pp. 167-171.

Stevenson, F.C., Légère, A., Simard, R.R., Angers, D.A., Pageau, D. & Lafond, J. (1998), Integrated pest management. Manure, tillage, and crop rotation: Effect on residual weed interference in spring barley cropping systems. Agron. J. 90, 496-504.

Swanton, C.J., Shrestha, A., Roy, R.C., Ball-Coelho, B.R. & Knezevic, S.Z. (1999), Effect of tillage systems, N, and cover crop on the composition of weed flora. Weed Science 47, 454-461.

Zanin, G., Otto, S., Riello, L. & Borin, M. (1997), Ecological interpretation of weed flora dynamics under different tillage systems. Agriculture, Ecosystems and Environment 66, 177-188.

Trivedi, S. & Tripathi, R.S. (1982), Growth and reproductive strategies of two annual weeds as affected by soil nitrogen and density levels. New Phytol. 91, 489-500.

Walter, A.M., Christensen, S. & Simmelsgaard, S.E. (2002), Spatial correlation between weed species densities and soil properties. Weed research, 42, 26-38.

Whitfield, C.P., Davison, A.W. & Ashenden, T.W. (1996), Interactive effects of ozone and soil volume on Plantago major. New Phytol. 134, 287-294.

van den Akker, J.J.H. (2002), Determination of the susceptibility of subsoil to compaction and ways to prevent subsoil compaction. In: Pagliai, M., Jones, R. (eds.) Sustainable land management – environmental protection: A soil physical approach. Advances in Geology 35, 291-304.

van Elsen, T. & Günther, H. (1992), Auswirkungen der Flächenstillegung auf die Ackerwildkraut-Vegetation von Grenzertrags-Feldern. Z. PflKrankh. PflSchutz, Sonderh. 13, 49-60.

Vleeshouwers, L.M., Modelling the effect of temperature, soil penetration resistance, burial depth and seed weight on pre-emergence growth of weeds. Annals of Botany 79, 553-563.

Use of Biological Methods to loosen Compacted Soil

Katrin Trükmann, Endla Reintam, Jaan Kuht, Virgo Rääts

Estonian University of Life Sciences
Department of Soil Science and Agrochemistry
Kreutzwaldi Str. 64, EE – 51014 Tartu, Estonia

Abstract

Soil compaction is potentially a major threat to agricultural productivity and the main form of soil degradation in Europe. The mechanical methods to eliminate compacted soil layers are expensive and energy consuming. One good alternative might be to use plants with vigorous root system to modify the compacted subsoil. The experiments with yellow lupine (*Lupinus luteus* L.), oil rape (*Brassica napus* ssp. *oleifera* Hertzg.), lucerne (*Medicago sativa* L.) and spring barley (*Hordeum vulgare* L.) were made on Estonian Agricultural University research field in 2004. The sandy loam Stagnic Luvisol was compacted by tractor MTZ-82 (total weight 4.9 Mg) by multiple tyre-to-tyre passing 1, 3 and 6 times. To evaluate the changes in the soils physical properties the bulk density and cone resistance of soil was measured. The root analyses were made by washed root measurement WinRhizo scanner and software. The best results showed yellow lupine in six times passed subsoil, where the penetration resistance was 1 MPa lower than the area planted with barley or other cultures. Annual cultures, such as yellow lupine and oil rape, also decreased soil bulk density on 3- and 6-times passed soil up to 0,3 Mg m^{-3}, respectively. At the same time growing of lupine and oil rape increased soil penetration resistance and bulk density on control treatment, compared to barley. Perennial cultures, such as lucerne, increased soil bulk density and penetration resistance mostly in topsoil in all investigated compaction treatments. Lupine had the lowest root volume and surface area but the thickest roots compared to the other cultures. Lucerne had the highest root volume and root surface area on uncompacted and 3-times passed soil. Compaction affected barley and oil rape rooting depth more than rooting depth of lupine and lucerne. As the effect of deep-rooted plants growing on soil properties will appear in the next year by growing next culture, when the roots are decayed, further experiments are needed to find a convenient culture to eliminate subsoil compaction.

Introduction

Compaction of agricultural soils is a global concern because of the negative effect on crop yields and the environment. Soil compaction is assessed to be responsible for the degradation alone of an area of 33 million ha in Europe (Van Ouwerkerk & Soane 1994).

One of the reasons for soil compaction is the use of heavy equipment during the tillage, which produces significant changes in the physical properties of the affected soils (Horn et al. 2000). Compaction is responsible for many adverse changes in soil bulk density, a porosity soil-water relationship and permeability and penetration resistance (Alakukku & Elonen 1994). The direct damages of soil compaction on the crop are impaired root growth and root function, caused by diminished soil aeration and increased mechanical resistance in the soil (Tardieu 1994). Either the plant root length is reduced or the root system develops superficially. As a result, the roots explore a relative small soil volume and their ability to observe water and nutrients is limited (Oussible et al. 1992). This can be the major constraint to plant growth and yields.

Some research has determined more than 20 plants species, which are able sufficiently to develop their root system in dense and strength soil (Materachera et al. 1991). That kind of penetration of plant roots into compacted soils has been described as a possible natural process which may ameliorate compacted soils (Dexter 1991). Crops with deep tap roots provide more channels for subsequent crop roots to follow and allow water to percolate more deeply into the soil.

The aim of this work was to investigate the effect of yellow lupine (*Lupinus luteus* L.), oil rape (*Brassica napus* ssp. *oleifera* Hertzg.), lucerne (*Medicago sativa* L.) and spring barley (*Hordeum vulgare* L.) growing on compacted soil. Therefore crops with vigorous root systems may provide a less expensive and less energy consuming alternative to tillage to reduce the effects of subsoil compaction on agricultural soils.

Materials and Methods

Data was collected from research field (58°23'N, 26°44'E) of Estonian Agricultural University with different levels of soil compaction (4 levels) on sandy loam soil in 2004. Soil compaction was carried out from 2001 every year in spring, before sowing. Compaction was generated using a tractor MTZ-82 provided with a front loader. A wheeled vehicle loaded with 2.25 Mg on the first axle and – 2.55 Mg on the rearward axle (total load is 4.9 Mg). Four different levels of soil compaction were produced using the following treatments: uncompacted – control, no pass of tractor wheels; compacted 1 time – one pass of tractor wheels; compacted 3 time – three passes of tractor wheels; compacted 6 time – six passes of tractor wheels. Data used in this paper was collected from three compaction treatments: uncompacted, passed 3 and 6 times. Seeds of yellow lupine were sown 50 kg ha^{-1}, seeds of oil rape 12 kg ha^{-1}, on different level of soil compaction. Seeds of lucerne 20 kg ha^{-1} were sown one year earlier (2003) as a subculture to spring barley. The seed rate of barley was 450 germination seeds per m^2 (250 kg ha^{-1}). Sowing was done in the middle of May. In the case of barley complex fertilizer (N20: P3.5: K10) was used in rate $N_{80}P_{14}K_{40}$ kg ha^{-1} before sowing. No herbicides or other pesticides were used in barley and other cultures. The samples of soil and plants were taken in earing phase of barley (middle of July). Root samples and soil bulk density were taken with 1131 cm^3 steel cylinders (h=15 cm, Ø=9.8 cm) in 15 cm layers up to 60 cm in four replications. Special samples were taken to measure soil moisture. Soil penetration resistance was measured with cone (30°) penetrometer from each 5 cm layers up to 60 cm. For the root analysis a scanner and software WinRhizo Version 2004a was used. Washed roots were spread out in about 15 mm depth of water in a scanner ray. The scanner was used to create a digital image of the roots. Total root volume, average root diameter and total root surface area were determined per sampling cylinder volume. In 2004 the weather was very rainy. During the vegetation period (April-September) the precipitation was 582 mm and the average air temperature 13°C. The distribution of precipitation was quite uneven. In June there were 143 mm of precipitation.

Results

In uncompacted soil the highest dry bulk densities were measured by growing lucerne and yellow lupine, lowest under oil rape and spring barley (Figure 1, A). Growing of lupine and oil rape decreased soil bulk density in 3- and 6-times compacted soil, up to 0,3 Mg m^{-3} compared to spring barley (Figure 1, B, C). The values of soil bulk density in 3- and 6-times passed soil were higher under lucerne and barley, compared to the lupine and oil rape.

Year 2004 was very rainy and the moisture content in soil was high during the sampling period (210 g kg^{-1}). For this reason there were no significant differences of penetration resistance between culture treatments in the upper part of the uncompacted soil (Figure 2, A). Below the plough layer soil was drier and significant differences between barley, lucerne and lupine were detected. Soil penetration resistance was 1.5 MPa higher by barley growing. In 3-times passed soil the highest values of penetration resistance were measured under lupine (Figure 2, B). Under barley penetration resistance was 1 MPa lower than under lupine and Lucerne in 3-times passed topsoil (0-30 cm). There were no significant differences in deeper soil. In 6-times passed soil lupine and lucerne decreased significantly soil penetration resistance compared to barley (Figure 2, C).

Fig. 1: The influence of barley (*Hordeum vulgare* L.), lucerne (*Medicago sativa* L.), lupine (*Lupinus luteus* L.) and oil rape (*Brassica napus ssp. Olifera* Hertzg.) growing on soil bulk density in three different levels of soil compaction (A, B, C)

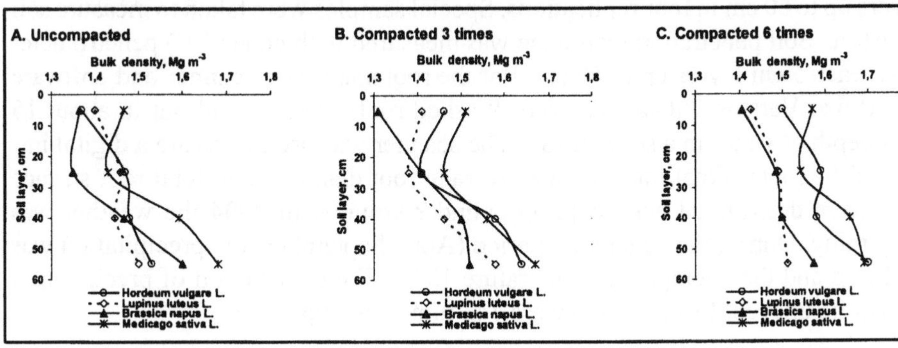

Fig. 2: The influence of barley (*Hordeum vulgare* L.), lucerne (*Medicago sativa* L.) and lupine (*Lupinus luteus* L.) growing in three different levels of soil compaction (A, B, C)

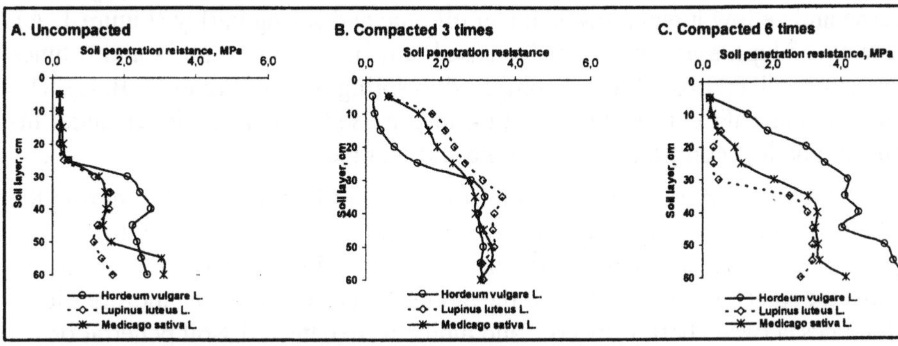

The highest root volume in uncompacted soil was detected on oil rape in the upper 15 cm soil layer, where it was two times higher than on barley and lucerne and 2.5 times higher than on lupine (Figure 3, A). In deeper soil layers volume of oil rape decreased suddenly and was the lowest compared to other investigated cultures. In deeper soil layers the highest root volume was detected on lucerne. Barley root volume was higher than root volume of lupine. In three times passed soil lucerne had the highest root volume and lowest again lupine (Figure 3, B). Three passes increased barley and lupine root volume in the upper 15 cm soil layer. Three passes affected barley, oil rape and lucerne rooting depth more than lupine. No differences between barley, lupine and oil rape root volumes were detected deeper than 15 cm. Also six passes increased lucerne root volume in the upper 15 cm soil layer, but also oil rape root volume (Figure 3, C). Lupine root volume was lowest in the same layer, but higher than in 3-times compacted soil. In the deeper soil lupine root volume was the highest.

Fig. 3: The total root volume of barley (*Hordeum vulgare* L.), lucerne (*Medicago sativa* L.), oil rape (*Brassica napus ssp. Olifera* Hertzg.) and lupine (*Lupinus luteus* L.) in three different levels of soil compaction (A, B, C)

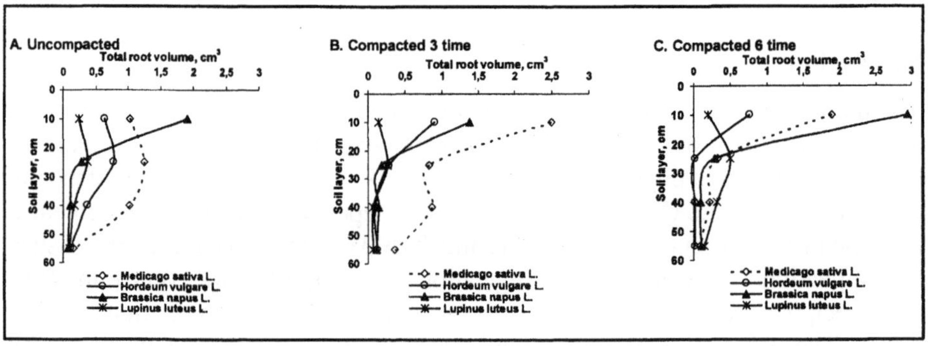

Differences in root surface area were similar to root volume. In uncompacted soil lucerne, barley and oil rape had bigger surface areas than lupine in the upper 15 cm soil layer and in the deeper soil lucerne and barley had bigger surface areas than lupine (Figure 4, A). Three and six passes increased root surface area of lucerne, oil rape and barley in the upper 15 cm soil layer and soil decreased it in deeper layers (Figure 4, B, C). Almost no changes in lupine surface area were detected due to compaction. In all compaction treatments lucerne had the biggest root surface area below 15 cm in soil.

Fig. 4: The total root surface area of barley (*Hordeum vulgare* L.), Lucerne (*Medicago sativa* L.), oil rape (*Brassica napus ssp. Olifera* Hertzg.) and lupine (*Lupinus luteus* L.) on penetration resistance in three different levels of soil compaction (A, B, C)

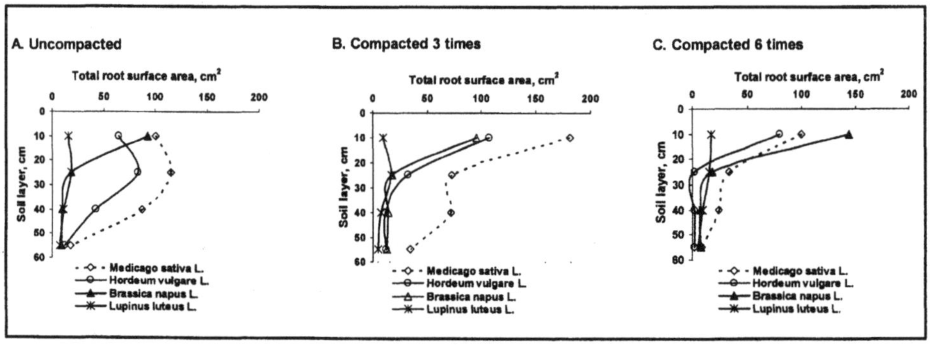

Though lupine had the lowest root volume and root surface area, lupine roots were the thickest. In all compaction treatments lupine root diameter was the highest compared to the other investigated cultures, except in uncompacted and 6-times passed soil in the upper 15 cm soil layer (Figure 5, A, B, C). If three passes decreased the average root diameter of all investigated cultures, than six passed increased it again. Compaction had almost no effect on barley root diameter.

Discussion

The unfavorable effect which occurs with compaction depends on soil moisture and weather conditions during compaction and vegetation period. To evaluate the changes in the soils physical properties the bulk density and cone resistance of soils were measured. Cone resistance is highly dependant on soil water content at the time of the measurements (Abu-Hamdeh 2003). Dry soils are more resistant to compaction than wet soil. The vegetation period in 2004 was very rainy (582 mm precipitation) and soil moisture content during the sampling period was 210 g kg^{-1}. The penetration resistance will decrease if the water content in the soil increases (Laboski et al. 1998). It is usually considered that any resistance over 2 MPa can significantly reduce root growth and development (Ishaq et al. 2001). The result of compaction is decreased root size and smaller rooting depth.

After three years of compaction the significant differences in soil bulk density and penetration resistance from the soil parameters were detected in topsoil and also in subsoil. Measurement of penetration resistance indicated the thickened layers in 12-20 cm and 30-40 cm depth in soil, especially in 3- and 6-time compacted variant (Reintam & Kuht 2004).

Fig. 5: The average root diameter of barley (*Hordeum vulgare* L.), lucerne (*Medicago sativa* L.), oil rape (*Brassica napus ssp. Olifera* Hertzg.) and lupine (*Lupinus luteus* L.) in three different levels of soil compaction (A, B, C)

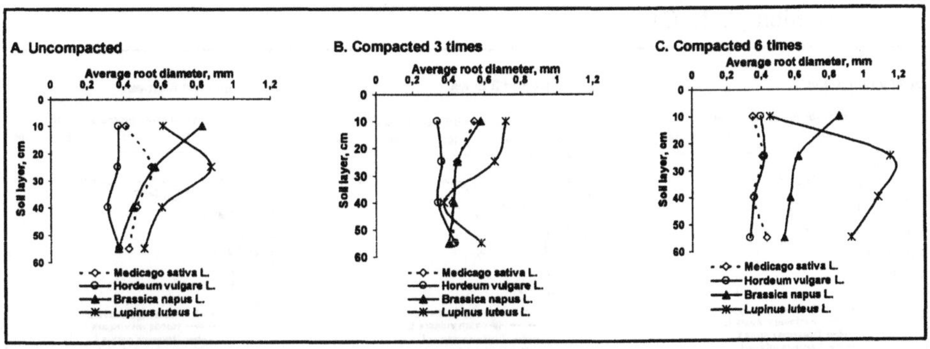

Reeves et al. (1984) have come to interesting conclusions by growing spring wheat in Australia. If the spring wheat was grown in soil with a bulk density of 1.52 g cm^{-3} in the 0-20 cm depth there were less roots than in soil with a bulk density of 1.32 g cm^{-3}. In our experiments the highest bulk density was measured by lucerne in 0-20 cm depth and it was 1.57 Mg m^{-3}. In an uncompacted area under lucerne the bulk density was 1.47 Mg m^{-3} but in both treatments the total root surface area was over 100 cm^{-2} (Figure 4).

If penetration resistance is lower than 2 MPa, then an increase in average root diameter is expected (Materachera et al. 1992). Rosolem and Takahashi (1998) observed an increase in soybean root diameter when the soil resistance to penetration was >1.2 MPa. In our experiments the average penetration resistance was 2.2 MPa by growing lupine and the average root diameter was predominantly the highest when comparing with other cultures (Figure 5). Compaction decreases the volume, size and continuity of pores in the soil (Hillel 1998).

In uncompacted soil higher soil bulk density by lupine and lucerne growing occurred. The root diameters are usually larger than most soil pores and soil particles are pushed aside and the density of soil near the root increases (Bruande et al. 1996). For root growth the existence of macropores in soil is essential, they have a greater diameter than roots and therefore it is a chance for undisturbed root growth to evade the zone where the mechanical impedance is high (Hatano et al. 1998; Glinski & Lipiec 1990). Goss (1991) found the ability of roots growing into existing pores and channels increases in deeper and stronger layers where they can be pathways for root growth. There are several studies that have shown that roots are able to perforate compact layers, creating accessible pathways for the roots of succeeding crops and leaving biopores which can increase water movement and gaseous diffusion (Unger & Kaspar 1994).

There are some instances of growing more horizontal proliferation, when roots find higher soil resistance to vertical penetration. The growth of many fine lateral roots leads to higher root length density in the more favorable soil zones (Rosolem et al. 2002). Some researchers have found that plants with numerous fine roots had better growth in compacted layers (Whitely & Dexter 1984). Materechera et al. (1992) have a different opinion, they have come to conclusion, that thicker roots would have a higher ability to grow in higher growth pressure. Bennie (1996) observed that the decrease in root elongation rate with increasing soil penetration resistance seems to be different in different species. In the current experiment compaction increased barley and oil rape root volume in the upper 15 cm soil layer and decreased below this depth.

Conclusions

The results of this investigation showed the positive effect of yellow lupine and lucerne growth on compacted Stagnic Luvisol. The root system of lupine was able to penetrate the compacted subsoil in all investigated soil compaction treatments, roots of lucerne in 3-times passes soil. As the effect of deep-rooted plants growing on soil properties will appear in the next year by growing the next culture, when the roots are decayed, further experiments are needed to find a convenient culture to eliminate subsoil compaction.

Acknowledgements

The study was supported by Estonian Science Foundation grand no. 5418.

References

Abu-Hamdeh, N.H. (2003), Soil Compaction and root distribution for ork as affected by tillage and vehicle parameters. Soil Tillage Res., 74, 25-35.

Alakukku, L. & Elonen, P. (1994), Finish experiment on subsoil compaction by vehicles with high axle load. Soil Tillage Res., 29, 151-155.

Bennie, A.T.P. (1996), Growth and mechanical impedance. In Waisel, Y., Eshel, A. & Kafkafi, U. (eds.), Plant roots: The Hidden Half, 2nd Edition. Marcel Dekker, New York, pp. 453-470.

Bruand, A., Cousin, I., Nicoullad, B., Duval, O. & Begon, J.C. (1996), Backscattered electron scanning images of soil porosity for analyzing soil compaction around roots. Soil Sci. Soc. Am. J. 60, 895-901.

Dexter, A.R. (1991), Ameliration of soil by neutrals processes. Soil Tillage Res., 20, 87-100.

Glinski, J. & Stepniewski, W. (1985), Soil Aeration and Its Role for Plants. CRC Press, Boca Raton, FL, USA, 250 pp.

Goss, M.J., Consequences of the activity of roots on soil. In Atkinson, A. (ed.), Plant Root Growth. An Ecological Perspective. Spec. Publ. Br. Ecol. Soc., vol. 10. Blackwell, Oxford, 171-186.

Hatano, R., Iwanaga, K., Okajima, H. & Sakuma, T. (1988), Relationship between the distribution of soil macropores and root elongation. Soil Sci. Plant Nutr. 34, 535-546.

Hillel, D. (1998), Environmental Soil Physics. Academic Press, London.

Horn, R., van den Akker, J.J.H. & Arvidsson, J. (eds.) (2000), Subsoil Compaction: Distribution, Processes and Consequences. Advance Geoecology, vol. 32. Catean Verlag, Reiskirchen, pp. 462.

Ishaq, M., Ibrahim, M., Hassan, A., Saeed, M. & Lal, R. (2001), Subsoil compaction effects on crops in Punjab, Pakistan. II. Root growth and nutrient uptake of wheat and sorghum. Soil Tillage Res., 60, 153-161.

Laboski, C.A.M., Dowdy, R.H., Allmaras R.R. & Lamb, J.A. (1998), Soil strength and water content influences on corn root distribution in a sandy soil. Plant Soil, 203, 239-247.

Reeves, T.J., Haines, P.J. & Coventry, R.R. (1984), Growth of wheat and subterranean clover on soil artificially compacted at various depths. Plant Soil, 89, 135-138.

Reintam, E. & Kuht, J. (2004), Soil compaction effect on soil, nutrient cycling and weeds in agroecosystem. Integrative Approaches towards sustainability in the Baltic Sea region (Leal Filho, W. & Ubelis, A., eds.), Peter Lang Scientific Publishers, pp. 519-527.

Rosolem, C.A. & Takahashi, M. (1998), Soil compaction and soybean root growth. In: Box JR. J.E. (eds.), Root Demographics and their Efficiencies in Sustainable Agricultur, Grasslands and Forrest Ecosystems. Kluwer Academic Publishers, Dordecht, pp. 295-304.

Rosolem, C.A., Foloni, J.S.S. & Tiritan, C.S. (2002), Root growth and nutrient accumulation in cover crops as affected by soil compaction. Soil Tillage Res., 65, 109-115.

Tardieu, F. (1994), Growth and functioning of roots and root systems subjected to soil compaction. Toward a system with multiple signaling? Soil Tillage Res., 30, 217-243.

Materachera, S.A., Alston, A.M., Kirby, J.M. & Dexter, A.R. (1992), Influence of root diameter on the penetration of seminal roots into a compacted subsoil. Plant Soil, 144, 297-303.

Materachera, S.A., Dexter, A.R. & Alston A.M. (1991), Penetration of very strong soils by seedling roots of different plant species. Plant Soil, 135, 31-41.

Oussible, M., Crookston, R.K. & Larson, W.E. (1992), Surface compaction reduces the root and shoot growth and grain yield of what. Agron. J., 84, 34-38.

Unger, E.C. & Kaspar, T.C. (1994), Soil compaction and root growth: a review. Agron. J., 86, 759-766

van Ouwerkerk, C. & Soane, B.D. (1994), In Conclusions and recommendations for furthers research on soil compaction in crop production. Soil Compaction in Crop Production. Dev. Agric. Eng., vol. 11. Elsevier, Amsterdam, 627-642.

Whitely, G.M. & Dexter, A.R. (1984), The behavior of root encountering cracks in soil. I Experimental methods and results. Plant Soil, 77, 141-149.

Physiological Activity of Sapropel Decomposition Products

Ina Alsiņa, Laila Dubova

Latvia University of Agriculture, Department of Plant Biology and Protection,
Strazdu1, LV – 3004 Jelgava, Latvia

Abstract

Experiments were carried out to investigate the activity of sapropel decomposition products. Sapropel was taken from two lakes in Latvia: Lake Lobe (Ogre district) and Lake Zebrus (Dobele district). Decomposition was done microbiologically in 750 mL flasks in a termostat at 40 °C over 5 months. Half of the flasks were kept under aerobic conditions, the other half anaerobic. Plant biotests and pot experiments were conducted at the Department of Plant Biology and Protection, Latvia University of Agriculture. The effect of decomposed sapropel on the germination of radish (*Raphanus sativus var sativus*) and cress salad *(Lepidium sativum)* seeds was tested. The decomposed sapropel was used at concentrations 100, 10 and 1 g L^{-1}. Three different controls were used; undecomposed sapropel without microorganisms, a solution of microorganisms without sapropel, and water respectively. In the germination of seeds, the length of the primary root depends on the culture. Stimulating effect was observed on radish seeds, in contrast to the inhibition of growth of cress salads in the early stages of decomposition. The inhibition decreases with the time of decomposition, after four months a growth stimulating effect was observed in cress salads.

Introduction

Sustainable crop production depends much on good soil health. By optimal combination and addition of organic and inorganic components the soil health can be maintained. Repeated use of chemical fertilizers destroys soil biota. In nature, there are a number of useful soil micro organisms which can help plants to absorb nutrients. Their utility can be enhanced with human intervention by selecting efficient organisms, culturing them and adding them to soils directly or through seeds. In this context it is of interest to investigate the effect of sapropel as biofertiliser and improver of soil health. [1, 2]

Sapropropel, is the sediment formed in fresh water reservoirs such as lakes, ditches and swamps over long periods of time. It consists of decomposing parts of dead plants and zoo-organisms, and is formed as a result of underwater biological and chemical processes at low oxygen concentrations. Sapropel, or mud in every day language, is available in large quantities in lowland regions. At the moment, one of the major consumers of sapropel is agriculture.

Sapropel consists of matter of mineral- and organic origin. The mineral part originates from clay- and sand particles and ash of decomposed organic material. The organic part consists of material of biological origin decomposing as a result of anaerobic biochemical processes and material formed by microorganisms by subsequent re-synthesis. [3]

Chemical analyses show that sapropel contains most nutrients required by plants, which no other single fertilizer can provide. Sapropel contains a significant quantity of huminic acids and their salts. These are resistant to biological disintegration and essentially enrich the humus fund of the soil and are carriers of many soil fertility components. [1, 2]

The aim of this work was to investigate the physiological activity of sapropel and its decomposition products.

Materials and Methods

Experiments for investigation of the effect of sapropel and its decomposition products on plants were carried out in 2004 and 2005 at the Department of Plant Biology and Protection, Latvia University of Agriculture. Sapropel was taken from two lakes in Latvia: Lake Lobe (Ogre district) and Lake Zebrus (Dobele district) (Fig. 1). The chemical composition of the two sapropels is shown at Table 1.

Fig. 1. Sapropel was taken from Lake Zebrus and Lake Lobe

The Sapropel decomposition was done microbiologically in 750 mL flasks in a termostat at 40 °C over 5 months. Sapropel was mixed with a solution of microorganisms in proportion 1: 1 or 10: 1. Half of the flasks were kept at aerobic conditions, $+O_2$, the other half anaerobic, $-O_2$.

Tab. 1: Elemental analysis of Sapropel

Elemental Analysis of Sapropel [mg L^{-1}]		
Element	**Lake Lobe (L)**	**Lake Zebruss (Z)**
N	45.00	44.00
P	105.00	25.00
K	80.00	13.00
Ca	20,750.00	91,2500
Mg	1,750.00	1,875.00
S	550.00	400.00
Fe	1,150.00	75.00
Mn	65.00	170.00
Zn	25.50	7.00
Cu	2.90	2.60
Mo	0.04	0.04
B	7.80	0.70
pH	7.10	7.60

Seed Germination Test

A seed germination test was designed for investigation of the effect of decomposed sapropel on the germination of radish (*Raphanus sativus var sativus*) and cress salad *(Lepidium sativum)* seeds. Ten seeds of each species were placed in Petri dishes with filter papers moistened with 10 ml of sapropel or its decomposition solution and placed in the thermostat at 20 °C. The control was germinated with water only. The following dilutions of sapropel or its decomposition products were used; 100 ml L^{-1}; 10 ml L^{-1}; 1 ml L^{-1}. After 2 and 4 month of sapropel incubation the length of the primary roots was determined. [4] Experiments were done in 3 replicas.

Vegetation Experiments

Vegetation experiments were placed in 2 L pots with neutralized peat in 4 replicas. Four variants and a control were chosen:

Control (c1): neutralised peat;

Variant 1 (S -2): sapropel mixed with peat in concentration 10 ml L^{-1};

Variant 2 (S -3)- sapropel mixed with peat in concentration 1 ml L^{-1};

Variant 3 (S 1: 1 -2): decomposed sapropel mixed with peat in concentration 10 ml L^{-1};

Variant 4 (S 1: 1 -3): decomposed sapropel mixed with peat in concentration 1 ml L^{-1};

Plants were fertilized with Kemira 10: 10: 20 1g per 1L of peat at sowing time.

The weight, the length, the number of leaves, and dry matter content were determined after one month of vegetation.

Results and Discussion

The germination experiments showed that seeds of different cultures had different germination abilities. No effect of sapropel or its decomposition products was observed on the germination of vegetables.

The effect on the length of primary roots of cultures is shown in Table 2. A stimulating effect was observed on radish. The length of primary roots of radish increased 17.8% on average as a result of sapropel treatment. The opposite effect was observed on cress salads; there the largest dose of sapropel (100 ml L^{-1}) inhibited growth of primary roots. The lettuce plants were less sensitive to sapropel concentrations. A statistically significant effect was not observed on these plants.

Tab. 2: The effect of sapropel on the length of primary root of radish, cress salads and lettuce [cm]

Concentration [ml L^{-1}]	**Radish**		**Cress salads**		**Lettuce**	
	Lobe	**Zebrus**	**Lobe**	**Zebrus**	**Lobe**	**Zebrus**
100	9.2**	8.5**	3.3*	2.6*	1.4	1.8
10	8.2**	7.7**	5.5	5.8	1.3	1.5
1	7.2	9.2**	5.0	5.7	1.8	1.1
H_2O	7.1			6.1	1.6	
$\gamma_{0.05}$ between lakes	0.5		1.3		0.3	
$\gamma_{0.05}$ between conc.	0.9		2.2		0.5	
* Result is significantly less than control (H_2O)						
** Result is significantly higher than control (H_2O)						

After 2 and 4 months of sapropel decomposition in the thermostat the similar stimulating effect on radish primary roots was observed (Table 3). The 30.4% longer primary roots of radish were established as a result of sapropel treatment. The 10-22% higher results were observed if decomposition was done in aerobic conditions. There were no significant differences between lakes and the concentration of microorganisms used for decomposition.

Tab. 3: The effect of sapropel decomposition products on the length of primary root of radish, cm.

Time of incubation		Concentration of sapropel decomposition products					
		100 ml L^{-1}		10 ml L^{-1}		1 ml L^{-1}	
		$+O_2$	**$-O_2$**	**$+O_2$**	**$-O_2$**	**$+O_2$**	**$-O_2$**
After 4 months of incubation	L 1: 1	9.4**	5.8**	7.3**	6.6**	4.9	4.7
	L 10: 1	6.6**	5.9**	4.5	5.8**	5.9**	4.3
	Z 1: 1	8.4**	6.7**	6.7**	6.5**	6.4**	5.3
	Z 10: 1	7.8**	4.8	7.4**	5.6	5.3	4.1
	ūdens	4.6	4.6	4.6	4.64	4.6	4.6
$\gamma_{0.05}$ between variants 1.0							
$\gamma_{0.05}$ between concentrations 1.0							
After 2 months of incubation	L 1: 1	13.9**	9.4	11.9**	13.4**	14.9**	16.4**
	L 10: 1	14.9**	12.9**	13.9**	12.7**	13.3**	11.0
	Z 1: 1	11.8**	10.7	13.3**	11.2	11.6**	13.4**
	Z 10: 1	15.9**	12.2**	12.6**	11.1	13.9**	12.9**
	ūdens	9.9	9.9	9.9	9.9	9.9	9.9
$\gamma_{0.05}$ between variants 1.5							
$\gamma_{0.05}$ between concentrations 1.5							
** Result is significantly higher than control (H_2O)							

Two months of incubation with microorganisms was too short a period to overcome the inhibition of primary root growth of cress salads (Table 4). On average the length of the primary root was 7% shorter in comparison to the control. After 4 months of incubation the stimulating effect had developed. The primary roots increase 14.9% on average in comparison to the control. The aerobic decomposition products stimulate growth more effectively than anaerobic decomposition (the increase of primary root of 8% on average was observed). No significant differences between lakes were seen.

Tab. 4: The effect of sapropel decomposition products on the length of primary root of cress salads, cm.

Time of incubation		**Concentration of sapropel decomposition products**					
		100 ml L^{-1}		**10 ml L^{-1}**		**1 ml L^{-1}**	
		$+O_2$	**$-O_2$**	**$+O_2$**	**$-O_2$**	**$+O_2$**	**$-O_2$**
After 4 months of incubation	L 1: 1	4.9**	4.7	5.6**	4.3	4.9**	5.5**
	L 10: 1	4.6	4.1	5.5**	4.6	4.4	4.5
	Z 1: 1	6.6**	4.9**	4.5	5.6**	5.1**	5.2**
	Z 10: 1	5.9**	4.5	6.4**	4.8	5.6**	4.8
	ūdens	4.4	4.41	4.4	4.4	4.4	4.4
$\gamma_{0.05}$ between variants 0.5							
$\gamma_{0.05}$ between concentrations. 0.5							
After 2 months of incubation	L 1: 1	7.3	5.6*	7.8	7.6	6.8*	6.6*
	L 10: 1	8.3	6.8*	7.5	7.3	6.8*	6.6*
	Z 1: 1	7.7	5.7*	7.8	8.2	6.3*	7.3
	Z 10: 1	7.9	7.2	7.2	6.9	6.4*	7.8
	ūdens	7.7	7.7	7.7	7.7	7.7	7.7
$\gamma_{0.05}$ between variants 0.8							
$\gamma_{0.05}$ between concentrations. 0.8							
* Result is significantly less than control (H_2O)							
** Result is significantly higher than control (H_2O)							

The pot experiment showed that different cultures had different sensitivity to sapropel and its decomposition products. The stimulating effect was observed on lettuce. The plant weight was 23.4% higher in comparison to the control. The best concentration (dose) of sapropel was 10 ml L^{-1}. There were no statistically significant differences between the sapropels from the two lakes.

Fig. 2: The effect of sapropel's decomposition products on the weight of cress salads, salads and radish

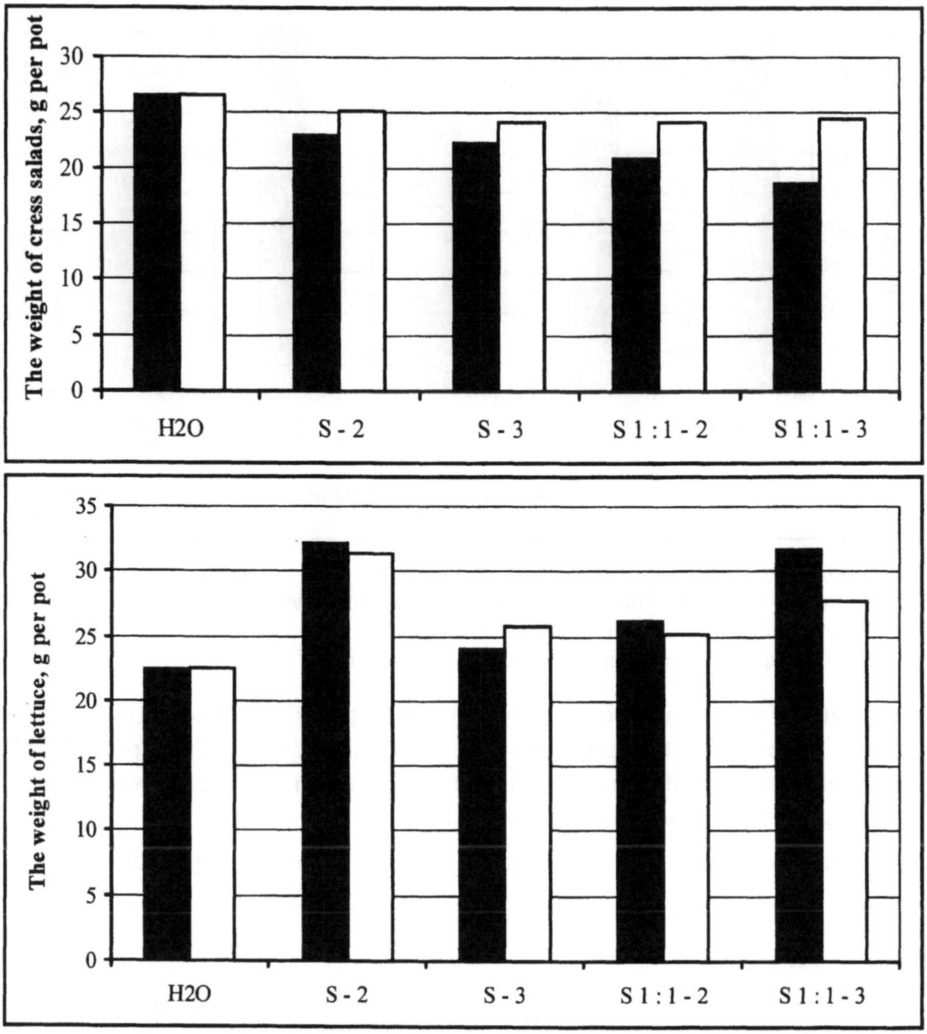

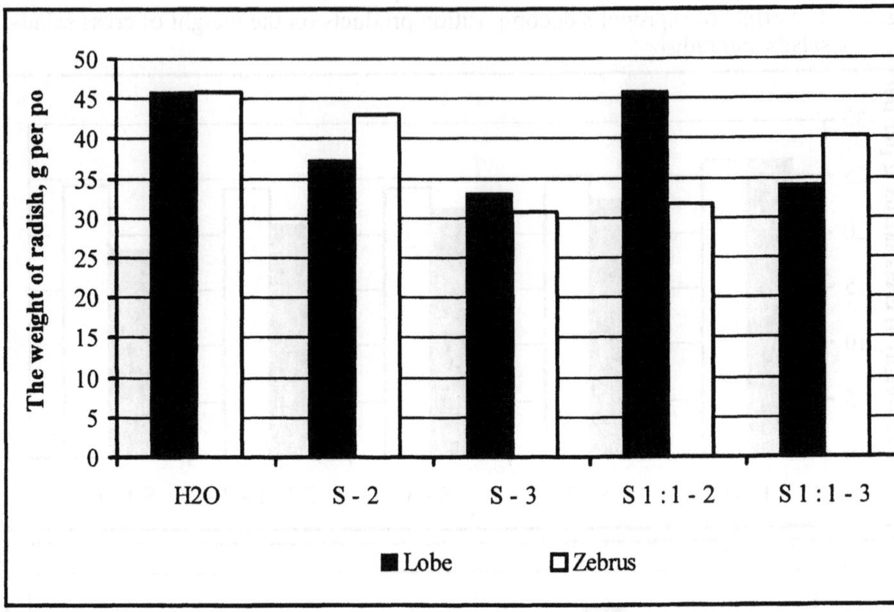

Legends:
S 1: 1 decomposed sapropel mixed with microorganisms in volumes 1: 1
S -2 sapropel mixed with peat in concentration 10 ml L^{-1}
S -3 sapropel sapropel mixed with peat in concentration 1 ml L^{-1}

The yields of radish and cress salads increase as the effect of addition of sapropel and its decomposition products. The cress salads grown with sapropel from Lake Lobe were smaller in comparison with Zebrus. These plants were significantly lighter in comparison with control too.

The radish grown with sapropel weighed 18% less in comparison to control.

Conclusions

The germination of seeds, the length of primary root and shoot depends on the culture. A stimulating effect was observed on radish seeds, in contrast to the inhibition of growth of cress salads at early stages of decomposition. The inhibition decreases with the time of decomposition, after four months a growth stimulating effect was observed.

The stimulating effect was observed in pot experiments with lettuce. Plants grown in pots with sapropel were 23.4% larger in comparison with control. The inhibiting effect was observed on radish and cress salads in pot experiment.

References

1. *Bakšiene, E.* (2004), Effect of lake sedements on changes sin andy loam cambisol properties and on crop productivity/Agronomy research 2(2), 111-119.
2. *Naumova, G.V., Younusova, L.Z., Ovchinnikova, T.F., Makarova, N.L., Chuduck, V.N. & Klyauze, I.V.* (1997), Resource-saving peat sapropel raw-material processing resulting in ecology-safe plant growth stimulators/Natural resources, no. 2., pp. 92-97.
3. *Andersons, P., Dudinsh, M. & Ositis, U.* (1997), Sapropel in animal nutrition 6th Symposium "Vitamins and Additives in the Nutrition of Man and Animal"/September, 24.-25. 1997, Jena (Germany)
4. *Гродзинский А.М. & Гродзинский Д.М.* (1973), Краткий справочник по физиологии растений, Киев, Наукова Думка, 591с.

Nutrient Transport Modelling in the Daugava River Basin

Andrea Wallin

Department of Earth Sciences, Uppsala University,
Villavägen 16, SE – 75236 Uppsala, Sweden

1. Introduction

During the 20th century and especially during the last 50 years, the loads of nutrients (nitrogen and phosphorus) in the Baltic Sea have increased dramatically. In order to master the problems of eutrophication in the Baltic Sea, proper water management is needed. A decision support system called Nest is presently being developed at Stockholm University within the frameworks of the scientific projects MARE (Marine Research on Eutrophication) and ELME (European Lifestyles and Marine Ecosystems). The aim of Nest is to help decision makers and managers working with Baltic Sea environmental issues to take the most cost effective measures against the eutrophication of the Baltic Sea. Nest consists of interlinked models that combine knowledge in ecology, physical oceanography, biogeochemistry, and economics (MARE 2005). However, the sub-model for prediction of nutrient loads from the river basins of the Baltic drainage area is not satisfactory. Therefore it has been decided that a better watershed model for the determination of nutrient loads to the Baltic Sea shall be incorporated into the system. The chosen model is GWLF (Generalized Watershed Loading Functions). A number of river basins have been selected in order to test this model in the Baltic Sea region, among these is the river basin of Daugava.

The purpose of this study was to apply the GWLF to the river basin of Daugava, which runs into the Baltic Sea, and thereby produce estimates of site specific parameters and input data needed by the model. The modelling was done using the Windows-based simulation system CatchmentSim. The purpose of the study is also to evaluate the model and the simulation system and point out strengths and weaknesses that should be considered when applying the model to the rest of the Baltic basin.

2. Materials and Methods

2.1 General Model Description

The Generalized Watershed Loading Functions (GWLF) is a simple mathematical model that estimates monthly loads of nitrogen and phosphorus from watersheds. Examples of required inputs to the model are climate data and information on land use and soil types in the basin. Hydrology is the driving force of the model. There are three types of water storage; soil water (unsaturated zone), shallow groundwater and deeper groundwater. The amount of water in these three storages characterises the system at each moment and determines the evapotranspiration, surface runoff and groundwater flows. The catchment is divided into land use categories with different characteristics. Surface runoff from each land use area and groundwater flow from the entire catchment to the river is predicted from data on daily precipitation and temperature. The user has to state the nutrient concentration in surface runoff, groundwater and in eroded material. The nutrient load is obtained by multiplication of the nutrient concentration by the water flow. Point sources of nutrients are added in order to obtain the total nutrient load. In the original GWLF model the nutrient load from septic systems in the catchment is included, but this part of the model has been excluded in the present study. A complete description of the model equations can be found in The BasinSim 1.0 Users Guide (Dai et al. 2000). For the simulations the computer program CatchmentSim has been used. The program is based on a modified version of the GWLF model and it has been developed at Stockholm University (Mörth 2004).

Fig. 1: The Daugava River Basin.

2.2 Site Description – the Daugava River Basin

The Daugava River is the third largest river in the Baltic Sea basin. The catchment area is 85 900 km^2 (Hannerz & Destouni 2005). 32% of the area belongs to Russia, 39% to Belarus and 27% to Latvia. The remaining 2% of the basin belong to Estonia and Lithuania.

About half of the basin area consists of agricultural land while 40% is covered by forest (Figure 1) (Hannerz & Destouni 2005). The population in the area is about two million, of which 1.4 million live in the Latvian part of the basin. The major cities in the basin are Riga (800 000 inhabitants) in Latvia and Vitsyebsk (400 000 inhabitants) in Belarus (Gooch et al. 2002).

2.3 Available Observation Data Used in the Study

Climate data (daily precipitation and mean air temperature) for the years 1980 to 2000 was taken from the BALTEX Hydrological Data Centre database administrated by SMHI. The value of daily temperature and precipitation is a spatial and temporal mean of several measuring points in the basin and several times of measurement during the day (Danielsson 2005).

The reported data on streamflow, total nitrogen load and total phosphorus load consists of monthly means from 1970-1999 (Wulff & Rahm 1990; Stålnacke 1996) with some data gaps during the 90's. Data from 1970-1990 were compiled by Stålnacke (1996) with help from data obtained from the Latvian Hydrometeorological Agency in Riga. Gaps in the data series were filled in by the use of statistical interpolation and extrapolation methods. Different sampling sites were used and it is not clear where the sampling sites are located, but they are probably situated a considerable distance from the mouth of the river. Since the areas of different land uses in the basin have been calculated from the mouth of the river this means that the drainage area used as a model input is overestimated with respect to the drainage area of the sample stations. Also, the impact of Riga is not included in reported data, since Riga is situated at the mouth of the Daugava. Considering the data from the 90's, there is little information on how the data series were compiled and where the sample station is located.

2.4 Calibration and Validation

Data from 1993-1999 was used for calibration. As a first step calibration was done for the hydrological part of the model. A few of the model parameters were used for calibration, for example the coefficient for evapotranspiration The calibration was a trade off between obtaining an average value over the seven years of the predicted streamflow as close to the observed average value as possible, minimising the average percentage error and reaching the observed top flows. The process was iterative and since there is no built-in calibration procedure in CatchmentSim the work had to be done manually. When the calibration was complete the validation was done on data from the 1980's.

When the hydrological part of the model was calibrated, the next step was to add nutrient concentrations in groundwater and surface runoff, erosion of solid nutrients and nutrient loads from point sources. The nutrient concentrations in the shallow and deeper groundwater were used as calibration parameters. Focus was placed on achieving the best fit for loads, but concentrations were also studied. The calibration was done in the same way as for the hydrological part. As for streamflow, validation was done on data from the 1980's.

Fig. 2: Reported and modelled values of yearly means of streamflow, nitrogen loads and phosphorus loads from the Daugava River for the calibration period (1993-2000) and the validation period (1980-1990).

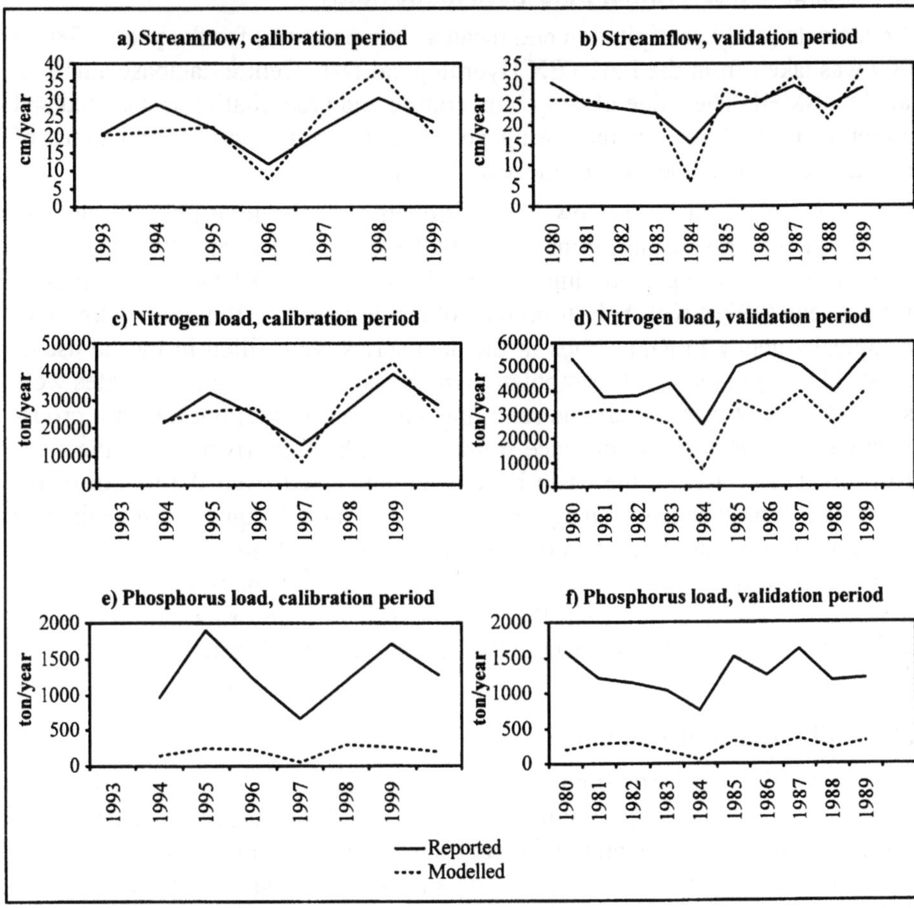

3. Results

The model output is monthly means of loads and streamflow, but since the purpose of the model is to estimate long-term loads it is of greater interest to see how well the model predicts yearly means. In Figure 7 the reported and modelled yearly means of streamflow and nutrient loads are presented. The model seems to estimate yearly streamflow quite well, both for the calibration period and the validation period. There are, however, tendencies towards a higher variance for modelled yearly flows than for observed yearly flows. Yearly nitrogen loads are well estimated by the model for the calibration period, but for the validation period nitrogen loads are underestimated.

Regressions lines were made between reported and modelled values. The R^2 value of the regression of modelled and reported yearly nitrogen loads is 0.78 and the corresponding slope of regression is 1.17. The R^2 value for yearly phosphorus loads is 0.5, but since the regression slope is only 0.14 the ability of the model to predict phosphorus loads must be considered to be very low.

4. Discussion and Conclusions

The aim of the model is to predict long-term nutrient loads from a watershed. The model predicts yearly nitrogen loads from the Daugava basin quite well. The mean absolute percentage error of estimations of yearly nitrogen loads during the years 1993-1999 is 18%, and the corresponding R^2-value is 0.78. Considering the simplicity of the model and the defectiveness of input data this can be considered a quite good result. For the 1980's (validation period) the model underestimates nitrogen loads by approximately 30% even though streamflow is quite well predicted during this period. A reason for this may be that the surface runoff and groundwater nitrogen concentrations were higher during the 1980's than during the 1990's. This is probable since the use of fertilisers, especially in the Latvian part of the basin, dropped dramatically in the beginning of the 1990's due to the collapse of the Soviet Union. There has been a negative trend in nitrogen loads from the Daugava during the years 1987-1998 (Stålnacke et al. 2003). Since the nutrient concentration in the GWLF model is constant, the model can never predict trends in nutrient concentration due to changes in the groundwater or surface runoff concentrations. In that sense the model is static and not suitable for long-term nutrient load predictions.

The GWLF model, as implemented, fails to predict phosphorus loads. The modelled phosphorus loads are about 90% less than reported loads. One reason could be that phosphorus concentrations in surface runoff and groundwater used were too low or that phosphorus loads from erosion were underestimated. Never the less, a more probable explanation for the underestimation of phosphorus loads is that the reported values of point source phosphorus loads in the basin are a lot less than the actual value and that phosphorus from septic systems, which are not included in the modelling, contributes substantially to the total load. According to HELCOM (2004), about 70% of diffuse phosphorus loads to the Gulf of Riga origin from other sources than agriculture and forestry, for instance from septic systems. Looking at Daugava the figure is probably something similar since water from the Daugava is the main contributor to the Gulf of Riga. Gren et al. (1997) state that the lack of sewage treatment in the Gulf of Riga region contributes to very large input of phosphorus to the sea. Since agriculture and forestry are the main sources of phosphorus according to the model and since septic systems are not included it is therefore not surprising that the model fails to predict phosphorus loads.

Nitrogen retention is not included in the model. The inability of the model to estimate nitrogen retention might affect the usefulness of the model for scenario analysis. For example, the construction of wetlands may help decrease nitrogen loads from water passing through the wetland since nitrogen is lost by denitrification, but in the GWLF the nitrogen load will not be affected by the increase in wetland areas. Also, since retention is not included in the model, the impact of land anywhere in the basin will have the same impact, a situation that is unlikely to occur in reality.

The quality of data is of major importance for the result of any modelling. In the Daugava basin it is difficult to find reliable data. The reported streamflow and nutrient loads can also be questioned. In order for modelling results of nutrient flows from the entire Baltic Sea basin to be trustworthy more reliable data are therefore needed. It is of utmost importance that such data are easily accessed and homogenous in order to support sustainable water management in the Baltic Sea Region (Hannerz et al. 2005).

In order to use the GWLF for long-term prediction of nutrient loads, modifications of the model are suggested. For instance, nutrient concentrations should be variable with time. For simulations over short time periods it is probable that the groundwater concentration remains more or less the same. However, considering long time spans or periods with sudden changes in management, it is not likely that the groundwater concentration will be constant. As stated above, it must be remembered that the model is static in this sense. Another change that could be made to the model is to divide the existing land cover categories into several subcategories depending on more site-specific land use. This way it would be easier to handle scenarios where new agricultural policies affect management practices.

Acknowledgements

I would like to thank my supervisors, Fredrik Hannerz and Jerker Jarsjö, Department of Physical Geography and Quaternary geology, Stockholm University, for their support and guidance during the project. Additionally I would like to thank Georgia Destouni, Department of Physical Geography and Quaternary Geology, Stockholm University, who contributed with guidance and expert knowledge. I would also like to thank Magnus Mörth, Department of Geology and Geochemistry, Stockholm University, for answering questions regarding the simulation system and the model in general.

5. References

BALTEX Hydrological Data Centre data base (2004), SMHI. URL. www.smhi.se/sgn0102/bhdc/(16-02-2004).

Dai, T., Wetzel, R.L., Christensen, T.R.L. & Lewis, E.A. (2000), The BasinSim 1.0 Users Guide. URL: www.vims.edu/bio/models/bsdownload.html (05-11-2004).

Danielsson, Åsa (2005), Institutionen för Tema, Linköpings universitet, Personal communication.

Gooch, G.D., Höglund, P., Roll, G., Lopman, E. & Aliakseyeva, N. (2002), Review of existing structures, models and practices for transboundary water management. Report no D1b. URL: www.mantraeast.org/pdf/D1b.pdf (05-11-2004).

Gren, I.-M., Söderqvist, T. & Wulff, F. (1997), Nutrient Reductions to the Baltic Sea: Ecology, Costs and Benefits. In: Journal of Environmental Management 51 (1997), pp. 123-143.

Hannerz, F. & Destouni, G. (2005), New definition and update of spatial data and statistics for the Baltic Sea drainage basins. Ambio. (in review)

Hannerz, F., Destouni, G., Cvetkovic, V., Frostell, B. & Hultman, B. (2005), A flowchart for integrated river basin planning with the Water Framework Directive. European Water Management Online. (in press)

HELCOM (2004), The forth Baltic Sea Pollution Load Compilation (PLC-4) Helsinki Commission, 2003, Pro no. 93. Erneko Painotuote Oy, Finland, ISSN 0357-2994.

MARE. URL: www.mare.su.se (05-01-05).

Mörth, Magnus (2004), Institutionen för geologi och geokemi, Stockholms universitet, Personal communication.

Stålnacke, P. (1996), Nutrient loads to the Baltic Sea. Linköping University, Department of Water and Environmental Studies.

Stålnacke, P., Grimvall, A., Libiseller, C., Laznik, M. & Kokorite, I. (2003), Trends in nutrient concentations in Latvian rivers and the response to the dramatic change in agriculture. In: Journal of Hydrology 283: 184-205.

Wulff, F. & Rahm, L. (1990), A database and its tools. Chapter 13. In: Wulff, F. (ed.), Large-scale environmental effects and ecological processes in the Baltic Sea. Research programme for the period 1990-1995 and background documents. SNV REPORT 3849. Stockholm. pp. 217-225.

Forest Fire Monitoring from Satellites

Edgars Andersons

University of Latvia
19 Rainis Blvd., LV – 1586 Riga, Latvia

Abstract

Referring to damage caused by recent forest fires in Latvia the paper considers the existing forest fire monitoring system and advancements based on possible solutions of the problems of fire risk assessment and monitoring provided by information from space-borne instruments.

Introduction

The limited resources of useful minerals and oil in Latvia are counterbalanced by a forest cover the area per capita of which is considerably larger than in many other places of the EU. Therefore, forest is the main national wealth of the country and has always been an important natural resource for the people of Latvia throughout its history. Forestry has become the most important and vital constituent of economic development ever since Latvia has regained its independence and, particularly, it is the basis for development of the emerging commitments and projects associated with recreational services and tourism. Presently 2.85 million ha – about 45% of the territory of Latvia is covered by forest [1]. 50% of the forest is managed by the State Forest Service, 42% – by private and 8% – by other owners.

Efficient prevention and protection from forest fires is one of important aspects of sustainable forestry and forest management. As shown in Table 1, forest fires cause a considerable loss of forest stocks every year.

Monitoring of Forest Fires

Forest fire monitoring in Latvia is based on 186 fixed observation towers with observers on top and is performed by the State Forest Service. The observation system, apart from other disadvantages (observation of smoke flumes is available only in daylight), does not cover all the forest area. Except for special cases, air-borne surveys are not practiced. Risk assessment of arising forest fires – the so-called "fire index" is absent. On the basis of current weather information the Latvian Environment, Geology and Meteorology Agency includes warning of arising fire hazards in weather forecasts for general public.

The forest fire monitoring system includes three conditional basic tasks: assessment of the risk of arising forest fire (determining the fire index), detection and determining location of the fire, and assessment of the damage (determining the area burned). It has to meet at least five requirements – it should be continuous, fast, reliable, precise, and informative. Alarm centres should receive full necessary information as soon as the fire is detected (within 15 minutes) [3]. Occasions of false alarms should be minimized (no more than 10%). The accuracy of determining the fire site should be within 300-500 metres from the actual location, and observing should provide detection of fire area that does not exceed 50 m^2. The information received at the alarm centre with time intervals of less than 1 hour, apart from precise location, should contain estimates of direction and speed the fire is spreading.

Tab. 1: Number of forests fires, burned area, and financial loss in 2000 to 2004 [1].

No.	Year	Number of forest fires	Burned area [ha]	Financial loss, LVL	Remarks
1.	2000	915	674	285,109	
2.	2001	271	200	84,602	wet summer
3.	2002	1,742	2,364	1,000,000	dry summer
4.	2003	900	663	280,456	
5.	2004	629	443	190,410	

Financial loss does not include environmental costs such as reduced absorption of CO_2 and costs of replanting the forest.

Remote Sensing Techniques

Presently about 15 satellites of 5 different types on low earth orbits (LEO) [7] are equipped with instruments providing information that could be used by a monitoring system. For example, the "Centre for Remote Imaging, Sensing and Processing" of the national University of Singapore successfully operates as a satellite ground station supplied with 6 m and 13 m X-band reception antennae acquiring data from 6 different types of LEO remote sensing satellites [5]. A space-borne forest fire monitoring system can provide fire-fighting organizations with an innovative and powerful tool for successful management of the forest fire problems particularly important and necessary in the case of Latvia.

The two main types of satellite instruments – the active Synthetic Aperture Radar (SAR) and the passive Advanced Very High Resolution Radiometer (AVHRR) can both be used for tasks of forest fire monitoring. The major advantage of SAR instruments is obtaining information regardless of weather conditions, cloud coverage, and sun illumination. Despite being inefficient for fire detection, the cloud penetration capability and spatial resolution of the order of 20 to 5 metres makes the SAR instruments potentially useful in identifying the dry forest areas most vulnerable to fire hazard, which is essential to assessment of fire risk and preventive measures. Since the intensity of backscattered radiation is dependent on the dielectric permeability of soil and vegetation is sensitive to moisture content the instrument has the advantage in determining the burnt areas and recovery of the moisture content reduced by fire. The passive sensors detect electromagnetic radiation of different wavelengths in visible and infrared. Detecting of infrared emissions from objects on earth can be successfully used for locating forest fires.

Assessment of the Risk of Arising of Forest Fires

The simplest way to assess the risk of fire hazard is provided by a fire danger rating system based on data about current temperature and relative humidity. A more complex system of fire risk assessment has been developed in Canada and U.S.A. [4] combining measurements of fuel, vegetation drought, topography, weather, and risk of ignition (both natural and caused by human activities). For conditions in Latvia it could be enough to begin with a system accounting for combined measurements of vegetation drought, temperature, wind, and relative humidity. The amount of living vegetation and its moisture content has a strong effect on propagation and severity of forest fires. Current assessment of the moisture of living vegetation relies on various methods of manual sampling. While the data thus obtained are quite accurate, covering of broad areas and providing continuous data is difficult and time-consuming.

The current polar orbiting meteorological satellites and other LEO satellites can provide a continuous data flow including information on greenness and other parameters (local temperature, local relative humidity, etc.) needed for updated calculation of fire index required for early warning of forest fire danger [4].

Fire Detection

Three major requirements for space-borne fire detection concern spectral sensitivity of sensors, spatial resolution, and period of observation. To discriminate hot events the sensors should cover the range from visible to thermal infrared [2]. To detect relatively small size fires it is necessary to combine a moderately high spatial resolution with good ground coverage. Repetition rate of periodic observations should provide a good enough temporal coverage.

Crude data flow from different satellites is continuously received and processed round the clock at the State Forest Agency. Evaluated and selected information is distributed to fire fighting centres.

Fig. 1: Block diagram of a forest fire monitoring system

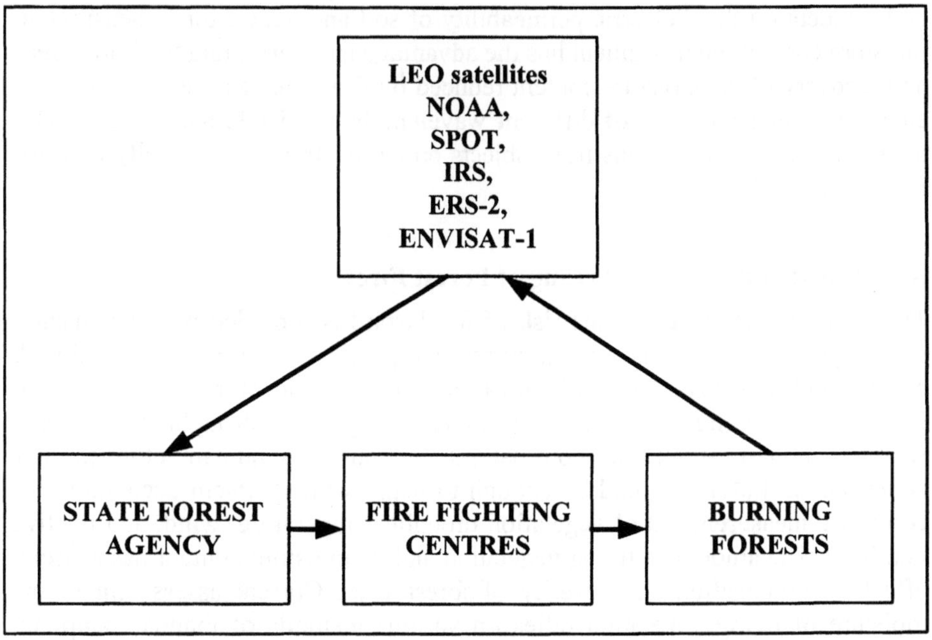

A number of orbiting satellites covering the territory of Latvia are equipped with sensors suitable for purposes of forest fire monitoring: five National Oceanic and Atmospheric Administration (NOAA) satellites carry very high resolution radiometers; ERS-2 is equipped with an along track scanning radiometer; advanced along track scanning radiometer is on board of environmental satellite ENVISAT-1; three SPOT satellites carry high resolution visible imaging instruments; five Indian research satellites (IRS) satellites have onboard wide-field and short-wave infrared sensors.

Equipment on board of the 15 LEO satellites, apart from the three basic tasks of fire monitoring, have the capacity for measurement of the change of carbon cycle and the temperature over urbanised areas.

The Monitoring System

To provide a favourable frequency of observations the active satellite-based forest fire-monitoring system should integrate different types of data from different types of available LEO satellites. Block diagram of a potential monitoring system is shown in Figure 1. Since the space-borne components of the system collecting the data already exist and function, only the ground components and networks of information exchange and dissemination need to be designed and set up to supplement and, if necessary, eventually reorganise the present network of fire fighting services.

Conclusion

Some fifteen satellites already launched in low earth orbits carry the instruments observing the territory of Latvia and collecting data that might be used for forest fire monitoring tasks and other purposes. On the other hand, the hardware and software necessary for data acquisition and processing is also available. Political will and skilled management is all the country needs to introduce a flexible and efficient space-borne monitoring system to prevent damage caused by fires to the main natural resource.

References

1. The Fire-Fighting and Rescue Services Department in Latvia – "Forest Fires Statistical Data from 2000 to 2004".
2. *Robinson, J.M.* (1991), "Fire from space: global fire evaluation using infra-red remote sensing". International Journal of Remote Sensing, 12, 3-24.
3. *Rauste, Y., Herland, E., Frelander, H., Soini, K., Kuoremaki, T. & Ruokari, A.* (1997), "Satellite-based forest fire detection for fire control in boreal forests", International Journal of Remote sensing, 18, no. 12, p. 2641- 2656.
4. *S.C. Liew, Lim, O.K., Lim, L.K. & H.* (1998), "A study of the 1997 forest fires in South East Asia using SPOT quicklook mosaics", Proc. 1998 Int. Geosci. Remote Sensing Symp., vol. 2, 879-881.
5. www.crisp.nus.edu.sg/.
6. www.fire-freiburg.de/programmes/und/idndr_ha.htm.,"Hazard assessment as the basis of risk analysis".
7. www.ioc.unesco.org/oceanteacher/OceanTeacher2/01 "Satellite Mission Information".

C. Social Corporate Responsibility and the Environment

Sustainable Community Development and the Role of Social Capital

Remigijus Čiegis, Dalia Gineitienė

Socio-Cultural Research Centre, Kaunas Faculty of Humanities,
Vilnius University,
3 Universiteto str., LT – 01513 Vilnius, Lithuania

The Problem

With the increase of environmental awareness and knowledge in society the way to look towards environmental problems has also changed. It became clear that all sectors of society are expected to carry a shared responsibility for the environment. Involvement of the local communities is an especially important task as most of environmental problems are caused by activities at a local level and are related to the ways of living and individual consumption. Social capital can be thought of as a resource to action, which may lead to a broad range of outcomes since the social capital enhances capacities for community problem solving. According to Grootaert et al. (2003), such social problems as crime, health, poverty and unemployment, have been linked empirically to a community's social capital or lack of it. Citizens are concerned that new forms of social capital must be constructed when older forms decline as a result of technological or demographic change.

The Research Objectives and Tasks

The main attention in the article is given to the social capital in the sustainable community development.

The Research Method

Logic abstraction encompassing generalisations on the problems of sustainable community development and social capital, according to the conclusions and reasoning of scientists from other countries was used in the article.

Forms of Capital Related to Sustainable Development

Speaking about the *sustainable development of community* it is relevant to discuss different types of capital, which are related to the solution of different environmental issues. The present need for sustainably managed systems resulted in the search for the *sustainability assessment*. It can be stated, that appropriate evaluation of capital is quite essential to the sustainability approach. In order to secure sustainable development of the society it is essential to preserve constant resources of *nature's capital (Kn)* (Pearce & Turner 1990). It is still not clear how all supplies of natural resources can be characterized by a single digital constant *(Kn)?*

We inevitably approach the nature's capital evaluating problem when the attempt is made to use the criterion of "constant nature's capital". There are two theoretical ways to measure the nature's capital: to analyse the *physical amount* of nature's capital or to evaluate this capital in *monetary units*. Using the first method, we should have tools to exercise the summing up of physical supplies but the lack of such tools inables us to calculate the physical quantities of different (unequal) origin. This argument could be a strong invitation to apply monetary expression in identifying nature's capital constant. But usually we do experience difficulties in the process of calculating "the right price" for nature's equity, which is not traded in markets; here we also face the fundamental problem of sustainability: we do not have any methods of calculating how future generations will evaluate certain components of natural resources. It is clear that when applying the monetary evaluation of nature's capital another problem arises, i.e. a sharp price take-off in the case of nature's capital decreasing power. This problem is strengthened by the general problem of resource economy: it is possible to have a theoretically optimal solution, which could evoke unsustainable consequences of environmental protection at the same time.

Based on the arguments stated above, we can conclude that the definition of nature's capital, serving as a useful metaphor for attracting attention to sustainable development, cannot be used as a practical instrument for scientific and political efforts to secure development within the ecologically sustainable content.

Interpretations of *strong and weak sustainability* can also be justified by studying the possibilities of substituting supplies of *nature's* and *economic capital* or complementing each other.

Strong sustainability requires both types of capital not to decrease for the benefit of one of the significant indicators. In other words, according to the law of *strong sustainability*, the aggregate physical quantity of nature's capital or general nature's capital and its value should not decrease and should be preserved for future generations. In addition, according to the criteria of strong sustainability, a supposition is made that nature's and economic capital are complementary in the production process rather than substituting (Costanza & Daly 1992). Actually, it is recognized that some natural resources and services cannot be totally substituted as these forms of nature's capital supply vital services for all life-supporting environmental systems.

The version of *weak sustainability* is more acceptable for dominating economic theories, oriented towards securing the status where "wealth does not decrease in the time lag" (Pearce 1993). The version of *weak sustainability* contains (making an unrealistic assumption about the perfect substitution of nature's and man-made capital) the sum of the forms of both the nature's and the economic capital or any other aggregated unit of measurement (e.g., the "green" GNP), and requires that it does not decline all the time. Thus, there is an orientation towards the stock of the capital, which we leave for the future generations expecting that the capital stocks should not be less than our generation has.

When we speak about *weak* and *strong sustainability*, it must be noted, that the use of weak sustainability started, when it was noticed that strong sustainability was impossible to use in practise. Tietenberg (1984) defined sustainability as *relentless utility*. Then, applying the concept of weak sustainability, sustainability might denote securing peoples welfare, which could thrive or at least remain stable. Considering such interpretation of sustainability, it equals the prerequisite to ensure that peoples wealth does not decrease.

When using the *ecological capital* for undertaking socio-economic activities, and bearing in mind the four dimensions of sustainability, the following factors have to be taken into consideration for achieving sustainable community development (Maciag 1996: 21):

- *Physical capacity* (the maximum level of activity that does not disturb the environmental balance);
- *Ecological capacity*, (the maximum amount of activities undertaken without destroying flora and fauna);
- *Socio-psychological capacity*, (the maximum amount of activities undertaken without destroying local communities and family ties);
- *Infrastructural capacity*, (the maximum amount of activities undertaken without deterioration of the infrastructure).

Community Development Paradigms

The idea of local sustainable development is closely related with the *socio-psychological capacity*. Social capital can be a very important resource for the economic growth of a certain territory stimulating the incentives of local actors, using local networks and resources for local sustainable development. *Social capital* as a concept describing social relations became an important issue in academic debate, public policy and the practice of sustainable community development. Social capital is now seen as one of the essential ingredients of sustainable communities, as it facilitates regional growth and enables community renewal. Thus it is relevant to discuss issues of community development and the role of social capital in detail.

Efforts to address the problems of communities have been guided by two competing paradigms. The rational planning paradigm based on scientific approaches to problem analysis and the evaluation of results. This paradigm was dominant in the 1960s. With the *rational community planning* typically the process begins with analysis of needs or problems, proceeds with formulation of a plan and is followed by its implementation. All these activities are taken by the external agent who acts rationally for the community needs while communities may participate but without a say.

Since 1970's a new paradigm emerged, based on the assumption that communities had to be empowered to define and to solve their own challenges. According to Wallis (1996: 35) this new paradigm called "community-making" is defined by six principles-of-practice: "*social capital, civic infrastructure, assets-orientation, collaboration, strategic/vision based action*, and *the arts of democracy*". *Social capital* consists of the obligations among people who share a common identity as members of a community. Social capital forms where people hold common values, where they are in contacts via formal and informal networks, and they share a sense of trust. *Civic infrastructure* consist of more formalized patterns of communication in a community. These are local organization e.g. church, clubs, etc. They can act as mediators and mobilize financial and human resources. While creating social capital community-making also strengthens civic infrastructure. Planning with *assets-orientation* principle seeks that the source of solutions to community problems is within the community itself. In comparison to the rational planning when elimination of deficits is considered, the question to be answered is "what we can do with the assets we have now?" (Wallis 1996: 37). In order to use community assets effectively *collaboration* should be built among local organizations. Under the paradigm of community-making, the comprehensive plan is replaced with *a vision* of community embodying shared values and followed with goals, tasks, and actions. The vision serves to generate goals and direct actions. The arts of democracy could be understood as one of the important principles of community-making is employing processes of consensus formation and conflict resolution (Wallis 1996). A community-making paradigm brings people together and helps to identify common values, strengthens existing and creates new networks and so on, so striving to achieve common purpose can build social capital, which can play an important role in community development. It should be stressed that the "community-making" paradigm could be successfully used in sustainable community development based on Local Agenda 21 principles. The role of social capital is important in this process so further the concept of "social capital" is analysed.

Concept of Social Capital

The differences in the definitions and concepts of social capital emerged in the sociological and political literature. The theories on social capital may be broadly divided into two subgroups (Raiser et al. 2001). Social capital by the definition of Bourdieu (1993) is accepted as a private good and can be converted into cultural capital or "symbolic capital" that indicates social status of an individual. Coleman (1988) also emphasises the benefits of social capital to the individual. The social capital is the "stock" of social relations and shared values that enables people to cooperate and "Like other forms of capital, social capital is productive, making possible the achievement of certain goals [...]. Unlike other forms of capital, social capital inheres in the structure of relations between actors" (Coleman cited in Greenley 1997).

The second approach to social capital is more common and refers to the nature and extent of one's involvement in various informal networks and formal civic organizations. According to Putnam (1993) social capital is a cultural phenomenon, among other things denoting the existence of social norms, promoting collective action and the degree of trust in public institutions. Putnam suggests that social capital has the properties of a public good and "refers to features of social organizations such as networks, norms and social trust that facilitate coordination and cooperation for mutual benefit" (Putnam 1993: 67). Networks of civic engagement, such as neighbourhood associations, sports clubs, and cooperatives, are an essential form of social capital, and the denser these networks, the more likely that members of a community will cooperate for mutual benefit.

Thus, according to Bourdieu (1993), Putnam (1993) and Coleman (1988), social capital can be understood as networks of social relations which are characterised by norms of trust and reciprocity and which lead to outcomes of mutual benefit. Defined in this way, social capital can be thought of as a resource to action, which may lead to a broad range of outcomes. Stocks of social capital enhance capacities for community problem solving.

Social capital is the glue that holds a community together. Potapchuk et al. (1997) notes that social capital "operating at the interpersonal, organizational, institutional, and political levels of community" actually is [...] "the people who carry forward the values and visions of the community along with the parents, the house of worship, and the other civic institutions. All of this is done in and through the relationships among persons" (p. 130). They also expressed an idea that social capital like financial capital increases with use. The more the community invests, the more it has.

Putnam (1993) distinguishes between two types of social capital, i.e. *localized* and *generalized* capital. Localized capital developed from various local societies and this could be seen as a more "private" localized form of social capital, which helps to foster trust and cooperation that make communities work. Generalized capital or "public" grew out of patterns of reciprocity, collaboration and community spirit found in the private sphere of society and it involves relationships between larger groups and organizations.

Though Putnam (1995) speaks about the decline of people involvement in various fields (organizational memberships, religious associations, labour unions, parent-teachers associations), which traditionally were popular, while other authors contradict these statements. However, indicated reasons for the decline of social capital could be noticed in many countries:

- The movement of women into the labour force;
- Mobility of residents;
- Demographic transformations (fewer marriages, fewer children, more divorces);
- Technological transformation of leisure (e.g. television and computer).

The importance of the social capital for the formation of the sustainable community and the civic society in general is stressed as well as its importance for the work of local governments as it strengthens cooperation and trust. Thus, it is relevant to analyse the role of social capital in community building as well as the need and ways of strengthening of the social capital.

Community Building and Development of Social Capital

Recently we are experiencing the re-emergence of interest in *community*. Potapchuk et al. (1997) presents "the ladder of community building" from the level of individuals to the whole community, which illustrates the role of social capital for collective community action and also shows community development.

Fig. 1: The ladder of community building

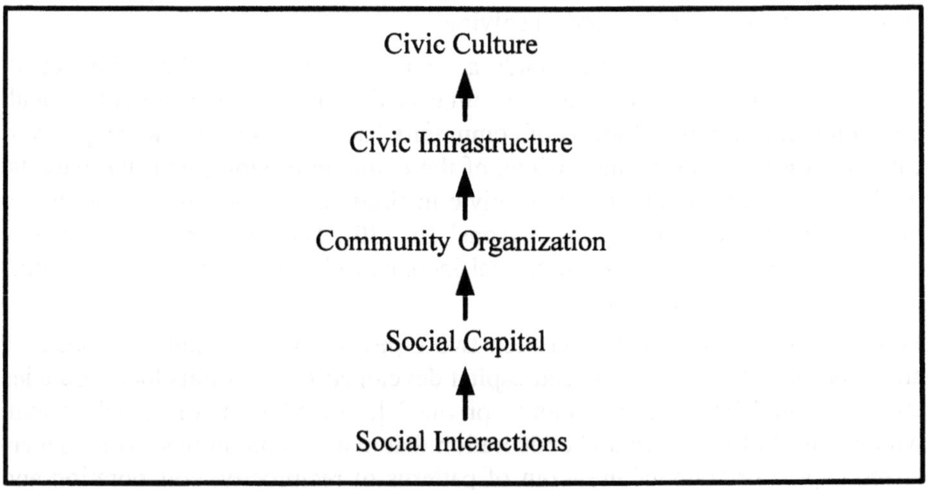

By *social interactions* the authors mean bonds among individuals produced by their interactions in daily activities. They are important for community but not always turn to social capital. *Social capital* is created by interaction of groups and individuals. *Community organization* refers to the formal and informal ways that individuals come together to create a community. When a community needs to deal with a range of social, environmental and other issues it has to mobilize its *civic infrastructures*. If community organizations in the form of local capital are strong and able to work together making generalized social capital it often determines success in problem solving initiatives. An example of the Not-In-My-Back-Yard (NIMBY) phenomenon when individuals come together to oppose the local placement of waste treatment plant illustrates the lack of generalized social capital, which is necessary to successfully solve community problems. *Civic culture* refers to the values and norms that characterize a community. In short, community is people in interaction, sharing common ideas, trust, goals with each other (social capital) and in a certain geographical area. Communities with their variety of forms and expressions make civil society.

Governmental institutions can play an important role in building social capital and communities. On the one hand, a strong government role is advocated to protect the varied interest of society, equitably distribute resources and balance inequality of different groups of society and, on the other hand, the market is believed to be the most efficient way to distribute resources, provide services. However, both these options do not see citizen and social organizations to collectively participate in making the decisions that directly affect their lives though participation of the social capital in neighbourhoods and communities reveals benefits for all.

The neighbourhoods represent the unit of social organization that is closest to citizens where they live and work. A civic infrastructure can bring neighbours together to work on shared problems and create the net of relationships that make up social capital. At the neighbourhood level, localized social capital is most clear. At the level of community, generalized social capital becomes necessary for collaboration essential for effective civil society. Successful development is determined by the ability of stakeholders to work together and this requires high levels of trust and cooperation, i.e. social capital. Without a generalized social capital and cooperation among community groups and organizations effective solution of local problems and the development is hardly possible (Potapchuk et al. 1997). It is possible to speak about community participants and their impact on activities and this could depend on age, education, sex, possibly material possessions as well as the role in formal organizations (membership) and informal networks that exist in a community. Not all community activities are expressed through formal organizations some occur rather spontaneously. Quite often people unites for common goal usually against some activities of private and public sector that may have impact on their lives, for example, protest against plan to build gas station in the neighbourhood. The following phases of community action could be distinguished:

- rise of interest;
- somebody takes leader's role;
- goal setting;

- gaining and maintaining participation;
- gaining support;
- carrying out activities.

Encouragement of participation in community activities could be perceived as an instrument for greater effectiveness as well as a new source of investment in building of social capital. Participatory process, which is stressed in Local Agenda 21, helps to avoid failures because of a close knowledge of the local reality, which external agents or governmental bureaucrats do not have. It should be noted that community networks are important for the success of achieving goals and that co-operation of organisations enables them to carry out developmental activities. Participation is justified because it expresses not only the will of the community towards sustainable development but also it is the way for the community to ensure that the important moral, humanitarian, social, cultural and economic objectives of a more human and effective development can be peacefully attained.

To foster the formation of the civil society through strengthening communities is important for a developing democratic country and there is no alternative. The only question may be how to do this. One of the ways to start is to strengthen and develop existing social capital. To build a social capital that can more effectively address complex problems with a wide variety of social actors will require greater capacities for participatory learning and assessment within many institutional arenas. It should be noted, as the literature demonstrates, that the creation of social capital is a complex process heavily influenced by social, political, and cultural factors as well as by the dominant types of economic activities (Grootaert 2003). There is a need to support through financial and organizational aids local sustainable development based on formation of networks between individual subjects and collective actors. Underlining the importance of the role of social capital for local sustainable development does not mean devaluation of the role of governmental institutions. The institutions at the non-local level could help "from above" the local community to mobilize its social capital for activities "from below". However, governmental institutions could have a negative impact on social capital when they start to undertake activities that should be better left to the civil society. If the governmental institutions start organizing everything, people will loose their spontaneous ability to work with one another.

Education and raising awareness is the area where the governmental institutions have one of the greatest possibilities to generate social capital and stimulate public participation. Though "investing" in social capital is more difficult than investing in human capital, where a number of time-tested approaches are available (building schools, training teachers, developing appropriate curricula, etc.), but educational institutions do not simply transmit human capital, they also pass on social capital in the form of social rules and norms. (Grootaert 2003). Environmentally responsible citizens accept responsibility for what happens in their community –

not only environmentally but also politically, and socio-economically. Therefore, a primary educational mission should be to teach citizens to be able to influence public decisions where environmental issues are extremely important. Recently Education for Sustainable Development is seen to play a key role in helping communities to adopt sustainability principles in their development. Thus, educating people not only about various public issues but also about their rights and how to use them is an important factor in raising level of public involvement and thus development of social capital.

Conclusions

1. There are two theoretical ways to measure the nature's capital: to analyse the physical amount of nature's capital or to evaluate this capital in monetary units.
2. Interpretations of strong and weak sustainability can also be justified by studying the possibilities of substituting supplies of nature's and economic capital or complementing each other.
3. When using the ecological capital for undertaking socio-economic activities and having in mind the four dimensions of sustainability, the following factors have to be taken into consideration for achieving sustainable community development: physical capacity, ecological capacity, socio-psychological capacity, infrastructural capacity, physical capacity, ecological capacity, socio-psychological capacity, infrastructural capacity.
4. Social capital is now seen as one of the essential ingredients of sustainable communities as it facilitates regional growth and enables community renewal. Thus it is relevant to analyse in detail issues of community development and the role of social capital.
5. The theories on social capital may be broadly divided into two subgroups and social capital can be understood as networks of social relations which are characterised by norms of trust and reciprocity and which lead to outcomes of mutual benefit.
6. It is possible to present community development as the ladder from the level of individuals to the whole community where the role of social capital for collective community action is very important.

References

Bourdieu, P. (1993), Forms of Capital. In: J. Richardson (ed.), Handbook of Theory and Research in the Sociology of Education, New York: Greenwood Press.

Coleman, J. (1988), Social Capital and the Creation of Human Capital, American Journal of Sociology, 94 (supplement): 94-120.

Costanza R. & Daly, H.E. (1992), Natural capital and sustainable development. Conservation Biology. No. 6, pp. 37-46.

Greenley, A. (1997), The Other Civic America. Religion and Social Capital. The American Prospect 32, 68-73.

Grootaert, Ch., Narayan, D., Jones, V.N. & Woolcock, M. (2003), Integrated Questionnaire for the Measurement of Social Capital (SC-IQ). The World Bank, http://poverty.worldbank.org/files/11998_SCIQfinal.doc.

Ladd, E.C. (1996), The Data Just Don't Show Erosion of America's "Social Capital". The Public Perspective. June/July. 5-22.

Maciag, J. (1996), Zrodla i perspectywy turystyki wiejskiej. Od wywczasiw do agroturystyki //Wies i Rolnictwo N 3.

Pearce, D.W. & Turner, R.K. (1990), Economics of Natural Resources and the Environment. Baltimore.

Pearce, D.W. (1993), Economic Values and the Natural World. London.

Putnam, R. (1993), Making Democracy Work. Civic Traditions in Modern Italy. Princeton: Princeton University Press.

– (1995), Bowling Alone, Revisited. The Responsive Community. Spring. 18-33.

– (1995), Bowling Alone: America's Declining Social Capital. Journal of Democracy. 6(1), 65-78.

– (1996), The Strange Disappearance of Civic America. American Prospect. 24.

Raiser, M., Haerfer, Ch., Nowotny, T. & Wallace, C. (2001), Social Capital in Transition: a First look at the Evidence. European Bank for Reconstruction and Development, Working paper, no. 61.

Tietenberg, T. (1984), Environmental and Natural Resource Economics. Glenview, IL.

Wallis, A.D. (1996), Toward a Paradigm of Community-making. National Civic Review. 85 (4). 36-47.

Corporate Responsibility at an International Level

–

Stakeholder Management in Multinational Firms as an Integrative Approach towards Sustainability

Agnes Haraszti

Centre for Sustainability Management,
University of Lüneburg,
Scharnhorststr. 1, D – 21335 Lüneburg, Germany

The social impact of a company's operations is a key area of business performance where competitive advantage can be either won or lost, and risks managed or mismanaged. In order to assure market success and hence maximise shareholder value over an uncertain time frame, multinational firms have a vested interest in the prudent management of all entities claiming a stake in their activities, i.e., forming part of their business environment. Therefore, professionally managed stakeholder relationships have a clear instrumental value and a strategic role.

To be able to pinpoint the characteristic value systems (governing values, espoused theories as well as theory-in-use) (Argyris 1992) and the influencing potential of stakeholders as well as possible ways to cope with them, companies must identify all relevant individuals and groups and create an appropriate segmentation framework by analysing the interest and power relations. On the basis of this research work, then, a coherent communication strategy and a reciprocal interest vindication system must be elaborated, balancing stakeholder satisfaction and stakeholder contribution.

Multinational firms with premises embedded in various political, legal and social structures, in particular, are expected to be responsive to stakeholder concerns on a global as well as on a local scale. Their stakeholder management work, thus, must be tailored to the specific local circumstances, with the prevailing cultural context being a crucial component. In order to better understand the complexity of stakeholder management against a multinational and, consequently, multicultural background, further, to develop indicators and metrics to measure and improve corporate social accountability and provide guidelines how to create genuinely stakeholder-balanced multinational firms, we explore what national and multinational firms can learn from one another by scrutinising best-practice examples within one chosen branch in a chosen country.

This endeavour is characterised by a high degree of interdisciplinarity and can therefore be regarded as a genuinely integrative approach towards sustainability. Stakeholder management in itself perfectly amalgamates the three vertices of the sustainability triangle by enriching the economic dimension of management with social as well as ecological aspects through the diverse stakeholder groups. By further introducing the international, i.e., multicultural perspective into this framework, we can claim that we are making a sincere contribution to spurring multinational companies on to making a significant and credible difference.

In order to duly highlight the manifold theoretical background of international, i.e., intercultural stakeholder management, we must first give an overview of the distinct fields this concept imbraces, namely management, stakeholder management, international management and culture, one at a time.

1. Management

Running a business entity is a reasonably complex task requiring management, leadership as well as governance skills and abilities. Managers must be capable of taking on a multitude of management roles like planning, control, organisation, direction, decision, co-ordination, controlling, performance evaluation, information management, wielding power, staffing, trouble-shooting, etc. (Mintzberg 1983). At the same time, they have to handle typical management situations, e.g. setting up an enterprise, organisation development, operating, change management, crisis management, expansion, entrepreneurship, strategic management, corporate governance, mergers and acquisitions, transformation and re-organisation, winding up a company, social responsiblity and environmental challenges, organisational learning, etc. And last but not least they have to be able to operate with a number of management tools amongst others in planning and monitoring, information systems, business policy, organisation design, leadership and management problem-solving, operations management, human resource management, decision-making, marketing, etc.) (Angyal 1999).

Managers, further, must be apt to balance self-management, contact management and operations management (Heller 2002) in order to maximise the competitiveness, and hence the overall success, of the company through a sound blend of their own personal success, successful relations and successful operations. It follows from the foregoing that the proper maintenance of relationships to the various stakeholder groups is just as crucial to the survival and the fruitful operation of a company as either of the other two factors. Nevertheless, managers tend to assign these contacts a lesser significance which this paper aims to make aware and critical of.

2. Stakeholder Management

Historically, companies were controlled by their owners through direct control of their managers, if not direct management, in pursuit of their economic objectives. However, with the rapid growth in company size, the economic activities of these companies tended to have increasing social consequences. Parallelly, they came more and more under the implicit control of their managers, and the notion of social responsibility (the voluntary consideration of public social objectives in addition to private economic ones) arose to endow their actions with the needed legitimacy (Mintzberg 1999).

Today, business entities face a complex and interrelated set of interests and influencing power. Individuals, groups and organisations claiming a stake in the activities of companies abound. Companies practically exist to serve the objectives of their stakeholders with these becoming their primary objectives. What the companies give to and take from their diverse stakeholder groups to achieve these primary objectives are their secondary objectives. Expecting the primary objectives to result, process stakeholders (employees and suppliers) design, implement and manage processes to realise the secondary objectives based on explicit and implicit contracts (also negotiated with environmental stakeholders; customers, owners, community, etc.) (Atkinson et al. 1997).

The recent corporate scandals have caused many to question the purely *shareholder value* theory of corporate governance. According to this approach, profitability outshines responsibility. Organisations are seen chiefly as instruments of their owners and their success as something measurable by share price, dividends and economic profit. The management of stakeholders is perceived rather as a means than as an end in itself. Social responsibility is believed not to be a matter for organisations, further, society is claimed to be best served by economic rationale, i.e., by companies pursuing enlightened self-interest and economic efficiency and maintaining market-based relations with all stakeholders. Recognising that it is instrumental (expedient) to pay attention to them does not mean that the purpose of a company should be to serve them, but to maximise shareholder value resulting in societal wealth being maximised (Rappaport 1998).

The prevailing view of companies and strategy-making nowadays tends to shift towards the stakeholder view. All businesses operate within a complex and interrelated system of interests and influences. The individuals and groups depending on the organisation to achieve their own goals and on whom, in turn, the organisational success depends, are its *stakeholders* (Recklies 2001). When identifying the stakeholders of a company, it is not enough to focus either (a) on stakeholders accepted as such by the company or (b) the formal structures of the organisation. The first lapse can be eluded by taking all individuals, groups and organisations claiming a stake in the activities of the company themselves into consideration. In other words, all persons who and organisations which are either affected by another organisation, or perceive that they have an interest in the activities of the organisation in question for whatever reason (Curtin-Jones 2000) should be taken into account. In order to avoid making the second mistake, then, it is necessary to scrutinise informal and indirect relationships and influences as well.

Having identified the stakeholders, the next step needed is to map how much impact or power they could exert, i.e., to what extent they could persuade, induce or coerce other stakeholders into following certain courses of action. This can be done by considering the sources and indicators (visible signs) of power. Besides the ability to act, it is also important to evaluate in what measure stakeholders are willing to exert their power, further, how they relate to certain issues of concern exactly (supporters, supporters on condition, opposed, opposed on condition or neutral). From another point of view, there are dependent stakeholders, affected stakeholders, unknown stakeholders, supportive stakeholders, third party advocates and the firm opposition (Curtin-Jones 2000). Based on the above, diverse segmentation frameworks of stakeholders can be set up which serve as tools of reasonable consequence in strategic management.

The *stakeholder value* perspective, as a matter of fact, emphasises responsibility over profitability and sees companies as interest coalitions to serve all parties involved. According to this theory, organisational success should be measured by the satisfaction among stakeholders, as a consequence, their management is perceived both as an end and a means at the same time. Social responsibility is handled as an organisational matter since society is best served by pursuing joint interests and economic symbiosis striving for common wealth. Stakeholder management refuses to give shareholders a higher moral claim on the company than providers of other resources. This thinking introduces values other than just financial value in the spectrum of what ought to be pursued by a company. Stakeholder management is normative, resulting in high levels of motivation and trust in the environment of the organisation (Donaldson-Preston 1995).

The above depicted normative approach can also be labelled as *intrinsic stakeholder commitment* where the interests of stakeholders have intrinsic value and enter the decision-making process of a firm prior to strategic considerations, hence forming a moral foundation for corporate strategy itself. Yet what much more often drives business in reality is the clear instrumental value of good stakeholder management since the prudent management of a company's operating environments, including stakeholder relationships, is part of sound management in general and since stakeholders control resources crucial to the survival and success of a company.

According to this *instrumental stakeholder management approach*, stakeholders are part of the business environment which must be managed in order to assure revenues, profits, and ultimately, returns to shareholders. Stakeholder management, in short, is a means to an end as strategic (instrumental) ethics enters the picture as an addendum to the imperative of corporate wealth maximisation for the manager-agent to follow (Janisch 1992). Consequently, stakeholder relations form part of corporate strategy, but by no means drive it. The two variants of this approach are the *direct effects model* (with stakeholder orientation having a direct effect on corporate financial success, independent of corporate strategy) and the *moderation model* (where stakeholder orientation moderates the relationship between the other two elements) (Berman et al. 1999, based on Freeman 1984).

In conclusion we must underline that the management of stakeholder relationships is a management field of strategic importance. Without the support of stakeholders with resources critical to the company at their disposal, no business activities can be pursued successfully in the long term. By balancing stakeholder contribution and stakeholder satisfaction (Cranfield University), however, reciprocally favourable relationships can contribute to lasting business excellence, and hence, financial reward.

3. International Management

In globalising business, companies incessantly strive for higher returns through exploring new opportunities, creating more favourable operating conditions and minimising resource costs. The traditional reasons for going international were securing raw material, expanding into new markets and hiring cheaper labour force abroad. Apart from these original factors, the ever increasing economies of scale and scope, soaring R&D costs as well as the radically shortening product life cycles also spur businesses nowadays to cross national borders. Furthermore, several countries attempt to entice firms investing abroad by making various (tax, customs, etc.) allowances for them. International companies also attain a whole range of strategic advantages by strengthening their internal operations and their global market position at the same time. Last, but not least, modern information and communication technology along with the emergence of sophisticated e-commerce solutions have opened inviting new perspectives for entrepreneurs.

Internationalisation is a multi-stage process. *Domestic* companies concentrating upon homogeneous national markets, hence, increasingly perceive the need and the lure to expand beyond the borders and become *international* firms, still producing locally, but developing ambitious export transactions (in general strictly within a distinct export division in the company). By expanding its production and service activities over a number of countries, a firm turns into a *multinational* one. Genuine *global* players, last, elaborate their corporate strategy and policies at a global level and have a truly multicultural flair to their corporate culture. (Another way to illustrate possible stepping stones from nationality to multinationality is through export-import, franchise and licence contracts, management contracts, turn-key contracts and direct capital investment) (Poór-Farkas 2001).

Multinational companies manage their local subsidiaries in different ways. The *ethnocentric* orientation tends to value home-country managers and production facilities over subsidiaries. *Polycentric* management, on the contrary, attaches great importance to local peculiarities and knowledge, but generally excludes local managers from central management positions. *Regionally oriented* multinational companies manage their subsidiaries grouped in regional units in which local managers perform comprehensive tasks and shoulder responsibility. The pure *global* orientation, however, has other top management selection criteria than national status and aims at establishing a firm leading position on a local as well as on a global scale.

International management, as such, can be defined as a set of functions and tools which promote the efficient operations of companies and institutions acting in an international environment, taking the diverse influencing factors (demography, economic development, resources, legal, political, cultural and competition environment) and objectives (sales expansion, corporate growth, resource aquisition, diversification, local advantages, globalisation) into consideration (Bartlett-Ghoshal 1998). International management has extended functions and fields compared to management within national boundaries, including communication across cultural borders, international corporate strategy, the organisation of international companies, international marketing, operations management on an international scale, international information management, managing international human resources and international industrial relations. International management is, then, more complex than the management of national companies. Consequently, successful managers of multinational companies must also dispose of specific knowledge and soft skills to tackle this additional dimension and live up to the multiple expectations.

This wide spectrum of expectations raised on management must be perceived as extended corporate responsibility. In other words, multinational companies must be seen as 'quangos', organisations practically functioning as public institutions (Ulrich 1977) owing to the immense effect they have on society and the environment through influencing consumption patterns, lifestyle, attitudes and beliefs – briefly, by influencing culture. These quasi-public multinational organisations, even more pronounced than national companies must be aware of the triple bottom line it all comes down to. They must be able to shoulder corporate social and environmental responsibility towards all their stakeholders and act as corporate citizens of our globalised world. This yields opportunities just as well as threats. However, those companies which take on this challenge and manage to fulfill the transparency and accountability requirements through reporting based on international standards (e.g. Global Reporting Initiative, SIGMA, etc.), sound codes of conduct reflecting genuine commitment (Claude-Feral 1998) and organically integrating sustainability with their strategic management tools for example by applying the Sustainability Balanced Scorecard (Schaltegger 2004) will create added value not just for society, but also for themselves in the strictly economic sense which will guarantee their business success in the long term (Bohnet-Joschko & Schiereck 2003).

4. Culture

International activities obviously involve contact with heterogeneous cultures. Stakeholders with different cultural backgrounds can make totally different demands on a multinational company which, in turn, must be able to recognise, respect and make good use of these differences. *Culture*, in general, can be defined as a set of attitudes towards the fundamental questions of life shared by a group of people and the values, norms, forms of behaviour, habits as well as the abstract and material products created and accumulated by this community. Culture, hence, consists of explicit and implicit elements (artifacts, values and basic assumptions, respectively) acquired and transmitted by symbols, embodied by heroes, experienced and strengthened through rituals (Hofstede 1980; Trompenaars 1993). This holds true for all levels of culture, including national culture as well as economic culture, business culture, professional culture and organisational (or corporate) culture. What makes international, i.e., intercultural management such a challenge is that all these cultural levels must be taken into account in one single concrete situation since, as it has been found, corporate culture neither erases nor diminishes national cultural differences, for example (Adler 1991). It might affect the aspects of culture connected to practices or artifacts, but not the level of the underlying values deep beneath (Laurent 1983).

This is why the basic assumptions of the single national cultures must be scrutinised in order to avoid possible misunderstandings and eventual conflicts, in other words, as a reliable footing for international management. The frameworks for cultural research which can be used to compare national cultures and cultural blocks abound, not unlike international comparative surveys on corporate culture types. The framework of Hofstede can be applied very well, for example, with the dimensions individualism vs. collectivism, power distance, uncertainty avoidance, masculinity vs. feminity and long- vs. short-term orientation (Hofstede 1980). The arising challenges of internatioanl management can easily be deduced from the single country scores. Other relevant research frameworks include that of Trompenaars, for example, examining value orientations: attitudes to people (universalism vs. particularism, individualism vs. collectivism, emotional neutrality vs. emotional saturation, specific vs. diffuse relationships, status-by-achievement vs. ascribed status), to time (past, present or future orientation, sequential approach or synchronicity) and to nature (supremacy of man vs. supremacy of nature) (Trompenaars 1993). A third possible reference can be made here to Hall who differentiates low and high cultural context, and monochrone and polichrone culture types (Hall 1983).

The above mentioned frameworks focus, evidently, upon differences in national cultures. Corporate culture types can be defined as mechanical or organic (Morgan 1986) or as instrumental (functional) or social organisations (Laurent 1983). Mole, further, differentiates individual and group leadership, directive and participatory leadership, autocratic and democratic leadership as well as supremacy and egalitarian leadership (Mole 1990). These are undeniably important dimensions of management and become possible pitfalls of international management if only insufficient attention is paid to the misinterpretations and conflicts stemming from not knowing or not respecting the differences between the parties involved.

When two (or more) cultures meet in the case of any international business activity (be it simple contracts or complex projects like mergers and acquisitions), all parties must be aware of their own cultural disposition, of that of the other party (or parties) and they must be prepared to co-operate with the other culture (or cultures). Knowledge and willingness can only bear fruit together. The common goals are sometimes easily achieved, in other cases, difficulties arise. The smooth operation of a multinational company, for example, depends upon the strength of local values and the perception of the attractiveness of the transferred global principles and practices (Veser 2005). If the local cultural values regarding the global principles and practices are weak and the new principles and practices perceived as unattractive, the reaction will be defensive and submission, basic symbolic adoption will follow. If the local values are weak, but the transferred principles and practices attractive, the response will be accomodative and assimilation can be expected (even if merely with limited entrenchment). Entrenchment will only be strong in the case of strong local values paired with attractive new principles and practices (proactive reaction and integration). However, if the local cultural values are strong, but the transferred principles and practices not attractive enough, the resistance will probably prove strong and adoption (if any) will be purely symbolic and ceremonial.

Consequently, in order to attain the targeted objectives in such a multicultural situation through acceptance and implementation, i.e., through integration and strong entrenchment, a multinational company must have overall information about the cultural circumstances, about the possible reactions of its stakeholders with different cultural backgrounds (including their communication practices, interest vindication culture, protest culture, conflict solution culture, etc.) and elaborate a suitable and comprehensive strategy to manage them.

5. International Stakeholder Management

On an international scale, there are two levels of analysis which are critical to multinational companies as far as stakeholder management is concerned, namely global and local stakeholders between which two groups a broad distinction can be made. Global stakeholders are for the most part worldwide initiatives and are accordingly active at the global level whereas local stakeholders have a primary interest in local issues directly affecting their community. Taking these two levels of stakeholder interest (and presumably, power as well), further, the two levels of a multinational company (namely the headquarters in the home country and the local subsidiaries in the given local environment) together, four directions of broad stakeholder claims can logically be taken for granted.

In order to interact with their global and local stakeholders, multinational firms can choose either a global, a multidomenstic, or a transnational stakeholder management strategy. In the framework of a *multidomestic strategy*, all stakeholder claims are forwarded to the local subsidiaries which are in charge of strategy implementation and hence interact with both stakeholder levels. When a *global stakeholder management strategy* is employed, on the other hand, it is the headquarters to which practically the whole stakeholder management work is assigned. Finally, in the case of a *transnational strategy*, a combination of the two mentioned in the foregoing, stakeholder management responsibility is shared and lies, according to the nature of stakeholder claims either with the headquarters in the home country or with the local subsidiaries (Veser 2005). Since these claims cover a rather wide spectrum and generally relate to issues concerning local as well as company levels, nevertheless, we can say that the transnational strategy is the only one of the three which leads to genuine sustainability across all worldwide operations by ensuring that stakeholders have the widest possible forum to express their concerns on certain matters and engage in real dialogue with all relevant actors within the company in question.

In the framework of the PhD research constituting the basis of this paper, stakeholder management practices in multinational companies are studied, with a particular focus on the influence of culture, with the aim of comparing the practices of national and multinational firms within the same branch in the very same country (i.e., national culture), primarily in order to be able to formulate practical management implications. On the basis of combined desk and field research, consequently, our goal is not only to present, analyse and evaluate the prevailing patterns, but to point out possible ways of improving these practices. Recent, for the most part environmental, scandals have shown how fragile corporate image and hence financial success is. By presenting best-practice examples and constructively criticising disputable approaches, our objective is to contribute to making powerful multinational companies vastly influencing society and the environment, aware of their responsibility and at least to some extent help them turn into stakeholder-balanced places of value creation in the long term.

In the focal point of the research work are the following questions: (1) Assuming that stakeholders with different cultural backgrounds expect different things from a multinational company, is there a significant difference in the way these treat their stakeholders? (2) Do they take cultural differences into account? (3) To what extent is their stakeholder management different compared to that of national companies? (4) How can a multinational company best live up to the claims of stakeholders coming from a number of national cultures? (5) How can the prevailing practices be improved? (6) How can the single management fields and corporate functions contribute to this? (7) How can such an effort be efficiently co-ordinated? (8) What supplementary skills and abilities are needed for managers to cope with these new challenges?

Based on the research work (in the first place through the analysis of case studies and surveys), the following hypotheses are expected to prove correct: (1) Stakeholders belonging to different cultural contexts have different claims on a multinational company. (2) Multinational companies do not respect these culture-based differences as much as the stakeholders would wish. (3) Consequently, multinational companies are more satisfied with their own stakeholder management work than their stakeholders are. (4) There are a number of lessons multinational companies could learn from national ones operating in the same branch in the same country, in order to develop a more sophisticated stakeholder management by introducing culture-based differentiation where necessary.

References

Adler, N.J. (1991), International Dimensions of Organizational Behavior. Boston: PWS-KENT Publishing Company.

Angyal, Á. (1999), A vezetés mesterfogásai (The Masterstrokes of Management). Budapest: Kossuth Kiadó.

Argyris, C. (1992), On Organizational Learning. Oxford: Blackwell Business Publishers.

Atkinson, A.A., Waterhouse, J.H. & Wells, R.B. (1997), A Stakeholder Approach to Strategic Performance Management. In: MIT Sloan Management Review Spring 1997, vol. 38, no. 3, pp. 25-37.

Bartlett, C.A. & Ghoshal, S. (1998), Managing Across Borders. The Transnational Solution. Boston, Mass.: Harvard Business School Press.

Berman, Wicks, Kotha & Jones (1999), Intrinsic Stakeholder Commitment vs. Strategic Stakeholder Management. In: Academy of Management Journal; Oct99, vol. 42 Issue 5.

Bohnet-Joschko, S. & Schiereck, D. (eds.) (2003), Socially Responsible Management. Impulses for Good Governance in a Changing World. Metropolis-Verlag.

Claude, A. & Feral, M. (2003), Codes de conduite des multinationales. Outils de progrès social ou coup de pub? Rennes: Peuples Solidaires.

Curtin, T. & Jones, J. (2000), Managing Green Issues. MacMillan Press Ltd.

Donaldson, T. & Preston, L. (1995), The Stakeholder Theory of the Corporation. Concepts, Evidence, and Implications. In: Academy of Management Review 20 (1995) 1.

Freeman, E.R. (1984), Strategic Management. A Stakeholder Approach. Marshfield, Mass.: Pitman.

Hall, E.T. & Hall, M.R. (1990), Understanding Cultural Differences. Yarmouth: Intercultural Press.

Heller, R. (ed.) (2002), Manager's Handbook. London: Dorling Kindersley Ltd.

Hofstede, G. (1980), Culture's Consequences. International Differences in Work-related Values. Beverly Hills: Sage.

Janisch, M. (1992), Das strategische Anspruchsgruppenmanagement. Vom Shareholder Value zum Stakeholder Value. Bamberg.

Laurent, A. (1983), The cultural diversity of Western conceptions of management. In: International Studies of Managament & Organization 13(1): 75-96. Measuring Stakeholder Value. Performance Prism. Cranfield University.

Mintzberg, H. (1983), Power in and around Organizations. Englewood Cliffs, NJ: Prentice-Hall.

– (1999), Who should control the corporation? In: Mintzberg, H., Quinn, J.B. & Ghoshal, S.: The Strategy Process. Harlow: Pearson Education Ltd.

Mole, J. (1990), Mind Your Manners. Managing Culture Clash in the Single European Market. London: Industrial Society.

Morgan, G. (1986), Image of Organisation. Beverly Hills: Sage.

Poór, J. & Farkas, F. (2001), Nemzetközi menedzsment (International management) KJK.

Rappaport, A. (1998), Creating Shareholder Value. A Guide for Managers and Investors. The Free Press.

Recklies, D. (2001), Stakeholder Management. Recklies Management Project GmbH.

Schaltegger, S. (2004), Sustainability Balanced Scorecard. Unternehmerische Steuerung von Nachhaltigkeitsaspekten. In: Controlling, Heft 8/9, August/ September 2004.

Trompenaars, F. (1993), Riding the Waves of Culture. Understanding Cultural Diversity in Business. London: Nicholas Brealey Publishing.

Ulrich, P. (1977), Die Grossunternehmung als quasi-öffentliche Institution. Stuttgart: Poeschel.

Veser, M. (2005), The Influence of Culture on Stakeholder Management. Social Policy Implementation in Multinational Corporations. Bern: Haupt.

Concepts for Education of Sustainable Development at the Faculty of Chemistry of University of Latvia

Andris A. Spricis

Associated Professor, Dr. chem.

University of Latvia, Faculty of Chemistry,
48 Kr.Valdemara Street, LV – 1013 Riga, Latvia

Summary

Concepts for education of sustainable development at the Faculty of Chemistry of University of Latvia are analysed. Integrated approaches to important sectors of economy, to chemical and environmental problems and to concepts of sustainable development at the study programmes of Faculty of Chemistry of the University of Latvia are described. An overview is given of the study programmes and study courses offered for education of issues of sustainable development linked with chemistry. It must be pointed out, that cooperation with International Baltic University has givien excellent possibilities for the University of Latvia and other universities for transfer of knowledge, for identification and studies of common concepts of sustainable development for the Baltic Sea region.

Introduction

Many important issues of resource management, of industrial production, of energy sector, of agriculture, of transport sector, of waste management, of material flows, of community development are closely linked with chemistry, with environmental sciences, with questions of possibilities for sustainable development.

Close links of all these sectors with chemical processes, with chemical environmental pollution, with global environmental problems, with greenhouse effect and climate change, with problems of resource management, with practical use of different chemicals and chemical products, with issues of sustainable development are basis for implementation of integrated approaches for these issues at the study courses and study programmes of the Faculty of Chemistry of University of Latvia.

Issues of Sustainable Development at Study Programmes and Study Courses of the Faculty of Chemistry of University of Latvia

Education of issues of sustainable development with emphasis on problems linked with chemistry is offered as part of the several educational programmes of the Faculty of Chemistry of the University of Latvia:

- On the Bachelor of science study programme in Chemistry;

- On the Higher professional education study programme in "Environmental Protection and Impact Assessment" [1],
- On the Master of science study programme in Chemistry with special studies at specialization in Environmental Chemistry.

The objective of the environmental study programmes at Faculty of Chemistry of the University of Latvia is to prepare highly qualified wide profile specialists in environmental protection with specialization in environmental protection sectors with chemistry profile.

The environmental study programmes at the Faculty of Chemistry of the University of Latvia have focus on education of principles of sustainable development.

Integrative approaches for issues of chemistry, of resource management, of industrial production, of energy production and of energy use, of agriculture, of transport sector, of water management, of waste management, of material flows and of sustainable development are delivered at several study courses with effective university – enterprise cooperation:

- At the BSc in Chemistry level the course Chemistry of Natural Resources and Applied Chemistry;
- At the level of Higher Professional study programme the course Environmental Protection;
- At the level of MSc in Chemistry the course Sustainable Development of the Baltic Sea Region and Environmental Problems in Latvia;
- At the level of MSc in Chemistry the course Environmental Pollution and Pollution Control;
- At the level of MSc in Chemistry the course Environmental Chemistry;
- At the level of MSC in Chemistry the course Food Packaging;
- At the level of MSc in Chemistry the course *Water Chemistry and Management.*

Special attention on these courses is devoted to problems associated with energy resources in the World and in Latvia, with other resources, with theoretical principles of treatment, management and use of different kind of resources, with processes of energy production and with processes of industrial productions of different kind of materials and chemicals, with safety and the environment.

Integrated Approaches for Energy and Chemistry Issues, Environmental Impacts and Sustainable Development

Energy and chemistry issues, environmental impacts and sustainable development are with special importance at the concepts of sustainability studies at the programmes of the Faculty of Chemistry of University of Latvia.

The epic importance of oil is recognised – no other substance has so changed the world and affected so many people in such a short time as oil [2].

A major part of the problems associated with oil are closely linked with chemistry, chemical processes, chemical industries, particularly with different types of fuels for energy production.

Oil has become a vital part of industry, agriculture, and the fabric of society at large. Oil by its derivatives, gasoline, kerosene, and naphtha, fuel more than 600 million vehicles worldwide. But oil is a finite resource, and we are using it at an exponential rate. There will soon be a post-petroleum paradigm [2].

There are several important issues linked with the energy sector, with chemistry and chemical processes, with problems of environmental protection and sustainable development included at the study courses of the study programmes of the Faculty of Chemistry of University of Latvia:

- Energy problem, world energy resources;
- Energy conversion processes;
- Theoretical principles of use of fossil fuels;
- Production and use of other fuel materials;
- Use of waste materials for energy production;
- Thermal energy production by combustion processes;
- Fossil fuels and the global atmosphere;
- Flows of energy and matter;
- Greenhouse gases and global climate;
- Energy production in fission nuclear power plants;
- Development of alternative energy sources;
- Development of fusion technologies;
- Nuclear power and radioactive pollution;
- Energy efficiency, saving;
- Energy transportation;
- Materials for energy sector;
- Autonomous energy sources – fuel cells, batteries and others;
- Energy sector, safety and environment.

Emphasis is given to the impacts of energy sector on the global atmosphere, to greenhouse gases and global climate, to consequences of climate change, to possibilities for reduction of atmospheric impacts by use of alternative energy systems, to renewable energy sources, to use of biomass, to methods of effective use of energy, to practical implementation of modern technologies and use of new environmentally more friendly materials.

It is analysed at the study courses, that many environmental problems result from the extraction, transportation, conversion and end-use of energy, that current energy systems are associated with a number of problems such as resource depletion and environmental pollution.

It is shown that energy related environmental problems include water pollution and land degradation from mining and extraction of fossil fuels, oil spills from transportation, air pollution from fuel combustion processes and the storage of radioactive waste from nuclear power plants.

It is emphasized that the energy sector is linked with environmental problems associated with major air pollutants: sulphur dioxide, nitrogen oxides, greenhouse gases and that a change of energy systems in a sustainable direction is inevitable if environmental and development goals as agreed by international bodies are to be fulfilled. Improvement of energy efficiency in all sectors of society is one important part of such a change. Energy efficiency improvements are a key components in any strategy for a more sustainable energy future. Another key component is a shift away from dependence on fossil fuels to renewable sources of energy.

Chemistry, Important Sectors of Economy, Environmental Problems and Sustainability

The use of oil has changed world economies, social and political structures, and lifestyles beyond the effect of any other substance in such a short time. But oil supplies are limited. The peak of world oil production and the beginning of the irreversible decline of oil is clearly in sight. World agriculture is now highly dependent on oil and natural gas for fertilizers and pesticides. Without these, agricultural productivity would markedly decline. As a base for the production of these materials, oil and natural gas are irreplacible. Lifestyles and affluence in the post-petroleum paradigm will be quite different from today. At that time concern will be much more centered on obtaining basic resources, especially agricultural, by which to survive [2].

Other fossil fuels – natural gas, coal and others as well many of mineral resources are limited also and chemical processing by using all these materials to produce and use different kind of products are of high importance.

Problems of very high priority are linked also with water management and chemistry, particularly with supply of good quality drinking water, with waste water treatment, with quality of surface water and groundwater.

Serious issues are associated also with waste management and chemistry problems, inclooding safe waste disposal at landfills, waste material treatment, reuse, recycling, waste inceneration and others.

Many different industries, agriculture sector, transport sector, municipal sector and others are associated with chemistry aspects and with questions of possibilities of sustainable development for respective sector.

Very important also are issues of different kind of environmental impacts linked with chemistry.

There are several important issues associated with different sectors of economy, with chemistry and chemical processes, with problems of environmental protection and sustainable development included at the study courses of the study programmes of the Faculty of Chemistry of University of Latvia:

- Technologies and processes of production of different materials, chemicals and products;
- Industrial production and environmental threats
- Special treatment of different kind of resources;
- Water chemistry and pollution;
- Drinking water treatment;
- Waste water treatment;
- Waste material treatment;
- Chemistry of sediments;
- Use and turnover of nutrients, eutrophication;
- Atmospheric chemistry and pollution;
- Combustion processes;
- Acid rain and acidification;
- Stratospheric ozone and UV radiation;
- Reducing atmospheric impact;
- Soil quality;
- Chemicals for technical purposes and pollution;
- Consumer products and pollution;
- Chemicals for agriculture and pollution;
- Toxic chemicals and pollution;
- Metal flows and environmental impact;
- Contaminating chemicals;
- Distribution of pollutants in the environment;
- Greening production processes;
- Environmental design;
- Environmental management systems;
- Controlling of pollutants, etc.

To strengthen practical knowledge of environmental education, linked with chemistry and sustainable development issues, study visits to different industrial plants, to power stations, to drinking water and waste water treatment plants, to waste landfill and to several institutions are performed.

Education in chemistry and sustainable development concepts at the Faculty of Chemistry of the University of Latvia is provided in close partnership with International Baltic University [3, 4].

The Baltic University course materials in English and TV series for courses are widely used for studies. Modern information technologies, in particular Internet, computer-, video- and audio- conferences are giving possibility to strengthen network learning [5].

Conclusions

Close links of important sectors of economy with chemistry and chemical processes, with chemical environmental pollution, with global environmental problems, with greenhouse effect and climate change, with problems of resource management, with practical use of different chemicals and chemical products are basis for implementation of issues of sustainable development at the study courses and study programmes of the Faculty of Chemistry of University of Latvia.

Integrated approaches for issues of resource management, of industrial production, of energy sector, of agriculture, of transport sector, of waste management, of material flows, of community development and others, closely linked with chemistry, with environmental sciences, with questions of possibilities for sustainable development are bases of concepts for education of sustainability at the study courses and study programmes of the Faculty of Chemistry of University of Latvia.

Cooperation with International Baltic University is giving excellent possibility for University of Latvia and other universities for transfer of knowledge for identification and studies of common concepts of sustainable development for the Baltic Sea region.

References

1. *Spricis, A.* (2001), The study programme "Environmental Protection and Impact Assessment" at the University of Latvia, it's focus on sustainability and the regional impacts of training offered. International Journal of Sustainability in Higher Education, ISSN 1467-6370, vol. 2, no. 4, pp. 339-348.
2. *Youngquist, W.* (1999), The Post-Petroleum Paradigm and Population. A Journal of Interdisciplinary studies, vol. 20, no. 4.
3. *Ryden, L., Migula, P. & Andersson, M.* (2003), Environmental Science. ISBN 91-970017-0-8, Uppsala, The Baltic University Press, 824 p.
4. *Salay, J.* (1997), A Sustainable Baltic Region. Session 2. Energy ISBN 91-7005-125-9. Uppsala, Uppsala University, 52 p.
5. *Spricis, A.A.* (2005), International Baltic University for Environmental Baltic Sea region Studies in Latvia. In: ICTE in Regional Development. Annual Proceedings of Vidzeme University College. Ed. by Blumel, E. & Ginters, E. ISBN 9984-633-01-2,Vidzeme University College, Valmiera, pp. 12-17.

Thematic Index

List of Participants

For Article	p.	e-mail
Uno Svedin – Sustainable Development – R&D Perspectives from Sweden and the European Union	13	Uno.Svedin@formas.se
Pierre Valette – Welcome Statement – Integrative Approaches towards Sustainability (SHARING)	21	Pierre.Valette@cec.eu.int
Joachim H. Spangenberg – The Economic Sustainability of the Economy – Assessing the Sustainability of a Complex System by Applying Orientor Theory	25	JoachSpangenberg@aol.com
Morihiro Kurushima – Challenges on New Technologies for Sustainable Energy System – Climate Change Policy Measures in Japan	43	Kurushimamrh@nedo.go.jp
Mickael Planasch, *Gernot Gwehenberger, Hans Schnitzer, Christoph Brunner* – From Cleaner Production to Zero Emissions	49	Planasch@TUGraz.at
Per G. Berg – Factor 5 Flow City – Future Images of Sustainable Urban Development in the Baltic Sea Region	81	Per.Berg@lpul.slu.se
Sergio Ulgiati, *Catia Cialani* – Environmental and Thermodynamic Indicators in Support of Fair and Sustainable Policy Making – Investigating Equitable Trade among Latvia, Denmark and Italy	101	ulgiati@unisi.it
Walter Leal Filho – Coastman – Integrative Approaches to Coastal Zone Management in the Baltic Sea	125	leal@tutech.de
C. Lagerberg-Fogelberg, *F. Fogelberg, G. Grankvist* – The Use of Environmental Information in Decision Making Promoting a Sustainable Food Chain	137	Charlotte.Lagerberg@cul.slu.se
Janis Gravitis – Green Biobased Chemistry Platform for Sustainability	145	jgravit@edi.lv
G. Geletukha, *T. Zhelyezna, Y. Matveev, N. Zhovmi* Bioenergy Development in Ukraine – State of the Art and Perspectives	161	geletukha@biomass.kiev.ua
Lars Rydén – Urban Sustainability Strategies – Is there a Common Structure?	177	lars.ryden@balticuniv.uu.se
Arnolds Ūbelis – Teaching for Sustainable Development and University-Municipality Partnerships (Project “Inversed Brain Drain”)	195	arnolds@latnet.lv

For Article	p.	e-mail
Bjarne Rasmussen – The Local Agenda 21 Change Agent – Why, What, Who, How?	205	bjr@npk.stam.dk
Bernd Delakowitz – Environmental Management (EM) and Sustainable Development (SD) in University's Strategic Management – Motivation and Experience from the Hochschule Zittau/Goerlitz – University of Applied Sciences (F.R.G.)	209	b.delakowitz@hs-zigr.de
Krzysztof Nawratek – City as a Vehicle of Education	221	nawratek@nams.arch.lv
Luigi Bistagnino, *Pier Paolo Peruccio, Carlotta Montanera* – A Challenge for the Design – The Systems Design Thinking	229	luigi.bistagnino@polito.it
Vladimir Stevanovic, ***Ivica Radovic***, *Slobodan Regner, Sreten Mandic, Zlatko Bulic, Aleksandar Joksimovic, Dejan Radovic* – Adriatic Sea and Coastal Zone of Serbia and Montenegro – Characteristics of Biodiversity and Possibilities for their Sustainable Exploitation	239	iradovic@bf.bio.bg.ac.yu
Hans-Peter Winkelmann – Towards an International e-Learning Didactic for Sustainable Development – The European Virtual Seminar on Sustainable Development (EVS)	255	hpw@copernicus-campus.org
Andrea Segrè, Luca Falasconi, Marina Mantini, Sabina Morganti, Elisabetta Nanni, Hajnalka Petrics, ***Matteo Vittuari*** – The UniAdrion Experience – A Virtual University for the Adriatic-Ionian Basin	269	matteo.vittuari@unibo.it
Stina Soewarta – The Contribution of Energy Policy to Sustainability	277	Stina.SOEWARTA@cec.eu.int
Hajnalka Petrics – Sustainable Development Concepts and Tools – The World Bank's Approach to Sustainable Development: Environmental Sustainability and Poverty Reduction	283	h.petrics@unibo.it
Raimonds Ernsteins – Local Agenda 21 Process Facilitation – Environmental Communication and Self-Experience Development in Latvia	305	Raimonds.Ernsteins@lu.lv
Dina Bērziņa – Climate Change Mitigation Initiatives – Energy Policy in Compliance with UNFCC and Kyoto Targets	319	dinab@latnet.lv
Markus Will – Foresight and Integrated Management Systems in Small and Medium sized Enterprises (SME) – Preliminary Considerations for a Tool-kit towards Corporate Sustainability	335	m.will@web.de

For Article	p.	e-mail
H. Schnitzer, G. Gwehenberger, ***Ch. Zwatz*** – Zero Emissions Research in Austria – A Case Study	345	christian.zwatz@tugraz.at
Heather Worosz*, Paulo Bento Maffei de Souza, Yury Kazhura* – Sustainable Community Development in the Baltic Sea Region	353	hktworosz@yahoo.com
Agrita Ozola – Integration of Cultural Heritage as a Resource for Sustainable Development in Municipality – A Case of Durbe Manor in Tukums	361	tukmuz@apollo.lv
Fredrik Fogelberg – Baltic Sea Algae as a Resource for Renewable Plant Nutrition and Bio-Fuel	373	fredrik.fogelberg@hihm.no
Janis Gravitis*, Janis Grabis, Oskars Bikovens, Andris Veveris* – European Rice Husks Studies with Focus on Utilization Opportunities	377	jgravit@edi.lv
Andrejs Lūsis – Removal of Hazardous Substances in Electronics – Processes and Techniques for SMEs	389	lusis@latnet.lv
G. Shulga*, O. Nestore, V. Shakels, B. Neiberte, A. Verovkins, T. Betkers, O. Belous, A. Žukauskaite, D. Ambrazaitene, M. Lukošius* – Collaboration Research on a Modified Biopolymer as a Soil Stabiliser and Amendment	397	shulga@junik.l
Sergio Ulgiati*, Silvia Bargigli, Marco Raugei* – Emergy Life Cycle Assessment within a Zero Emission Framework	403	ulgiati@unisi.it
G. Vabolienė*, A.B. Matuzevičius* – The Research of Biological Phosphorus and Nitrogen Removal from the Waste Water using new Wastewater Treatment Technology	419	gie-dre@one.lt
Péter Sipos – Modelling the Behavior of Heavy Metals Releasing from Sewage Sludge to Soils	433	sipos@geochem.hu
Konstantin Drozd*, Georgiy Geletukha* – Feasibility Study of Wood-fired Boiler House Construction for Heat and Hot Water Supply of Slavutich City	443	drozd@biomass.kiev.ua
L. Kliučininkas*, D. Martuzevičius, J. Mockuvienė* – Ambient Air Quality Mapping for Integrated Urban Traffic Flows Management	451	linas.kliucininkas@ktu.lt
Maxim Slenkin*, Nikolay Zhovmir* – Development of Ecologically Friendly Technology for Gasification of MSW	461	maxst@biomass.kiev.ua
Agnė Kazlauskienė*, Pranas Baltrėnas* – Chloride and Sodium Concentrations in Snow Cover on Roadsides of Highway Panevėžys – Šiauliai in Lithuania	467	kazlauskas@post.skynet.lt

For Article	p.	e-mail
V. Biziks, *B. Andersons, Z. Cakste, G. Zakis, B. Neiberte* – Utilization of Pine Bark as Waste Wood for Soil Improvement	473	orkaans79@inbox.lv
J. Rizhikovs, *M. Puke, J. Zandersons, N. Vedernikovs, G. Dobele, A. Tardenaka, B. Spince* – Impact of Water-Soluble Substances of Birch Lignocellulose on the Mechanical and Sorption Properties of Granular Activated Carbon	479	rijans@inbox.lv
Kristina Ložienė, *Petras R. Venskutonis* – Variation of Chemical Composition of Essential Oils in *Thymus serpyllum L. subsp. serpyllum* growing wild in Lithuania	487	kristina.l@botanika.lt
J. Pukite, *S. Kühl, W. Wilms-Grabe, C. von Friedeburg, T. Deutschmann, S. Sanghavi, J. Hollwedel, S. Beirle, C. Frankenberg, M.F. Khokhar, M. Grzegorski, T. Marbach, B. Kirchhof, S. Kraus, U. Platt, T. Wagner* – SCIAMACHY – Limb Measurements as a New Tool for Stratospheric Ozone Studies	493	Janis.Pukite@iup.uni-heidelberg.de
Pille Meier, *T. Kaps, U. Kallavus* – Behaviour of Pinewood (*Pinus Sylvestris*) in Various Mixtures	503	meier@cc.ttu.ee
Adrienn Tóth – A Method Supporting Sustainable Landuse Management Applying Soil Erosion Models and Ecological Investigations	511	tot8371@helka.iif.hu
Dainis Ozols – Landform Map and Sediments of North Vidzeme	519	biodain@latnet.lv
E. Reintam, *J. Kuht, K. Trükmann, J. Puust* – Composition of Weed Community Depending on Soil Compaction in Barley Field	529	endla.reintam@emu.ee
Katrin Trükmann, ***Endla Reintam***, *Jaan Kuht, Virgo Rääts* – Use of Biological Methods to loosen Compacted Soil	541	endla.reintam@emu.ee
Ina Alsiņa, ***Laila Dubova*** – Physiological Activity of Sapropel Decomposition Products	551	lailad@cs.llu.lv
Andrea Wallin – Nutrient Transport Modelling in the Daugava River Basin	561	andrea.wallin@mdh.se
Edgars Andersons – Forest Fire Monitoring from Satellites	569	edgarz@lja.bkc.lv
Remigijus Čiegis, ***Dalia Gineitienė*** – Sustainable Community Development and the Role of Social Capital	577	dalia.gineitiene@vukhf.lt

For Article	p.	e-mail
Agnes Haraszti – Corporate Responsibility at an International Level – Stakeholder Management in Multinational Firms as an Integrative Approach towards Sustainability	587	harasztiagnes@hotmail.com
Andris A. Spricis – Concepts for Education of Sustainable Development at the Faculty of Chemistry of University of Latvia	599	spricis@latnet.lv

Umweltbildung, Umweltkommunikation und Nachhaltigkeit
Environmental Education, Communication and Sustainability

Herausgegeben von/Edited by Walter Leal Filho

Band 1 Walter Leal Filho (Hrsg.): Umweltschutz und Nachhaltigkeit an Hochschulen. Konzepte-Umsetzung. 1998.

Band 2 Walter Leal Filho / Farrukh Tahir (Eds.): Distance Education and Environmental Education. 1998.

Band 3 Walter Leal Filho (ed.): Environmental Engineering: International Perspectives. 1998.

Band 4 Walter Leal Filho / Mauri Ahlberg (Eds.): Environmental Education for Sustainability: Good Environment, Good Life. 1998.

Band 5 Walter Leal Filho (ed.): Sustainability and University Life. 1999. 2nd, revised ed. 2000.

Band 6 Wout van den Bor / Peter Holen / Arjen Wals / Walter Leal Filho (eds.): Integration Concepts of Sustainability into Education for Agriculture and Rural Development. 2000.

Band 7 Manfred Oepen / Winfried Hamacher (Eds.): Communicating the Environment. Environmental Communication for Sustainable Development. 2000.

Band 8 Walter Leal Filho (ed.): Communicating Sustainability. 2000.

Band 9 Walter Leal Filho (ed.): Environmental Careers, Environmental Employment and Environmental Training. International Approaches and Contexts. 2001.

Band 10 Rolf Jucker: Our Common Illiteracy. Education as if the Earth and People Mattered. 2002.

Band 11 Walter Leal Filho (ed.): Teaching Sustainability at Universities. Towards curriculum greening. 2002.

Band 12 Walter Leal Filho (ed.): International Experiences on Sustainability. 2002.

Band 13 Lars Wohlers (Hrsg.): Methoden informeller Umweltbildung. 2003.

Band 14 Ulisses Azeiteiro / Fernando Conçalves / Walter Leal Filho / Fernando Morgado / Mário Pereira (eds.): World Trends in Environmental Education. 2004.

Band 15 Walter Leal Filho / Arnolds Ubelis (eds.): Integrative approaches towards sustainability in the Baltic Sea Region. 2004.

Band 16 Walter Leal Filho / Michael Littledyke (eds.): International Perspectives in Environmental Education. 2004.

Band 17 Daniela Krumland: Beitrag der Medien zum politischen Erfolg. Forstwirtschaft und Naturschutz im Politikfeld Wald. 2004.

Band 18 Walter Leal Filho / Bernd Delakowitz (Hrsg.): Umweltmanagement an Hochschulen: Nachhaltigkeitsperspektiven. 2005.

Band 19 Walter Leal Filho / Volker Lüderitz / Gunther Geller (Hrsg.): Perspektiven der Ingenieurökologie in Forschung, Lehre und Praxis. 2006.

Band 20 Walter Leal Filho (ed.): Handbook of Sustainability Research. 2005.

Band 21 Walter Leal Filho / Dieter Greif / Bernd Delakowitz (eds.): Sustainable Chemistry and Biotechnology – A Contribution to Rivers Management. 2006.

Band 22 Walter Leal Filho / David Carpenter (eds.): Sustainability in the Australasian University Context. 2006.

Band 23 Walter Leal Filho / Arnolds Ubelis / Dina Berzina (eds.): Sustainable Development in the Baltic and Beyond. 2006.

www.peterlang.de